Earthquakes and Multi-hazards Around the Pacific Rim, Vol. I

Edited by
Yongxian Zhang
Thomas Goebel
Zhigang Peng
Charles A. Williams
Mark R. Yoder
John B. Rundle

Previously published in *Pure and Applied Geophysics* (PAGEOPH),
Volume 174, No. 6, 2017

Editors
Yongxian Zhang
Earthquake Prediction Division
China Earthquake Networks Center
Beijing, China

Thomas Goebel
Seismological Laboratory
Earth and Planetary Sciences
University of California-Santa Cruz
Santa Cruz, CA, USA

Zhigang Peng
School of Earth and Atmospheric Sciences
Georgia Institute of Technology
Atlanta, GA, USA

Charles A. Williams
Tectonophysics Dept.
GNS Science
Lower Hutt, New Zealand

Mark R. Yoder
Department of Physics
University of California-Davis
Davis, CA, USA

John B. Rundle
Departments of Physics and Earth & Planetary Science
University of California-Davis
Davis, CA, USA

ISSN 2504-3625
ISBN 978-3-319-71564-3

Library of Congress Control Number: 2017960904

Cover illustration: Historical seismicity within the Farallon transform fault region, Gulf of California, Mexico. Taken from Castro, R.R., Stock, J.M., Haukoon, E., and Clayton, R.W., Pure Appl. Geophys. 174 (2017), 2239-2256

Cover design: Deblik, Berlin

Printed on acid-free paper

This book is published under the trade name Birkhäuser (www.birkhauser-science.com)
The registered company is Springer International Publishing AG
The registered company address is: Gewerbestrasse 11, 6330 Cham, Switzerland

Contents

Pure Appl. Geophys. 174 (2017), 2195–2198
© 2017 Springer International Publishing
DOI 10.1007/s00024-017-1580-4

| Pure and Applied Geophysics

Earthquakes and Multi-hazards around the Pacific Rim, Vol. 1: Introduction

Yong-Xian Zhang,[1] Thomas Goebel,[2] Zhigang Peng,[3] Charles Williams,[4] Mark Yoder,[5] and John Rundle[5]

The seismic belt along the Pacific Rim is the greatest earthquake zone in the world, generating more than 80% of the world's largest earthquakes (https://earthquake.usgs.gov/learn/topics/megaqk_facts_fantasy.php). It is also susceptible to tsunamis and volcanic eruptions, which could generate serious multi-hazards. Since the beginning of the twenty-first century, most countries along the Pacific Rim have suffered from tremendous multi-hazards, especially earthquakes and tsunamis. For example, the 2004 Sumatra $M9.1$ earthquake in Indonesia and the 2011 Tohoku $M9.0$ earthquake in Japan triggered mega-tsunami and caused significant damages and human casualties. An improved understanding of the underlying physical processes and potential interactions of these multi-hazards, and better simulation and forecasting of their occurrences are needed for better disaster prevention/mitigation.

The APEC Cooperation for Earthquake Simulation (ACES) (http://www.aces.org.au/), endorsed by APEC (Asia–Pacific Economic Cooperation) in 1997, has been focusing on understanding, forecasting, and mitigating the effects of earthquakes and other natural disasters for about 20 years. It links the complementary strengths of the earthquake research programs of individual APEC member economies via collaborations toward the development of earthquake simulation models and creates the research infrastructure to enable large-scale simulations and to assimilate data into the models. Since 1997, twelve workshops, including nine international workshops and three working group meetings on earthquake simulations (http://www.aces.org.au/), have been held by ACES: (1) Inaugural ACES Workshop, January 31–February 5, 1999, Brisbane and Noosa, Queensland, Australia; (2) 2nd ACES Workshop, October 15–20, 2000, Tokyo and Hakone, Japan; (3) 2nd ACES Working Group Meeting, July 29–August 3, 2001, Maui Supercomputer Center, USA; (4) 3rd ACES Workshop, May 5–10, 2002, Maui, Hawaii, USA; (5) 3rd ACES Working Group Meeting, June 2–6, 2003, Melbourne and Brisbane, Australia; (6) 4th ACES Workshop and iSERVO colloquium July 9–14, 2004, Beijing, China; (7) 5th ACES International Workshop April 4–6, 2006, Hawaii, USA; (8) 6th ACES International Workshop May 11–16, 2008, Cairns, Australia; (9) 7th ACES International Workshop October 3–8, 2010, Hokkaido, Japan; (10) ACES Workshop on Advances in Simulation of Multihazards, Maui, Hawaii, May 1–5, 2011; (11) 8th ACES International Workshop on Advances in Simulation of Multihazards, Maui, Hawaii, October 23–26, 2012; (12) 9th ACES International Workshop on Advances in Simulation of Multihazards, Chengdu, China, August 10–16, 2015. As a result of ACES, much progress has been achieved on Lattice Solid particle simulation Model (LSM), Australian Computational Earth Systems Simulation (ACcESS), Earth Simulator of Japan, Geotechnical Finite Element Analysis (GeoFEM), Geophysical Finite Element Simulation Tool (GeoFEST), Earthquake Simulator (QuakeSIM), Solid Earth Virtual Research Observatory Institute (SERVO), International Solid Earth Virtual Research Observatory Institute (iSERVO), Load–Unload Response Ratio (LURR),

[1] China Earthquake Networks Center, Beijing 100045, China. E-mail: yxzhseis@sina.com
[2] Seismological Laboratory, Earth and Planetary Sciences, University of California, Santa Cruz, Santa Cruz, CA 95064, USA. E-mail: tgoebel@ucsc.edu
[3] Georgia Institute of Technology, Atlanta, GA 30332, USA. E-mail: zpeng@gatech.edu
[4] GNS Science, Lower Hutt 5040, New Zealand. E-mail: c.williams@gns.cri.nz
[5] Department of Physics, UC Davis, Davis, CA 95616, USA. E-mail: mryoder@ucdavis.edu; jbrundle@ucdavis.edu

Pattern Informatics (PI), Critical Sensitivity, Earthquake Critical Point Hypothesis, the Virtual California model (VC), Relative Operating Characteristic (ROC), Multiscale Finite-Element Model (MFEM), the Uniform California Earthquake Rupture Forecast (UCERF), etc. In the late 2000s, multihazards became a theme of ACES, and the ACES Workshop on Advances in Simulation of Multihazards was held in Maui, Hawaii, May 1–5, 2011, soon after the Tohoku M9.0 earthquake and tsunami.

Special Issues have been published after each ACES workshop, with themes related to the themes of the workshop (Donnellan et al. 2004, 2015; Fukuyama et al. 2013; Matsu'ura et al. 2002; Mora et al. 2000; Yin et al. 2006). This special issue, named 'earthquakes and multi-hazards around the Pacific Rim,' contains many of the results presented at the 9th ACES International Workshop on Advances in Simulation of Multihazards, Chengdu, China, August 10–16, 2015 (http://www.csi.ac.cn/ACES2015/Home/index.html), as well as additional related topics.

The 9th workshop included seven regular sessions (microscopic simulation; scaling physics; macroscale simulation: earthquake generation and cycles; macroscale simulation: dynamic rupture and wave propagation; computational environment and algorithms; data assimilation and understanding; model applications) and four special sessions. The special sessions were highlighted by 'Global Navigation Satellite System (GNSS) Tsunami Early Warning System: Models, Simulations, Data and Technology.' The research results of the April 24, 2015, Magnitude 7.8 Lamjung (Kathmandu), Nepal Earthquake were also discussed in a special session. The other two special sessions were 'Earthquake simulation and forecasting in China: State-of-the-art and future prospective,' and 'The lure of LURR—Celebration of Professor Yin's 80th Birthday.'

This topical issue is divided into two volumes. The first volume (Vol. I) includes 16 papers, which are further divided into three sections. Papers on earthquake physics are presented first, followed by papers on earthquake simulation and data assimilation. The final section covers multi-hazard assessment and earthquake forecasting models.

In the first section, S. W. Hao et al. present experimental results that examine the evolution properties of rocks during the secondary, nearly constant strain rate stage of deformation. Their experiments indicate that a lower ratio in the slope of the secondary stage with respect to the average rate of the entire lifetime implies more brittle failure modes. Y. Urata et al. examine the influence of loading velocity, cumulative slip and gouge on constitutive parameters within a rate-and-state friction law, using experiments and a spring-slider model. The authors find that conventional rate-and-state formulations cannot capture the observed evolution of frictional parameters with cumulative slip. R. R. Castro et al. determine source parameters and quality factors for an earthquake sequence that occurred in the Gulf of California in October 2013. They resolve a mainshock stress drop of 1.7 MPa and the frequency dependence of seismic wave attenuation. In order to study the stress distribution near the seismic gap between the M8.0 Wenchuan and M7.0 Lushan Earthquakes, Y. H. Yang et al. determine the focal mechanisms of 228 earthquakes with magnitude $M \geq 3$ from January 2008 to July 2014 near the seismic gap along the Longmenshan Fault Belt by using a full waveform inversion method and then apply a damped linear inversion method to derive the regional stress field based on the determined source parameters. Their results suggest that from west to east across the three main imbricated faults of the Longmenshan fault system, the faulting types change from thrust in the westernmost region to strike-slip in the central part and to normal and mixed faulting at the east end.

In the second section, K. W. Schultz et al. (2017) introduce a new slip-weakening friction law used in the Virtual Quake Simulator. They explore the effects of the frictional law parameters on seismicity rate simulations with the UCERF3 California fault model. They find that the new model extends the magnitude ranges where earthquake rates from simulations match observations in California, and improves the agreement between the simulated and observed scaling relations. M. Wilson et al. develop a statistical method to evaluate earthquake simulators based on observed seismicity data. J. Parker et al. discuss automated methods to locate and measure surface fault slip using deformation measurements from the NASA Uninhabited Aerial Vehicle Synthetic Aperture Radar (UAVSAR) data. UAVSAR is similar to satellite-based SAR, but can facilitate higher mission flexibility and superior pixel resolution, of approximately 7 m. The presented methods use freely available data products from the UAVSAR mission. T. H. W. Goebel et al. examine the variability of seismic stress drops, looking specifically at the San Gorgonio Pass and Ventura Basin in southern California. They improve the resolution of their method by stacking large numbers of source spectra. They find that events in the Ventura Basin (high loading rates) have lower stress drops than those in the San Gorgonio Pass (slow loading rates). Y. Y. Kropivnitskaya et al. analyze anthropogenic data from social media sources like Twitter in combination with contemporary physical sensor data, to estimate local earthquake intensity. Their algorithm and results show that combining social media-based metrics with data from physical type sensors produces intensity maps with more complete coverage, improved accuracy, and higher resolution than maps using either data source separately.

In the third section, H. W. Li et al. conduct a probabilistic tsunami hazard assessment in the South China Sea and neighboring basins. They perform a thorough review of historic earthquake and tsunami events followed by evaluations of the upper and lower bounds of tsunami hazard based on different corner magnitudes. They suggest that multi-disciplinary studies with seismic, geodetic, tectonic, and tsunami generation are needed to improve tsunami hazard evaluation in this region. Z. L. Wu et al. summarize the approach in China since the last 1.5 decade of using apparent stress for time-dependent seismic hazard assessment or earthquake forecasting. The research results show that this approach, seemingly uniquely carried out on a large scale in mainland China, provides the earthquake catalogues for the predictive analysis of seismicity with an additional degree of freedom, deserving a systematic review and reflection. H. Z. Yu et al. use a combination of spatial and temporal earthquake forecast methods and suggest that using this ensemble of methods may help improve short to intermediate term forecasts. S. F. Zhang et al. investigate methods to determine false alarm rates in annual earthquake forecasts for China and suggest a method to reduce such false alarms. Y. X. Zhang et al. present and validate an earthquake forecast, based on the Pattern Informatics (PI) method, of the Tibetan Plateau region during the time period 2008–2014, including the 2008 M8.0 Wenchuan earthquake. The forecast is verified using a receiver operating characteristic (ROC) metric and R score. They show that the PI metric significantly outperforms a random forecast and that some models with a larger grid and longer time window have higher forecasting efficacy. K. Katsumata analyzes earthquake catalogues provided by the International Seismological Center (ISC) to detect earthquake quiescence in and near Japan by using a simple scanning technique (ZMAP). They de-cluster earthquake swarms and aftershocks by a stochastic de-clustering method based on the Epidemic-Type Aftershock Sequence model (ETAS). Their results show that 11 significant quiescences of more than 9 years occurred during 1964–2012, and 3 of them were followed by $M_w \geq 8.25$ earthquakes, while there were 4 $M_w \geq 8.25$ earthquakes during this time period. A. Hawkins et al. examine the statistics of induced and triggered seismicity at the Geysers geothermal field in California. Their results support the idea that these earthquakes are caused by a reduction in friction on the associated faults as a result of injected fluid. They also find that the induced seismicity obeys Gutenberg–Richter (GR) scaling, and that aftershocks of the induced earthquakes obey GR scaling as well.

We thank the contributors to this and previous topical volumes, especially the authors, reviewers, Birkhauser personnel to make this topical volume happen. We would like to thank Dr. Renata Dmowska, the Editor-in-Chief, for Topical Issues of PAGEOPH, for her support and patience during the editing process. Special thanks to Prof. X. C. Yin's contribution to ACES for nearly 20 years. Finally, we would like to acknowledge our sponsors of the 9th ACES workshop, the China Earthquake Administration, the Ministry of Science and Technology, and the Ministry of Finance of the People's Republic of China. The China Earthquake Networks Center hosted the ACES workshop together with the Institute of Earthquake Science, the China Earthquake

Administration, the Sichuan Earthquake Administration, the Computer Network Information Center, the Chinese Academy of Sciences, and the State Key Laboratory of Nonlinear Mechanics, Institute of Mechanics, Chinese Academy of Sciences.

REFERENCES

Donnellan, A., Mora, P., Matsu'ura M. & Yin, X.-C. (Eds.). (2004). Computational earthquake science, part I and II. Pure and Applied Geophysics, 161 (9/10 & 11/12), (Birkhauser, 2004).

Donnellan, A., Williams, C. & Pierce, M. (2015). Multihazard simulation and cyberinfrastructure. *Pure and Applied Geophysics, 172*(8), 2083–2085. doi:10.1007/s00024-015-1074-1.

Fukuyama, E., Rundle, J.B. & Tiampo, K.F. (Eds.) (2013), Earthquake hazard evaluation. Pure and Applied Geophysics, 170 (1/2), 560 pp. (Springer, Basel, 2013).

Matsu'ura, M., Mora, P., Donnellan, A. & Yin, X-C. (Eds.) (2002), Earthquake processes: physical modelling, numerical simulation, and data analysis, part I and II. Pure and Applied Geophysics, 159(9/10), (Birkhauser, 2002).

Mora, P., Matsu'ura, M., Madariaga, R. & Minster, J-B. (Eds.) (2000), Microscopic and macroscopic simulation: towards predictive modelling of the earthquake process. Pure and Applied Geophysics, 157 (11/12), (Birkhauser Verlag, 2001).

Yin, X.-C., Mora. P., Donnellan, A. & Matsu'ura, M. (Eds.). (2006). Computational earthquake physics: simulations, analysis and infrastructure, part I and II. Pure and Applied Geophysics, 163(9 & 11–12) (Birkhauser, 2006 and 2007).

(Published online June 1, 2017)

Pure Appl. Geophys. 174 (2017), 2199–2215
© 2017 Springer International Publishing
DOI 10.1007/s00024-017-1523-0

Pure and Applied Geophysics

CrossMark

Scaling law of average failure rate and steady-state rate in rocks

SHENGWANG HAO,[1,2] CHAO LIU,[1] YINGCHONG WANG,[1] and FUQING CHANG[1]

Abstract—The evolution properties in the steady stage of a rock specimen are reflective of the damage or weakening growth within and thus are used to determine whether an unstable transition occurs. In this paper, we report the experimental results for rock (granite and marble) specimens tested at room temperature and room humidity under three typical loading modes: quasi-static monotonic loading, brittle creep, and brittle creep relaxation. Deformed rock specimens in current experiments exhibit an apparent steady stage characterized by a nearly constant evolution rate, which dominates the lifetime of the rock specimens. The average failure rate presents a common power–law relationship with the evolution rate in the steady stage, although the exponent is different for different loading modes. The results indicate that a lower ratio of the slope of the secondary stage with respect to the average rate of the entire lifetime implies a more brittle failure.

Key words: Steady stage, time-to-failure, failure mode, rock.

1. Introduction

When a rock specimen is loaded quasi-statically in the laboratory, the accelerating deformation that leads to eventual failure always follows an apparent, constant strain rate stage. This stage is called the steady, or secondary stage, and can be observed in three types of experiments: monotonic loading by controlling the crosshead of the testing machine moving at a constant velocity (Hao et al. 2013), brittle creep testing (Brantut et al. 2014), and creep relaxation testing (Hao et al. 2014). Creep failure in rocks is typically classified within three temporal stages: primary creep, secondary creep, and accelerating tertiary creep (Scholz 1968; Okubo et al. 1991; Lockner 1993). The primary and tertiary stages

receive the most attention because researchers believe they are more closely associated with failure. At the primary creep stage, the strain rate $\dot{\varepsilon}(t)$ decays as a power law $\dot{\varepsilon}(t) \sim t^{-p_c}$, called the Andrade's law (Andrade 1910), with time following the application of the stress. The exponent $p_c \approx 2/3$ (Andrade 1910). The power law decaying behavior for the rate of damage events (Amitrano and Helmstetter 2006) in the primary stage is similar to the modified Omori's law (Omori 1894; Utsu 1961) $\dot{n}(t) \sim (c + t)^{-p_n}$, which describes the rate $\dot{n}(t)$ of earthquakes decaying with time after the main shock. The exponent p_n could differ from 1, though it is usually almost equal to 1. These relations could suggest a clue for understanding and predicting the delayed failure triggered by the main shock or other causes.

The tertiary creep stage represents rapid, unstable growth, and thus it should provide insight into the failure process. Power–law creep acceleration behavior during tertiary creep was revealed by researchers (Voight 1988, 1989; Guarino et al. 2002; Nechad et al. 2005). Voight's relation (Voight 1988, 1989) $\dot{\Omega}^{-\theta}\ddot{\Omega} - A = 0$ has been widely accepted as the predominant method for describing the behavior of a material in the terminal stage of failure, where A and θ are the constants of experience and Ω is a measurable quantity such as strain. The dot refers to differentiation with respect to time. Similar power–law accelerations were also observed for natural structures, such as landslides (Saito and Uezawa 1961; Saito 1969; Petley et al. 2002), volcanoes (Voight 1988), or cliff collapses (Amitrano et al. 2005). Kilburn (2012) proposed a model to extend analyses to deformation under increasing stress and suggested an alternative relation between fracturing and stress. Hao et al. (2013) compressed granites and marbles in the laboratory by controlling the crosshead of the testing machine moving at a constant velocity.

[1] School of Civil Engineering and Mechanics, Yanshan University, Qinhuangdao, China. E-mail: hsw@ysu.edu.cn

[2] The State Key Laboratory of Nonlinear Mechanics, Institute of Mechanics, Chinese Academy of Science, Beijing, China.

They defined a response function as the change of the sample's deformation with respect to the displacement of crosshead. The results showed that following a pseudo-steady stage, the response function increased rapidly as a power law relationship with displacement. Hao et al. (2016) presented a systematical analysis of this critical accelerating behavior and suggested a new relation $\dot{\Omega}^{-1}\ddot{\Omega} \sim (t_f - t)$ to predict failure. where t_f represents the failure time. Efforts on describing the primary and tertiary stages improved our knowledge on the mechanism of failure and its prediction. The results also imply that the secondary stage as a stable stage between the primary and tertiary stages should reflect the specific properties of a specimen and its lifetime. Especially, the secondary stage reflects how a sample evolves from the decelerating stage to the accelerating stage.

It should be noted that damage, or weakening growth, in the secondary stage determines the unstable transition to the accelerating tertiary stage, and thus its evolution properties should contain information linked to the time to failure. In the laboratory tests, the secondary stage always dominates the lifetime of rock samples. Likewise in earthquake cycles, it is stated that throughout the interseismic period, the secondary stage is the dominant process (Perfettini and Avouac 2004). The slow speed of pre-earthquake deformation and strain accumulation (Chen et al. 2000; Shen et al. 2005; Meade 2007; Zhang 2013) in the Longmen Shan fault zone, which hosted the 2008 Wenchuan Mw 7.9 earthquake, China, led to an incorrect assessment of this hazard event. Therefore, revealing the relationship between the secondary stage and ultimate failure is critical to predicting failure and understanding the underlying mechanisms for it.

The strain rate in secondary stage creep is strongly dependent on the applied stress (Amitrano and Helmstetter 2006). It was shown that the time to failure for a rock decreases with increasing mean stress (Scholz 1968; Kranz et al. 1982; Boukharov et al. 1995; Baud and Meredith 1997; Lockner 1998), and that the average time to failure and the applied stress exhibits an exponential relationship (Das and Scholz 1981). The experiments (Hao et al. 2013, 2014) indicated that the monotonic and creep-relaxation experiments also showed a pseudo-steady

stage similar to the creep experiments. The dependence of the secondary creep rate and time to failure on the applied stress suggests that there should be a possible relationship between the secondary creep stage and the time to failure (Hao et al. 2014).

In this paper, we aim to establish an empirical relationship between the eventual failure and the evolution properties in the secondary stage in a rock sample under quasi-static tests. Monotonic (quasi-static constant displacement rate), brittle creep, and brittle creep relaxation experiments are the three typical laboratory experiments performed to investigate the dependency of the time-to-failure and failure modes on the secondary stage property.

2. Experimental methodology and material

Granite and marble, of the two major types of rocks found in the earth's crust, are tested in this experiment. Some of the physical properties of these two types of rocks are listed in Table 1. The rocks were sampled from a depth of ~ 10 m from Beijing, China, and cut into prismatic blocks (40 mm in height and 16 mm $\times$ 20 mm in cross-section). The surfaces of the specimens were cut for parallelism and perpendicularity between the faces, with particular emphasis on ensuring that the two ends were exactly perpendicular to the longitudinal axis of the sample. The specimens were intact, but had many randomly distributed intrinsic and natural microfractures. Figure 1 shows two microscopic images of crack pattern measured by SEM (scanning electron microscope) for two investigated rocks under loading to give an insight into their microstructures and fracture behaviors.

We performed three types of quasi-static experiments that were designed to simulate the three typical types of fault loading occurring in the upper crust. Figure 2 illustrates the experimental set-up and loading processes: monotonic loading (type 1), brittle creep (Amitrano and Helmstetter 2006; Heap et al. 2011; Brantut et al. 2013, 2014) (type 2), and brittle creep relaxation (Hao et al. 2013) (type 3). In laboratory tests, the loading system consists of a load apparatus and a deformed sample. The load apparatus is always modeled by an analogous "elastic spring"

Table 1

Physical properties of rocks

Rocks	Bulk density (g/cm³)	Water absorption (%)	Uniaxial compressive strength (Mpa)	Type 1 experiments	Type 2 experiments	Type 3 experiments
Granite	~2.5 to 2.7	0.16	222.5	✔	✔	✔
Marble	~2.6 to 2.8	0.46	162.4	✔	–	✔

Water absorption is the mass's ratio of water absorbed in the atmosphere by unit volume rock with respect to the mass of the dry rock

(a)   **(b)**

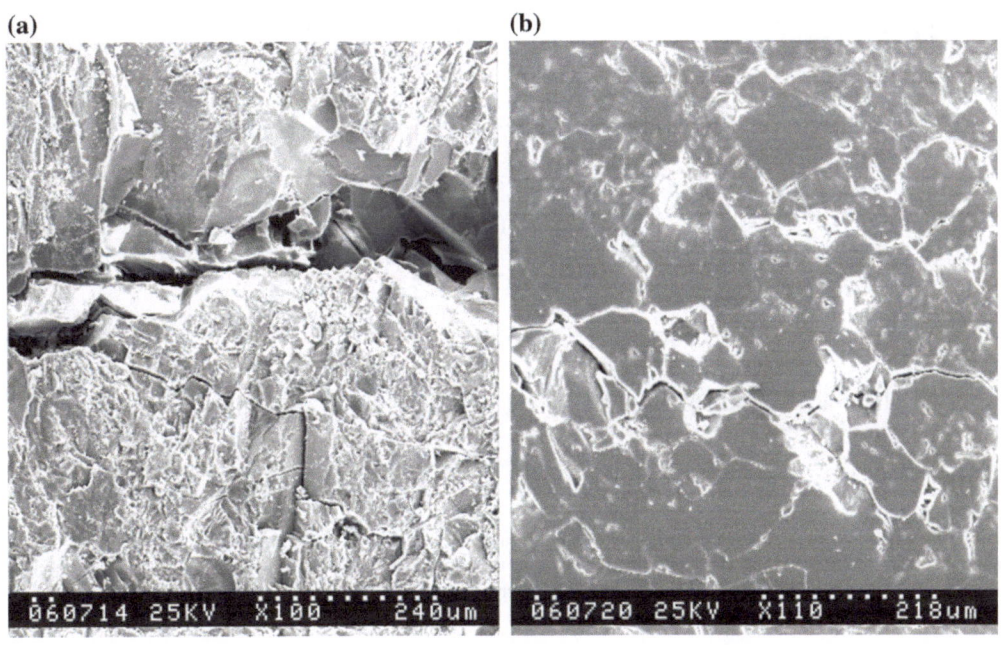

Figure 1
SEM images of crack patterns under loading for **a** granite. **b** Marble

as shown in Fig. 2. The displacement, U, of the crosshead of the testing machine includes the deformations of the loading apparatus, and that, u of the rock samples. In the first type of experiment (type 1), a rock specimen is compressed by monotonically and quasi-statically moving the crosshead of the testing machine at a constant rate (i.e. U is in linear relation with time). Brittle creep experiments (Amitrano and Helmstetter 2006; Heap et al. 2011; Brantut et al. 2013, 2014) were the second type of experiment (type 2) performed. In these experiments, a rock specimen was first loaded to a prescribed initial stress, which was then maintained at a constant value to observe the evolution of the deformation. The evolution of the deformation relates to the responses under applied

and invariant stresses, (Benioff 1951; Scholz 1968; Singh 1975; Lockner 1993; Du and McMeeking 1995; Lienkaemper et al. 1997; Heap et al. 2011) and could be possibly analogous to some processes of fault weakening that may lead to a seismic rupture. Brittle creep relaxation (Hao et al. 2014) was the third type of experiment (type 3) performed. In this experiment, a rock specimen was first loaded to an initial state by imposing an initial displacement to the crosshead of the testing machine. Then, the displacement was held at a constant value. The rock specimen then undergoes a combination process in which it deforms but the stress relaxes because the crosshead of the testing machine is held constant. This experiment simulated the process in which the

7

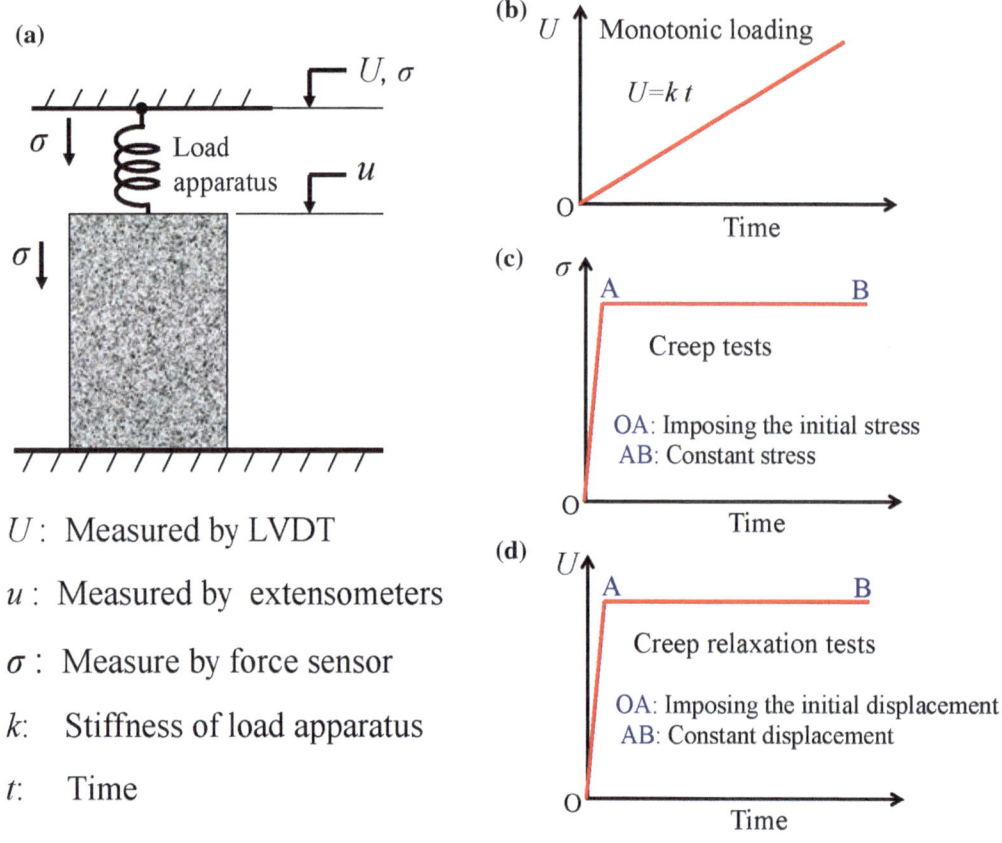

Figure 2
Sketch of the experimental set-up and loading processes. **a** Sketch of the experimental set-up; **b** the monotonic loading process; **c** the brittle creep test loading process; **d** the creep relaxation test loading process

elastic energy release of the surroundings drives the damage propagation, or weakening of a fault zone after a main earthquake (Hao et al. 2014).

In the experiments, the rock specimens were compressed uniaxially along the 40-mm axis at room temperature, ranging from 10 to 30 °C, and room humidity, with average relative humidity ~62%. Uniaxial compression was achieved using a screw-driven crosshead, which is a universal electromechanical testing machine equipped with a load cell with an offset load of 1 kN. The deformation, u, of a specimen was measured using 1 μm resolution extensometers located on the sides of the specimen. The displacement, U, of the crosshead was continuously measured using a linear variable differential transformer with a resolution 1 μm.

In the type 1 experiment, the rock specimens were compressed by monotonically increasing the crosshead displacement at a rate of 0.02 mm/min (leading to a strain rate of approximately 8.3×10^{-6} s^{-1}) until failure (i.e., no-hold step). The experimental protocols for type 1, type 2, and type 3 are illustrated in Figs. 3, 4 and 5. In brittle creep tests (type 2), the rock specimen was first loaded to an initial stress, (AB portion in Fig. 4) which was held at a constant value (AB portion in Fig. 4) while the deformation was measured. In the brittle creep relaxation experiment (type 3), the rock specimen was first rapidly loaded to the initial deformation state (OA part in Fig. 5) with a crosshead speed of 1.5 mm/min for the subsequent relaxation test. The crosshead was then held at this constant position (AB portion in Fig. 5) and the deformation and stress of the sample were measured as it relaxed. Hao et al. (2014) have detailed this loading process.

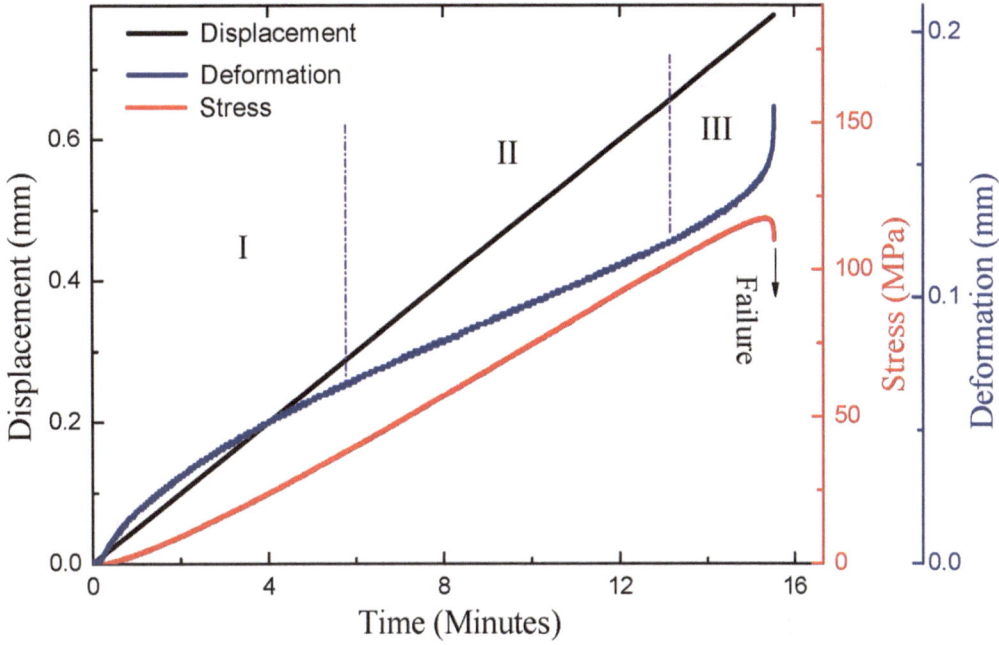

Figure 3
Deformation–time and stress–time plots from a static monotonic experiment illustrating the experimental protocol. The displacement, U, of the crosshead of the testing machine is controlled to increase linearly with respect to time and is a combination of the deformation of the loading apparatus and the deformed rock sample. The rock specimen failed catastrophically at the post-failure portion after the peak stress

In this paper, the symbol t_f denotes the failure time and t_0 represents the start time of the creep phase, or creep-relaxation phase, in brittle creep or creep-relaxation experiments. Thus, $(t_f - t_0)$ is the creep time or creep-relaxation time in these two experiments. The experimental parameters of all specimens tested in three types of experiments are listed in Tables 2, 3 and 4. It should be mentioned that there was an event of fracture that occurred during the testing of specimens SC-G-80-2 and SC-G-85-3. An audible sound was emitted by the fracture but the specimen did not fail completely. This small event induced a small jump in the curves (see Fig. 7a, b). Thus, the steady stage has a more direct relation with this small event than the macroscopic failure, and then we select the u_f and t_f corresponding to this small event to calculate the value of the average creep rate $\mu = \frac{\mu_f - \mu_0}{t_f - t_0}$.

We calculated the rate of deformation (or stress) du/dU, $d\sigma/dt$ or $d\varepsilon/dt$ by using the finite difference method. This constant rate stage is defined as the secondary stage in the present paper, and correspondingly, the value of the rate plateau is

determined as the slope, λs, of the steady stage in the deformation (or stress) curves.

3. Results

3.1. Stages of evolution to failure

Let us have a close examination of the evolution of the response variables, such as stress or deformation, in the three experiments. Figure 3 shows a typical result for rock specimens tested in the type 1 experiment. The solid blue line plots the axial deformation (u) against time, and the solid red squares plot the stress–time curve. It can be seen that at the early stage, the stress–time curve is slightly convex upwards (also see Rudnicki and Rice 1975; Jeager et al. 2007; Hao et al. 2007, 2013) and the deformation growth is characterized by an initial convex upward phase of decreasing strain rate. Later, an almost linear stress–time relation, as well as a nearly linear $u \sim t$ relation, follows. Finally, the deformation, u,

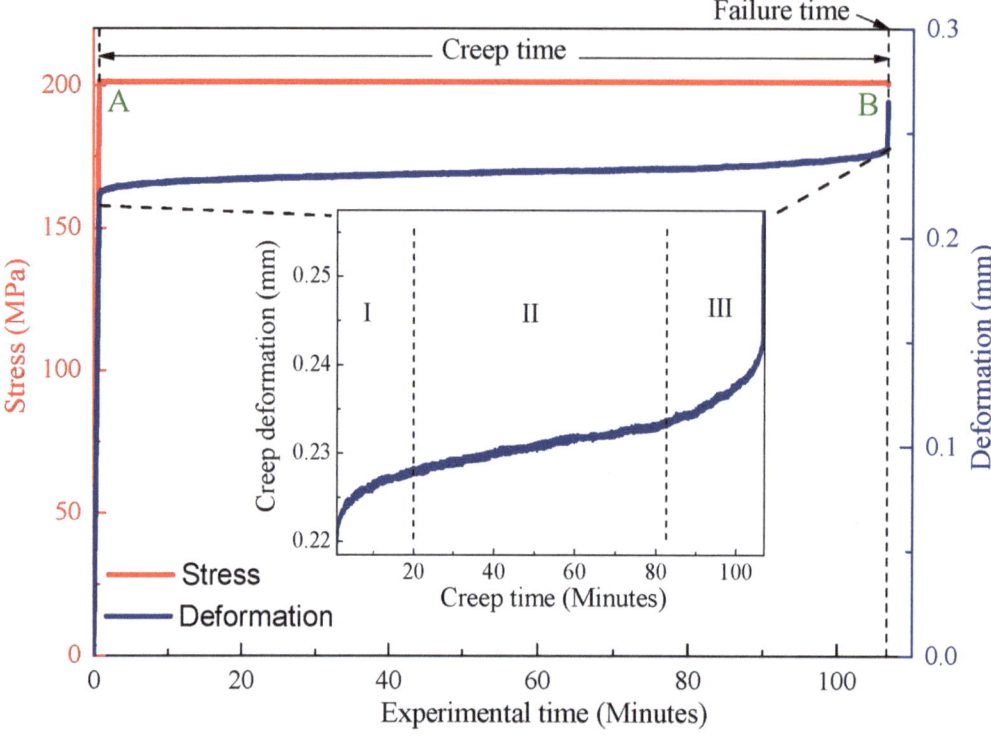

Figure 4
Stress–time and deformation–time curves in a brittle creep experiment on rocks. The experiment is divided into two phases: a phase (0A) of imposing an initial stress on rock samples, and a constant stress (creep) phase (*AB*). The *inset* shows a zoomed-in view to the creep phase. Three creep phases (*I, II, III*) are labeled on the *inset* and a long steady stage with almost constant creep rate is indicated

of the sample increases rapidly to catastrophic failure, which is associated with a jump of stress (Fig. 3). It has been demonstrated that the catastrophic failure is induced by the elastic energy release from the loading apparatus, which occurs at some point in the strain softening section after peak stress (Salamon 1970; Hundson et al. 1972; Labuz and Biolzi 1991; Bai et al. 2005; Jeager et al. 2007; Hao et al. 2007, 2010, 2013). Figure 6 plots the deformation–displacement curves for specimens. It shows that specimens exhibit the typical three-staged behavior of primary, secondary, and terminal accelerated evolution to failure. In the primary stage, the deformation–displacement curve is concave downward and the slope of the deformation–displacement decreases with increasing displacement. Later, an almost linear deformation–displacement relation, i.e. a pseudo-steady stage follows. In the tertiary stage, the curve is concaved upwards and the deformation increases rapidly.

In brittle creep tests, rock specimens were loaded with different, constant applied stresses. Figure 4 shows a typical result of the type 2 experiment to illustrate the complete process of brittle creep testing. It can be seen that at the creep phase (AB portion) after imposing the initial stress (OA portion), the stress was held well at a constant value. Figure 7a–d shows the curves of strain against time for all samples during the creep phase, and the applied stresses are indicated in the corresponding figures. It can be seen that the deformation vs. time curves show typical brittle creep behavior characterized by the three-stage behavior as seen in previous studies (Amitrano and Helmstetter 2006; Heap et al. 2011; Brantut et al. 2013, 2014). Each primary creep stage is characterized by an initially high strain rate that decreased over time to reach an almost constant secondary stage strain rate, which is often interpreted as steady creep. Finally, the samples entered a tertiary phase characterized by an accelerated increase in strain. This

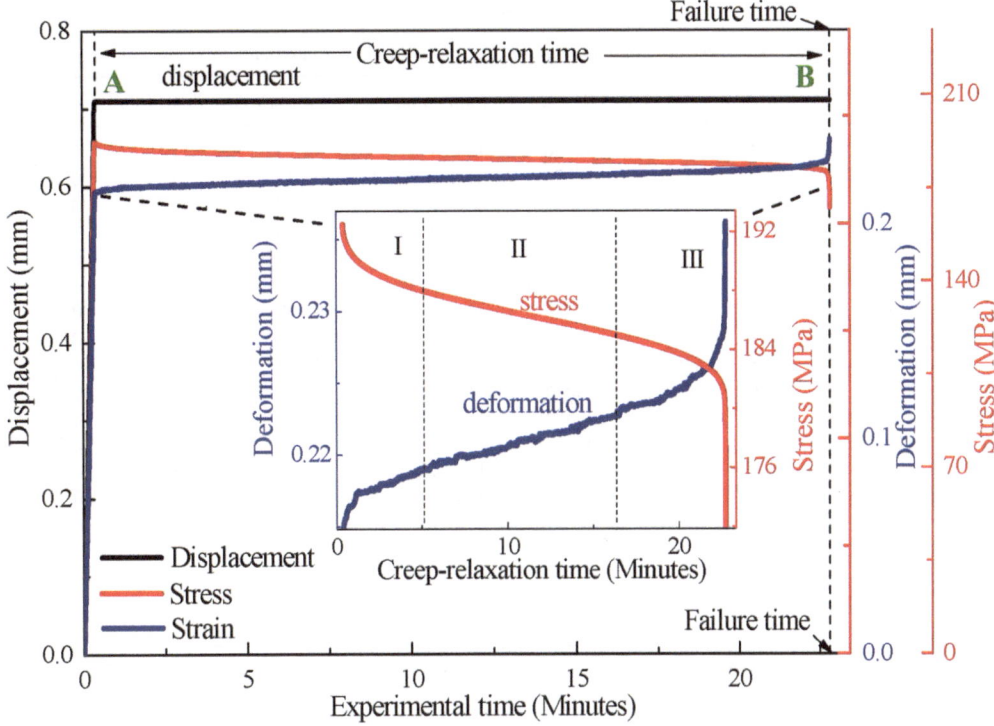

Figure 5
An example of the creep-relaxation curves observed in the experiments. The experiment is divided into two phases: a phase (0A) of applying an initial displacement on the crosshead, and a constant displacement (creep-relaxation) phase (*AB*). The *inset* shows a zoomed-in view of the creep-relaxation phase. Three phases (*I, II, III*) are labeled on the *inset* and a long steady stage with a nearly constant creep-relaxation rate is indicated

eventually resulted in macroscopic failure of the rock specimen.

Figure 5 illustrates a typical brittle creep relaxation result to demonstrate the complete process. As shown, the displacement was held constant (AB portion) well after the crosshead had reached the initial position (OA portion) (Fig. 5). The evolution curves of the deformation and stress at the constant displacement (AB) portion shown in the inset indicate that stress relaxation accompanied the increase in the deformation of the specimen. Figure 8 shows the results at the creep relaxation phase. It is clear that a typical stress relaxation process can be described as the rapid initial relaxation of stress, followed by a pseudo-steady stage in which the stress decreases at constant rate (Hao et al. 2014). This constant rate of stress relief terminates abruptly by an acceleration of stress loss and deformation that often culminates in catastrophic failure (Fig. 5).

In the brittle creep and creep relaxation experiments, some specimens immediately failed upon application of the initial stress or displacement. These experiments were discarded as they produced no data (the typical curves in type 2 and type 3 experiments are shown in Figs. 15 and 16 in Appendix, respectively). Some specimens did not fail during the available time window of 4 days. These experiments were also discarded (the exemplary curves in type 2 and type 3 experiments are shown in Figs. 15 and 16, respectively).

3.2. Scaling law of average failure rate and steady-state rate

It can be seen that the monotonic and creep-relaxation experiments showed a pseudo-steady stage similar with the creep experiments. To further characterize the three stages of the evolution of the

Table 2

Experimental conditions and results for all type 1 experiments

Sample number	σ_{max} (MPa)	σ_f (MPa)	u_f (mm)	U_f (mm)	Sample number	σ_{max} (MPa)	σ_f (MPa)	u_f (mm)	U_f (mm)
Tl-M-1	189.2	189.0	0.221	0.550	Tl-G-1	77.76	77.89	0.271	0.596
Tl-M-2	183.7	182.6	0.262	0.573	Tl-G-2	58.45	58.73	0.258	0.548
Tl-M-3	180.9	179.3	0.272	0.570	Tl-G-3	69.79	69.97	0.243	0.585
Tl-M-4	182.5	180.1	0.272	0.560	Tl-G-4	59.85	60.10	0.252	0.563
Tl-M-5	110.4	108.6	0.220	0.397	Tl-G-5	39.179	40.146	0.328	0.423
Tl-M-6	127.6	126.4	0.205	0.397	Tl-G-6	61	61.40	0.304	0.610
Tl-M-7	125.6	124.4	0.241	0.396	Tl-G-7	61.88	61.98	0.228	0.542
Tl-M-8	113.3	112.4	0.260	0.407	Tl-G-8	42.46	42.63	0.216	0.431
Tl-M-9	133.5	132.5	0.211	0.398	Tl-G-9	49.05	49.46	0.311	0.526
Tl-M-10	111.5	110.1	0.236	0.375	Tl-G-10	41.19	41.76	0.303	0.502
Tl-M-11	122.7	121.4	0.233	0.384	Tl-G-11	46.61	47.77	0.292	0.509
Tl-M-12	174.5	173.2	0.335	0.542	Tl-G-12	62.96	63.36	0.213	0.515
Tl-M-13	121.0	119.9	0.216	0.386	Tl-G-13	68.54	68.7	0.206	0.559
Tl-M-14	196.1	196.1	0.240	0.582	Tl-G-14	71.69	71.96	0.233	0.581
Tl-M-15	174.4	174.4	0.186	0.481	Tl-G-15	74.51	74.91	0.277	0.583
Tl-M-16	172.6	172.5	0.183	0.466					
Tl-M-17	88.8	87.9	0.289	0.395					
Tl-M-18	91.3	90.8	0.287	0.395					
Tl-M-19	182.4	181.3	0.298	0.584					

T1 type 1 experiments, *M* marble, *G* granite. Load rate: $dU/dt = 0.02$ mm/min

properties of rock failure, the first derivatives of the deformation with respect to the displacement (du/dU) for the monotonic experiments were plotted and two such plots are shown in Fig. 9 as examples. For the brittle creep experiments, the first derivatives of the deformation–time curves (i.e., the strain rate against time) are calculated and shown in Fig. 10. Figure 11 presents the curves of the strain and stress rates against time for the creep relaxation tests.

It can be seen that the responses showed a common three-stage behavior for all three experiments. All of these plots demonstrated that there are long segments with an almost constant slope, as indicated by the horizontal sections of the curves. It is clear that the secondary stage dominated the lifetime of the specimens in all experiments under the three loading modes. The evolution characteristics of the secondary stage determine the transition from the steady state to the unstable state in the tertiary stage and thus determine whether macroscopic failure will occur.

For the brittle creep relaxation experiments and creep experiments, the resultant creep deformation is given by $u_f - u_0$, and hence, $\mu = \frac{\mu_f - \mu_0}{t_f - t_0}$ represents the average creep deformation rate. Similarly, for the

monotonic loading experiments, the average deformation rate is $\mu = \frac{\mu_f}{U_f}$. Here, μ_f is the resultant deformation and μ_f is the resultant loading displacement (or a proxy measure of lifetime because U has a linear relationship with time). For a brittle rock material, the initiation, propagation, interaction, and coalescence of cracks are the main mechanism of deformation. Therefore, the deformation can be a proxy measurement of the change in damage during sample deformation. The ratio λ_s/μ represents proportion of damage rate in the secondary state. A larger value of λ_s/μ implies a smaller proportion of duration in the secondary stage, and consequently, a larger part of the damage is developed in the tertiary stage.

Therefore, as shown in Figs. 6, 7 and 8, a steep slope of the steady stage implies a short lifetime. The double-logarithm plots of the lifetime and λ_s, are shown in Figs. 12, 13 and 14. The linear relationship in the double-logarithm plots indicates that μ follows a power law

$$\mu = A\lambda_s^\alpha \tag{1}$$

with the secondary creep rate λ_s.

Table 3

Experimental conditions and results for all type 2 experiments

Sample number	σ_{max} (MPa)	u_0 (mm)	u_f (mm)	$t_f - t_f$ (s)
T2-G-80-1	181.2	0.208	0.262	570
T2-G-80-2	181.2	0.216	0.267	6355
T2-G-80-3	181.2	0.208	0.232	5775
T2-G-85-1	190.6	0.227	0.265	133
T2-G-85-2	190.6	0.226	0.255	397
T2-G-85-3	190.6	0.216	0.227	2051
T2-G-90-1	203.1	0.247	0.271	56
T2-G-90-2	203.1	0.222	0.248	19
T2-G-90-3	203.1	0.237	0.269	115
T2-G-9(M	203.1	0.239	0.267	122
T2-G-90-5	203.1	0.228	0.252	290
T2-G-90-6	203.1	0.231	0.262	539
T2-G-90-7	203.1	0.232	0.254	830
T2-G-90-8	203.1	0.227	0.252	9929
T2-G-90-9	203.1	0.227	0.245	1317
T2-G-90-10	203.1	0.232	0.250	378
T2-G-90-11	203.1	0.224	0.257	4569
T2-G-90-12	203.1	0.232	0.253	481
T2-G-95-1	212.5	0.293	0.321	23
T2-G-95-2	212.5	0.289	0.310	20
T2-G-95-3	212.5	0.251	0.276	32
T2-G-95-4	212.5	0.250	0.279	153
T2-G-95-5	212.5	0.244	0.266	230
T2-G-95-6	212.5	0.236	0.269	559

It should be mentioned that there was an event of fracture that occurred during the testing of specimens SC-G-80-2 and SC-G-85-3. We heard a sound emitted by the fracture but the specimen did not fail completely. This small event induced a small jump in the curves (see Fig. 7a, b). Thus, the stable stage has a more direct relation with this small event than the macroscopic failure, and then we select the u_f and t_f corresponding to this small event to calculate the value of the average creep rate μ

T2 type 2 experiments, *G* granite

Table 4

Experimental conditions and results for all type 3 experiments

Sample number	U (mm)	σ_0 (MPa)	σ_f (MPa)	u_0 (MPa)	u_f (MPa)	$t_f - t_f$ (s)
T3-G-1	0.770	238.4	227.9	0.241	0.231	347
T3-G-2	0.750	222.2	237.1	0.214	0.232	797
T3-G-3	0.760	2040	229.1	0.174	0.193	497
T3-G-4	0.730	228.0	209.3	0.222	0.243	2471
T3-G-5	0.710	2140	193.6	0.214	0.238	1345
T3-G-6	0.682	169.7	120.9	0.107	0.182	125
T3-G-7	0.753	214.7	183.1	0.169	0.219	285
T3-G-S	0.753	225.4	198.3	0.147	0.189	208
T3-G-9	0.741	203.1	170.0	0.220	0.281	593
T3-G-10	0.721	202.7	149.8	0.233	0.324	926
T3-M-1	0.760	225.8	214.8	0.226	0.237	593
T3-M-2	0.783	186.0	157.8	0.160	0.239	83
T3-M-3	0.756	191.9	152.9	0.543	0.382	129
T3-M-4	0.756	166.9	127.2	0.342	0.456	64

T2 type 3 experiments, *M* marble, *G* granite

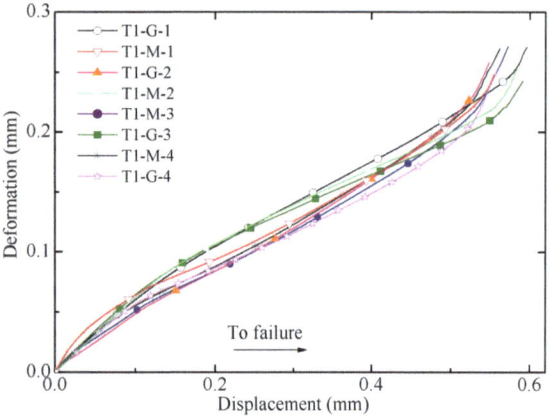

Figure 6
Deformation curves in monotonic tests. The *curves* exhibit the apparent tri-stage behaviors. *T1* type 1 experiments, *G* granite, *M* marble

The power law exponent was approximately 0.70 [0.72 ± 0.04 for the marble specimens (Fig. 12a) and 0.70 ± 0.04 for the granite specimens (Fig. 12b)] in the monotonic loading experiments. The exponent α is 0.92 ± 0.04 in the brittle creep experiments (Fig. 13), which is almost a linear relationship between μ and λ_s. On the other hand, the exponent α is about 0.85 [0.84 ± 0.03 (Fig. 14a) for stress relaxation and 0.86 ± 0.03 (Fig. 14b) for deformation evolution] in the brittle creep relaxation experiments. Our results indicate that the power law relation (1) is fitted well for the three types of experiments.

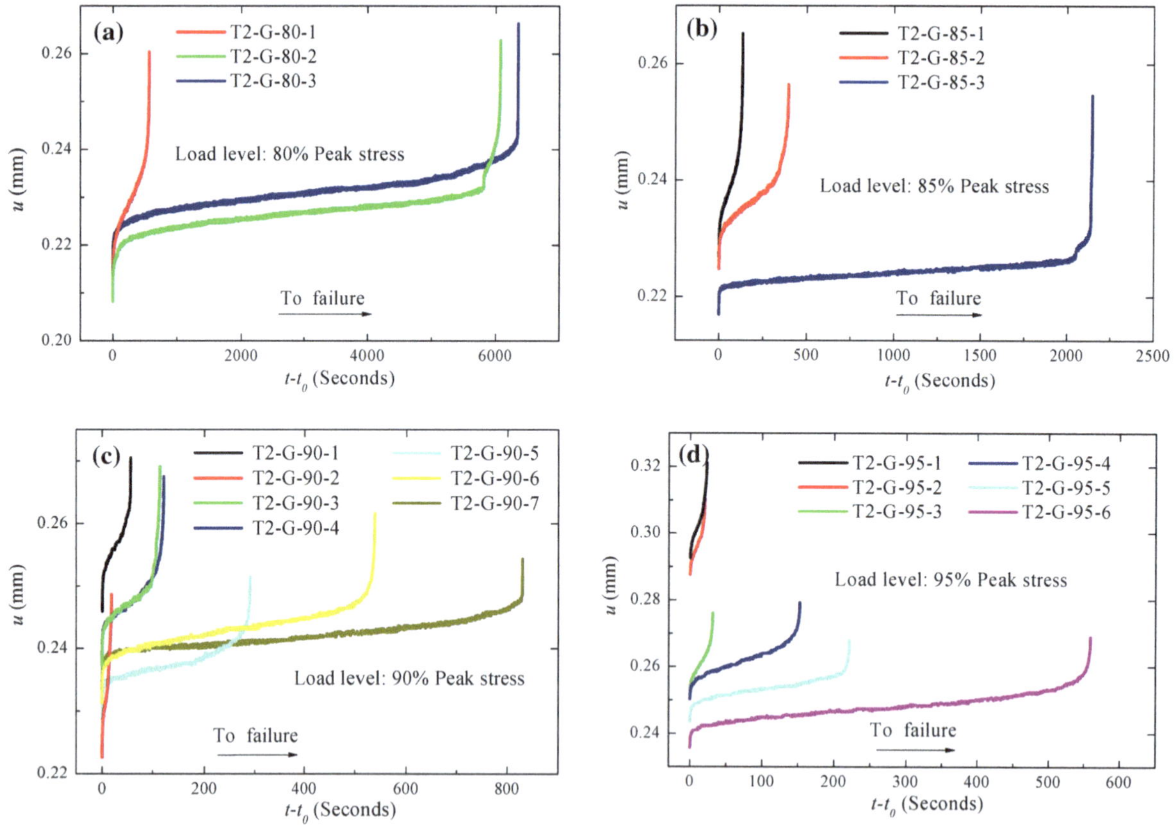

Figure 7
Creep curves for different applied stress levels. **a** Load level: 80% peak stress; **b** load level: 85% peak stress; **c** load level: 90% peak stress; **d** load level: 95% peak stress. The *curves* exhibit the apparent tri-stage behaviors. The specimens present different failure times and strain under the same stress level. *T2* type 2 experiments, *G* granite, *M* marble

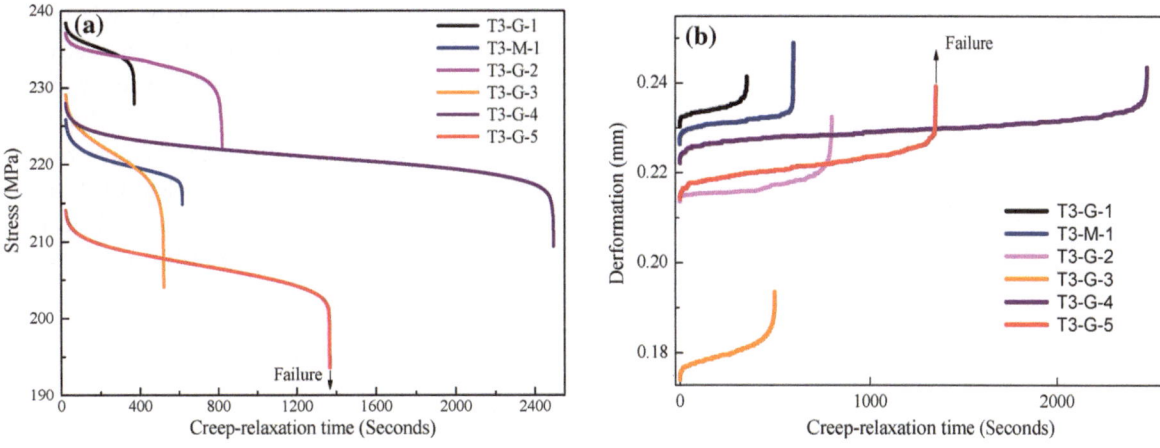

Figure 8
Creep-relaxation curves for six rock samples. Temporal evolution of **a** the stress and **b** deformation. The *curves* exhibit the apparent tri-stage behaviors. *T3* type 3 experiments, *G* granite, *M* marble

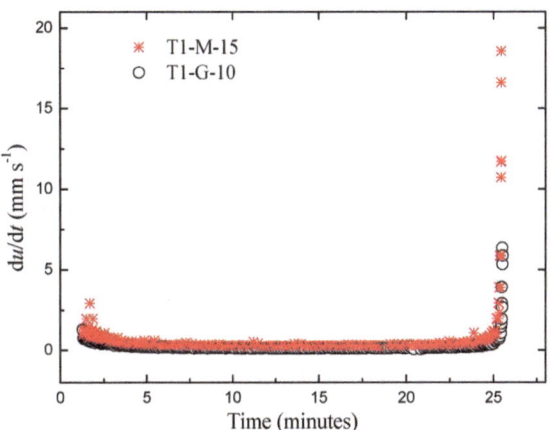

Figure 9
Deformation rate-displacement curves in the monotonic test. *G* granite, *M* marble

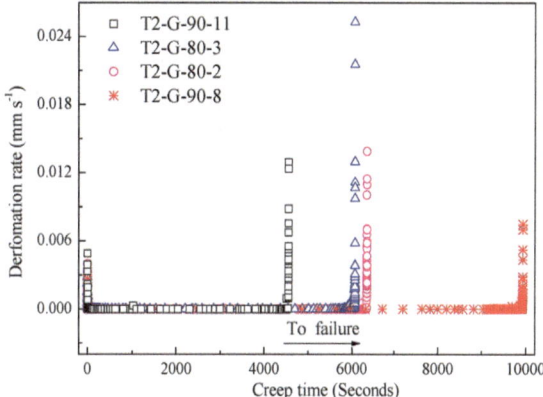

Figure 10
Deformation rate–time curves in the brittle creep experiments. *G* granite

4. Discussion

The subcritical growth of micro cracks, i.e. crack growth, can occur when the stress intensity factor K is lower than its critical value (also called as facture toughness). K_c, is suggested to be the main mechanism responsible for brittle creep of rocks (Scholz 1972; Atkinson and Meredith 1987; Lockner 1993; Amitrano and Helmstetter 2006). Subcritical crack growth can be caused by several competing mechanisms, including: stress corrosion, diffusion, dissolution, ion exchange and microplasticity (Atkinson 1984; Atkinson and Meredith 1987). The

theory of stress corrosion postulates that the reaction between strained bonds and the environmental agent produces a weakened (an activated) state that can then be broken at lower stresses than the unweakened bonds. The model of diffusion assumes that mass transport can be the dominant mechanism of subcritical crack growth. The dissolution model suggests that the growth rate of cracks is controlled by the silica dissolution rate. According to the theory of ion exchange, if the chemical environment contains species which can undergo ion exchange with species in the solid phase and if there is a gross mismatch in the size of these different species, then lattice strain can result from ion exchange which can facilitate

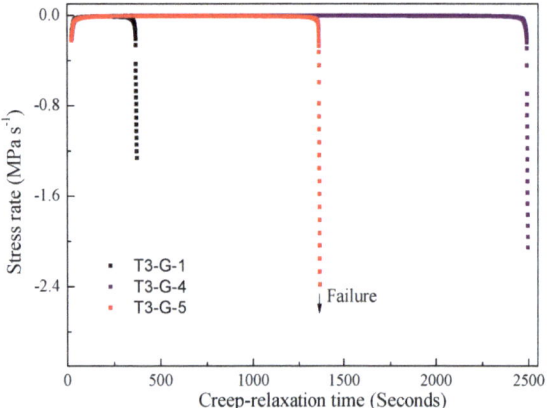

Figure 11
Stress rate–time curves in the creep-relaxation experiments.
G granite

crack growth. The theory of micro-plasticity suggests that the dislocation activity allowing plastic flow may be a significant mechanism of crack propagation. It is assumed that stress corrosion is the main mechanism of subcritical growth in shallow crustal conditions (Atkinson 1984; Atkinson and Meredith 1987).

The crack velocity fits a power law with stress intensity factor (Charles 1958; Atkinson 1984; Atkinson and Meredith 1987; Amitrano and Helmstetter 2006)

$$\frac{V}{V_0} = \left(\frac{K}{K_c}\right)^q \exp\left(\frac{-H}{RT}\right) \quad K_0 < K < K_c \quad (2)$$

where V is the crack growth velocity, R is the gas constant, T is the temperature, H is the activation energy, and K_0 is the threshold value below which no crack propagation is observed. Based on Eq. (2), a power–law relationship of the time-to-failure with the applied stress (Charles 1958; Cruden 1974; Kranz 1980; Lockner 1993; Amitrano and Helmstetter 2006) has been derived analytically and observed experimentally

$$t_f = t_* \left(\frac{\sigma}{\sigma_*}\right)^{-p} \quad (3)$$

The constants t_* and σ_* depend on rock properties and ambient conditions (Scholz 1972). An exponential relation (Wiederhorn and Bolz 1970; Das and Scholz 1981; Amitrano and Helmstetter 2006)

$$t_f = t_* \exp\left(-b\frac{\sigma}{\sigma_*}\right) \quad (4)$$

has also been suggested. These two empirical relations (3) and (4) are equivalent in terms of correlation coefficient (Amitrano and Helmstetter 2006) and indicate the stress dependence of the time-to-failure.

A theoretical derivation presented by Main (2000) gives a relationship between the strain rate $\dot{\varepsilon}$ and stress

$$\dot{\varepsilon} = C\sigma^m, \quad (5)$$

where C is a constant. Then, Eqs. (3) and (5) suggest a possible power law relation

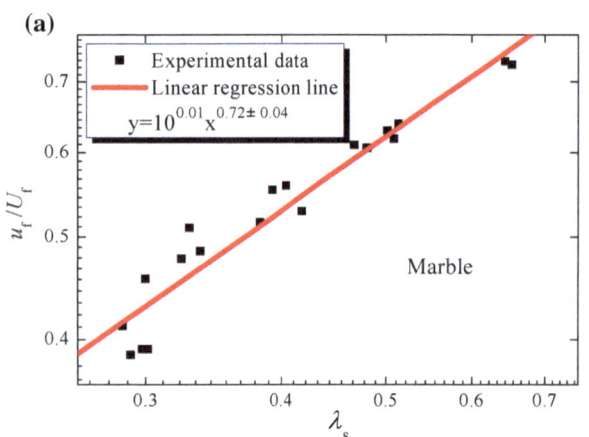

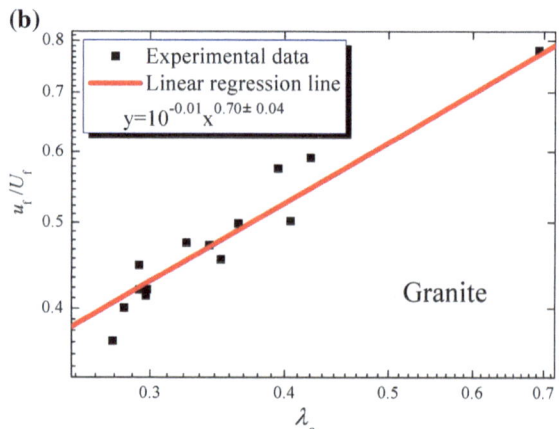

Figure 12
Double-logarithm plots of μ and λs in the monotonic loading tests. a Marble, b granite. The data fit well a power law. The two types of rock exhibit few differences with the exponents: marble is 0.72 and granite is 0.70

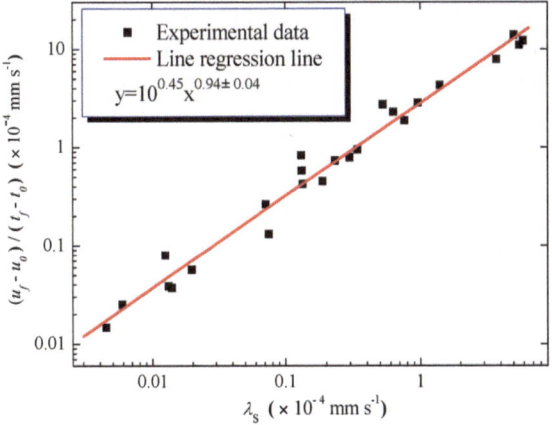

Figure 13
The double-logarithm plots of μ and λs in brittle creep tests. The data fit well a power law

$$\frac{K}{t_f} = \dot{\varepsilon}^\beta \qquad (6)$$

between the strain rate and the time-to-failure similar with the relation (1). It should be noted that if $\kappa = (\varepsilon_f - \varepsilon_0)$, then expression (6) reproduces our empirical relation (1).

Under quasi-static loading experiments, the damage evolution processes in rocks could present some common properties. In three types of experiments, the first stage is either related to the elastic closing of the pre-existing cracks and pores, or to the condition where the crack-closure rate exceeds the

crack-opening and propagation rates (Rudnicki and Rice 1975; Heap et al. 2009; Hao et al. 2013). This leads to an effect of strain hardening in the type 1 (Hao et al. 2013) and type 2 experiments (Lockner and Byerlee 1980; Lockner 1993). The secondary stage corresponds to the stage where damage to rock samples occurs randomly, and thus the spatial distribution of the strain field presents weak fluctuations (Hao et al. 2007, 2010). This occurs so that the macroscopic physical quantities (e.g., average strain or stress) that describe the global average mechanical responses evolve steadily. The diffused distribution of acoustic emission events has also been observed throughout secondary stage in the creep experiments and quasi-constant acoustic emission rate experiments (Lockner et al. 1991; Lockner 1993). The accelerating evolution in the tertiary stage always results from the localization and coalescence of cracks observed in monotonic loading experiments (Hao et al. 2007, 2010), creep experiments, and in quasi-constant acoustic emission rate experiments (Lockner et al. 1991).

To further demonstrate the relationship between failure and the secondary stage, let us focus on the brittle creep relaxation experiments. For a plastic material, stress relaxation is achieved through a process in which the initial imposed elastic strain is replaced over time by an inelastic strain. However, for heterogeneous brittle materials such as rocks, a

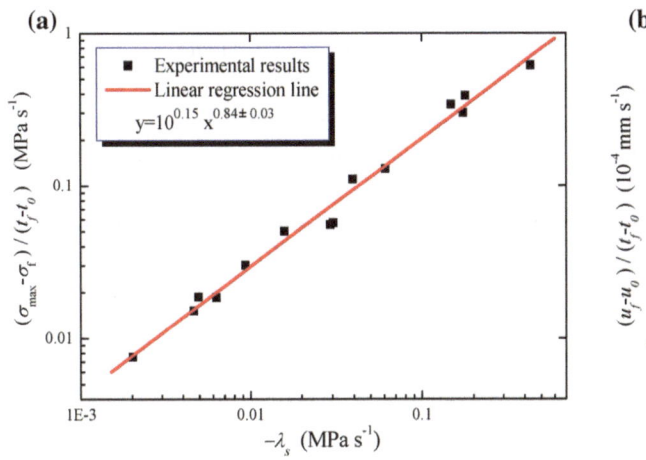

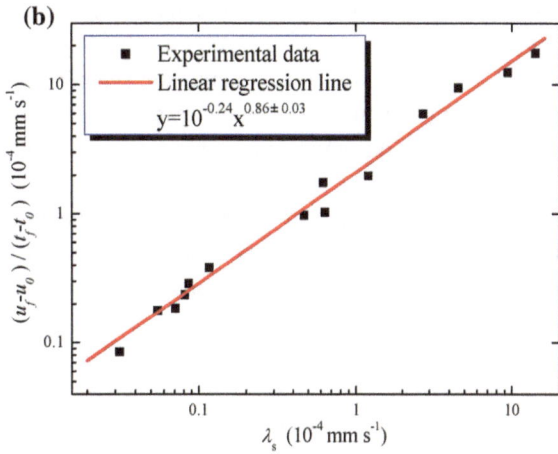

Figure 14
The double-logarithm plots of μ and λs in brittle creep-relaxation tests. **a** $(\sigma_{\max} - \sigma_f)/(t_f - t_0)$ versus $-\lambda_s$ and **b** $(u_f - u_0)/(t_f - t_0)$ versus λ_s. The data fit well to a power law

S. Hao et al. — Pure Appl. Geophys.

drop in the stress is induced by the development of damage, which includes the initiation, propagation, coalescence, and growth of cracks and defects.

It should be mentioned that the evolution of the macroscopic response depends on the microphysical of the rock specimen, which induces sample-specificities of rock specimens. The evolution properties (e.g., the slope λ_s of the secondary stage) of the response (strain or stress) curves and the eventual failure are determined by the properties of damage evolution for a specific specimen and the loading condition. In type 1 and type 3 experiments, the elasticity of the testing machine (the stiffness of the loading apparatus is ~130 kN/mm) played an extremely important role in promoting failure. The condition of catastrophic rupture in type 1 experiments is controlled by two main factors (Bai et al. 2005; Jeager et al. 2007; Hao et al. 2007, 2010, 2013): the stiffness ratio between the load apparatus and the rock sample, and the damage evolution properties of the rock. The condition of catastrophic rupture in type 3 experiments is controlled by three factors (Hao et al. 2013): the stiffness ratio between the load apparatus and the rock sample, the initial applied displacement, and the damage evolution properties of rock. In the type 2 experiment, failure is determined mainly by the initial stress level and the specific damage evolution properties of a rock specimen. Therefore, it is difficult to decouple these combined effects of loading level, testing machine, and sample-specificities of rock specimens. Both the slope of the secondary stage and failure time are commonly dependent on these conditions. As a result, the variability of these factors do not change the scaling law between the two main macroscopic responses (λ_s and μ), but lead to the differences in λ_s, power exponent and failure time among the specimens.

The ratio λ_s/μ between the average deformation rate λ_s of the secondary stage and the average rate λ_s for whole lifetime reflects the contribution of the secondary stage to the whole failure.

5. Conclusions

In the three types of our quasi-static experiments, rock specimens presented an apparent long steady (or pseudo-steady) stage followed by a rapid deformation stage, leading to a macroscopic failure. The steady stage dominates almost the entire lifetime of a rock specimen. A steep slope of the steady stage implies a short lifetime. These properties promise a unified description of the relationship between the time-to-failure and failure with the evolution properties in the steady stage.

The present experimental results show that the lifetime of rock specimens can be commonly expressed as a power–law relationship with the slope of the steady stage [cf. Eqs. (1) and (6)]. The power law exponent is approximately 0.70 in the monotonic loading experiments, 0.92 ± 0.04 in the brittle creep experiments, and approximately 0.85 in the brittle creep relaxation experiments. Further investigations are needed to determine the reason for the different exponent values. Under the creep experiments, the approximate linear relationship of the lifetime of rock specimens with a steady stage slope provides a potential method for predicting time-to-failure by using a linear extrapolation.

λ_s/μ, which is the ratio of the evolution rate in the steady stage with respect to the average rate of the total lifetime, is proposed to describe the failure mode. A larger value of λ_s/μ means a smaller damage proportion occurred in the steady stage, and consequently, a larger portion of damage occurred in the tertiary stage, corresponding to more brittle failure.

Acknowledgements

This work is supported by National Natural Science Foundation of China (Grant 11672258), National Basic Research Program of China (Grant 2013CB834100) and Natural Science Foundation of Hebei Province (Grant D2015203398). We acknowledge useful comments of two anonymous reviewers.

Appendix: results for specimens that failed immediately and did not fail after a long time loading

See Figs. 15 and 16.

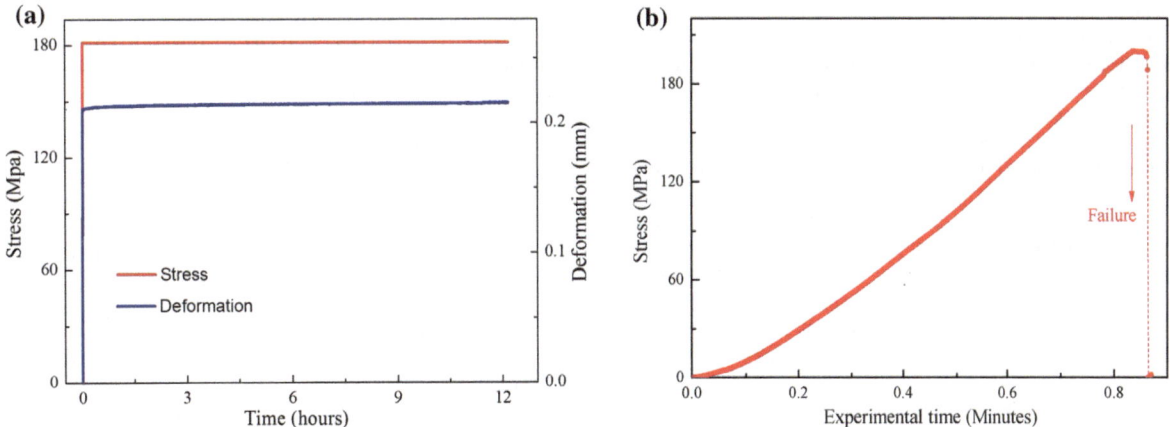

Figure 15
Results of the brittle creep tests: one specimen failed immediately while trying to hold the stress and one did not fail after a long load time. **a** The results of a specimen that did not fail after a long load time. After approximately 12.16 h of testing, the data are too large to be stored. **b** The *curves* of stress vs. time of a specimen that failed immediately while trying to hold the stress

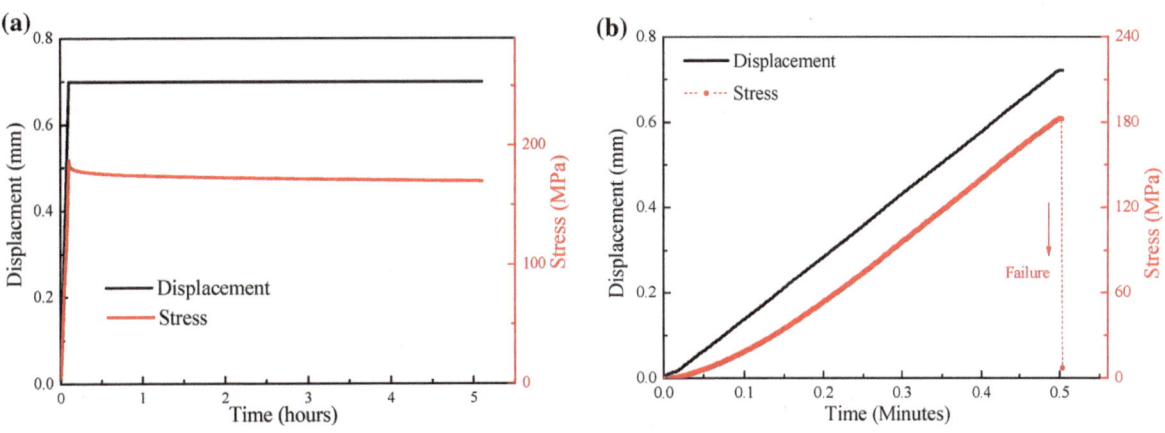

Figure 16
Results in creep relaxation tests: one specimen failed immediately when trying to hold the displacement and one that did not fail after a long time loading. **a** The results of a specimen that did not fail after a long time loading. After about 5.1 h in the tests, the data is too large to be stored. **b** The results of a specimen that failed immediately when try to hold the displacement

REFERENCES

Amitrano, D., Grasso, J. R., & Senfaute, G. (2005). Seismic precursory patterns before a cliff collapse and critical point phenomena. *Geophysical Research Letters, 32*(5), L08314. doi:10.1029/2004GL022270.

Amitrano, D., & Helmstetter, A. (2006). Brittle creep, damage and time to failure in rocks. *Journal of Geophysical Research, 111*, 1–17, B11201. doi:10.1029/2005JB004252.

Andrade E. N. da C. (1910). On the viscous flow in metals and allied phenomena. *Proceedings of the Royal Society of London. Series A, 84*, 1.

Atkinson, B. K. (1984). Subcritical crack growth in geological materials. *Journal of Geophysical Research, 89*, 4077–4114.

Atkinson, B., Meredith, P. (1987) The theory of subcritical crack growth with applications to minerals and rocks. In: Fracture mechanics of rocks (pp. 111–166). New York: Academic Press.

Bai, Y. L., Wang, H. Y., Xia, M. F., & Ke, F. J. (2005). Statistical mesomechanics of solid, liking coupled multiple space and time scales. *Applied Mechanics Review, 58*, 372–388.

Baud, P., & Meredith, P. (1997). Damage accumulation during triaxial creep of darley dale sandstone from pore volumetry and acoustic emission. *International Journal of Rock Mechanics and Mining Sciences and Geomechanics Abstracts, 34*(3–4), 1–8.

Benioff, H. (1951). Earthquake and rock creep. *Bulletin of the Seismological Society of America, 41*(1), 31–62.

Boukharov, G., Chanda, M., & Boukharov, N. (1995). The three processes of brittle crystalline rock creep. *International Journal of Rock Mechanics and Mining Sciences and Geomechanics Abstracts, 32*(4), 325–335.

Brantut, N., Heap, M. J., Baud, P., & Meredith, P. G. (2014). Rate- and strain-dependent brittle deformation of rocks. *Journal of Geophysical Research, 119*, 1818–1836. doi:10.1002/2013JB010448.

Brantut, N., Heap, M. J., Meredith, P. G., & Baud, P. (2013). Time-dependent cracking and brittle creep in crustal rocks: a review. *Journal of Structural Geology, 52*, 17–43.

Charles, R. (1958). The static fatigue of glass. *Journal of Applied Physics, 29*, 1549–1560.

Chen, Z., et al. (2000). Global Positioning System measurements from eastern Tibet and their implications for India/Eurasia intercontinental deformation. *Journal of Geophysical Research, 105*, 16215–16227.

Cruden, D. (1974). The static fatigue of brittle rock under uniaxial compression. *International Journal of Rock Mechanics and Mining Sciences and Geomechanics Abstracts, 11*, 67–73.

Das, S., & Scholz, C. (1981). Theory of time-dependent rupture in the Earth. *Journal of Geophysical Research, 86*, 6039–6051.

Du, Z. Z., & McMeeking, R. M. (1995). Creep models for metal matrix composites with long brittle fibers. *Journal of the Mechanics and Physics of Solids, 43*, 701–726.

Guarino, A., Ciliberto, S., Garcimartin, A., Zei, M., & Scoretti, R. (2002). Failure time and critical behavior of fracture precursors in heterogeneous materials. *European Physical Journal B: Condensed Matter and Complex Systems, 26*, 141–151.

Hao, S. W., Liu, C., Lu, C. S., & Elsworth, D. (2016). A relation to predict the failure of materials and potential application to volcanic eruptions and landslides. *Scientific Reports, 6*, e27877. doi:10.1038/srep27877.

Hao, S. W., Rong, F., Lu, M. F., Wang, H. Y., Xia, M. F., Ke, F. J., et al. (2013). Power-law singularity as a possible catastrophe warning observed in rock experiments. *International Journal of Rock Mechanics and Mining Sciences, 60*, 253–262.

Hao, S. W., Wang, H. Y., Xia, M. F., Ke, F. J., & Bai, Y. L. (2007). Relationship between strain localization and catastrophic failure. *Theoretical and Applied Fracture Mechanics, 48*, 41–49.

Hao, S. W., Xia, M. F., Ke, F. J., & Bai, Y. L. (2010). Evolution of localized damage zone in heterogeneous media. *International Journal of Damage Mechanics, 19*(7), 787–804.

Hao, S. W., Zhang, B. J., Tian, J. F., & Elsworth, D. (2014). Predicting time-to-failure in rock extrapolated from secondary creep. *Journal of Geophysical Research Solid Earth, 119*, 1942–1953. doi:10.1002/2013JB010778.

Heap, M. J., Baud, P., Meredith, P. G., Bell, A. F., & Main, I. G. (2009). Time-dependent brittle creep in Darley Dale sandstone. *Journal of Geophysical Research Solid Earth, 114*, B07203. doi:10.1029/2008JB006212.

Heap, M. J., Baud, P., Meredith, P. G., Vinciguerra, S., Bell, A. F., & Main, I. G. (2011). Brittle creep in basalt and its application to time-dependent volcano deformation. *Earth and Planetary Science Letters, 37*(1–2), 71–82.

Hundson, J. A., Crouch, S. L., & Fairhurst, C. (1972). Soft, stiff and servo-controlled testing machines: a review with reference to rock failure. *Engineering Geology, 6*, 155–189.

Jeager, J. C., Cook, N. G. W., & Zimmerman, R. (2007). *Fundamentals of rock mechanics* (4th ed.). London: Wiley-Blackwell.

Kilburn, C. R. J. (2012). Precursory deformation and fracture before brittle rock failure and potential application to volcanic unrest. *Journal of Geophysical Research, 117*, B02211. doi:10.1029/2011JB008703.

Kranz, R. (1980). The effect of confining pressure and difference stress on static fatigue of granite. *Journal of Geophysical Research, 85*, 1854–1866.

Kranz, R., Harris, W., & Carter, N. (1982). Static fatigue of granite at 200°C. *Geophysical Research Letters, 9*(1), 1–4.

Labuz, J. F., & Biolzi, L. (1991). Class I vs class II stability: a demonstration of size effect. *International Journal of Rock Mechanics and Mining Sciences and Geomechanics Abstracts, 28*(2/3), 199–205.

Lienkaemper, J. J., Galehouse, J. S., & Simpson, R. W. (1997). Creep response of the Hayward fault to stress changes caused by the Loma Prieta earthquake. *Science, 276*, 2014–2016.

Lockner, D. A. (1993). The role of acoustic emission in the study of rock fracture. *International Journal of Rock Mechanics and Mining Sciences and Geomechanics Abstracts, 30*(7), 883–899.

Lockner, D. A. (1998). A generalized law for brittle deformation of westerly granite. *Journal of Geophysical Research, 103*(B3), 5107–5123.

Lockner, D. A., & Byerlee, J. D. (1980). Development of fracture planes during creep in granite. In H. R. Hardy, F. W. Leiton, (Eds.) *Proceedings of 2nd conference on acoustic emission/microseismic activity in geological structures and materials* (pp 11–25). Clausthal-Zellerfeld, Germany: Trans Tech Publications.

Lockner, D. A., Byerlee, J. D., Kuksenko, V., Ponomarev, A., & Sidorin, A. (1991). Quasi-static fault growth and shear fracture energy in granite. *Nature, 350*(7), 39–42.

Main, I. (2000). A damage mechanics model for power–law creep and earthquake aftershock and foreshock sequences. *Geophysical Journal International, 142*, 151–161.

Meade, B. J. (2007). Present-day kinematics at the India–Asia collision zone. *Geology, 35*, 81–84.

Nechad, H., Helmstetter, A., Guerjouma, R. E., & Sornette, D. (2005). Andrade creep and critical time-to-failure laws in heterogeneous materials. *Physical Review Letters, 94*, 045501.

Okubo, S., Nishimatsu, Y., & Fukui, K. (1991). Complete creep curves under uniaxial compression. *International Journal of Rock Mechanics and Mining Sciences and Geomechanics Abstracts, 28*(1), 77–82.

Omori, F. (1894). On the aftershocks of eathquakes. *Journal of the College of Science, Imperial University of Tokyo, 7*, 111–120.

Perfettin, H., & Avouac, J. P. (2004). Postseismic relaxation driven by brittle creep: A possible mechanism to reconcile geodetic measurements and the decay rate of aftershocks, application to the Chi-Chi earthquake, Taiwan. *Journal of Geophysical Research, 109*, B02304. doi:10.1029/2003JB002488.

Petley, D., Bulmer, M., & Murphy, W. (2002). Patterns of movement in rotational and translational landslides. *Geology, 30*(8), 719–722.

Rudnicki, J. W., & Rice, J. R. (1975). Conditions for the localization of deformation in pressure-sensitive dilatant materials. *Journal of the Mechanics and Physics of Solids, 23*, 371–394.

Saito, M. (1969). Forecasting time of slope failure by tertiary creep. In Proc. 7th Int. Conf. Soil Mechanics and Foundation Engineering, Mexico City (Vol. 2, pp. 677–683).

Saito, M., Uezawa, H. (1961) Failure of soil due to creep. In Proc. 5th Int. Conf. Soil Mechanics and Foundation Engineering, Montreal (Vol. 1, pp. 315–318).

Salamon, M. D. G. (1970). Stability, instability and design of pillar workings. *International Journal of Rock Mechanics and Mining Sciences, 7*(6), 613–631.

Scholz, C. (1968). Mechanism of creep in brittle rock. *Journal of Geophysical Research, 73*(10), 3295–3302.

Scholz, C. (1972). Static fatigue of quartz. *Journal of Geophysical Research, 77,* 2104–2114.

Shen, Z. K., Lu, J. N., Wang, M., & Burgmann, R. (2005). Contemporary crustal deformation around the southeast borderland of the Tibetan Plateau. *Journal of Geophysical Research, 110,* 11409. doi:10.1029/2004JB003421.

Singh, D. P. (1975). A study of creep of rocks. *International Journal of Rock Mechanics and Mining Sciences and Geomechanics Abstracts, 12,* 271–276.

Utsu, T. (1961). Statistical study on the occurrence of aftershocks. *Geophysical Magazine, 30,* 521–605.

Voight, B. (1988). A method for prediction of volcanic eruption. *Nature, 332,* 125–130.

Voight, B. (1989). A relation to describe rate-dependent material failure. *Science, 243,* 200–203.

Wiederhorn, S. M., & Bolz, L. H. (1970). Stress corrosion and static fatigue of glass. *Journal of the American Ceramic Society, 50,* 543–548.

Zhang, P. Z. (2013). Beware of slowly slipping faults. *Nature Geoscience, 6,* 323–324.

(Received February 1, 2016, revised March 2, 2017, accepted March 10, 2017, Published online March 21, 2017)

Pure Appl. Geophys. 174 (2017), 2217–2237
© 2016 The Author(s)
This article is published with open access at Springerlink.com
DOI 10.1007/s00024-016-1422-9

| Pure and Applied Geophysics

Apparent Dependence of Rate- and State-Dependent Friction Parameters on Loading Velocity and Cumulative Displacement Inferred from Large-Scale Biaxial Friction Experiments

YUMI URATA,[1] 🔘 FUTOSHI YAMASHITA,[1] EIICHI FUKUYAMA,[1] HIROYUKI NODA,[2] and KAZUO MIZOGUCHI[3]

Abstract—We investigated the constitutive parameters in the rate- and state-dependent friction (RSF) law by conducting numerical simulations, using the friction data from large-scale biaxial rock friction experiments for Indian metagabbro. The sliding surface area was 1.5 m long and 0.5 m wide, slid for 400 s under a normal stress of 1.33 MPa at a loading velocity of either 0.1 or 1.0 mm/s. During the experiments, many stick–slips were observed and those features were as follows. (1) The friction drop and recurrence time of the stick–slip events increased with cumulative slip displacement in an experiment before which the gouges on the surface were removed, but they became almost constant throughout an experiment conducted after several experiments without gouge removal. (2) The friction drop was larger and the recurrence time was shorter in the experiments with faster loading velocity. We applied a one-degree-of-freedom spring-slider model with mass to estimate the RSF parameters by fitting the stick–slip intervals and slip-weakening curves measured based on spring force and acceleration of the specimens. We developed an efficient algorithm for the numerical time integration, and we conducted forward modeling for evolution parameters (b) and the state-evolution distances (L_c), keeping the direct effect parameter (a) constant. We then identified the confident range of b and L_c values. Comparison between the results of the experiments and our simulations suggests that both b and L_c increase as the cumulative slip displacement increases, and b increases and L_c decreases as the loading velocity increases. Conventional RSF laws could not explain the large-scale friction data, and more complex state evolution laws are needed.

Key words: Rate-and-state friction, large-scale experiment, stick–slips, numerical simulation, spring-slider model.

1. Introduction

An earthquake cycle involves a very wide range of slip velocities, from orders of magnitude slower than a plate motion to as fast as a slip velocity at a rupture front during an earthquake. As earthquakes occur on a fault repeatedly, the internal structure of the shear zone and its mechanical properties are considered to evolve with increasing cumulative slip displacement (e.g., Beeler et al. 1996). The modeling of a sequence of earthquakes over the geologically long time scale probably requires a fault constitutive law which can comprehensively describe the mechanical properties of faults over the wide range of slip velocities and cumulative displacement.

The rate- and state-dependent friction (RSF) laws have been widely used to simulate earthquake sequences (e.g., Hori et al. 2004; Lapusta and Liu 2009; Noda and Lapusta 2013). These laws were originally proposed to model laboratory experimental data (Dieterich 1978, 1979; Ruina 1983), and the RSF parameters have been investigated using biaxial loading apparatuses at the low slip velocity from $\sim$0.01 μm/s to $\sim$1 cm/s, in which the cumulative displacement was of the order of cm at most (e.g., Mair and Marone 1999).

To achieve higher slip velocity and larger cumulative displacement, rotary shear apparatuses were developed (e.g., Tullis and Weeks 1986; Tsutsumi and Shimamoto 1997). Beeler et al. (1996) estimated the RSF parameters for large cumulative displacement at the slip velocity of 1–10 μm/s. Since a rotary shear apparatus is capable of producing high slip velocity up to a seismic rate, steady-state friction coefficients of various rock types have been investigated at a wide range of slip velocities, and a

---
[1] National Research Institute for Earth Science and Disaster Resilience (NIED), 3-1 Tennodai, Tsukuba, Ibaraki 305-0006, Japan. E-mail: urata@bosai.go.jp
[2] Disaster Prevention Research Institute, Kyoto University, Kyoto, Japan.
[3] Central Research Institute of Electric Power Industry (CRIEPI), Abiko, Japan.

remarkable velocity-weakening property of rock friction was revealed (e.g., Di Toro et al. 2011). Although a rotary shear apparatus enables the investigation of rock friction properties with a wide range of slip velocities and large displacement as described above, the apparatus would not be suitable to investigate stick–slip behavior, which could be considered analogous to a sequence of earthquakes on natural faults (Brace and Byerlee 1966). To study the effect of cumulative displacement, it is important to consider the history of slip velocity in nature. In usual friction experiments at low slip rates, steady-state sliding of a fault is simulated, and the shear zone internal structure is developed under such circumstances. Natural fault hosting a sequence of earthquakes experiences quite different deformation conditions (e.g., repeated transients in the slip velocity with stress concentration at rupture fronts), and the resulting internal structure should be different from what is developed under steady-state sliding. Since the evolution of the internal structure causes evolution in the parameters in RSF (e.g., Beeler et al. 1996), it is important to study them in experiments with stick–slips to better understand behavior of seismogenic faults.

In addition to the limitations of the slip velocity and the cumulative displacement, conventional studies used small (on the order of 10 cm at most) rock specimens to estimate the constitutive friction parameters (e.g., Dieterich 1972; Marone and Cox 1994; Beeler et al. 1996). The constitutive friction parameters estimated for the small rock specimens may be different from those for the large rock specimens, as Yamashita et al. (2015) suggested that rock friction in meter-sized rock specimens starts to decrease at a work rate (the product of the shear stress and the slip rate) one order of magnitude smaller than that in centimeter-sized rock specimens.

Many previous studies stated above obtained the RSF parameters by the method of step changes in load point velocity. The RSF parameters can be estimated also from stick–slip behaviors (Mitchell et al. 2015). Mitchell et al. (2015) performed the inversions of experimental data for unstable sliding using a spring-slider model, but they ignored the inertia in their numerical simulations [see their Eq. (7)], which may lead to inaccurate estimation of

the RSF parameters because in the quasi-static system, finite amplitude periodic oscillations are observed for very limited parameters and the slip velocity becomes infinite in unstable sliding regimes (Gu et al. 1984) where the inertia makes the slip and stress evolution completely different (Rice and Tse 1986).

In this study, we estimated the RSF constitutive parameters for the data obtained in experiment data by large-scale (on the order of meters) biaxial rock friction experiments conducted by Fukuyama et al. (2014) to investigate the dependence of the parameters on the loading velocity and cumulative displacement. For the estimation, we performed fully dynamic simulations of a single-degree-of-freedom spring-slider model. For efficient calculation, we developed a new algorithm of numerical simulations which tremendously reduces the calculation time relative to conventional methods such as embedded Runge–Kutta method.

2. Large-Scale Biaxial Rock Friction Experiments

2.1. Experimental Procedure

Fukuyama et al. (2014) constructed a large-scale biaxial friction apparatus using a large-scale shaking Table (15 m wide and 14.5 m long) at the National Research Institute for Earth Science and Disaster Resilience (NIED) in Japan. Figure 1a shows a schematic diagram of the apparatus. A pair of rock specimens made of Indian metagabbro (see Fukuyama et al. 2016 for its mineral composition) was used. The lower specimen moved with the shaking table and the upper specimen was fixed to the outer base of the shaking table by a reaction force bar. The simulated fault between the specimens was 1.5 m long and 0.5 m wide, slid at the nominal loading velocity ($\bar{v}_L$) of either 0.1 or 1.0 mm/s in a single experiment. The shaking table was instructed to move at the constant loading velocity, but the loading velocity oscillated slightly, which will be detailed in Sect. 2.2.2. The slip displacement was approx. 40 and 400 mm for experiments with $\bar{v}_L$ of 0.1 and 1.0 mm/s, respectively. The specimens were reused in a series of experiments.

(a)

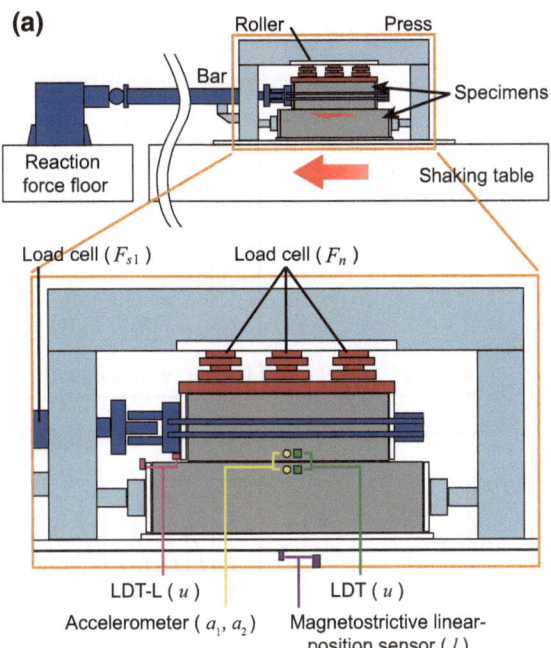

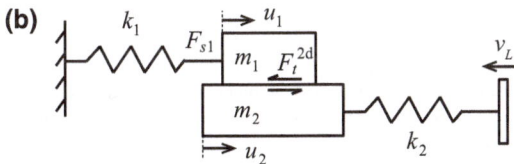

(b)

$$m_1 \quad F_t^{2d}$$

Figure 1

a A schematic diagram and **b** an equivalent mechanical model of a
large-scale rock friction apparatus. The spring force F_{s1} is
measured by a horizontal load cell, and the normal force F_n is
measured by three vertical load cells. The relative displacement of
the sliding surfaces u is measured by two laser displacement
transducers with different measurement ranges: short-range (LDT,
green squares) and long-range (LDT-L, *magenta squares*). In this
study, the LDT-L was installed on the side plate attached to the end of
the lower specimen and its target was attached at the edge of the
upper specimen. The LDT and its target were attached at the center
of the lower and upper specimens, respectively. Acceleration (a_1
and a_2) was measured in the upper and lower specimens at 20 mm
from the slip interface by two accelerometers installed there
(*yellow circles*). The table displacement l was measured by a
magnetostrictive linear-position sensor installed at the bottom of
the shaking table (*purple squares*)

Shear force on the simulated fault was produced
by the movement of shaking table and was sustained
by the reaction bar. The constant normal stress of
1.33 MPa was applied by three jacks and was
measured using three load cells serially connected
to each jack (Fig. 1a). Before the first experiment, the
sliding surfaces were flattened within 0.01 mm
undulation by a large-scale surface grinder. After

Table 1

Conditions of analyzed experiments

Experiment ID	$\bar{v}_L$ (mm/s)	Slip displacement (mm)	Gouge removal before experiment
LB01-127	0.1	40	Yes
LB01-134	0.1	40	No
LB01-142	1.0	400	No

each experiment, we sometimes removed and other
times left gouge particles produced during the
previous experiments. It should be noted that even
when the gouge particles were not removed, detailed
gouge structure might not be preserved because we
have to unload the normal stress and separate two
sliding surfaces at each end of the experiment due to
the limitation of shaking table operation. The detailed
conditions of experiments analyzed in this study are
shown in Table 1.

The relative displacement of the sliding surfaces
was measured by two laser displacement transducers
with different measurement ranges: a long-range
transducer (LDT-L) and a short-range transducer
(LDT). The LDT-L was installed on the side plate
attached to the end of the lower specimen, and its
target was attached at the edge of the upper specimen
(magenta squares in Fig. 1a). The LDT and its target
were attached at the center of the lower and upper
specimens, respectively (green squares in Fig. 1a).
Acceleration was measured by two accelerometers
installed in the upper and lower specimens at 20 mm
from the slip interfaces (yellow circles in Fig. 1a).
The force applied by the reaction force bar F_{s1} was
measured by a load cell (Fig. 1a).

The apparatus can be considered as a coupled
two-degree-of-freedom model (Fig. 1b), in the same
manner that Shimamoto et al. (1980) and Noda and
Shimamoto (2009) did. The force applied by the
reaction force bar corresponds to the spring force of
the upper spring F_{s1}. The loading velocity applied by
the shaking table is represented by v_L. The equations
of motion in this system are:

$$m_1 a_1 = -k_1 u_1 - F_t^{2d}$$
$$= F_{s1} - F_t^{2d} \qquad (1)$$

$$m_2 a_2 = -k_2 u_2 - k_2 v_L t + F_t^{2d}, \qquad (2)$$

where m is mass of the rock specimens, a is the acceleration of the specimens, k is the spring stiffness, u is the displacement of the specimens, subscripts 1 and 2 stand for the upper and lower specimens/springs, respectively, t is time, and F_t^{2d} is the shear force between the two specimens. It should be noted that LDT and LDT-L measured u ($=u_1 - u_2$), and that F_t^{2d} could not be measured directly.

2.2. Experimental Results

2.2.1 Behavior of Stick–Slips

Figures 2 and 3 show the experimental results for the friction ($\mu'_{ob} = F_{s1}/F_n$) obtained by a shear load F_{s1} divided by a normal load F_n (Fig. 1a). We applied a 4-kHz Butterworth-type low-pass filter to the load

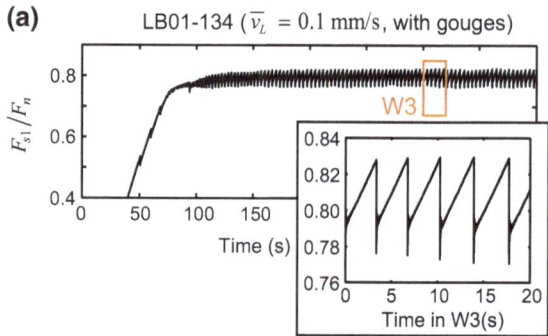

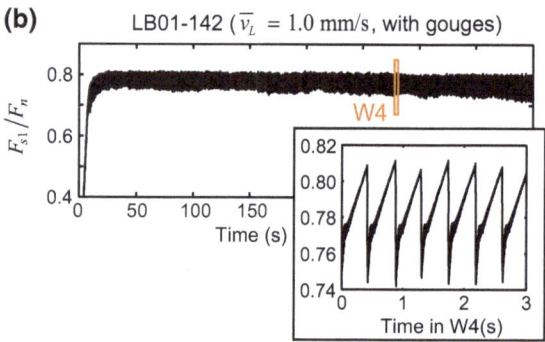

Figure 3
Time history of F_{s1}/F_n for experiments before which several experiments were conducted without gouge removal. $\bar{v}_L$ was **a** 0.1 mm/s (LB01-134) and **b** 1.0 mm/s (LB01-142). The *insets* in **a** and **b** show the details for time windows W3 (20 s) and W4 (3 s), respectively. The origin times of the *insets* in **a** and **b** correspond to 301 and 278 s in their *panels*, respectively

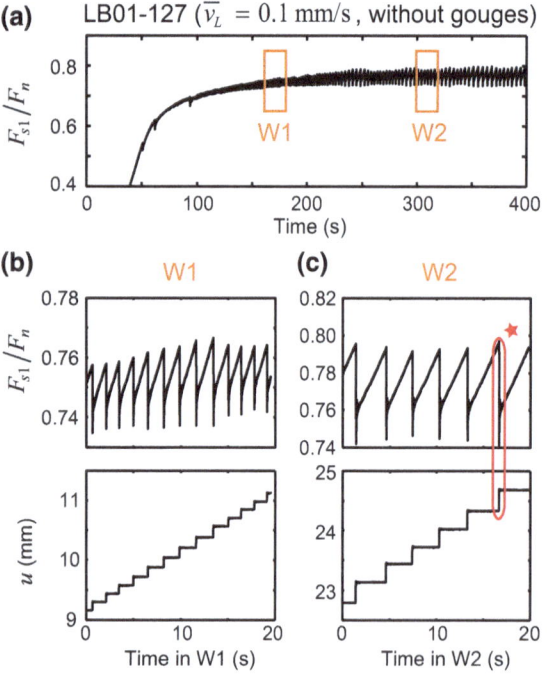

Figure 2
Results of an experiment (LB01-127) that started after the removal of gouges on the sliding surface. The loading velocity $\bar{v}_L$ was 0.1 mm/s. **a** Time history of reaction force F_{s1} normalized by the normal force F_n. **b, c** *Upper* and *lower panels* show the time histories of F_{s1}/F_n and cumulative displacement u, respectively, for time windows **b** W1 and **c** W2 in **a**. The origin times of **b** and **c** correspond to 161 and 299 s in **a**, respectively. The *red star* is referred to in Fig. 5

cell data to remove high-frequency noise. Many stick–slip events occurred during the experiments, and they had the following features.

The recurrence interval (ΔT_{ob}), friction drop ($\Delta \mu'_{ob}$), and displacement during a stick–slip event increased with cumulative slip displacement within a single experiment (LB01-127, Fig. 2a) if gouges on the surface were removed before it. The average ΔT_{ob} increased from approx. 1.5 s at short cumulative displacement (~ 10 mm, window W1, Fig. 2b) to approx. 3.1 s at long cumulative displacement (~ 24 mm, window W2, Fig. 2c). The average $\Delta \mu'_{ob}$ increased from approx. 0.022 at short cumulative displacement to approx. 0.046 at long cumulative displacement. The slip amount during a stick–slip event increased from approx. 0.14 mm at short cumulative displacement to approx. 0.32 mm at long cumulative displacement (bottom panels in Fig. 2b, c; black solid circles in Fig. 4). In contrast, ΔT_{ob} and $\Delta \mu'_{ob}$ became almost constant throughout the

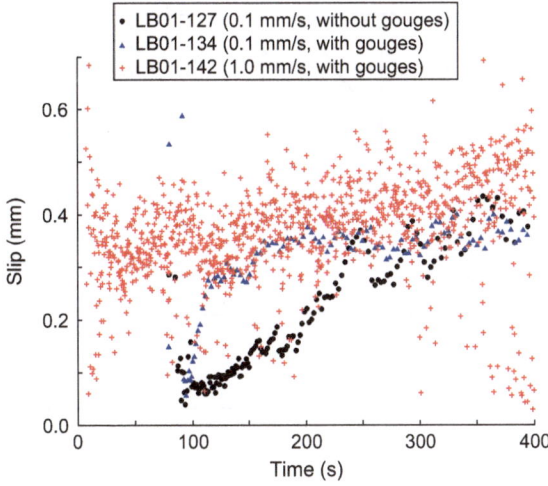

Figure 4
Temporal variation of slip amount during a stick–slip event in LB01-127 (*black circles*), LB01-134 (*blue triangles*), and LB01-142 (*red crosses*). It should be noted that in this plot, the slips due to slow slips and foreshocks are included

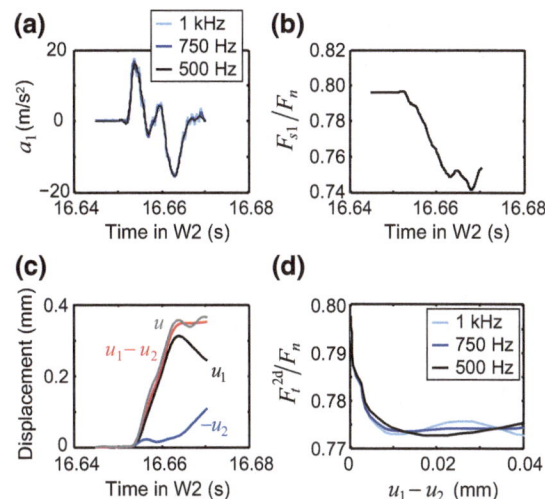

Figure 5
Estimation of a slip-weakening curve for the event marked by the *red star* in Fig. 2c. **a** Time history of the acceleration of the upper rock specimen a_1. Three different low-pass filtered accelerations are shown: 1 kHz (*light blue curve*), 750 Hz (*blue curve*), and 500 Hz (*black curve*). **b** Time history of F_{s1}/F_n. **c** Time history of the displacement of the upper (*black curve*) and lower (*blue curve*) specimens estimated by double integration of the original acceleration waveforms. The relative displacement by acceleration (*red curve*) was consistent with that of the LDT (*gray line*). **d** Slip-weakening curves obtained using the acceleration data with a 1-kHz (*light blue*), 750-Hz (*blue*), and 500-Hz (*black*) low-pass filter applied

experiment when the gouge particles were left on the fault prior to the experiment (Fig. 3a; blue triangles in Fig. 4).

In addition to the dependence on the cumulative slip, those characteristics of stick–slips depended on $\bar{v}_L$; the slip amount during an event and $\Delta\mu'_{ob}$ were larger and ΔT_{ob} was shorter for the experiments with faster $\bar{v}_L$. The average ΔT_{ob} was approx. 3.6 s for slow $\bar{v}_L$ ($\bar{v}_L = 0.1$ mm/s, LB01-134, window W3 in Fig. 3a) and approx. 0.44 s for fast $\bar{v}_L$ ($\bar{v}_L = 1.0$ mm/s, LB01-142, window W4 in Fig. 3b). The average $\Delta\mu'_{ob}$ was approx. 0.052 for slow $\bar{v}_L$ (window W3) and approx. 0.060 for fast $\bar{v}_L$ (window W4). The slip amounts per event were approx. 0.35 mm for slow $\bar{v}_L$ (window W3) and 0.39 mm for fast $\bar{v}_L$ (window W4) (blue triangles and red crosses in Fig. 4).

We estimated the shear force F_t^{2d} as a function of slip on the fault $u_1 - u_2$ during the stick–slip events. Figure 5 shows an example for the event indicated by a red star in Fig. 2c. From Eq. (1), the shear force F_t^{2d} is

$$
\begin{aligned}
F_t^{2d} &= -m_1 a_1 - k_1 u_1 \\
&= -m_1 a_1 + F_{s1}.
\end{aligned}
\tag{3}
$$

Following this equation, we can obtain F_t^{2d} from the values of m_1 calculated from rock density (2980 kg/

m^3) and mass volume, a_1 measured by the accelerometer (Fig. 5a), and $F_{s1} = -k_1 u_1$ measured by the load cell (Fig. 5b).

For the estimation of $u_1 - u_2$, we conducted a double time-integration of $a_1 - a_2$ because LDT-L did not have enough resolution. Figure 5c shows a comparison of $u_1 - u_2$ obtained by the double time-integration of $a_1 - a_2$ and u measured by LDT, indicating that we can estimate the short-term slip displacement from the accelerograms. Examples of estimated slip-weakening curves are shown in Fig. 5d. In this estimation, we corrected the timing of the recording system as pointed out by Fukuyama et al. (2014). We applied a 400-Hz Butterworth-type low-pass filter to F_{s1}. We examined the contribution from high-frequency waves by applying a 500-, 750-Hz, and 1-kHz low-pass Butterworth filter to the acceleration data, and we computed the slip-weakening curves (black, blue, light blue curves, respectively, in Fig. 5). Since we did not observe any significant differences in the slip-weakening curves, we confirmed that the high-frequency waves did not contribute to the estimated slip-weakening curves and used a 400-Hz cutoff. We will show the results using the acceleration data in which the 750-Hz low-pass filter was applied below.

We estimated the peak slip velocity during stick–slip events by time-integration of $a_1 - a_2$. The peak slip velocity was approx. 0.02 m/s for time window W1, 0.04 m/s for time windows W2 and W3, and 0.05 m/s for time window W4.

2.2.2 Behavior of Experimental Apparatus

Since we used a shaking table system, which was originally developed for the vibration experiments of large-scale constructions, we needed to carefully examine the behavior of the shaking table during the friction experiments. The shaking table was instructed to move at the constant loading velocity $\bar{v}_L$ (instructed loading velocity to the apparatus), but the loading velocity oscillated slightly. This is due to the delay of servo controller response of the actuator of the shaking table. As can be seen in Fig. 6, we observed that the shaking table moved faster than $\bar{v}_L$ immediately after a stick–slip event occurred. This might have been the result of dynamic overshoot of shear stress observed by the friction apparatus during a stick–slip event. The shaking table moved slightly slower than $\bar{v}_L$ just after the fast movements. Note that the total movement of the shaking table was consistent with $\bar{v}_L t$ as a long time period average.

We estimated the fast and slow loading velocity, v_{Lf} and v_{Ls}, respectively, and the duration of the fast movement, d_f, from the observed movement of the shaking table (l), as follows. We divided each time window from W1 to W4 (Figs. 2, 3) into the shorter time windows for large and small slopes of l, and we estimated l for each time window by a straight line, as shown in Fig. 6c. The average values of the large and small slopes and the duration of the large slopes correspond to v_{Lf}, v_{Ls}, and d_f, respectively. Table 2 lists the values for each time window from W1 to W4.

We estimated the stiffness k_1 for each time window from W1 to W4. We measured the increasing rate of the spring force ($\dot{F}_{s1}$) during each stick from Figs. 2 and 3, and obtained $k_1 = \dot{F}_{s1}/v_{Ls}$. The

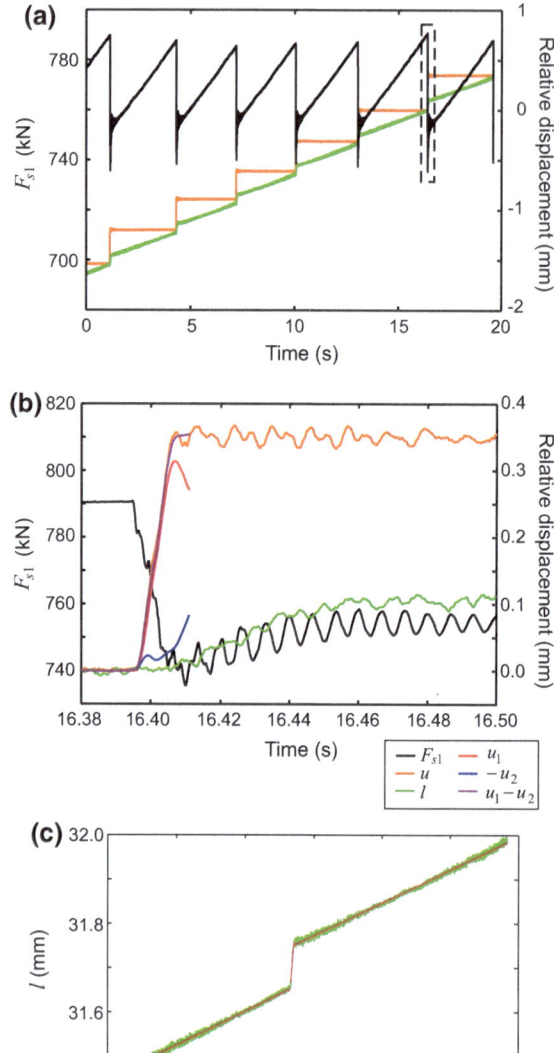

Figure 6
a, b Observed force (F_{s1}), laser displacement (u), table displacement (l), displacements by accelerometers (u_1 and u_2) and their differences. **c** Observed table displacement (l) and estimated straight line (*red line*). **a** Longer time frame for F_{s1}, u, and l. **b** Shorter time window indicated by *broken square* in **a**. Note that the stick–slip event in **b** is the same as that shown in Fig. 5, although the origin time for plotting was different

average values of k_1 for each time window were almost the same, as shown in Table 3.

The fast recovery of the friction immediately after the sharp friction drop observed at a stick–slip event would be related to the oscillation of the shaking

Table 2

Estimated loading velocity

Experiment ID	Time window	v_{Ls} (mm/s)	v_{Lf} (mm/s)	d_f (s)
LB01-127 (short cumulative disp.)	W1	0.0753	0.9622	0.039
LB01-127 (long cumulative disp.)	W2	0.0750	1.9717	0.042
LB01-134	W3	0.0740	2.3608	0.036
LB01-142	W4	0.7356	3.3027	0.046

Table 3

Estimated stiffness

Experiment ID	Time window	Stiffness k_1 (10^8 N/m)	Normalized critical stiffness κ	
			Slip law	Aging law
LB01-127 (short cumulative disp.)	W1	1.42	2.6×10^{-2}	2.4×10^{-2}
LB01-127 (long cumulative disp.)	W2	1.48	3.2×10^{-2}	1.6×10^{-2}
LB01-134	W3	1.48	7.4×10^{-2}	7.4×10^{-2}
LB01-142	W4	1.48	6.5×10^{-3}	7.7×10^{-3}

table. This is because the ratio of $\dot{F}_{s1}$ during the fast recovery of the friction and during stick was similar to the ratio of v_{Lf} and v_{Ls}, that is, the stiffness k_1 was almost constant through our experiments.

3. Method for Numerical Simulations

We applied a spring-slider model with mass with one-degree-of-freedom (Fig. 7) to explain the stick–slips because we found that the displacement of the lower specimen is approx. 10% of that of the upper specimen during dominant slip (16.653–16.664 s in Fig. 5c) and can be ignored. The equation of motion is

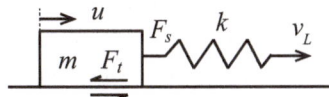

Figure 7

The one-degree-of-freedom spring-slider model used in our numerical simulations

$$\frac{dv}{dt} = (F_s - F_t)/m, \qquad (4)$$

where v is the slip velocity, t is time, F_s is spring force, F_t is the shear force on the fault, and m is mass. We perform dynamic simulations (i.e., accounting for the inertial effects), in contrast to the previous studies which ignored the inertia at low slip velocity (e.g., Rice and Tse 1986; Bizzarri 2011). F_s is

$$F_s = F_0 + k(v_L t - u), \qquad (5)$$

where F_0 is the steady-state shear force at a reference slip velocity of the friction law v_0, k is the spring stiffness, v_L is the load point velocity, and u is the fault displacement. F_t is assumed to obey the RSF law. We examine both Slip law (Ruina law)

$$F_t = F_0 + aF_n \ln(v/v_0) + \theta \qquad (6)$$

$$\frac{d\theta}{dt} = -\frac{v}{L_c}(\theta + bF_n \ln(v/v_0)) \qquad (7)$$

and Aging law (Slowness law)

$$F_t = F_0 + aF_n \ln(v/v_0) + bF_n \ln(v_0\theta/L_c) \qquad (8)$$

$$\frac{d\theta}{dt} = 1 - \frac{v\theta}{L_c}, \qquad (9)$$

where a and b are parameters representing the direct and evolution effects, respectively, θ is the state variable, and L_c is the critical slip distance (e.g., Marone 1998a). We calculated m from rock density and dimensions (Table 4). We used the estimated values of k_1 (Table 3) as k, which was constant in our simulations because k_1 was almost constant through the experiments, as stated in Sect. 2.2.2. The parameters used in the simulations are shown in Table 4.

If we solve Eqs. (4)–(7) for the Slip law and Eqs. (4), (5), (8), and (9) for the Aging law by the Runge–Kutta method with adaptive step-size control (Press et al. 1992), it takes too long to integrate in

Table 4

Simulation parameters

Property	Symbol	Value
Mass	m	1.1×10^3 kg
Direct effect parameter	a	0.008
Reference velocity	v_0	10^{-5} m/s

time domain during interseismic periods for some of the parameter sets. Therefore, we newly developed an efficient algorithm for the numerical time integration which is an example of the exponential time differencing method (e.g., Cox and Matthews 2002) similar to that used by Noda and Lapusta (2010), as described in "Appendices 1 and 2".

We confirmed for the Slip law that the time-integration method provided results that were identical to those obtained by the Runge–Kutta method, but required only for 1/10,000 of the calculation time for $a = 0.008$, $b = 0.0092$, $L_c = 0.5$ μm, and the constant v_L of 0.1 mm/s (Fig. 18).

The loading velocity slightly fluctuated in our experiments, as stated in Sect. 2.2.2. Therefore, we assumed in our simulations that v_L is the faster loading velocity v_{Lf} for the duration of d_f after a stick–slip event finishes and v_L is the slightly slower loading velocity v_{Ls} at other times. We defined that the event occurs when $v > 10v_{Ls}$ and it finishes when $v \leq v_{Ls}$. Table 2 lists the values of v_{Lf}, d_f, and v_{Ls}, which were decided from the observed movement of the shaking table (Sect. 2.2.2).

Using the new time-integration method, we conducted many numerical simulations with various combinations of the evolution parameter b and the state-evolution distance L_c while keeping the direct effect parameter a constant. We then estimated the combinations of b and L_c which reproduce the recurrence time and the friction drop consistent with ΔT_{ob} and $\Delta \mu'_{ob}$. We also calculated the slip-weakening curve with each combination, because we could not determine the optimal parameters only by the recurrence time and the friction drop as described in the next section.

4. Results

4.1. Slip Law

4.1.1 Dependence of Constitutive Parameters on Cumulative Displacement

We estimated the combinations of constitutive parameters b and L_c to reproduce the stick–slip events that occurred at short and long cumulative

displacement in the experiment LB01-127 with $\bar{v}_L = 0.1$ mm/s (windows W1 and W2 in Fig. 2).

First, we modeled the observed recurrence time of the stick–slip events (ΔT_{ob}) and the observed friction drop ($\Delta \mu'_{ob}$). Figure 8a, b shows the computational results of the recurrence time of the stick–slip events (ΔT_{sy}) and friction drop ($\Delta \mu'_{sy}$), respectively, for various combinations of b and L_c values for the long cumulative displacements (window W2). The colored and gray circles indicate the parameters with which the system reached the limit cycle and the stable sliding, respectively. The crosses indicate parameters which provide the combination of plural recurrence time. The multiple recurrence time does not appear in the simulations for the constant v_L; it comes from the changes in v_L in our simulations. If ΔT_{sy} for v_{Lf} is shorter than d_f (duration of v_{Lf}), the stick–slip events occur during v_L of v_{Ls} and v_{Lf}, and the multiple recurrence time arises. Our simulations are not suitable in these cases because v_L decreases from v_{Lf} to v_{Ls} after duration d_f, whether or not a stick–slip event occurs during v_{Lf}. However, we need not consider these cases because ΔT_{ob} is much larger than d_f. The diamonds show the parameter sets (b, L_c) that provide ΔT_{sy} within $\Delta \bar{T} \pm \sigma_T$ and $\Delta \mu'_{sy}$ within $\Delta \bar{\mu}' \pm \sigma_\mu$, where $\Delta \bar{T}$ and $\Delta \bar{\mu}'$ are the average ΔT_{ob} and $\Delta \mu'_{ob}$ for window W2, respectively, and σ_T and σ_μ are the standard deviation of ΔT_{ob} and $\Delta \mu'_{ob}$ for window W2, respectively. The parameter sets providing ΔT_{sy} within $\Delta \bar{T} \pm \sigma_T$ are similar to those providing $\Delta \mu'_{sy}$ within $\Delta \bar{\mu}' \pm \sigma_\mu$. To model ΔT_{ob} and $\Delta \mu'_{ob}$ simultaneously, we calculated the following evaluation function

$$J_1 = \frac{\left(\Delta T_{sy} - \Delta \bar{T}\right)^2}{\left(\sigma_T\right)^2} + \frac{\left(\Delta \mu'_{sy} - \Delta \bar{\mu}'\right)^2}{\left(\sigma_\mu\right)^2}, \quad (10)$$

as shown in Fig. 8c. The diamonds in Fig. 8c indicate the parameter sets with which $J_1 \leq 5.99146$. If the recurrence time and the friction drop exhibit normal distributions and they are independent of each other, J_1 follows the Chi-squared distribution with two-degrees-of-freedom, and $J_1 \leq 5.99146$ corresponds to the 95% confidence regions. Although this assumption is inadequate because the recurrence time correlates with the friction drop, the small J_1 indicates that ΔT_{sy} and $\Delta \mu'_{sy}$ are consistent with ΔT_{ob} and

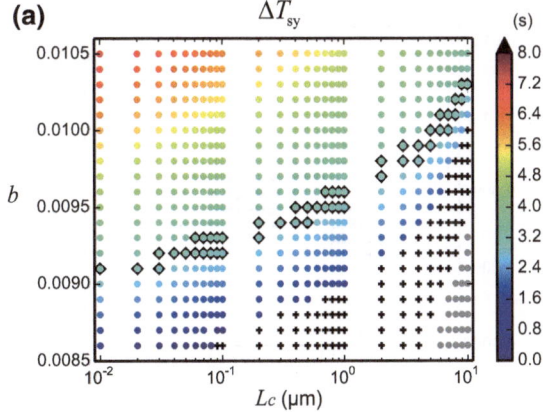

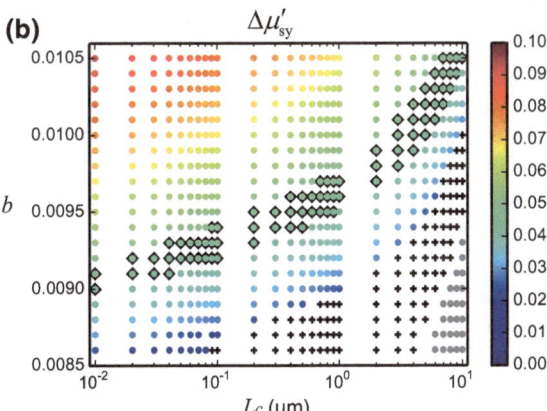

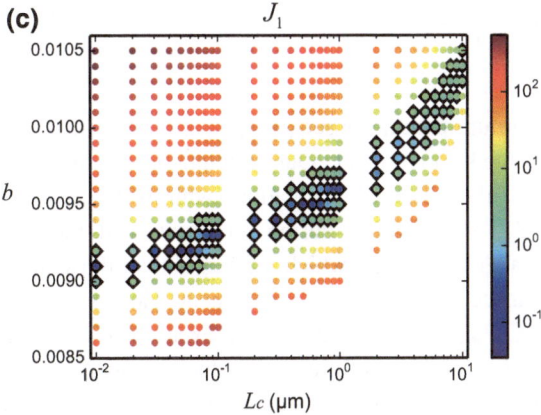

Figure 8

Synthetic **a** recurrence intervals ΔT_{sy} and **b** friction drop $\Delta\mu'_{sy}$, and **c** evaluation function J_1 plotted as a function of the constitutive parameters b and L_c in numerical simulations for long cumulative displacement of the LB01-127 (Fig. 2c, window W2). *Gray circles* show parameters with which the system reaches stable sliding. *Crosses* indicate parameters which provide multiple recurrence time. *Diamonds* in **a** and **b** show the parameters which provide ΔT_{sy} within $\Delta\bar{T} \pm \sigma_T$ and $\Delta\mu'_{sy}$ within $\Delta\bar{\mu}' \pm \sigma_\mu$, respectively. *Diamonds* in **c** indicate $J_1 \leq 5.99146$

$\Delta\mu'_{ob}$. That is, b and L_c values shown by the diamonds in Fig. 8c can reasonably reproduce ΔT_{ob} and $\Delta\mu'_{ob}$. Figure 8c shows that ΔT_{ob} and $\Delta\mu'_{ob}$ have little information on L_c, and thus we cannot constrain the parameter set of b and L_c uniquely only from the recurrence time and friction drop data.

To further constrain the possible range of L_c, we tried to fit the slip-weakening curves (Fig. 5d). We examined the following evaluation function

$$J_2(i) = \min_{\mu_0} \left[\sum_{j=1}^{Nj(i)} \left\{ \mu_{sy}(i,j) - \mu_{ob}(i,j) \right\}^2 \right] \bigg/ Nj(i)$$

(11)

for event i, where $Nj(i)$ is the number of data for the event i, and μ_{sy} and μ_{ob} are the synthetic and observed shear force divided by the normal force, respectively. For this calculation, we applied the linear interpolation to the numerical result. Note that the initial friction coefficient $\mu_0 = F_0/F_n$ affects only the absolute level of μ_{sy}. We analyzed all of the events for window W2. Figure 9a shows the average $\bar{J}_2 = \sum_{i=1}^{N_i} J_2(i)/N_i$, where N_i is the number of the events, for many computations with parameter sets (b, L_c). The diamonds show $\bar{J}_2$ less than the minimum value of $\bar{J}_2 + \sigma_J$ in all simulations, where σ_J is the standard deviation of $J_2(i)$ for each simulation. The parameter set with small value of $\bar{J}_2$ can reproduce the observed slip-weakening curves, but the parameter set with large value of $\bar{J}_2$ cannot, as demonstrated in Fig. 9b.

For a joint inversion of the recurrence time, the friction drop, and the slip-weakening curves, we evaluated the following function

$$J = J_1 + \frac{\alpha_1}{\alpha_2} \bar{J}_2,$$

(12)

as shown in Fig. 10a, where α_1/α_2 is a weight coefficient, α_1 is the average of J_1 in the range of $J_1 \leq 5.99146$ (diamonds in Fig. 8c), α_2 is the average of $\bar{J}_2$ less than the minimum value of $\bar{J}_2 + \sigma_J$ (diamonds in Fig. 9a). The minimum of J (star in Fig. 10a) gives the optimum parameter set $(b, L_c) = (0.0094, 0.3 \,\mu\text{m})$, and $J \leq 2\alpha_1$ (diamonds in Fig. 10a) can be a possible range of b and L_c. The comparison of the observed and synthetic time histories for 20 s of displacement and of the spring force

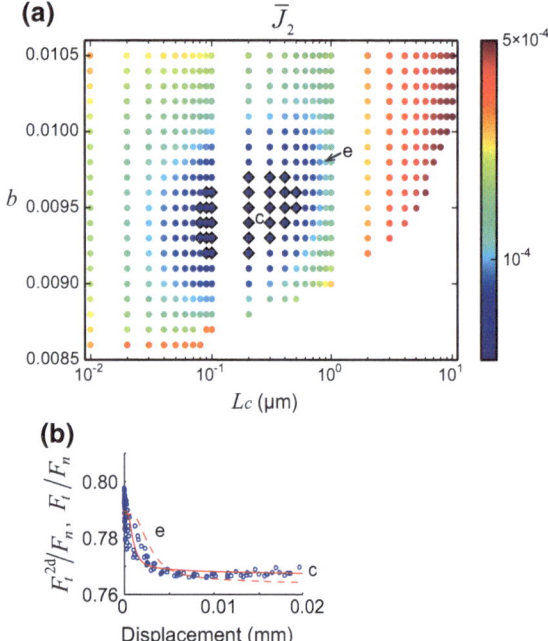

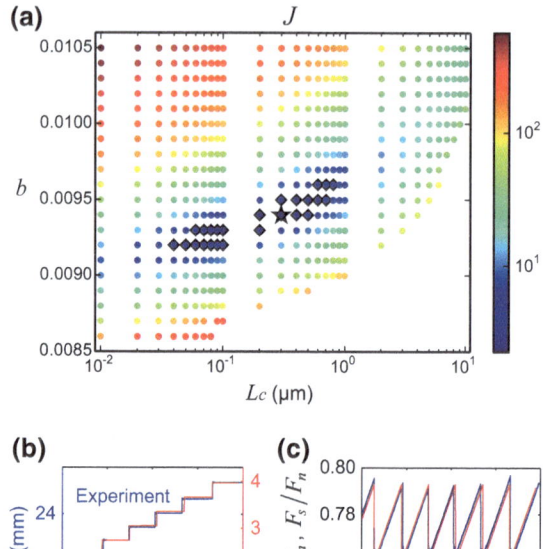

Figure 9

a $\bar{J}_2$ plotted as a function of the constitutive parameters b and L_C and **b** comparison of observed (*blue circles*) and synthetic (*red lines*) slip-weakening curves for long cumulative displacement of LB01-127 (Fig. 2c, window W2). *Diamonds* in **a** indicate $\bar{J}_2$ less than the minimum value of $\bar{J}_2 + \sigma_J$. *Small letters* 'c' and 'e' in **a** correspond to those in **b**. *Small letter* 'c' corresponds to the *star* in Fig. 10a. We plotted the observed slip-weakening curves for all of the events in window W2 (*blue circles* in **b**)

Figure 10

a J plotted as a function of the RSF parameters b and L_c and time history of **b** the cumulative displacement u and **c** spring force (F_{s1}/F_n and F_s/F_n) for the large cumulative displacements of LB01-127 (Fig. 2c, window W2). **a** *Star* and *diamonds* show the optimum and the possible parameter sets, respectively. The method of the calculation of the possible range is stated in the text. **b, c** *Blue lines* show the experimental data and *red lines* show the synthetic data for the optimum parameter set (*star* in **a**). The observed and synthetic data are plotted by the *same scales*. **c** The initial coefficient of friction μ_0 is set arbitrarily for plotting

is shown in Fig. 10b, c. Our simulation could reproduce the very sharp friction drop and the subsequent fast recovery of the friction observed at a stick–slip event. These behaviors in our simulation result from the variable v_L (Table 2) because the features do not appear in the simulations with the constant v_L. In addition, the synthetic recurrence time, slip amount during an event, cumulative slip for 20 s, friction drop, and the slip-weakening curves for the optimum parameter set are similar to the observations, as demonstrated in Fig. 10b, c and by the red solid line in Fig. 9b.

We estimated the combinations of b and L_c to reproduce the stick–slip events that occurred at the short cumulative slip displacement (window W1), in the same manner as at the long cumulative slip displacement (window W2). By a joint inversion of the recurrence time, the friction drop, and the slip-weakening curves (Eq. (12)), we obtained the

optimum parameter set $(b, L_c) = (0.0085, 0.09\,\mu\text{m})$, as shown in Fig. 11. Our simulations cannot reproduce the observed fluctuations with events of ΔT_{ob}, $\Delta\mu'_{\text{ob}}$, slip amount during an event, and the slip-weakening curves, which are larger than those at the long cumulative slip displacement. However, ΔT_{sy}, $\Delta\mu'_{\text{sy}}$, slip amount during an event, and the slip-weakening curves for a large value of J are out of the observed fluctuations. Therefore, our estimation is reasonable.

Possible parameter sets (b, L_c) are summarized in Fig. 12. For the short cumulative slip displacement case (window W1), $(b, L_c) = (0.0085, 0.09\,\mu\text{m})$ was the best of the examined cases (i.e., J was the minimum, Fig. 11a), and the possible ranges for b and L_c were $b = 0.0085$ and $0.04 \leq L_c \leq 0.1\,\mu\text{m}$, respectively (triangles in Fig. 12). For the long

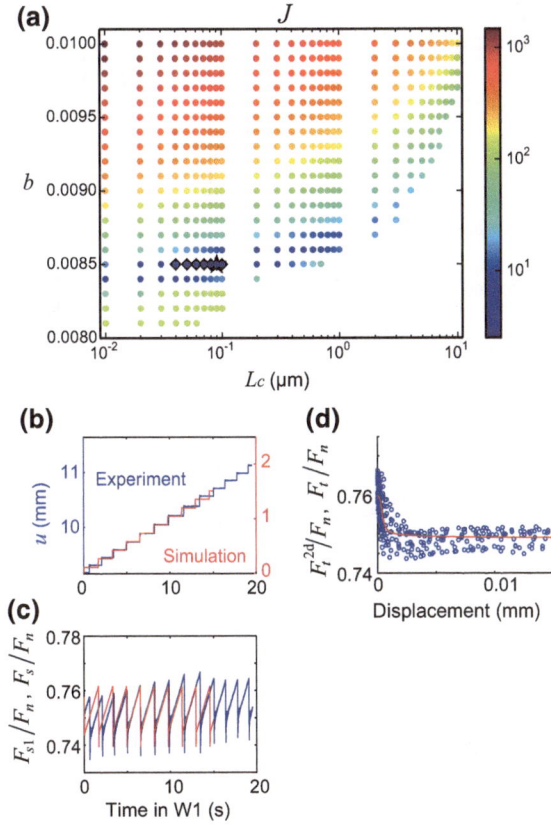

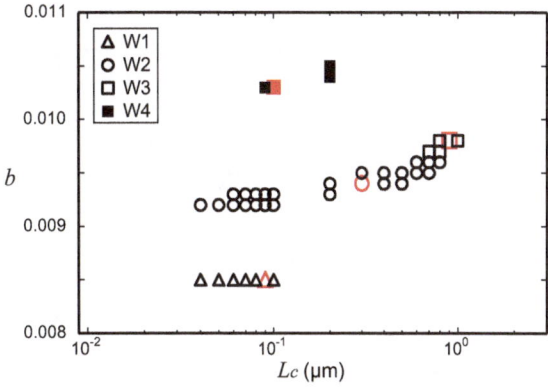

Figure 12

Parameter sets that provided stick–slip behaviors consistent with the experimental data in cases with the Slip law. *Triangles* and *circles* show the parameter sets for small and large cumulative displacement, respectively (slip dependence, Sect. 4.1.1). *Squares* show the rate-dependence (Sect. 4.1.2). *Open* and *solid symbols* correspond to $\bar{v}_L$ of 0.1 and 1.0 mm/s, respectively. *Red symbols* indicate the parameter sets with which J is the minimum

Figure 11

a J plotted as a function of the RSF parameters b and L_C and **b–d** comparison of the observed and synthetic data for the optimum parameter set (*star* in **a**) for the small cumulative displacements of LB01-127 (Fig. 2b, window W1). **a–d** are drawn in the same manner as in Figs. 9b and 10, respectively

cumulative slip displacement case (window W2), $(b, L_c) = (0.0094, 0.3\,\mu\text{m})$ was the best (Fig. 10a), and the possible ranges of b and L_c were $0.0092 \leq b \leq 0.0096$ and $0.04 \leq L_c \leq 0.8\,\mu\text{m}$, respectively (circles in Fig. 12).

These results suggest that the evolution-related constitutive parameters (b and L_c) increased as the cumulative displacement increased, even in a single experiment. By comparing the triangles and the circles in Fig. 12, we can say that the b value is significantly different. The L_c value might be different between these two time windows, but the error range was too large to judge the differences.

We compared ΔT_{sy} at limit cycles with ΔT_{ob} not at limit cycles. The synthetic stick–slips reach a limit cycle after several stick–slips in the simulations with the possible parameters. However, the observed

stick–slips did not reach a limit cycle even after many stick–slips occurred. Therefore, other processes such as changes in the state of the fault surfaces may have occurred in the experiment, which were not taken into account in synthetic model. This will be discussed in Sect. 5.

4.1.2 Dependence of Constitutive Parameters on Loading Velocity

We estimated the constitutive parameters b and L_c to reproduce the stick–slip events observed in experiments LB01-134 with the slow $\bar{v}_L$ ($\bar{v}_L = 0.1$ mm/s) and LB01-142 with the fast $\bar{v}_L$ ($\bar{v}_L = 1.0$ mm/s) (Fig. 3). Throughout each of these experiments, ΔT_{ob} and $\Delta \mu'_{ob}$ were almost constant (blue triangles and red crosses in Fig. 4). We estimated the constitutive parameters b and L_c in the same manner as that described in Sect. 4.1.1. Figures 13 and 14 show the results of the joint inversion of the recurrence time, the friction drop, and the slip-weakening curves for the slow and fast $\bar{v}_L$, respectively. The observed fluctuations of ΔT_{ob}, $\Delta \mu'_{ob}$, and the slip-weakening curves with events are smaller than those in LB01-127 (Sect. 4.1.1), and thus the synthetic slip-weakening curves fitted the observations better and we constrained the L_c value better.

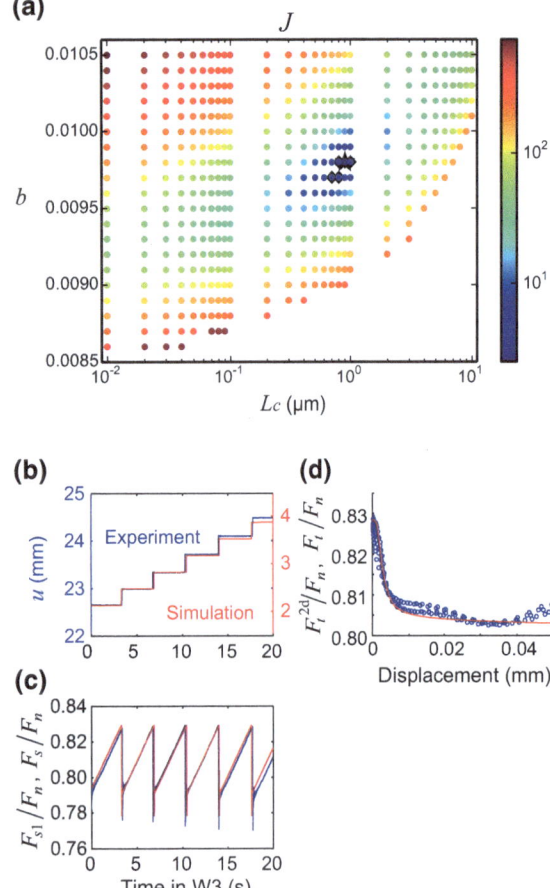

Figure 13

a J plotted as a function of the RSF parameters b and L_c and **b–d** comparison of the observed and synthetic data for the optimum parameter set (*star* in **a**) for LB01-134 (slow $\bar{v}_L$, Fig. 3a, window W3). **a–d** are drawn in the same manner as in Figs. 9b and 10, respectively

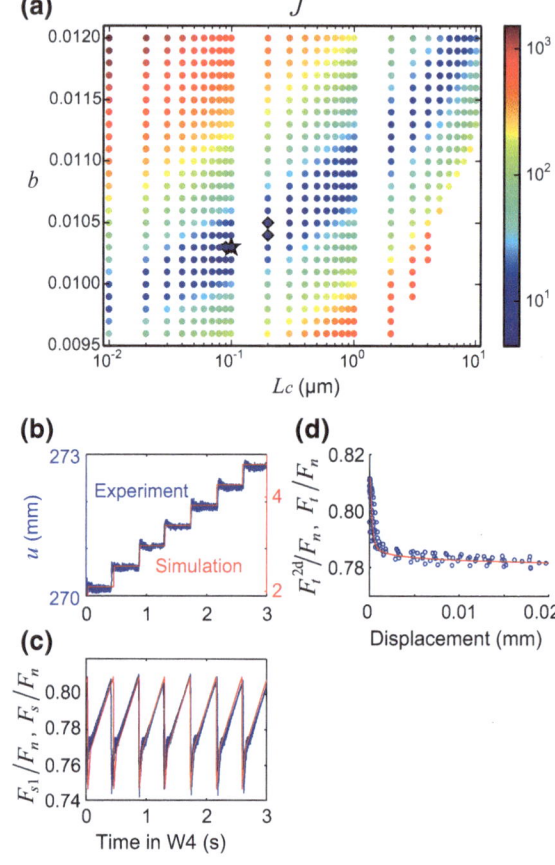

Figure 14

a J plotted as a function of the RSF parameters b and L_c and **b–d** comparison of the observed and synthetic data for the optimum parameter set (*star* in **a**) for LB01-142 (fast $\bar{v}_L$, Fig. 3b, window W4). **a–d** are drawn in the same manner as in Figs. 9b and 10, respectively

For the case of the slow $\bar{v}_L$, the best parameter set was $(b, L_c) = (0.0098, 0.9 \ \mu m)$, and the possible b and L_c values were $0.0097 \leq b \leq 0.0098$ and $0.7 \leq L_c \leq 1.0 \ \mu m$, respectively (open squares in Fig. 12). For the case of the fast $\bar{v}_L$, the best parameter set was $(b, L_c) = (0.0103, 0.1 \ \mu m)$, and the possible b and L_c values were $0.0103 \leq b \leq 0.0105$ and $0.09 \leq L_c \leq 0.2 \ \mu m$, respectively (solid squares in Fig. 12). Therefore, the constitutive parameters show clear dependence on $\bar{v}_L$; b increases and L_c decreases as $\bar{v}_L$ increases. By comparing the open and solid squares in Fig. 12, we can say that the b and L_c values are significantly different.

4.2. Aging Law

We examined the Aging law, in the same manner as in Sect. 4.1 for the Slip law. The estimated constitutive parameters b and L_c and the possible ranges are summarized in Fig. 15. The results for the short and long cumulative displacements (triangles and circles in Fig. 15) suggest that the evolution-related constitutive parameters (b and L_c) increase as the cumulative displacement increases in a single experiment. The b value is significantly different. The L_c value can be different between these two time windows, but the error range was too large to judge the differences. The comparison of the results for the slow and fast $\bar{v}_L$ (open and solid squares) suggest that

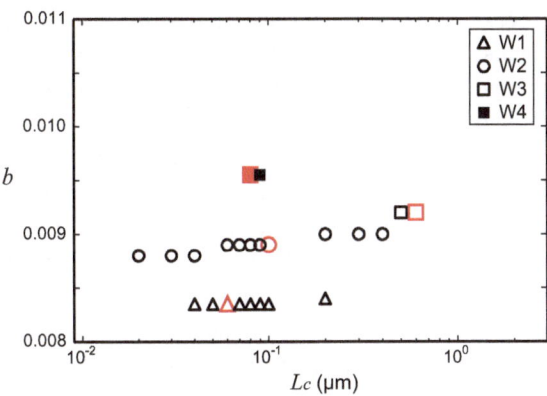

Figure 15

Parameter sets which provide stick–slip behaviors consistent with the experimental data in cases with the Aging law. This figure is drawn in the same manner as Fig. 12

Table 5

Values of evaluation functions for optimum parameter sets

Time window	Slip law			Aging law		
	J_1	$\bar{J}_2$ $(\times 10^{-5})$	J	J_1	$\bar{J}_2$ $(\times 10^{-5})$	J
W1	0.180	0.932	2.52	0.108	0.930	1.97
W2	0.0710	5.21	2.00	0.0414	5.21	1.99
W3	0.0743	2.96	2.53	0.0268	5.68	4.16
W4	1.28	5.77	4.44	1.29	6.89	4.86

b increases and L_c decreases as $\bar{v}_L$ increases and that the b and L_c values are significantly different.

The values of the evaluation functions J_1, J_2, and J (Eqs. (10)–(12)) for the estimated parameter sets were slightly different between the Slip and Aging laws, but we did not see any superiority of one over the other (Table 5). The b and L_c values estimated for the Aging law are smaller than for the Slip law (Fig. 12) in all of the examined cases (windows from W1 to W4). However, our results for the Aging law stated above are the same as for the Slip law in Sect. 4.1.

5. Discussion

The results of this study suggest that when a friction experiment starts without gouges on the fault, both the b and L_c values increase as the cumulative slip increases, as stated in Sect. 4.1.1. Beeler et al. (1996) also reported the decrease in $a − b$ in some initially bare surface experiments. On the other hand, Leeman et al. (2016) suggested that decreases in both $a − b$ and L_c with the cumulative slip. They constructed 3-mm-thick layers of powdered silica to simulate granular fault gauges, which might cause the contradiction between their results and ours on L_c. Our results on the slip dependence of the RSF parameters can be partly explained by the production of gouges, since the state-evolution distance L_c increases with increasing gouge layer thickness as suggested by Marone and Kilgore (1993). In fact, many gouge particles were produced during the present friction experiments as described by Fukuyama et al. (2014). From the estimated gouge production rate, we calculated 5.013×10^{-7} and 1.303×10^{-6} m as averaged thicknesses of the gouge layers for W1 and W2 in LB01-127, respectively. See "Appendix 3" for detail of this estimation. Note that we estimated these thicknesses assuming that the produced gouge materials are uniformly distributed over the fault surface. Actually, the gouge materials were locally produced in and around the generated grooves as revealed by Yamashita et al. (2015). Therefore, these thicknesses could be minimum estimates. The estimated b and L_c values were larger in experiment LB01-134 (open squares in Figs. 12, 15) than in experiment LB01-127 (triangles and circles in Figs. 12, 15), which is consistent with the slip dependence described in Sect. 4.1.1, because LB01-134 was conducted after several experiments following LB01-127 without the removal of gouges. The conventional RSF laws with a single set of the RSF parameters were not sufficient to explain the results of the long cumulative displacement experiments, and more complex state evolution laws accounting for gouge production are needed to comprehensively describe the large-scale experimental data.

It is important to note that most earthquake cycle simulations (e.g., Hori et al. 2004; Lapusta and Liu 2009; Noda and Lapusta 2013) used the conventional RSF law with a single state-variable. Based on the present findings, however, the evolution of friction as a function of slip during evolution of the internal structure of the shear zone (e.g., accumulation of wear material) could not be well expressed by the conventional RSF law. To account for the large-scale behavior in which the apparent RSF parameters evolve with changes in the internal structures of the shear zone, a different framework is required.

We found an increase in b as $\bar{v}_L$ increased, as stated in Sect. 4.1.2. This indicates a positive correlation between $\bar{v}_L$ and the friction drop. Some previous studies, however, suggested that the friction drops of the stick–slip events decreased with the increase of the loading velocity (Karner and Marone 2000; Mair et al. 2002; Anthony and Marone 2005; McLaskey et al. 2012). The negative correlation between $\bar{v}_L$ and the friction drop shown by the previous studies can be interpreted to result from contact aging associated with frictional healing during the inter-seismic period of the seismic cycle. Our results do not deny the effect of the frictional healing because peak friction was slightly higher at lower $\bar{v}_L$ in our experiments as shown in Fig. 3. Instead, our results may suggest that the velocity-weakening effect is stronger than that expected from the conventional RSF law with a single set of the RSF parameters. The estimated peak slip velocity was approx. 0.02–0.05 m/s and higher for the faster $\bar{v}_L$ in our experiments (Sect. 2.2.1). In this velocity range, friction weakens as a function of slip velocity (e.g., Di Toro et al. 2011); therefore, larger friction drop would occur for higher slip velocity. Kato et al. (1991) obtained similar data in experiments with a granite specimen as well as a composite specimen of granite and marble. The correlation between the friction drop and the loading velocity might relate to the characteristics of the apparatus used in the experiments.

The estimated L_c values except for time window W3 are smaller than those obtained by the previous studies (0.7 μm or longer, e.g., Dieterich 1979; Marone et al. 1990). The small L_c could result from the thin gouge layer. As described above, Marone and Kilgore (1993) proposed a scaling relation that L_c is proportional to the gouge thickness. The gouge layer thicknesses estimated in this study are two orders of magnitude smaller than those in the previous experiments (e.g., Marone et al. 1990; Marone and Kilgore 1993); therefore, L_c could be smaller in our experiments than in the previous studies. The small L_c might be also related to the velocity weakening

processes. This is because the previous studies obtained L_c by velocity step change tests, while we obtained L_c from stick–slips which involve high slip velocity (0.05 m/s at most).

Unstable (seismic) slip may occur for spring stiffness smaller than a critical value, as theoretically shown by Ruina (1983). Leeman et al. (2016) showed that the behaviors of stick–slips and stable sliding are related to normalized critical stiffness $\kappa = k/k_c$ where k is loading system stiffness, $k_c = \sigma_n(b-a)/L_c$ is critical stiffness of a fault, and σ_n is the normal stress. From the stiffness and the RSF parameters obtained in Sects. 2.2.2 and 4, we estimated κ as shown in Table 3. Many stick–slip events were observed for $\kappa \ll 1$, which are consistent with Leeman et al. (2016) and the theoretical works (e.g., Ruina 1983). However, stick–slip behaviors cannot be explained only by κ in our experiments. For example, $\Delta\mu'_{ob}$ of stick–slip events and slip amount per event were high in the order of W4, W3, W2, and W1, but κ was small in the different order.

We assumed that a value is constant since a is considered as the material property. However, a might depend on temperature change (e.g., Blanpied et al. 1995; Nakatani 2001) and on the loading velocity because Marone (1998b) showed that the static friction increases with the loading rate (1–10 μm/s) by double-direct shear experiments. To investigate the dependency of a on the estimation of b and L_c, we conducted the same computation using $a = 0.005$ and $a = 0.011$ for window W2. We estimated the constitutive parameters (b, L_c) and their possible ranges, in the same manner as in Sect. 4.1.1, summarized in Fig. 16. The estimated L_c values were slightly different among the examined cases. On the other hand, the estimated $b - a$ values increased as a values decrease because ΔT_{sy} and $\Delta\mu'_{sy}$ depended on not only b value but also a value. The values of the evaluation functions J_1, J_2, and J (Eqs. (10)–(12)) for the estimated parameter sets were similar in the cases with different a values, as shown in Table 6. This means that the optimum parameter set (a, b, L_c) cannot be determined uniquely. Thus, the L_c values estimated in Sect. 4 do not depend on the choice of a value, but the dependence of b values on the cumulative displacement and on $\bar{v}_L$ could be explained by the dependence of a values.

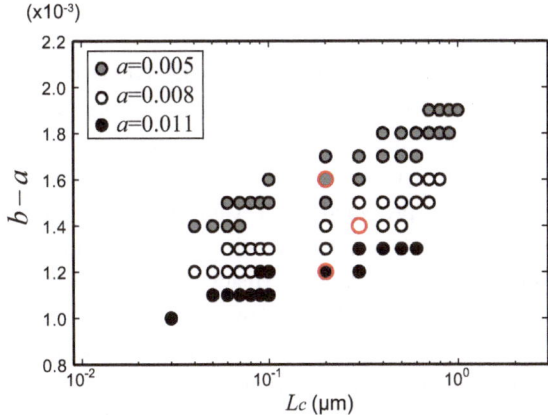

Figure 16

Optimum parameters (*red symbols*) and possible ranges for long cumulative displacement in experiment LB01-127 (window W2) for various *a* values. *Gray, open*, and *black circles* show the possible ranges when *a* is equal to 0.005, 0.008, and 0.011, respectively. *Open circles* are the same as those in Fig. 12

Table 6

Values of evaluation functions for optimum parameter sets for a of 0.005 and 0.011

a values	J_1	$\bar{J}_2$ $(\times 10^{-5})$	J
0.005	0.0287	4.69	1.56
0.011	0.0321	5.05	1.99

Fukuyama et al. (2014) observed that a stick–slip event was initiated by the nucleation of a dynamic rupture, which propagated on the simulated fault surface. In this study, we used a one-degree-of-freedom spring-slider model (Fig. 7) and did not take into account the finiteness of the slip area. The minimum nucleation size for the Aging law for anti-plane strain,

$$L_b = \frac{GL_c}{(1-v)b\sigma_n},$$
(13)

where G is the shear modulus and v is the Poisson's ratio, was estimated by Ampuero and Rubin (2008). G and v for Indian metagabbro used in our experiments are 39.3 and 0.31 GPa, respectively. From the parameters obtained in Sect. 4.2, L_b is estimated to be 0.31, 0.48, 2.8, and 0.36 m for windows W1 (short cumulative slip displacement), W2 (long cumulative slip displacement), W3 (slow $\bar{v}_L$), and W4 (fast $\bar{v}_L$),

respectively. The increase in L_b for windows from W1 to W3 could represent the gouge production. The L_b values are smaller than the sample size, except for window W3; therefore, a rupture can propagate dynamically as a stick–slip event. It matters whether or not the sample size is larger than the nucleation size, and the effect of dynamic rupture propagation on the overall friction deserves future experimental and theoretical investigation.

There is evidence that the stick–slip behavior may be related to the damage of sliding surface as described below. Unfortunately, in the series LB01 presented here, we did not conduct the stick–slip experiment at the beginning. However, we observed the dependence of the stick–slip behavior on the slip surface damage in the LB09 series, in which the width of the lower rock specimen was reduced to 0.1 m to increase the normal stress to 6.7 MPa. The rock was Indian metagabbro, same as in the present experiments described above. The fault surface was repeatedly slid for 900 s at $\bar{v}_L$ of 0.01 mm/s, and the gouges were removed after each experiment. We compared the stick–slip behaviors at the same time window (600–700 s) to avoid the effect of cumulative slips from gouge removal.

We found that ΔT_{ob} and $\Delta \mu'_{ob}$ were small in the first experiment of the LB09 series (Fig. 17a), whereas both became larger and almost constant in subsequent experiments (Fig. 17b). These observations support the idea that the friction drops and the recurrence time depend on the damage on the fault surface in addition to $\bar{v}_L$ and the cumulative displacement demonstrated in this paper. The effects of the damage on the fault surface will be further investigated in future works.

6. Conclusions

We estimated the constitutive parameters in the RSF law (for both Slip and Aging laws) by fitting numerical simulations to stick–slip experiments with a large-scale biaxial rock friction apparatus at the NIED. During the friction experiments, many stick–slip events were observed, and their features are summarized as follows. (1) The friction drop and recurrence time of the stick–slip events increased

(a)   **(b)**

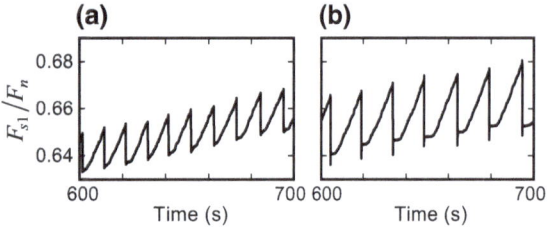

Figure 17

Time history of F_{s1}/F_n **a** for the first experiment with the rock specimens whose surfaces were prepared using a surface grinder and never used in any experiments (LB09-001) and **b** for the second experiment with the same specimens under the same conditions (LB09-002). Note that the gouges produced in the first experiment were removed before the second experiment

with cumulative displacement within a single experiment when gouges were removed before the experiment. (2) The friction drop and recurrence time became more or less constant throughout the experiment when the experiment was done after several experiments without removing gouges. (3) The friction drop became lager and the recurrence time was shorter for an experiment with faster loading velocity. We estimated the slip-weakening curves during the stick–slip events from measured spring force and the accelerations of the specimens. We applied a one-degree-of-freedom spring-slider model with mass to explain the observed stick–slips. We developed an efficient algorithm for numerical time integration, and we conducted many numerical simulations with various b and L_c values while keeping a constant. We then identified the values of b and L_c that provided a consistent recurrence time, friction drops, and slip-weakening curves during the stick–slip events.

The results of our analyses suggest that (1) both b and L_c increase as the cumulative displacement increases, and (2) b increases and L_c decreases as the loading velocity increases, for both Slip and Aging laws. Therefore, the conventional RSF laws with an invariable single set of the RSF parameters cannot explain the whole of the experimental data. More complex state evolution laws are needed to comprehensively describe the experimental data and to consider an earthquake cycle involving a wide range of slip velocities and a sequence of earthquakes over geologically long times during which the evolution of the internal structure of the shear zone is significant.

Acknowledgements

This research was supported by the NIED research project entitled 'Development of Earthquake Activity Monitoring and Forecasting' and the JSPS KAKENHI, Grant No. 23340131. Assistance for the experiments provided by Tetsuhiro Togo, Hironori Kawakata, Nana Yoshimitsu, Tadashi Mikoshiba, Makoto Sato, Chikahiro Misawa, Toshiyuki Kanezawa, Hiroshi Kurokawa, Toya Sato, and Toshihiko Shimamoto is greatly appreciated. Anonymous reviewers' comments were quite valuable in improving our manuscript. Friction experiment data are available upon request.

Appendix 1: Exponential Time Differencing Method for the SLIP Law

Governing Equations

A spring-slider system with one-degree-of-freedom consists of Eqs. (4)–(7) in the main text:

$$m\dot{v} = F_s - F_t \tag{14}$$

$$\dot{F}_s = k(v_L - v) \tag{15}$$

$$F_t = F_0 + A \ln\left(\frac{v}{v_0}\right) + \theta \tag{16}$$

$$\dot{\theta} = \frac{v}{L_c}\left(-B \ln\left(\frac{v}{v_0}\right) - \theta\right), \tag{17}$$

where dots on the top represent derivatives with respect to time t, $A = aF_n$ and $B = bF_n$ are parameters representing the direct and evolution effects, respectively, and the definitions of the other characters are the same as those in the main text. The time-derivative of Eq. (16) is:

$$\dot{F}_t = A\frac{\dot{v}}{v} + \dot{\theta}. \tag{18}$$

Now we normalize these equations by introducing nondimensional parameters. The slip rate shall be normalized by the steady-state velocity:

$$w = v/v_0. \quad (19)$$

The nondimensional time shall be defined as:

$$\tau = v_0 t/L_c. \quad (20)$$

The nondimensional frictional stress difference from the steady-state is:

$$g = \frac{F_t - F_0}{A}, \quad (21)$$

and the nondimensional spring force difference from the steady-state is:

$$h = \frac{F_s - F_0}{kL_c}. \quad (22)$$

The nondimensional state variable is:

$$\Psi = \theta/B. \quad (23)$$

The nondimensional equations then take the form of:

$$w' = Ch - Dg \quad (24)$$

$$h' = E - w \quad (25)$$

$$g' = C\frac{h}{w} - D\frac{g}{w} - \beta w(\ln(w) + \Psi) \quad (26)$$

$$\Psi' = w(-\ln(w) - \Psi), \quad (27)$$

where primes represent derivatives with respect to the nondimensional time. The parameters in the nondimensional equations are:

$$\beta = B/A \quad (28)$$

$$D = \frac{AL_c}{mv_0^2} \quad (29)$$

$$C = \frac{kL_c^2}{mv_0^2} = \frac{kL_c}{A}\frac{AL_c}{mv_0^2} = \kappa_1 D \quad (30)$$

$$E = v_L/v_0. \quad (31)$$

D represents the significance of the direct effect with respect to the inertia, κ_1 is the nondimensional spring constant, and C represents the significance of spring stiffness with respect to inertia. Note that $C^{1/2}$ is proportional to the natural angular frequency of the harmonic oscillator.

There are four Eqs. (24)–(27), but we have one constraint from the friction law:

$$w = \exp(g - \beta\Psi). \quad (32)$$

Therefore, the system is a three-dimensional ordinary differential equation. Since Eq. (32) describes the relation between w, g, and Ψ, we have only to integrate Eq. (25) and additional two equations among Eqs. (24), (26), and (27).

Exponential Time Differencing Method

By integrating Eqs. (25)–(27) with (32), we obtained the following second-order accurate (in terms of τ_1) expressions at τ_1, supposing we have h_0, g_0, and Ψ_0 at $\tau = 0$:

$$h_1^* = h_0 + (E - w_0)\tau_1 \quad (33)$$

$$g_1^* = g_0 \exp(-\tau_1/\tau_{g0}) + g_{ss0}(1 - \exp(-\tau_1/\tau_{g0})) \quad (34)$$

$$\Psi_1^* = \Psi_0 \exp(-\tau_1/\tau_{\Psi 0}) + \Psi_{ss0}(1 - \exp(-\tau_1/\tau_{\Psi 0})), \quad (35)$$

where

$$w_0 = \exp(g_0 - \beta\Psi_0) \quad (36)$$

$$g_{ss0} = \frac{C}{D}h_0 - \frac{\beta}{D}w_0^2(\ln(w_0) + \Psi_0) \quad (37)$$

$$\tau_{g0} = w_0/D \quad (38)$$

$$\Psi_{ss0} = -\ln(w_0) \quad (39)$$

$$\tau_{\Psi 0} = 1/w_0. \quad (40)$$

Note that g_{ss0} and Ψ_{ss0} are (pseudo-)steady-state values which would be achieved if g and Ψ were only variables in Eqs. (26) and (27), respectively. τ_g and τ_Ψ are decay time constants. Then we can estimate w at τ_1 as:

$$w_1^* = \exp(g_1^* - \beta\Psi_1^*). \quad (41)$$

Adopting those starred values leads to a first-order integration scheme, but we can iterate this scheme to increase the order of accuracy (see Noda and Lapusta 2010). In the second-order scheme, we integrate the first half time-step using values at $\tau = 0$:

$$h_{1/2} = h_0 + (E - w_0)\tau_1/2 \quad (42)$$

$$g_{1/2} = g_0 \exp(-\tau_1/2\tau_{g0}) + g_{ss0}(1 - \exp(-\tau_1/2\tau_{g0}))$$
$$(43)$$

$$\Psi_{1/2} = \Psi_0 \exp(-\tau_1/2\tau_{\Psi0})$$
$$+ \Psi_{ss0}(1 - \exp(-\tau_1/2\tau_{\Psi0}))$$
$$(44)$$

and then the latter half using the starred values estimated above

$$h_1^{**} = h_{1/2} + (E - w_1^*)\tau_1/2 \qquad (45)$$

$$g_1^{**} = g_{1/2} \exp(-\tau_1/2\tau_{g1}^*) + g_{ss1}^*$$
$$(1 - \exp(-\tau_1/2\tau_{g1}^*)) \qquad (46)$$

$$\Psi_1^{**} = \Psi_{1/2} \exp(-\tau_1/2\tau_{\Psi1}^*)$$
$$+ \Psi_{ss1}^*(1 - \exp(-\tau_1/2\tau_{\Psi1}^*)) \qquad (47)$$

where

$$g_{ss1}^* = \frac{C}{D}h_1^* - \frac{\beta}{D}w_1^{*2}(\ln(w_1^*) + \Psi_1^*) \qquad (48)$$

$$\tau_{g1}^* = w_1^*/D \qquad (49)$$

$$\Psi_{ss1}^* = -\ln(w_1^*) \qquad (50)$$

$$\tau_{\Psi1}^* = 1/w_1^*. \qquad (51)$$

We then estimate w at τ_1 as:

$$w_1^{**} = \exp(g_1^{**} - \beta\Psi_1^{**}). \qquad (52)$$

Variable Time Step for the Exponential Time Differencing Method

To decrease the calculation time, we use variable time steps determined by the following equation:

$$\Delta\tau = \min(\Delta\tau_{max}, \Delta\tau_{cnst}v_{max}/v). \qquad (53)$$

The time interval is $\Delta\tau_{max}$ when the slip rate of the block v is very small. When v is nearly v_{max}, the time interval is smaller than the time interval $\Delta\tau_{cnst}$ for appropriate calculations with the constant time interval. We set $\Delta\tau_{max} = 0.01$, $\Delta\tau_{cnst} = 10^{-5}$, and $v_{max} = 0.05$ m/s, which is consistent with the maximum slip rate in the large-scale experiments. Note that the slip increment for a time step is a fixed fraction of L_c because τ is equal to $v_0 t/L_c$ and $\Delta\tau$ is inversely proportional to v.

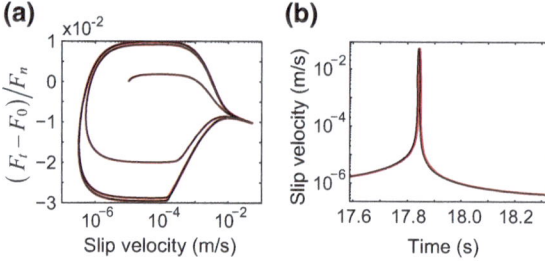

(a) x10⁻² ... $(F_t - F_0)/F_n$ ... Slip velocity (m/s)

(b) Slip velocity (m/s) ... Time (s)

Figure 18

Comparison of stick–slip behaviors obtained by the Runge–Kutta method (*black curves*) and our exponential time differencing method (*red curves*). **a** Changes in the friction coefficient $(F_t - F_0)/F_n$ as a function of slip velocity. **b** Time history of slip velocity for a stick–slip event

Comparison with the Runge–Kutta Method

The parameters of an example problem are $v_L = 0.1$ mm/s, $b = 0.0092$, and $L_c = 0.5$ μm. Simulations were carried out until $t = 20$ s. Figure 18 shows the stick–slip behaviors obtained by the Runge–Kutta method and our exponential time differencing method. They are almost identical; our numerical method works properly. The calculation time necessary for our method was approx. 1/10,000 of that needed to use the Runge–Kutta method.

Appendix 2: Exponential Time Differencing Method for the AGING Law

We developed the same method for the Aging law as for the Slip law. The definitions of the characters in the following equations are the same as those in "Appendix 1". We also used the same variable time step stated in "Appendix 1.3".

Governing Equations

A spring-slider system with one-degree-of-freedom consists of Eqs. (4), (5), (8), and (9) in the main text:

$$m\dot{v} = F_s - F_t \qquad (54)$$

$$\dot{F}_s = k(v_L - v) \qquad (55)$$

$$F_t = F_0 + A\ln(v/v_0) + B\ln(v_0\theta/L_c) \qquad (56)$$

$$\dot{\theta} = 1 - \frac{v\theta}{L_c}, \tag{57}$$

where the definitions of the other characters are the same as those in "Appendix 1" and the main text. The time-derivative of Eq. (56) is:

$$\dot{F}_t = A\frac{\dot{v}}{v} + B\frac{\dot{\theta}}{\theta}. \tag{58}$$

Now we normalize these equations by introducing nondimensional parameters. Nondimensional slip rate, time, frictional stress difference from the steady-state, and spring force difference from the steady-state shall be defined as Eqs. (19)–(22). The nondimensional state variable is:

$$\Psi = v_0\theta/L_c. \tag{59}$$

The nondimensional equations then take the form of Eqs. (24), (25) and

$$g' = C\frac{h}{w} - D\frac{g}{w} - \beta w + \beta\frac{1}{\Psi} \tag{60}$$

$$\Psi' = 1 - w\Psi. \tag{61}$$

The parameters in the nondimensional equations, β, D, C, and E, are Eqs. (28)–(31).

There are four Eqs. (24), (25), (60) and (61), but we have one constraint from the friction law:

$$w = \exp(g - \beta\ln(\Psi)). \tag{62}$$

Therefore, the system is a three-dimensional ordinary differential equation. Since Eq. (62) describes the relation between w, g, and Ψ, we have only to integrate Eq. (25) and additional two equations among Eqs. (24), (60), and (61).

Exponential Time Differencing Method

By integrating Eqs. (25), (60), (61) with (62), we obtained the second-order accurate (in terms of τ_1) Eqs. (33)–(35), where τ_{g0} and $\tau_{\Psi0}$ are Eqs. (38) and (40), respectively, and

$$w_0 = \exp(g_0 - \beta\ln(\Psi_0)) \tag{63}$$

$$g_{ss0} = \frac{C}{D}h_0 + \frac{\beta}{D}\frac{w_0}{\Psi_0} - \frac{\beta}{D}w_0^2 \tag{64}$$

$$\Psi_{ss0} = 1/w_0. \tag{65}$$

Note that g_{ss0} and Ψ_{ss0} are steady-state values which would be achieved if only g and Ψ were variables in Eqs. (60) and (61), respectively. τ_g and τ_Ψ are decay time constants. Then we can estimate w at τ_1 as:

$$w_1^* = \exp(g_1^* - \beta\ln(\Psi_1^*)). \tag{66}$$

Adopting those starred values leads to a first-order integration scheme, but we can iterate this scheme to increase the order of accuracy (see Noda and Lapusta 2010). In the second-order scheme, we integrate the first half time-step using values at $\tau = 0$, Eqs. (42)–(44), and then the latter half using the starred values estimated above, Eqs. (45)–(47), where τ_{g1}^* and $\tau_{\Psi1}^*$ are Eqs. (48) and (50), respectively, and

$$g_{ss1}^* = \frac{C}{D}h_1^* + \frac{\beta}{D}\frac{w_1^*}{\Psi_1^*} - \frac{\beta}{D}w_1^{*2} \tag{67}$$

$$\Psi_{ss1}^* = 1/w_1^*. \tag{68}$$

We then estimate w at τ_1 as:

$$w_1^{**} = \exp(g_1^{**} - \beta\ln(\Psi_1^{**})). \tag{69}$$

Appendix 3: Estimation of Gouge Production Rate

To evaluate effect of the increasing gouge layer thickness, we estimated the gouge production rate from the volume of collected gouge material and amount of mechanical works done during the experiments. Mass of the gouge materials collected after the experiment LB01-111 was 12.0009 g. This gouge was produced by four frictional experiments, LB01-104, -106, -108, and -111. See Fukuyama et al. (2014) for the details of the experimental conditions. From the measured mass, we estimated the volume of the produced gouge material to be 6.330×10^{-6} m^3 under the assumption that effective density of the gouge material of metagabbro is equal to 1896 kg/m^3 following Yamashita et al. (2015). Total amount of mechanical works was calculated as 1.086×10^5 J from F_{s1} integrated over the entire slip distances in the four experiments. As the result, we estimated the gouge production rate to be 5.828×10^{-11} m^3/J.

From this estimated rate, we can calculate the averaged thicknesses of the gouge layer for W1 and W2 in LB01-127 as 5.013×10^{-7} and 1.303×10^{-6}

m, respectively. Note that we estimated these thicknesses assuming the produced gouge materials are uniformly distributed over the fault surface. Actually, the gouge material was locally produced in and around the generated grooves as revealed by Yamashita et al. (2015). Therefore, these thicknesses could be minimum estimates.

REFERENCES

Ampuero, J.-P., & Rubin, A. M. (2008). Earthquake nucleation on rate and state faults—aging and slip laws. *Journal of Geophysical Research, 113*, B01302. doi:10.1029/2007JB005082.

Anthony, J. L., & Marone, C. (2005). Influence of particle characteristics on granular friction. *Journal of Geophysical Research, 110*, B08409. doi:10.1029/2004JB003399.

Beeler, N. M., Tullis, T. E., Blanpied, M. L., & Weeks, J. D. (1996). Frictional behavior of large displacement experimental faults. *Journal of Geophysical Research, 101*(B4), 8697–8715. doi:10.1029/96JB00411.

Bizzarri, A. (2011). Temperature variations of constitutive parameters can significantly affect the fault dynamics. *Earth and Planetary Science Letters, 306*, 272–278.

Blanpied, M. L., Lockner, D. A., & Byerlee, J. D. (1995). Frictional slip of granite at hydrothermal conditions. *Journal of Geophysical Research, 100*(B7), 13045–13064. doi:10.1029/95JB00862.

Brace, W. F., & Byerlee, J. D. (1966). Stick–slip as a mechanism for earthquakes. *Science, 153*(3739), 990–992. doi:10.1126/science.153.3739.990.

Cox, S. M., & Matthews, P. C. (2002). Exponential time differencing for stiff systems. *Journal of Computational Physics, 176*(2), 430–455. doi:10.1006/jcph.2002.6995.

Di Toro, G., Han, R., Hirose, T., De Paola, N., Nielsen, S., Mizoguchi, K., et al. (2011). Fault lubrication during earthquakes. *Nature, 471*, 494–498. doi:10.1038/nature09838.

Dieterich, J. H. (1972). Time-dependent friction in rocks. *Journal of Geophysical Research, 77*(20), 3690–3697. doi:10.1029/JB077i020p03690.

Dieterich, J. H. (1978). Time-dependent friction and the mechanics of stick–slip. *Pure and Applied Geophysics, 116*(4–5), 790–806. doi:10.1007/BF00876539.

Dieterich, J. H. (1979). Modeling of rock friction: 1. Experimental results and constitutive equations. *Journal of Geophysical Research, 84*(B5), 2161–2168. doi:10.1029/JB084iB05p02161.

Fukuyama, E., Yamashita, F., & Mizoguchi, K. (2016). Voids and rock friction at subseismic slip velocity. *Journal of Geophysical Research* (under Rev).

Fukuyama, E., et al. (2014). Large-scale biaxial friction experiments using a NIED large-scale shaking table—design of apparatus and preliminary results—. *Report of the National Research Institute for Earth Science and Disaster Prevention, 81*, 15–35.

Gu, J.-C., Rice, J. R., Ruina, A. L., & Tse, S. T. (1984). Slip motion and stability of a single degree of freedom elastic system with rate and state dependent friction. *Journal of the Mechanics and Physics of Solids, 32*(3), 167–196. doi:10.1016/0022-5096(84)90007-3.

Hori, T., Kato, N., Hirahara, K., Baba, T., & Kaneda, Y. (2004). A numerical simulation of earthquake cycles along the Nankai Trough in southwest Japan: Lateral variation in frictional property due to the slab geometry controls the nucleation position. *Earth and Planetary Science Letters, 228*, 215–226. doi:10.1016/j.epsl.2004.09.033.

Karner, S. L., & Marone, C. (2000). Effects of loading rate and normal stress on stress drop and stick–slip recurrence interval. In J. B. Rundle, D. L. Turcotte, & W. Klein (Eds.), *Geocomplexity and the Physics of Earthquakes* (pp. 187–198). Washington, D. C.: American Geophysical Union.

Kato, N., Kusunose, K., Yamamoto, K., & Hirasawa, T. (1991). Slowly propagating slip events in a composite sample of granite and marble. *Journal of Physics of the Earth, 39*(2), 461–476.

Lapusta, N., & Liu, Y. (2009). Three-dimensional boundary integral modeling of spontaneous earthquake sequences and aseismic slip. *Journal of Geophysical Research, 114*, B09303. doi:10.1029/2008JB005934.

Leeman, J. R., Saffer, D. M., Scuderi, M. M., & Marone, C. (2016). Laboratory observations of slow earthquakes and the spectrum of tectonic fault slip modes. *Nature Communications, 7*, 11104. doi:10.1038/ncomms11104.

Mair, K., Frye, K. M., & Marone, C. (2002). Influence of grain characteristics on the friction of granular shear zones. *Journal of Geophysical Research, 107*(B10), 2219. doi:10.1029/2001JB000516.

Mair, K., & Marone, C. (1999). Friction of simulated fault gouge for a wide range of velocities and normal stresses. *Journal of Geophysical Research, 104*(B12), 28899. doi:10.1029/1999JB900279.

Marone, C. (1998a). Laboratory-derived friction laws and their application to seismic faulting. *Annual Review of Earth and Planetary Sciences, 26*, 643–696. doi:10.1146/annurev.earth.26.1.643.

Marone, C. (1998b). The effect of loading rate on static friction and the rate of fault healing during the earthquake cycle. *Nature,*. doi:10.1038/34157.

Marone, C., & Cox, S. J. D. (1994). Scaling of rock friction constitutive parameters—the effects of surface-roughness and cumulative offset on friction of gabbro. *Pure and Applied Geophysics, 143*(1–3), 359–385. doi:10.1007/BF00874335.

Marone, C., & Kilgore, B. (1993). Scaling of the critical slip distance for seismic faulting with shear strain in fault zones. *Nature, 362*, 618–621. doi:10.1038/362618a0.

Marone, C., Raleigh, C. B., & Scholz, C. H. (1990). Frictional behavior and constitutive modeling of simulated fault gouge. *Journal of Geophysical Research, 95*(B5), 7007–7025.

McLaskey, G. C., Thomas, A. M., Glaser, S. D., & Nadeau, R. M. (2012). Fault healing promotes high-frequency earthquakes in laboratory experiments and on natural faults. *Nature, 491*, 101–104. doi:10.1038/nature11512.

Mitchell, E. K., Fialko, Y., & Brown, K. M. (2015). Frictional properties of gabbro at conditions corresponding to slow slip events in subduction zones. *Geochemistry, Geophysics, Geosystems, 16*(11), 4006–4020. doi:10.1002/2015GC006093.

Nakatani, M. (2001). Conceptual and physical clarification of rate and state friction: Frictional sliding as a thermally activated rheology. *Journal of Geophysical Research, 106*(B7), 13347–13380. doi:10.1029/2000JB900453.

Noda, H., & Lapusta, N. (2010). Three-dimensional earthquake sequence simulations with evolving temperature and pore pressure due to shear heating: Effect of heterogeneous hydraulic diffusivity. *Journal of Geophysical Research, 115*, B12314. doi:10.1029/2010JB007780.

Noda, H., & Lapusta, N. (2013). Stable creeping fault segments can become destructive as a result of dynamic weakening. *Nature, 493*, 518–521. doi:10.1038/nature11703.

Noda, H., & Shimamoto, T. (2009). Constitutive properties of clayey fault gouge from the Hanaore fault zone, southwest Japan. *Journal of Geophysical Research, 114*, B04409. doi:10.1029/2008JB005683.

Press, W. H., Flannery, B. P., Teukolsky, S. A., & Vetterling, W. T. (1992). *Numerical recipes* (2nd ed.). New York: Cambridge University Press.

Rice, J. R., & Tse, S. T. (1986). Dynamic motion of a single degree of freedom system following a rate and state dependent friction law. *Journal of Geophysical Research, 91*(B1), 521–530. doi:10.1029/JB091iB01p00521.

Ruina, A. (1983). Slip instability and state variable friction laws. *Journal of Geophysical Research, 88*(B12), 10359–10370. doi:10.1029/JB088iB12p10359.

Shimamoto, T., Handin, J., & Logan, J. M. (1980). Specimen-apparatus interaction during stick–slip in a tri axial compression machine: A decoupled two-degree-of-freedom model. *Tectonophysics, 67*, 175–205.

Tsutsumi, A., & Shimamoto, T. (1997). High-velocity frictional properties of gabbro. *Geophysical Research Letters, 24*(6), 699–702. doi:10.1029/97GL00503.

Tullis, T. E., & Weeks, J. D. (1986). Constitutive behavior and stability of frictional sliding of granite. *Pure and Applied Geophysics, 124*(3), 383–414. doi:10.1007/BF00877209.

Yamashita, F., Fukuyama, E., Mizoguchi, K., Takizawa, S., Xu, S., & Kawakata, H. (2015). Scale dependence of rock friction at high work rate. *Nature, 528*(7581), 254–257. doi:10.1038/nature16138.

(Received May 10, 2016, revised October 27, 2016, accepted October 31, 2016, Published online November 18, 2016)

Pure Appl. Geophys. 174 (2017), 2239–2256
© 2016 Springer International Publishing
DOI 10.1007/s00024-016-1346-4

Pure and Applied Geophysics

Source Functions and Path Effects from Earthquakes in the Farallon Transform Fault Region, Gulf of California, Mexico that Occurred on October 2013

RAÚL R. CASTRO,[1] JOANN M. STOCK,[2] EGILL HAUKSSON,[2] and ROBERT W. CLAYTON[2]

Abstract—We determined source spectral functions, Q and site effects using regional records of body waves from the October 19, 2013 ($M_w = 6.6$) earthquake and eight aftershocks located 90 km east of Loreto, Baja California Sur, Mexico. We also analyzed records from a foreshock with magnitude 3.3 that occurred 47 days before the mainshock. The epicenters of this sequence are located in the south-central region of the Gulf of California (GoC) near and on the Farallon transform fault. This is one of the most active regions of the GoC, where most of the large earthquakes have strike–slip mechanisms. Based on the distribution of the aftershocks, the rupture propagated northwest with a rupture length of approximately 27 km. We calculated 3-component P- and S-wave spectra from ten events recorded by eleven stations of the Broadband Seismological Network of the GoC (RESBAN). These stations are located around the GoC and provide good azimuthal coverage (the average station gap is 39°). The spectral records were corrected for site effects, which were estimated calculating average spectral ratios between horizontal and vertical components (HVSR method). The site-corrected spectra were then inverted to determine the source functions and to estimate the attenuation quality factor Q. The values of Q resulting from the spectral inversion can be approximated by the relations $Q_P = 48.1 \pm 1.1 f^{0.88 \pm 0.04}$ and $Q_S = 135.4 \pm 1.1 f^{0.58 \pm 0.03}$ and are consistent with previous estimates reported by Vidales-Basurto et al. (Bull Seism Soc Am 104:2027–2042, 2014) for the south-central GoC. The stress drop estimates, obtained using the ω^2 model, are below 1.7 MPa, with the highest stress drops determined for the mainshock and the aftershocks located in the ridge zone. We used the values of Q obtained to recalculate source and site effects with a different spectral inversion scheme. We found that sites with low S-wave amplification also tend to have low P-wave amplification, except for stations BAHB, GUYB and SFQB, located on igneous rocks, where the P-wave site amplification is higher.

Key words: Earthquakes in the Gulf of California, source and path effects, Farallon transform fault.

1. Introduction

The plate boundary between North America and the Pacific plates cuts across the Gulf of California. The present style of rifting started at ~6 Ma (Atwater and Stock 1998), and the deformation in this region is governed by oblique faults in the north and transform faults in the south (Fenby and Gastil 1991; Nagy and Stock 2000).

The October 19, 2013 ($M_w = 6.6$) earthquake was located by the National Seismological Service of Mexico (SSN) at 90 km east of Loreto, Baja California Sur, Mexico, near the Farallon transform fault (Fig. 1). This event and eight aftershocks with magnitudes (M_w) ranging from 1.9 to 4.4 occurred in one of the most seismically active regions of the Gulf of California. The aftershocks were located in the Carmen basin, between the Carmen and Farallon transform faults. The focal mechanism of the mainshock (Fig. 2), taken from the GCMT catalog, exhibits right-lateral strike–slip motion on a NW-striking fault (strike = 222°, dip = 86°, rake = 6°). Most of the aftershocks (Figs. 1, 2) are located within 27 km northwest of the epicenter of the mainshock, suggesting unilateral rupture to the northwest. The location of the centroid, where the maximum slip of the fault occurs, is also to the northwest (azimuth = 334°) of the epicenter reported by the SSN.

The large earthquakes in this region typically accommodate strike–slip movement (Goff et al. 1987; Castro et al. 2011a), but normal faulting events occur near the ends of the spreading centers where they connect to transform faults. Previous events on the Farallon transform fault include the $M_w = 7.0$ of December 9, 1901 (Pacheco and Sykes 1992); the $M_w = 6.2$ of August 28, 1995 (Tanioka and Ruff

---
[1] Departamento de Sismología, CICESE, Carretera Ensenada-Tijuana 3918, 22860 Ensenada, Baja California, Mexico. E-mail: raul@cicese.mx
[2] Seismological Laboratory, Caltech, 1200 E. California Blvd., Pasadena, CA 91125, USA.

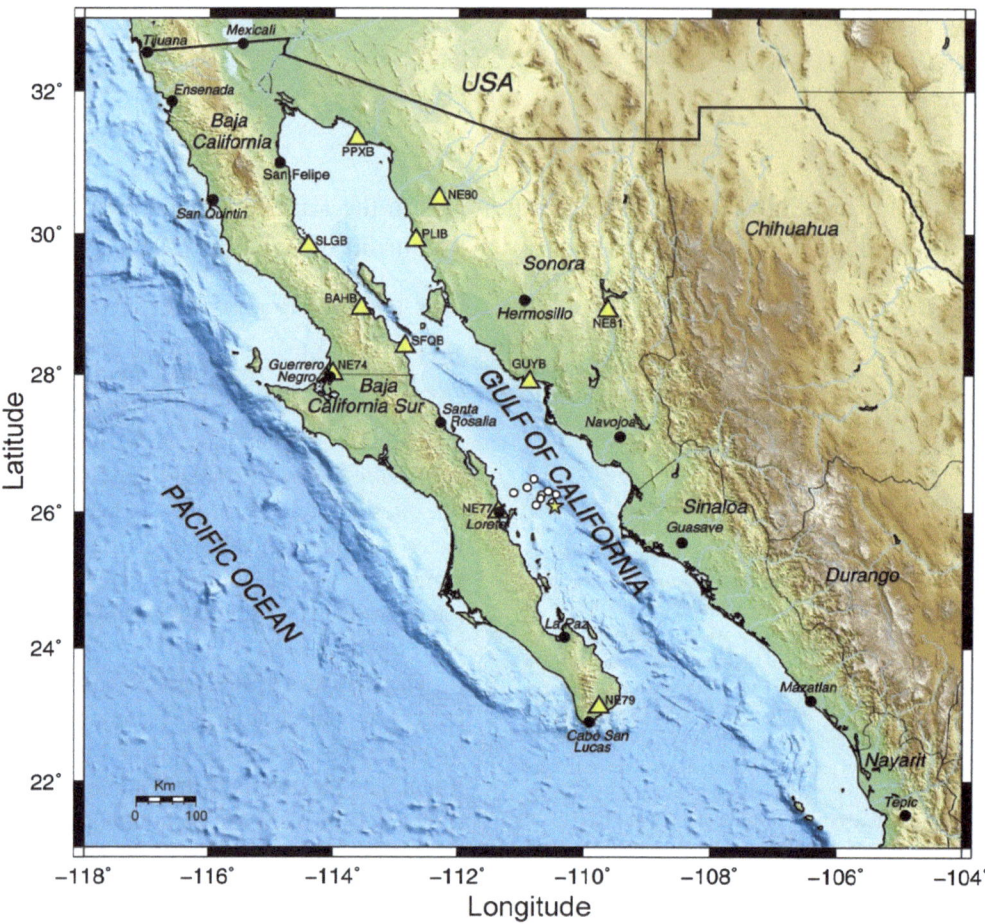

Figure 1
Distribution of stations of the RESBAN network (*triangles*), location of main event (*star*) and aftershocks analyzed (*circles*). This map was generated using Generic Mapping Tools (Wessel and Smith 1998)

1997); and the February 22, 2005 event, a moderate magnitude earthquake ($M_w = 5.5$) that occurred 48 km southeast of the October 19, 2013 epicenter. Rodriguez-Lozoya et al. (2008) calculated the focal mechanism of this event and found a right-lateral strike–slip fault plane solution. The southern region of the Gulf of California (GoC), where these earthquakes were located, is characterized by thin oceanic crust (Zhang et al. 2007), which suggests that longer rupture lengths may be expected for a given magnitude, than those from events in other tectonic environments with thicker crust and deeper rupture.

A few studies of source parameters have been made from earthquakes in the GoC, for instance, Munguía et al. (1977) studied the aftershock sequence of the July 8, 1975 Canal de Ballenas ($M_S = 6.5$) earthquake; Goff et al. (1987) calculated earthquake source mechanisms of events that occurred on transform faults; Rebollar et al. (2001) estimated source parameters of a $M_S = 5.5$ earthquake that occurred in the Delfin basin, northern GoC; and López-Pineda and Rebollar (2005) determined the source characteristics of the March 12, 2003 ($M_w = 6.2$) Loreto earthquake and 20 other earthquakes with magnitude greater than five that occurred in different parts of the GoC. More recently, Castro et al. (2011b) studied the August 3, 2009 ($M_w = 6.9$) Canal de Ballenas earthquake and its aftershocks. In general, the stress drops of the events in the GoC tend to be low; for instance, the average

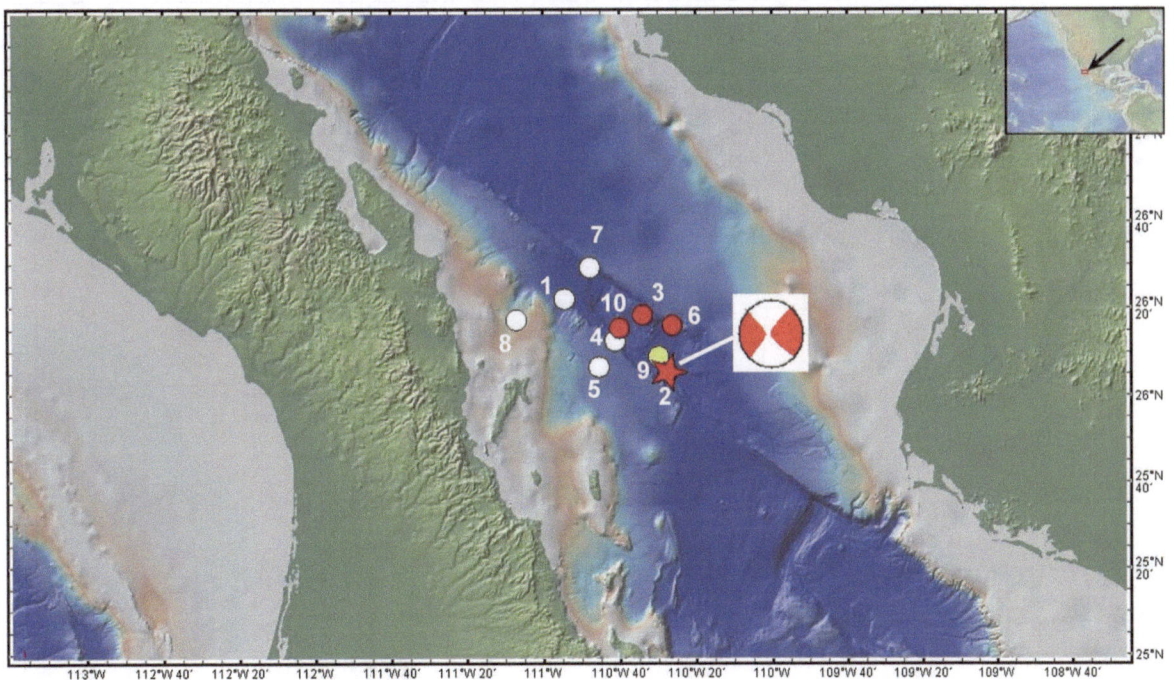

Figure 2

Epicenters of the 2013 sequence. The focal mechanism shown is of the main event and was taken from the GCMT catalog. The *white circle* marked with the number 1 is an $M = 3.3$ foreshock recorded on September 2, 2013 (47 days before the main event). *Red circles* indicate events with stress drop ($\Delta\sigma$) greater than 1.0 MPa, *yellow circles* events with $0.5 \leq \Delta\sigma < 1.0$ MPa, and *white circles* events with $\Delta\sigma < 0.5$ MPa. The topography and bathymetry are from GeoMap App

stress drop of the earthquakes analyzed by López-Pineda and Rebollar (2005) is 0.25 MPa. For the Canal de Ballenas region, Munguía et al. (1977) estimated a stress drop of 0.4 MPa for the July 1975 ($M_S = 6.5$) event, and Castro et al. (2011b) calculated a stress drop of 2.2 MPa for the August 2009 ($M_w = 6.9$) event.

Here, we determine source parameters and path characteristics of the October 19, 2013 ($M_w = 6.6$) earthquake sequence and a $M_w = 3.3$ earthquake that occurred 47 days before the main event in the rupture area.

2. Tectonic Setting

The Gulf of California is an oblique rift system with short spreading centers connected by transform faults. The peninsula of Baja California, located on the west side of the GoC, moves with the Pacific plate

with approximately 48 mm/year of spreading across the GoC (Lizarralde et al. 2007). Between about 12 and 3.5 Ma Baja California was a rigid micro-plate bounded by the displacement between the Pacific and North America plates. The dominant extensional faults in the Gulf Extensional Province strike NNW, but the exact direction of extension is unknown (Stock and Hodges 1989). In the central-south GoC, south of Canal de Ballenas (below 29°N), the plate boundary consists of a series of en echelon transform fault zones. The transform faults between Delfin and Carmen basins have generated 60 % of the plate boundary earthquakes with $M_S > 6$ (Goff et al. 1987).

Several significant earthquakes have occurred along the Farallon–Carmen transform faults system in the past. Figure 3 shows the location of earthquakes with $M_S > 5.8$ reported in this region. The 1901 $M_S = 7.0$ earthquake (Pacheco and Sykes 1992) is the biggest event instrumentally located in the GoC, and the epicenter is on the southeastern end of

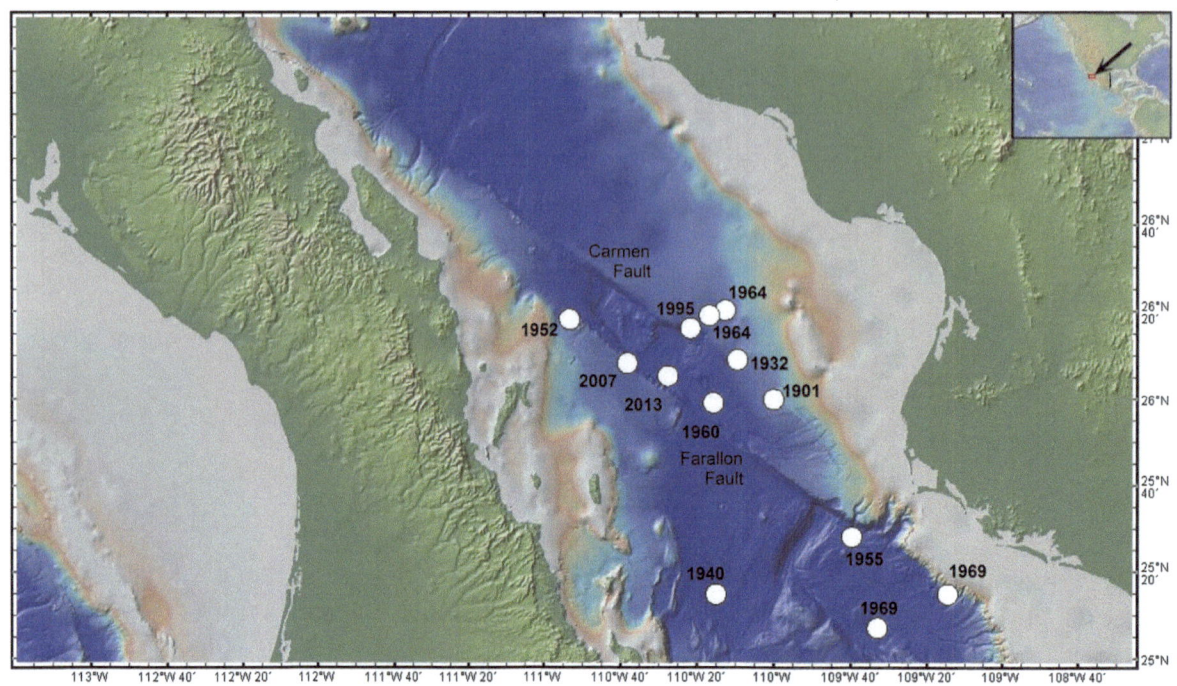

Figure 3
Historical seismicity within the Farallon transform fault region, only events with $M_S > 5.8$ are shown (Table 1). The 1901 earthquake, the biggest of all, has a magnitude $M_w = 7.0$ (Pacheco and Sykes 1992). The location of the 2013 ($M_w = 6.6$) event was taken from the Mexican National Seismological Service (SSN) catalog. The topography and bathymetry are from GeoMap App

Table 1

Historical earthquakes in the Farallon transform fault region taken from Pacheco and Sykes (1992) (reference 1), the ISC catalog (reference 2) and Goff et al. (1987) (reference 3)

Date	Time	Lat	Lon	Depth	Mag (M_S)	References
1901-12-09	02:12	26.00	−110.00	–	7.0	1
1932-07-12	19:24	26.15	−110.16	10.0	6.8	2
1940-06-03	18:05	25.25	−110.25	35.0	6.2	2
1952-11-07	20:55	26.30	−110.89	15.0	6.3	2
1955-04-05	15:09	25.47	−109.66	15.0	6.5	2
1960-03-31	19:56	25.99	−110.26	15.0	5.8	2
1964-07-05	19:08	26.34	−110.21	–	6.3	3
1964-07-06	02:14	26.32	−110.28	–	6.6	3
1969-08-17	20:13	25.25	−109.24	–	6.5 (M_w)	3
1969-08-17	20:15	25.12	−109.55	–	6.6	3
1995-08-28	10:46	26.27	−110.36	12.2	6.5	2
2007-03-13	02:59	26.14	−110.64	10.0	5.8	2

Carmen fault. Tanioka and Ruff (1997) studied the source function of the 1995 ($M_w = 6.2$) earthquake, which is also located on the Carmen fault together with the 1964 and 1932 earthquakes (Fig. 3). The earthquakes that occurred in 1952, 1955, 1960, 1969,

2007 and 2013 (Table 1) are located on the Farallon fault. The location of the 2013 ($M_w = 6.6$) event in Fig. 3 was taken from the Mexican National Seismological Service catalog. The epicenter of the 1940 earthquake was probably mislocated, because it is more than 30 km away from the closest active fault, and the epicenters in the GoC reported by ISC can have errors in the order of ~50 km (Sumy et al. 2013). It is likely that the 1940 event is on the Pescadero fault together with the 1969 event.

A more detailed review of current bathymetric data helps to place the previous seismicity, and the 2013 earthquake sequence, in context. The 2013 earthquakes are within and near the Del Carmen Basin area of the Pacific–North America plate boundary in the Gulf of California (Figs. 2, 3). This spreading segment lies between the wider and better studied Guaymas Basin and Farallon Basin segments. In this region, the seismic velocity structure is likely to be complicated due to the irregular juxtaposition of continental and oceanic blocks. The crust of these central Gulf basins is inferred to comprise a variable mix of sediments and nascent oceanic crust lacking the abyssal hill fabric and marine magnetic reversal patterns characteristic of the southern Gulf of California basins and East Pacific Rise (EPR) (Lonsdale 1989). From crustal-scale seismic profiles of the Guaymas, Alarcon, and northern EPR segments, Lizarralde et al. (2007) identified the Continent–Ocean transition as a gradient in Moho depth and a lateral increase in Vp. Their results indicate that in the Guaymas basin, a 280-km-width of new igneous crust formed since ca. 6 Ma. The width of newly accreted oceanic crust in the Del Carmen and Farallon segments is not well constrained, but is probably <280 km, because these segments include a wider zone of extended, partially submerged, continental crust in the Baja California continental borderland (Nava-Sánchez et al. 2001). In addition, the small Del Carmen basin is a younger offset in the plate boundary system; it appears to have formed along the transform offset separating the Guaymas and Farallon basins after those segments were already well established.

To relate the seismicity to the modern plate boundary geometry, we obtained multibeam data files from the IEDA Marine Geoscience Data System (Carbotte et al. 2004; http://www.marine-geo.org) for cruises AT15-31, DANA07RR, MOCE05MV, AT03-L46, and EW0210. We combined these multibeam data files using the software MB-System (Caress and Chayes 2014) (http://www.mbari.org/data/mbsystem) to produce a detailed bathymetric map of the study area, and used it to make the following measurements and observations.

Along an NW–SE profile parallel to plate motion, the Del Carmen basin is a 70-km-wide bathymetric low with basin shoulders at ca. 1700 m water depth. This low contains three sub-basins: two in the center and on the SE side of the basin reaching 2300 m water depth and a third one, on the NW side of the basin, with a 14-km-long, 3-km-wide flat-bottomed trough reaching 2800 m water depth. Multibeam bathymetry suggests that this narrow trough, trending N32°E, represents the modern plate boundary, connecting at its northern end to the Carmen transform fault which has a strike here of N61°W. Southeastward from this trough, the Farallon transform fault system extends 130 km SE to the Farallon basin, but in two different segments with changing character along strike: an irregular zone in the northwestern 60 km and, then, a straight, more linear scarp for the southeastern 70 km. In detail, the irregular northwestern section of the Farallon transform fault system is characterized by a zigzag pattern of bathymetric scarps trending N47°W, N24°W, N50°W, and N25°W. The overall oblique orientation, compared to the N61°W Carmen transform fault, indicates that there are likely multiple active structures, and a broad zone of deformation, accommodating extension within and outside of the region of the Del Carmen basin. The 2013 earthquake sequence appears to be confined to the irregular northern zone of the Farallon transform fault and to the adjacent Del Carmen basin.

Sumy et al. (2013) relocated events in the Guaymas–Del Carmen basins from October 2005 to October 2006 using wave arrivals from an array of ocean-bottom seismographs deployed as part of the Sea of Cortez Ocean Bottom Array (SCOOBA) experiment and onshore stations of the Network of Autonomously Recording Seismographs (NARS)-Baja array. Figure 4 shows only epicenters relocated by Sumy et al. (2013) between 24.5°N and 28°N, in the Guaymas basin and south. These well-located epicenters seem to concentrate near the spreading zones between the transform faults.

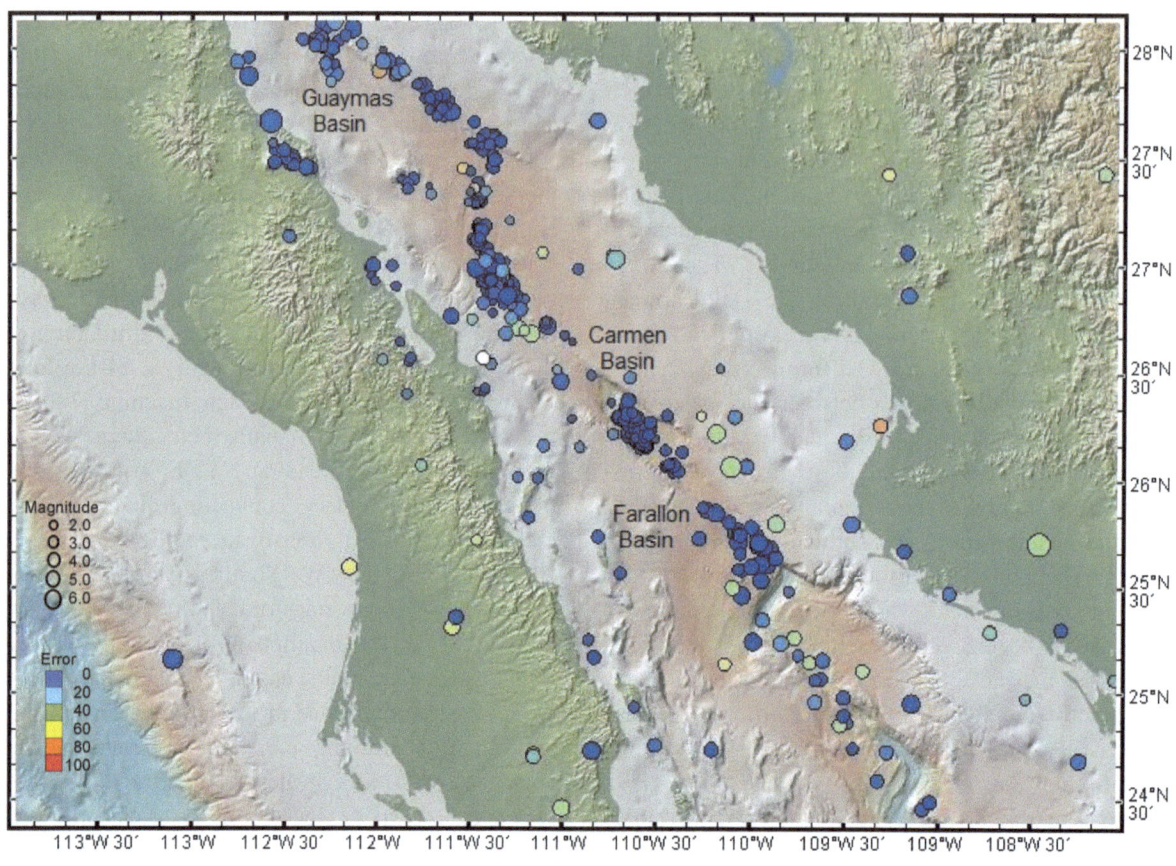

Figure 4
Epicenters relocated by Sumy et al. (2013) in the Guaymas—Del Carmen—Farallon basins from October 2005 to October 2006 using an array of ocean-bottom seismographs deployed as part of the Sea of Cortez Ocean Bottom Array (SCOOBA) experiment and onshore stations of the Network of Autonomously Recording Seismographs (NARS)-Baja array. The topography and bathymetry are from GeoMap App. The topography and bathymetry are from GeoMap App. The size of the circle is proportional to the magnitude of the event and the color is proportional to the location error (*blue* low error; *red* high error)

Table 2

Earthquake coordinates of the 2013 seismic sequence reported by the National Seismological Service (SSN) of the Instituto de Geofísica, UNAM

Event no.	Date Y/M/D	Origin time hour:minute	Latitude N	Longitude W	Depth (km)
1	2013/09/02	19:50	26.37	110.91	16
2	2013/10/19	17:54	26.09	110.46	14
3	2013/10/19	22:14	26.26	110.67	16
4	2013/10/19	23:32	26.21	110.69	16
5	2013/10/20	04:19	26.11	110.76	–
6	2013/10/20	06:20	26.27	110.44	–
7	2013/10/20	08:18	26.49	110.80	16
8	2013/10/20	08:32	26.29	111.12	15
9	2013/10/20	12:01	26.15	110.50	16
10	2013/10/31	15:08	26.31	110.57	23

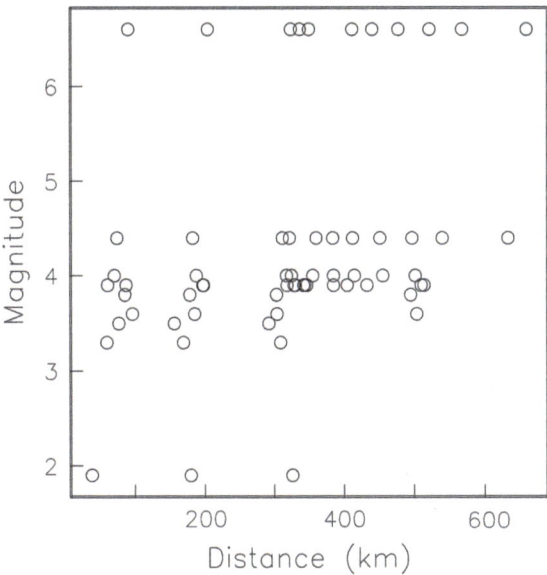

Figure 5
Epicentral distance distribution versus magnitude of the earthquakes analyzed

3. Data and Method

The dataset consists of 3-component waveforms from 11 stations (Fig. 1) of the broadband network RESBAN (*Red Sismologica de Banda Ancha del Golfo de California*) that recorded 10 moderate-size earthquakes (Table 2). The events analyzed have M_w magnitudes ranging from 1.9 to 6.6 and epicentral distances between 30 and 670 km (Fig. 5). The RESBAN stations are equipped with Guralp CMG-40T or CMG-3ESP digital three-component seismometers with a GPS and 24-bit Guralp digitizer. These broadband instruments are set to record earthquakes at a rate of 20 samples per second. Two of the stations (BAHB and GUYB) have Streckeisen STS-2 sensors and Reftek DAS 130 recorders that sample 100 samples per second. Figure 6 shows a sample of horizontal component waveforms from the mainshock (event 2 in Table 2). We consider that the 20 Hz sampling (10 Hz Nyquist) is high enough for our analyses, because only one of the earthquakes analyzed (event 8) has a magnitude $M < 3$. During the 2010–2012 Canterbury earthquake sequence, Van Houtte et al. (2014) show that events with $M > 3$ tend to have corner frequencies $f_c < 10$ Hz. Transform fault earthquakes with magnitudes between 5.5 and

7.1 tend to have even lower corner frequencies, $0.07 < f_c < 0.2$ Hz (Stein and Pelayo 1991).

All the records were corrected for instrument response and baseline-corrected by subtracting the mean, and we choose time windows containing clear P- and S-wave arrivals to calculate Fourier acceleration spectra of the signals. We selected the beginning of the window about 1 s before the first phase arrival and the end a few seconds after the peak amplitude. For the P wave, the time window ends before the S-wave arrival and for the S wave, before the surface waves arrive. The window lengths vary depending on the epicentral distance and are typically 4–38 s for P waves and 6–112 s for S waves. The beginning and the end of the time windows were tapered with a 5 % cosine taper before the Fourier transform was calculated. The spectral amplitudes were smoothed using a variable frequency band of ±25 % over 21 predefined central frequencies between 0.10 and 10.00 Hz, equidistant on a logarithmic scale. The spectral amplitudes at the central frequency selected are the average amplitude within the corresponding frequency band with the constraint that the total energy of the original spectrum is conserved (e.g., Castro et al. 1990). For further analysis, we inspected the spectral amplitudes to select frequency bands above the noise level, and we use spectral amplitudes with signal-to-noise ratio above a factor of two. We differentiated the records in the frequency domain to obtain acceleration spectra. Figure 7 displays the S-wave acceleration spectra of the main event obtained for each station. This figure illustrates the attenuation effect on the S-wave amplitude spectra. At 1 Hz, for instance, the amplitude decreased four orders of magnitude between the station at epicenter distance of 205 km and the station at 659 km, in a 454-km interval. For detailed analysis, we calculated the geometric mean $(a(f) = \sqrt{a_{NS}^2(f) + a_{EW}^2(f)})$ of the horizontal components of the S-wave amplitude spectra and the vertical component for the P wave.

We estimated site effects at the stations by calculating average spectral ratios between horizontal and vertical components of S waves (HVSR method). Since most stations of the array are located on hard-rock sites, we do not expect important amplifications, and the HVSR method can give a first-order estimate

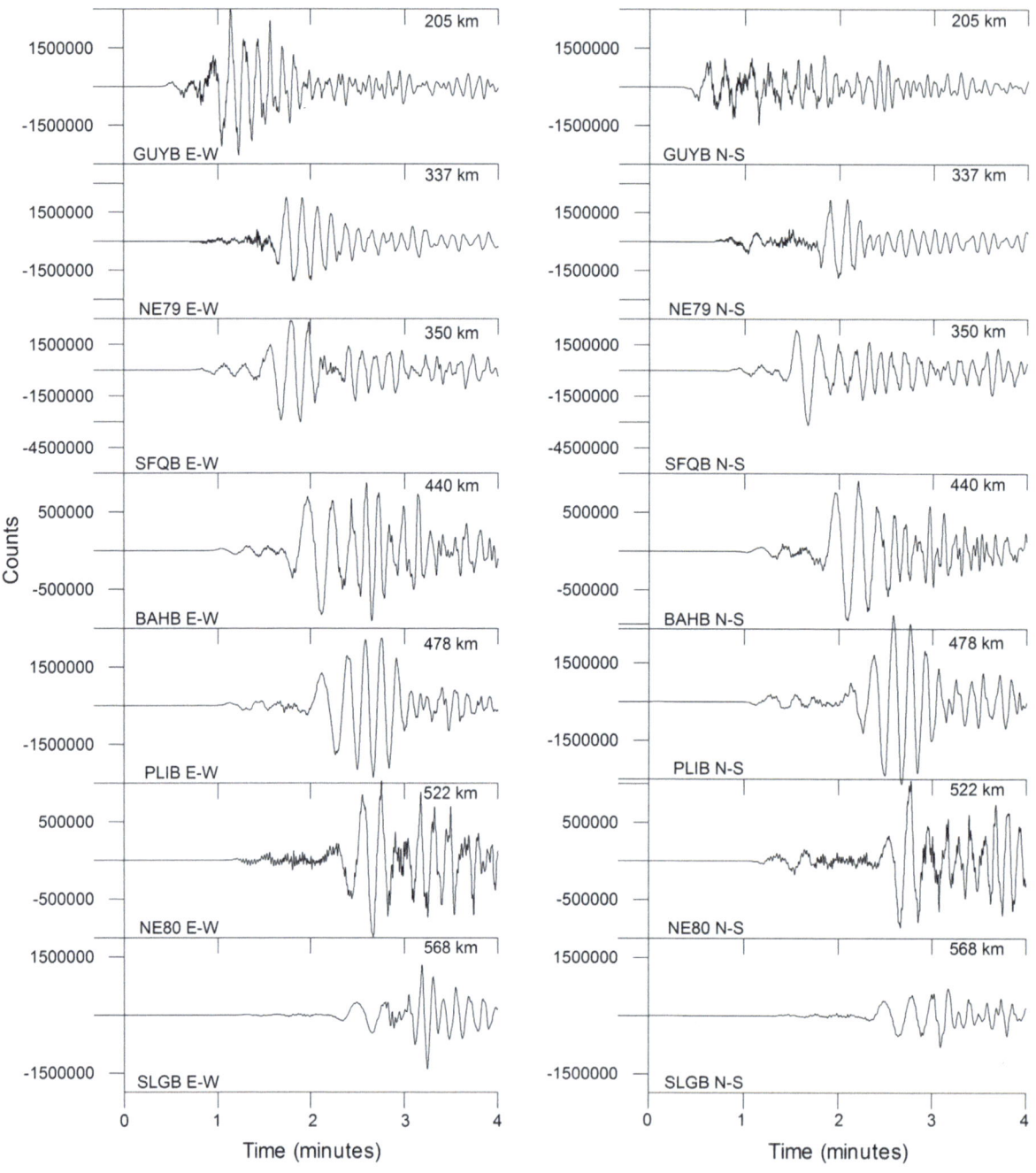

Figure 6
Horizontal component seismograms of main event of October 19, 2013 ($M_w = 6.6$). The *left column* is the east–west component, while the *right column* is the north–south component of the RESBAN stations

of the site response. We used waveforms from the 2013 sequence and events recorded by the RESBAN array and the Network of Autonomously Recording Seismographs of Baja California, NARS-Baja,

(Trampert et al. 2003; Clayton et al. 2004) between 2002 and 2006, and analyzed previously by Avila-Barrientos and Castro (2015). We calculated for each station the spectral ratio between the horizontal and

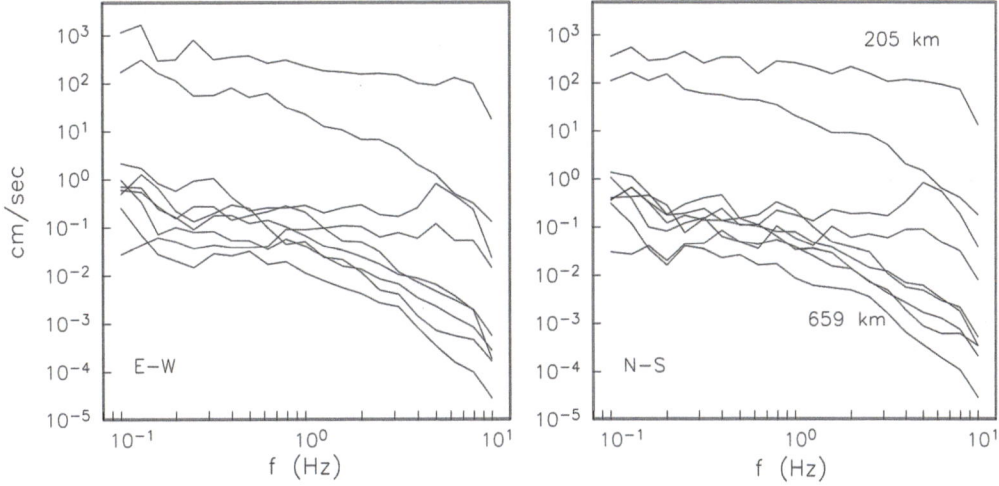

Figure 7
S-wave acceleration spectra of the main earthquake. *Left frame* are east–west components and *right frame* north–south components

vertical components (HVSR method) of each event. Then, we computed the average of all events and the geometric mean of both horizontal components. We assumed that the resulting average HVSR represents the site response of the corresponding station.

The HVSR method was originally proposed by Nakamura (1989) and extended by Lermo and Chávez-García (1993) to estimate the site amplification of the horizontal component of ground motion. This method assumes that the vertical component of the ground motion is insensitive to site amplification (Langston 1977; Nakamura 1989), so that when the spectral ratio between the horizontal and the vertical component is calculated, the source effect and any other path effect common for both components cancel out. The original S-wave spectra were corrected by site effect using the corresponding site response calculated with the HVSR method. The site-corrected spectra were then inverted to determine the source functions and to estimate the attenuation quality factor Q according to the following model:

$$U_i(r,f) = S_i(f) \cdot \frac{N}{r^n} \exp\left[-\frac{\pi f}{Qv}(r - N)\right], \quad (1)$$

where $U_i(r,f)$ is the observed spectral amplitude after site effect correction at frequency f from event i recorded at hypocenter distance r. $S_i(f)$ is the acceleration source function of event i, $\frac{N}{r^n}$ is the geometrical spreading function, with $N = 10$ km

being a normalization factor. Vidales-Basurto et al. (2014) calculated the geometrical spreading exponent n in the GoC for both P and S waves and found that this exponent varies between 0.8 and 1.1 for P waves and between 0.8 and 1.0 for S waves. We make $n = 0.9$ in Eq. (1). v is the wave velocity, we used a value of 3.77 km/s for S and 6.71 km/s for P waves. Q is the wave quality factor that accounts for both intrinsic and scattering attenuation. The estimates of Q are conditioned to the geometrical spreading function selected. However, the value of $n = 0.9$ is very close to the expected theoretical value for body waves ($n = 1$). Thus, the geometrical spreading function that we defined is a reasonable reference to estimate Q.

For a given frequency f and source-station distance r_j Eq. (1) can be linearized by taking logarithms:

$$u_{ij} = s_i + cQ^{-1} \quad (2)$$

where:

$$u_{ij} = \mathrm{Log}U_i(r,f) + 0.9\,\mathrm{Log}r_j - 1.0 \quad (3)$$

$$s_i = \mathrm{Log}S_i(f) \quad (4)$$

$$c = -\frac{\pi}{v}(r_j - 10)\mathrm{Log}e \quad (5)$$

The estimates of Q resulting from the inversion of Eq. (2) can be used, in a second step, to recalculate the S-wave site responses and to make new estimates of P-wave site responses. We corrected the original spectral records for attenuation effect with the same geometrical spreading function in Eq. (1) and the estimates of Q obtained solving Eq. (2). Thus, the observed spectral amplitude after attenuation effect correction at frequency f from event i at site j can be expressed as:

$$d_{ij} = s_i + z_j \, , \qquad (6)$$

where s_i is the acceleration source function of event i, as in Eq. (4), and $z_j = LogZ_j(f)$ represents the site

response of the station j. Equation (6) is solved by a least-squares inversion using the singular value decomposition technique (e.g. Castro et al. 1990).

A similar inversion technique has been used before by Phillips and Aki (1986) and Andrews (1986) to determine site effects and source functions in California. To eliminate the linear dependence between site and source terms, we used as a reference site the station that shows minimum HVSR values (station PPXB). The site and source terms in Eq. (6) are calculated independently at each frequency. The advantage of this method is that the stability of the

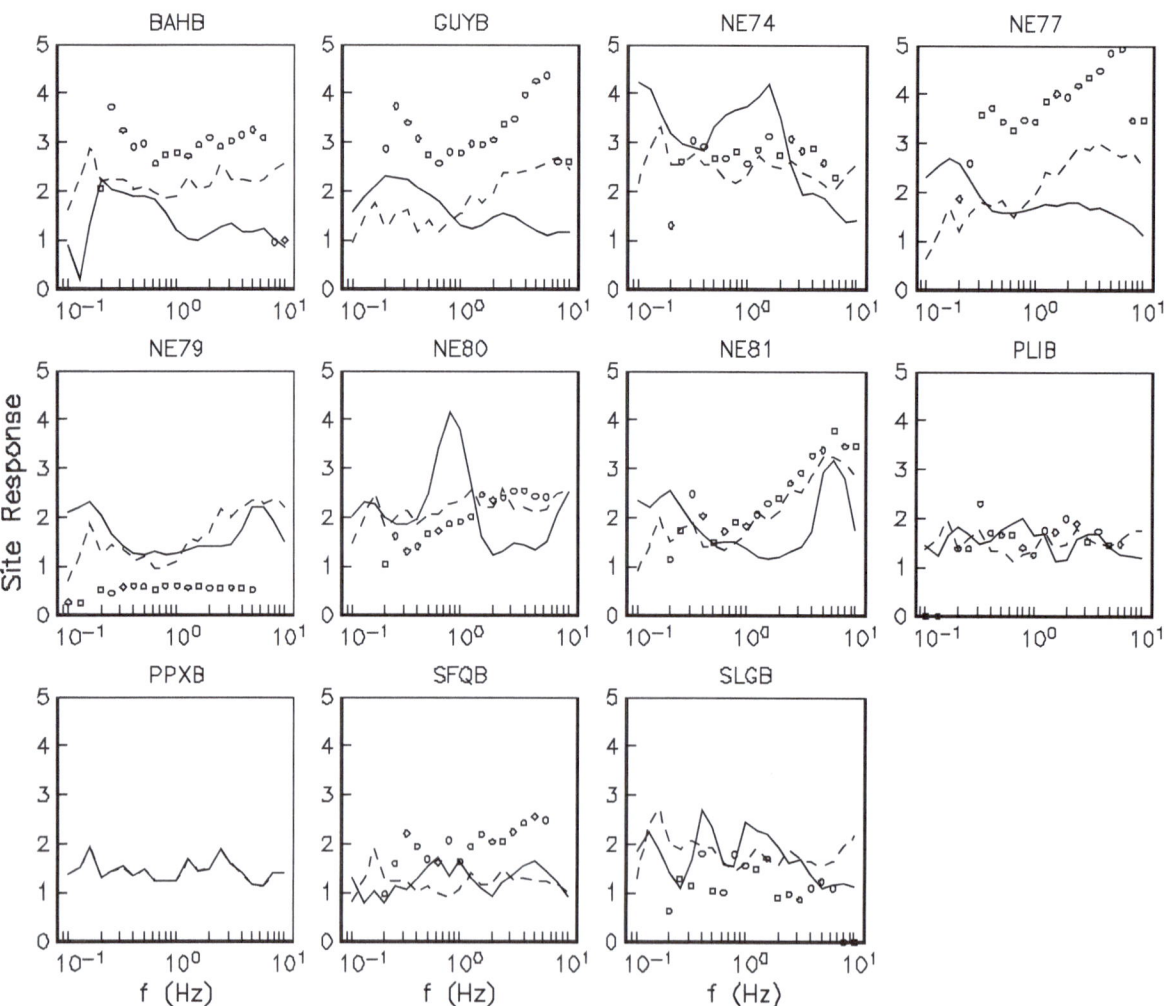

Figure 8
Site response of the stations analyzed. *Solid* and *dashed lines* are the S-wave responses obtained using horizontal-to-vertical spectral ratios (HVSR method) and the spectral inversion, respectively. The *small circles* are the P-wave responses resulting from the spectral inversion

Table 3

Station coordinates and site geology of the RESBAN network

Code	Latitude	Longitude	Elevation (m)	Group class	Site geology
BAHB	28.943	−113.561	35	I	Igneous intrusive: granodiorite-tonalite
PPXB	31.335	−113.632	10	I	Igneous intrusive: granite-granodiorite
PLIB	29.915	−112.694	40	I	Igneous intrusive: granodiorite
GUYB	27.899	−110.871	50	II	Igneous breach with fractures filled with calcium carbonate
NE74	28.008	−114.014	21	III	Eolian deposits unconsolidated composed of fine sands
NE77	26.016	−111.361	40	II	Sedimentary deposits and conglomerates
NE79	23.119	−109.756	225	I	Igneous intrusive: granite-granodiorite
NE80	30.500	−112.320	225	III	Unconsolidated sandstone
NE81	28.918	−109.636	295	III	Unconsolidated conglomerate
SFQB	28.405	−112.861	50	I	Igneous intrusive: granite
SLGB	29.830	−114.404	15	II	Volcanic breccia

Group I correspond to stations located on intrusive volcanic rocks with low degree of weathering; Group II are extrusive igneous rocks with moderate weathering; Group III are stations located on poorly consolidated conglomerates or soil

solutions is independent from the fall-off at high frequencies, since the shape of the source functions is unconstrained (e.g. Castro et al. 2013).

4. Site Effects

The site functions calculated with the HVSR method (solid lines in Fig. 8) show small amplifications (less than a factor of two) at most frequencies. Stations NE74 and NE80, that are located on sand sediments (Table 3), have site amplifications that reach a factor of four at 1.6 and 0.7 Hz, respectively. Avila-Barrientos and Castro (2015) studied the site response at the stations of the RESBAN and NARS-Baja networks and classified the sites in three groups according to the near-surface geology below the stations. Group I corresponds to stations located on intrusive volcanic rocks with low degree of weathering; Group II are extrusive igneous rocks with moderate weathering; and Group III are stations located on poorly consolidated conglomerates or soil. Stations NE74, NE80 and NE81 are in the Group III, NE81 has its peak site amplification of three at 6 Hz, the likely natural frequency of resonance of this site. PPXB (Group I) is on a hard-rock outcrop (granite), and its site response function is approximately flat for the whole frequency band analyzed (0.10–10.0 Hz). We used PPXB as the reference site to solve Eq. (6) and the other site functions (solid lines in Fig. 8) to

correct the observed spectral records for site amplification effect and to determine Q and source functions solving Eq. (2). Figure 8 also shows the site responses obtained from the observed P-wave spectra (small circles) and from S-wave (dashed lines) solving Eq. (6). The site response from P waves is similar to that obtained from S waves solving Eq. (6), except for sites BAHB, GUYB and SFQB where the P-wave site amplification is higher. Sites with low S-wave amplification also tend to have low P-wave amplification, for instance NE79, PLIB and SLGB. Comparing the S-wave site responses obtained from the two methods (HVSR and Eq. (6)), we can see that Eq. (6) tends to give larger amplifications at high frequencies ($f > 1$ Hz) at sites BAHB, GUYB, NE77, NE80 and SLGB. This may be an effect of using different site reference and number of events. For the HVSR method, there were more available events and records, and the possible azimuthal effects tend to average out better. Since the HVSR method assumes zero amplification on the vertical component, the lower S-wave amplification obtained with the HVSR between 1–10 Hz, compared to that obtained with the spectral inversion (Eq. (6)), could be due to S-wave amplification on the vertical component. For instance, Castro et al. (2004) observed in central Italy that sites located in sedimentary basins with thick sediments show important vertical amplifications at low frequencies. However, in general, HVSR gives a good estimate of the site

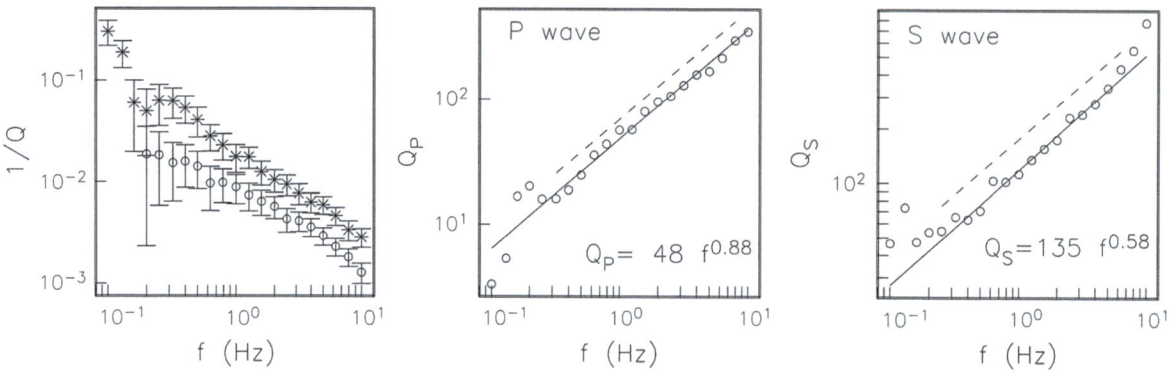

Figure 9
Estimates of the quality factor Q (*left frame*) obtained from the spectral inversion (*asterisks* for P and *circles* for S waves). The *continuous line* in the *middle* and *right frames* is the resulting linear regression of the Q estimates and the *dashed lines* the functions $Q_P = 69\ f^{0.87}$ and $Q_S = 176\ f^{0.61}$ obtained by Vidales-Basurto et al. (2014) for P and S waves, respectively, in the south-central region of the Gulf of California

response although tends to sub-estimate the amplifications at high frequencies. Similar results were obtained by Edwards et al. (2013) in Switzerland using regional seismicity.

5. P- and S-Wave Attenuation in the Gulf of California

We solved Eq. (2) for each of the 21 frequencies that sample the observed spectral records to estimate values of Q for P and S waves (Fig. 9). The values of Q for P waves (Q_P) tend to be lower than those for S waves (Q_S), indicating higher attenuation for compressional waves (left frame of Fig. 9) compared to shear waves. The frequency dependence of Q can be approximated with the functional form:

$$Q = Q_0(f/f_0)^a , \qquad (6)$$

where Q_0 is the value of Q at $f_0 = 1.0$ Hz.

We found the following Q-frequency relations for P and S waves:

$$Q_P = 48.1 \pm 1.1 f^{0.88 \pm 0.04} \qquad (7)$$

$$Q_s = 135.4 \pm 1.1 f^{0.58 \pm 0.03} \qquad (8)$$

Seismic attenuation in the south-central region of the GoC was previously studied by Vidales-Basurto et al. (2014) with records from the NARS-Baja and RESBAN arrays, and from an array of ocean-bottom seismographs (OBS) deployed as part of the Sea of Cortez Ocean Bottom Array experiment (SCOOBA)

(Sumy et al. 2013). We compare in Fig. 9 their results (dashed lines) with those obtained in this study. Although the data set used by Vidales-Basurto et al. (2014) covers a bigger area, their Q-frequency functions show similar trends. However, their Q values are slightly higher for both P and S waves. It is also interesting to compare these values of Q with those reported inland, north of the GoC. For the Imperial Valley, California region, Singh et al. (1982) found that $Q_S = 20f$. Rebollar et al. (1985) estimated Q from coda waves in northern Baja California, and they found that $Q_C = 37f^{0.87}$. Raoof et al. (1999) found in southern California that $Q_S = 180\ f^{0.45}$ for frequencies between 0.25 and 5.0 Hz. Adams and Abercrombie (1998) used data from the Cajon Pass borehole to study Q, and they found that from 1 to 10 Hz, Q_S is frequency-dependent with an average total (intrinsic and scattering) $Q_S = 1078$.

The ratio Q_P/Q_S (Fig. 10) is below 1.0 in the whole frequency band 0.1–10 Hz. Experimental studies made in rocks indicate that when the rocks are partially saturated, $Q_P/Q_S < 1$ and when the rocks are totally saturated, $Q_P/Q_S > 1$ (Winkler and Nur 1982). These results suggest that the body-wave amplitudes are affected by partially saturated rocks probably associated to magmatic bodies located near the ridge zones. Hauksson and Shearer (2006) found that $Q_S/Q_P > 1$ in southern California, suggesting that the crust is partially fluid-saturated in that region. They estimated that $Q_P \sim 500$–900 and $Q_S \sim 600$–1000 with a mean value of $Q_P/Q_S = 0.77$. In the Salton

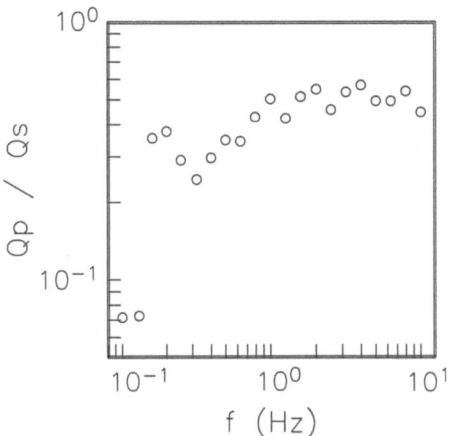

Figure 10
Q_P/Q_S ratios calculated with the estimates of Q shown in Fig. 9

Trough region, Schlotterbeck and Abers (2001) reported that $Q_P/Q_S = 1.43$.

The seismic attenuation in the GoC has been also evaluated by Castro and Avila-Barrientos (2015) who calculated the spectral decay parameter kappa (κ). κ measures the decay of high-frequency ground-motion amplitudes and is used for evaluating seismic risk and hazard. κ, as introduced by Anderson and Hough (1984), is controlled by the attenuation along the path and at the site. Castro and Avila-Barrientos (2015) found that κ near most sites of the RESBAN array have a value of $\kappa_0 = 0.03$ s, except for station GUYB (Guaymas) which is located on a less-consolidated soil and has a $\kappa_0 = 0.05$ s. They also found that at short distances (50–60 km), $\kappa = 0.04$ for a station located in the middle of the array (NE76). Converting $\kappa = 0.04$ s to Q_S ($Q_S = \frac{r}{\beta\kappa}$), for $r = 50$ km and $\beta = 3.77$ km/s, $Q_S = 332$. This value of Q_S corresponds to a frequency of 4.7 Hz, from Eq. (8). The near-surface attenuation parameter κ_0 is specific to each site (Anderson 1991) and can mask the source parameters, particularly the corner frequency. We account for the near-surface attenuation in this study by determining the site response function of each station used.

6. Source Functions

The acceleration source functions ± 1 SD resulting from the inversion of the observed acceleration spectra, after removing site effects (Eq. (2)) are shown in Fig. 11. Source 1 is the foreshock that occurred on September 2, 2013, the main event of the sequence is source 2, and the rest of the sources are displayed chronologically according to the time of occurrence. As expected, the amplitudes of the acceleration source functions tend to increase at low frequencies (1–2 Hz), although the source amplitudes of the main event (source 2) are approximately constant for the whole frequency band. At high frequencies ($f > 8$ Hz), the source amplitudes show a sudden decay, probably due to the near-source attenuation. Because the low sampling rate of the data and the narrow frequency band of this attenuation effect, we cannot quantify it. However, it is also possible that the observed high-frequency decay may be caused by the anti-alias filter that begins just before the Nyquist frequency of 10 Hz. We plotted the source functions between 0.1 and 8.0 Hz, to avoid the sampling rate effect, and observed that for events 2, 7, 9 and 10, the high-frequency decay persists. This suggests that a near-source attenuation effect is a possible explanation to the observed high-frequency decay of the source functions.

To estimate source parameters, we converted these functions into far-field source acceleration spectra $f^2 \dot{M}_0(f)$ (e.g. Boore 1986)

$$f^2 \dot{M}_0(f) = \frac{\rho \beta^3 r}{1.4\pi\Re} S(f) , \qquad (9)$$

where $\dot{M}_0(f)$ is the moment time derivative, $\rho = 2.8$ g cm^{-3}, $\beta = 3.77$ km s^{-1}, $r = 10$ km (the reference distance) and an average radiation pattern $\Re = 0.6$ for S waves. The factor of 1.4 accounts for the energy partition into two components and the free surface amplification.

The resulting far-field source acceleration spectra are displayed in Fig. 12 with solid lines. We fit these observed functions to the ω^2 model (Aki 1967; Brune 1970):

$$f^2 \dot{M}_0(f) = \frac{f^2 M_0}{1 + \left(\frac{f}{f_c}\right)^2} \qquad (10)$$

The low-frequency level was fixed with the seismic moment (M_0) calculated with the magnitude reported by the ISC catalog for each event, and the corner frequency (f_c) was obtained by trial and error

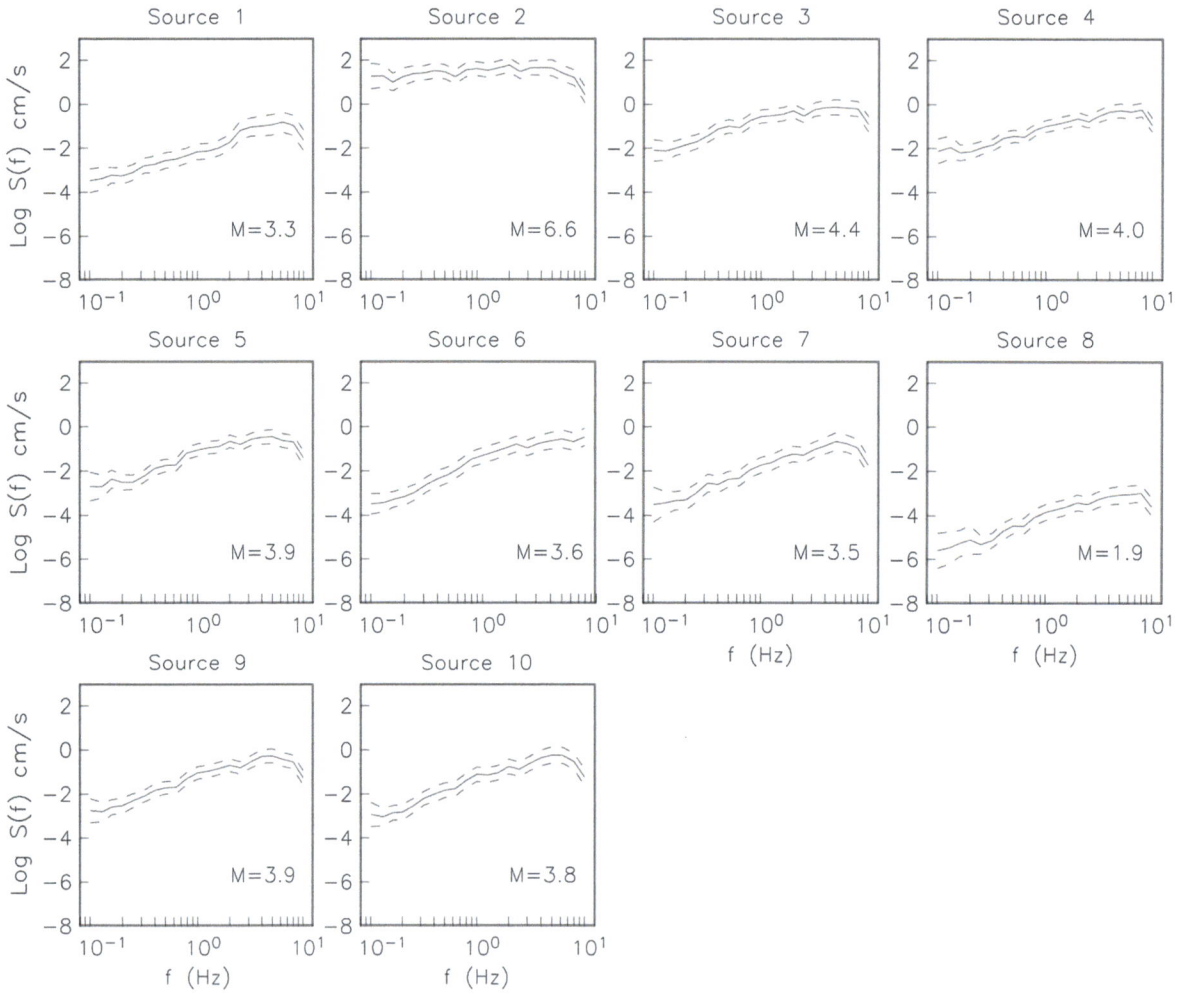

Figure 11
Source functions (±1 SD) resulting from the spectral inversion

testing all possible values from 0.1 to 10.0 Hz with increments of 0.001 Hz. The magnitudes reported by ISC for the sources 2, 4 and 9 fit well the low-frequency part of the spectra. For event 3, we used the magnitude (M 4.4) reported by the Mexican Seismological Service (SSN), which is in between the magnitudes reported by the ISC catalog. The SSN estimated a magnitude $M = 3.8$ for event 8 but is not listed in the ISC catalog, suggesting that perhaps the actual magnitude of this event is below the minimum magnitude of completeness of the ISC catalog. The low-frequency level of the source functions of events 1, 5, 6, 7 and 8 cannot be fit with the ω^2 model with

the M_0 estimated with the reported magnitudes. We calculated M_0 of these earthquakes with the observed low-frequency level Ω_0, and the corner frequency was recalculated.

We used the corner frequency that gives the smaller residual between the ω^2 model and the observed source spectra to estimate the stress drop with the Brune (1970) model:

$$\Delta\sigma = \frac{7}{16}\frac{M_0 f_c^3}{(0.37\beta)^3} \qquad (11)$$

We estimate a stress drop of 1.7 MPa for the mainshock and smaller values for the aftershocks and

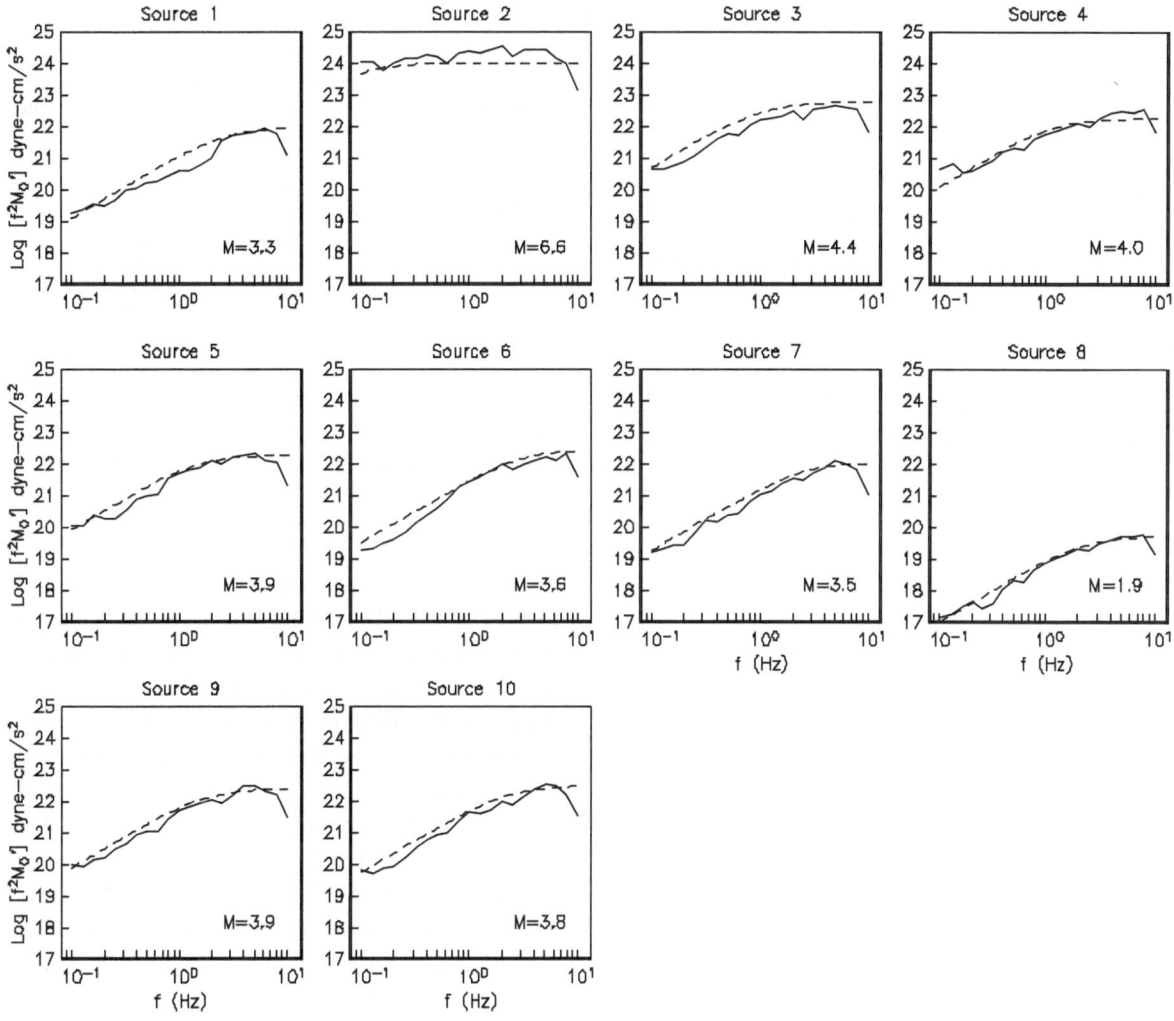

Figure 12
S-wave far-field acceleration source functions (*continuous line*) and the best fits of the Brune's ω^2 model (*dashed lines*)

the foreshock (Table 4). In general, the stress drops of this sequence are between 0.34 and 1.7 MPa, except for event 8, the smallest of all (M_w 1.9) that has a stress drop of 0.002 MPa (Fig. 13). We expect values of f_c in the range of 5–10 Hz for events with magnitudes $M_w \sim 2.0$ (e.g. Van Houtte et al. 2014). Because the low magnitude of this event, it is possible that the corner frequency may be close or above 10 Hz, outside the useful frequency band of the data, and our estimate of the stress drop is likely inaccurate for this event. We prefer to ignore the results from event 8 for further analyses.

7. Discussion and Conclusions

Although the foreshock, the main event and most aftershocks of the sequence occurred on the Farallon transform fault, several aftershocks are located near the ridge that separates Carmen and Farallon transform faults (Fig. 2). The foreshock (event 1) is located on the northern end of the Farallon fault and the mainshock 55 km to the southeast and SE of the rift. The rupture propagated NW for approximately 27 km, triggering earthquakes on the ridge zone and on the Carmen transform fault, 17 km north of the

Table 4

Source parameters estimated with the ω^2 model

Event no.	Fault	M_0 (N-m)	f_c (Hz)	Stress drop (MPa)	r_s (km)	M	Refs.
1	Farallon	0.1234E15	2.68	0.38	0.52	3.3	1
2	Farallon	0.8921E19	0.11	1.71	13.25	6.6	2
3	Ridge	0.4955E16	1.08	1.00	1.30	4.4	3
4	Farallon	0.1245E16	1.20	0.34	1.17	4.0	2
5	Farallon	0.8264E15	1.50	0.45	0.94	3.9	1
6	Ridge	0.3141E15	2.95	1.29	0.48	3.6	1
7	Carmen	0.1766E15	2.47	0.43	0.57	3.5	1
8	–	0.9931E12	2.28	0.002	0.62	1.9	1
9	Farallon	0.7889E15	1.77	0.71	0.79	3.9	2
10	Ridge	0.5585E15	2.34	1.16	0.60	3.8	1

M_0 is the seismic moment, f_c the corner frequency and r_s the source radius. The last column (Ref) indicates the source of the magnitude values: 1 from the seismic moment calculated with the observed low-frequency spectral level Ω_0; 2 from the ISC catalog; 3 from the Mexican Seismological Service catalog

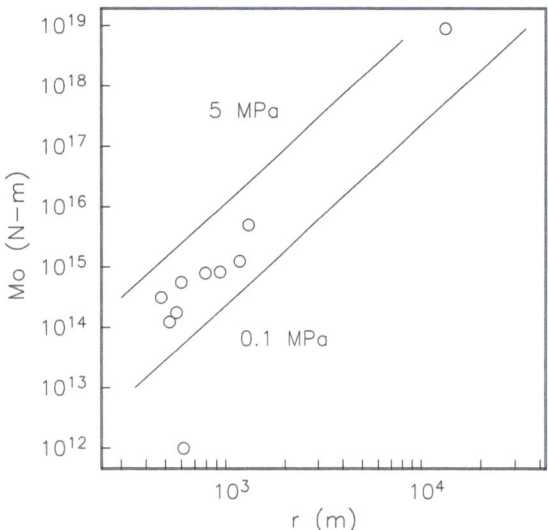

Figure 13
Seismic moment versus source radius obtained fitting the far-field source functions with the Brune's model. We estimated a stress drop of 1.7 MPa for the main event (M_w 6.6) and 0.002 MPa for the smaller aftershock (M 1.9)

Farallon fault trace. The bathymetry and the location of one of the aftershocks (event 6 in Fig. 2 and Table 2) suggest that the Carmen fault continues (SE) beyond the ridge for at least 23 km, contrary to the expected geometry of a typical en echelon transform fault system where the ridge zone marks the end of the fault and jumps to the other end of the ridge. The character of the Farallon transform fault looks smoother to the SE of the 2013 earthquake sequence,

and that is probably why the 2013 sequence was located on the NW section, where the transform fault is more geometrically complicated, and it likely has more off-fault deformation, which would account for the epicentral distribution of the aftershocks.

The estimates of site response obtained with the HVSR method indicate that most stations of the RESBAN network have low site amplifications; however, stations NE74, NE80 and NE81 show significant site response amplifications that are above a factor of 3 (Fig. 8). The S-wave site responses obtained with HVSR tend to give larger amplifications at high frequencies ($f > 1$ Hz) at sites BAHB, GUYB, NE77, NE80 and SLGB compared with the S-wave responses estimated with the source-site spectral inversion (solving Eq. (6)). This may be an effect of azimuthal dependence of the site response. The site responses from P waves are similar to those obtained from S waves, except for sites BAHB, GUYB and SFQB where the P-wave site amplification is higher. These three stations are located on hard-rock sites (igneous rocks) and show small S-wave site amplifications (Fig. 8).

The frequency dependence of the quality factor of S waves can be approximated with the relation $Q_s = 135.4 \pm 1.1 f^{0.58 \pm 0.03}$ ($0.1 \leq f \leq 10$), which is consistent with previous estimates by Vidales-Basurto et al. (2014) obtained for the south-central region of the Gulf of California. Similarly, for P waves, we find the relation $Q_P = 48.1 \pm 1.1 f^{0.88 \pm 0.04}$. The ratio $Q_P/Q_S < 1.0$ found in the frequency band 0.1–10 Hz

(Fig. 10) suggests that partially saturated rocks may be present along the wave propagation paths. These zones of low Q_P/Q_S may be related to magmatic bodies located near the ridges.

We estimate that the October 19, 2013 (M_w 6.6) earthquake has a stress drop of 1.7 MPa. Aftershocks 3, 6 and 10 (Table 4), located near the ridge, also show the highest stress drops (Fig. 2). The seismicity reported by Sumy et al. (2013) in the Farallon–Carmen regions (Fig. 4) tends to concentrate near the ridges, suggesting that the interaction of the Farallon and Carmen transform faults may concentrate stress in between, near the ridge, where the higher stress drop earthquakes of the 2013 sequence occurred.

Acknowledgments

The operation of the RESBAN network has been possible thanks to the financial support of the Mexican National Council for Science and Technology (CONACYT) (projects CB-2011-01-165401(C0C059), G33102-T and 59216). This paper was prepared while the first author (RRC) was on sabbatical year in Caltech. We thank Prof. Gurnis for the support provided. Dr. Lenin Avila-Barrientos facilitated part of the spectral records used to calculate the site functions. Antonio Mendoza Camberos pre-process the data from the RESBAN network and Arturo Perez Vertti maintains and operates the stations. We thank Dr. Edwards and the anonymous reviewer for their careful revisions, comments and suggestions which help us to improve the manuscript. We also acknowledge the Editor, Dr. Thomas H.W. Goebel.

References

Adams, D. A., & Abercrombie, R. E. (1998). Seismic attenuation above 10 Hz in southern California from coda waves recorded in the Cajon Pass borehole. *Journal Geophysical Research, 103*, 24257–24270.

Aki, K. (1967). Scaling law of seismic spectrum. *Journal Geophysical Research, 72*, 1217–1231.

Anderson, J. G. (1991). A preliminary descriptive model for the distance dependence of the spectral decay parameter in southern California. *Bulletin of the Seismological Society of America, 81*, 2186–2193.

Anderson, J. G., & Hough, S. E. (1984). A model for the shape of the Fourier amplitude spectrum of acceleration at high frequencies. *Bulletin of the Seismological Society of America, 74*, 1969–1993.

Andrews, D.J. (1986). Objective determination of source parameters and similarity of earthquakes of different size. In S. Das, J. Boatwright, C. H. Sholz (Eds.) *Earthquake source mechanics*. Washington, DC: American Psychological Union. doi:10.1029/GM037p0259

Atwater, T., & Stock, J. (1998). Pacific-North America plate tectonics of the Neogene Southwestern United States: An update. *International Geologiy Review, 40*, 375–402.

Avila-Barrientos, L., & Castro, R. R. (2015). Site response of the NARS-Baja and RESBAN broadband networks of the Gulf of California, Mexico. *Geofísica International, 55*, 131–154.

Boore, D. M. (1986). Short-period P- and S-wave radiation from large earthquakes: implications for spectral scaling relations. *Bulletin of the Seismological Society of America, 76*, 43–64.

Brune, J. N. (1970). Tectonic stress and the spectra of seismic shear waves from earthquake. *Journal Geophysical Research, 75*, 4997–5009.

Carbotte, S. M., Arko, R., Chayes, D. N., Haxby, W., Lehnert, K., O'Hara, S., et al. (2004). New integrated data management system for Ridge 2000 and MARGINS research. *EOS Transactions, AGU, 85*(51), 553. doi:10.1029/2004EO510002.

Caress, D. W., & D. N. Chayes. (2014). MB-System: Mapping the Seafloor. https://www.mbari.org/data/mbsystem and http://www.ldeo.columbia.edu/res/pi/MB-System. Accessed 4 Jan 2016

Castro, R. R., Anderson, J. G., & Singh, S. K. (1990). Site response, attenuation and source spectra of S waves along the Guerrero, México, subduction zone. *Bulletin of the Seismological Society of America, 80*, 1481–1503.

Castro, R. R., & Avila-Barrientos, L. (2015). Estimation of the spectral parameter kappa in the region of the Gulf of California, Mexico. *Journal of Seismology 19*, 809–829. doi:10.1007/s10950-015-9496-x.

Castro, R. R., Pacor, F., Bindi, D., Franceschina, G., & Luzi, L. (2004). Site response of strong motion stations in the Umbria, Central Italy, region. *Bulletin of the Seismological Society of America, 94*, 576–590.

Castro, R. R., Pacor, F., Puglia, R., Ameri, G., Letort, J., Massa, M., & Luzi, L. (2013). The 20 May, 2012 Emilia earthquake, Italy and the main aftershocks: S-wave attenuation, acceleration source functions, and site effects. *Geophysical Journal International,*. doi:10.1093/gji/ggt245.

Castro, R. R., Pérez-Vertti, A., Mendez, I., Mendoza, A., & Inzunza, L. (2011a). Location of moderate size earthquakes recorded by the NARS-Baja array in the Gulf of California region between 2002 and 2006. *Pure and Applied Geophysics, 168*, 1279–1292.

Castro, R. R., Valdes-Gonzalez, C., Shearer, P., Wong, V., Astiz, L., Vernon, F., et al. (2011b). The 3 August 2009 M_w 6.9 Canal de Ballenas region, Gulf of California, earthquake and its aftershocks. *Bulletin of the Seismological Society of America, 101*, 929–939.

Clayton, R. W., Trampert, J., Rebollar, C. J., Ritsema, J., Persaud, P., Paulssen, H., et al. (2004). The NARS-Baja array in the Gulf of California R ift Zone. *Margins Newsletter, 13*, 1–4.

Edwards, B., Michel, C., Poggi, V., & Fah, D. (2013). Determination of site amplification from regional seismicity: Application to the Swiss National Seismic Networks. *Seismological Research Letters, 84*, 611–621. doi:10.1785/0220120176.

Fenby, S. S., & Gastil, R.G. (1991). Geologic-Tectonic map of the Gulf of California and surrounding áreas. In J. P. Dauphin & B. T. Simoneit (Eds.), *The Gulf and Penninsular provinces of the Californias* (Vol. 47, pp. 79–83). Tulsa:American Association of Petroleum Geologist.

Goff, J. A., Bergman, E. A., & Solomon, S. C. (1987). Earthquake source mechanism and transform fault tectonics in the Gulf of California. *Journal Geophysical Research, 92*, 10485–10510.

Hauksson, E., & Shearer, P. M. (2006). Attenuation models (Q_P and Q_S) in three dimensions of the southern California crust: Inferred fluid saturation at seismogenic depths. *Journal Geophysical Research, 111*, B05302. doi:10.1029/2005JB003947.

Langston, C. A. (1977). Corvallis, Oregon, crustal and upper mantle receiver structure from teleseismic P and S waves. *Bulletin of the Seismological Society of America, 67*, 713–724.

Lermo, J., & Chávez-García, F. J. (1993). Site effect evaluation using spectral ratios with only one station. *Bulletin of the Seismological Society of America, 83*, 1574–1594.

Lizarralde, D., Axen, G. J., Brown, H. E., Fletcher, J. M., Gonzalez-Fernandez, A., Harding, A. J., et al. (2007). Variation in styles of rifting in the Gulf of California. *Nature,*. doi:10.1038/nature06035.

Lonsdale, P. F. (1989). Geology and tectonic history of the Gulf of California. *The Eastern Pacific Ocean and Hawaii, Decade of North American geology* (Vol. N, pp. 499–521). Denver, CO, USA: Geol. Soc. Am.

López-Pineda, L., & Rebollar, C. J. (2005). Source characteristics of the M_w 6.2 Loreto earthquake of 12 march 2003 that occurred in a transform fault in the middle of the Gulf of California, Mexico. *Bulletin of the Seismological Society of America, 95*, 419–430.

Munguía, L., Reichle, M., Reyes, A., Simons, R., & Brune, J. (1977). Aftershocks of the 8 July, 1975 Canal de las Ballenas, Gulf of California, earthquake. *Geophysical Reseach Letters, 4*, 507–509.

Nagy, E. A., & Stock, J. M. (2000). Structural control on the continent-ocean transition in the Northern Gulf of California. *Journal Geophysical Research, 105*, 16251–16269.

Nakamura, Y. (1989). A method for dynamic characteristics estimation of subsurface using microtremor on the ground surface. *Report Railway Technical Research Institute, 30*, 25–33.

Nava-Sánchez, E. H., Gorsline, D. S., & Molina-Cruz, A. (2001). The Baja California peninsula borderland: structural and sedimentological characteristics. *Sedimentary Geology, 144*, 63–82. doi:10.1016/S0037-0738(01)00135-X.

Pacheco, J. F., & Sykes, L. R. (1992). Seismic moment catalog of large shallow earthquakes, 1900 to 1989. *Bulletin of the Seismological Society of America, 82*, 1306–1349.

Phillips, W. S., & Aki, K. (1986). Site amplification of coda waves from local earthquakes in central California. *Bulletin of the Seismological Society of America, 76*, 627–648.

Raoof, M., Hermann, R. B., & Malagnini, L. (1999). Attenuation and excitation of three-component ground motion in southern California. *Bulletin of the Seismological Society of America, 89*, 888–902.

Rebollar, C. J., Quintanar, L., Castro, R. R., Day, S. M., Madrid, J., Brune, J. N., et al. (2001). Source characteristics of a 5.5 magnitude earthquake that occurred in the transform fault system of the Delfin Basin in the Gulf of California. *Bulletin of the Seismological Society of America, 91*, 781–791.

Rebollar, C. J., Traslosheros, C., & Alvarez, R. (1985). Estimates of seismic wave attenuation in northern Baja California. *Bulletin of the Seismological Society of America, 75*, 1371–1382.

Rodriguez-Lozoya, H. E., Quintanar, L., Ortega, R., Rebollar, C. J., & Yagi, Y. (2008). Rupture process of four medium-size earthquakes that occurred in the Gulf of California. *Journal Geophysical Research, 113*, B10301. doi:10.1029/2007JB005323.

Schlotterbeck, B. A., & Abers, G. A. (2001). Three-dimensional attenuation variations in southern California. *Journal Geophysical Research, 106*, 30719–30735.

Singh, S. K., Apsel, R. J., Fried, J., & Brune, J. N. (1982). Spectral attenuation of SH waves along the Imperial fault. *Bulletin of the Seismological Society of America, 72*, 2003–2016.

Stein, S., & Pelayo, A. (1991). Seismological constraints on stress in the oceanic lithosphere. *Philosophical Transactions of the Royal Society of London A, 337*, 53–72.

Stock, J. M., & Hodges, K. V. (1989). Pre-Pliocene extension around the Gulf of California and the transfer of Baja California to the Pacific plate. *Tectonics, 8*, 99–115.

Sumy, D. F., Gaherty, J. B., Kim, W.-Y., Diehl, T., & Collins, J. A. (2013). The mechanism of earthquakes and faultingin the southern Gulf of California. *Bulletin of the Seismological Society of America, 103*, 487–506.

Tanioka, Y., & Ruff, L. (1997). Source time functions. *Seismological Research Letters, 68*, 386–400.

Trampert, J., Paulsen, H., Van Wettum, A., Ritsema, J., Clayton, R., Castro, R., et al. (2003). New array monitors seismic activity near the Gulf of California in México, EOS. *Transactions American Geophysical Union, 84*, 29–32.

Van Houtte, C., Ktenidou, O. J., Larkin, T., & Holden, C. (2014). Hard-site κ_0 (Kappa) calculations for Christchurch, New Zealand, and comparison with local ground motion prediction models. *Bulletin of the Seismological Society of America, 104*, 1899–1913.

Vidales-Basurto, C. A., Castro, R. R., Huerta, C. I., Sumy, D. F., Gaherty, J. B., & Collins, J. A. (2014). An attenuation study of body-waves in the south-central region of the Gulf of California, México. *Bulletin of the Seismological Society of America, 104*, 2027–2042.

Wessel, P., & Smith, W. H. F. (1998). New, improved version of generic mapping tools released. *EOS Transactions, AGU, 79*(47), 579.

Winkler, K. W., & Nur, A. (1982). Seismic attenuation: effects of pore fluids and frictional sliding. *Geophysics, 47*, 1–15.

Zhang, X., Paulsen, H., Lebedev, S., & Meier, T. (2007). Surface wave tomography of the Gulf of California. *Geophysical Reseach Letters, 34*, L15305. doi:10.1029/2007GL030631.

(Received January 25, 2016, revised June 27, 2016, accepted June 29, 2016, Published online July 9, 2016)

Pure Appl. Geophys. 174 (2017), 2257–2267
© 2016 Springer International Publishing
DOI 10.1007/s00024-016-1360-6

Stress Distribution Near the Seismic Gap Between Wenchuan and Lushan Earthquakes

Yihai Yang,[1,2] Chuntao Liang,[1,2] Zhongquan Li,[3,4] Jinrong Su,[5] Lu Zhou,[1,2] and Fujun He[1,2]

Abstract—The Wenchuan M_S 8.0 earthquake and Lushan M_S 7.0 earthquake unilaterally fractured northeastward and southwestward, respectively, along the Longmenshan fault belt. The aftershock areas of the two earthquakes were separated by a gap with a length of nearly 60 km. We have determined the focal mechanisms of 471 earthquakes with magnitude $M \geq 3$ from Jan 2008 to July 2014 near the seismic gap using a full waveform inversion method. Normal, thrust and strike-slip focal mechanisms can be found in northern segment. But in a significant contrast, focal mechanisms of the earthquakes in the southern segment are dominated by thrust faulting. Based on the determined source parameters, we further apply a damped linear inversion method to derive the regional stress field. The southern segment is characterized by an obvious thrust faulting stress regime with a nearly horizontal maximum compression that orients in SE–NW direction. The stress environment in the northern segment is a lot more complicated. The maximum compressional stresses appear to rotate around the "asperity" near west of the Dujiangyan city. Stress field also shows strong variation with time and depth. Before 2009, the seismic activities are more concentrated on the Pengxian–Guanxian fault and Yingxiu–Beichuan fault with dominant strike-slip faulting and normal faulting, while after 2009, the seismic activities are dominated by thrust faulting from north to south, while the activities are more concentrated on the Wenchuan–Maoxian fault in northern segment and Pengxian–Guanxian fault in southern segment. The maximum compressional stresses vary in different depths from north to south, thus may imply the decoupled movement in shallow and in depth.

Key words: Wenchuan earthquake, Lushan earthquake, seismic gap, focal mechanism, stress field.

[1] State Key Laboratory of Geohazard Prevention and Geoenviroment Protection, Chengdu University of Technology, Chengdu 610059, China. E-mail: liangchuntao12@cdut.cn

[2] Key Laboratory of Earth Exploration and Information Technique of Education Ministry of China, Chengdu University of Technology, Chengdu 610059, China.

[3] State Key Laboratory of Oil and Gas Reservoir Geology and Exploitation, Chengdu University of Technology, Chengdu, Sichuan 610059, China.

[4] Key Laboratory of Tectonic Controlled Mineralization and Oil Reservoir, Ministry of Land and Resources, Chengdu University of Technology, Chengdu, Sichuan 610059, China.

[5] Earthquake Administration of Sichuan Province, Chengdu 610041, China.

1. Introduction

Nearly 5 years after the devastating Wenchuan M_s 8.0 earthquake in 2008, the April 20, 2013 Lushan M_s 7.0 earthquake stroke along the NE–SW striking Longmenshan fault (LMSF) zone. The epicenter of this earthquake was located approximately 84 km southwest of the epicenter of the Wenchuan earthquake. Field investigation suggests that the 2013 Lushan earthquake strikes on a blind thrust fault in the LMSF zone (Xu et al. 2013), which is different from the Yingxiu–Beichuan fault, the seismogenic fault of Wenchuan earthquake. The Wenchuan and Lushan earthquakes unilaterally fractured northeastward and southwestward, respectively, along the Longmenshan fault. The distribution of earthquakes in this area was clearly separated by a gap with a length of nearly 60 km (Fig. 1). Chen et al. (2013) estimated if this section of seismic gap breaks completely in an event, the moment magnitude can be as large as 6.8.

The Longmenshan range is under a horizontal compression from the eastward extrusion of crustal materials from Bayankala block, the central part of Tibetan plateau, against rigid Sichuan basin in the western part of Yangtze craton. Its uplift can be dated back to pre-Cenozoic and accelerated by the collision of Eurasian and Indian plates since ∼50 Ma (Xu et al. 2008). The Longmenshan fault system is composed of a series of imbricated thrust faults, including Pengxian–Guanxian fault (F1 in Fig. 1), Yingxiu–Beichuan fault (F2 in Fig. 1) and Wenchuan–Maoxian fault (F3 in Fig. 1). The Pengguan complex, Baoxing complex and Kangding complex form the basement complex belt, and many nappes composed of limestone overridden on the Sichuan foreland basin at the front of these complexes lie at the eastern

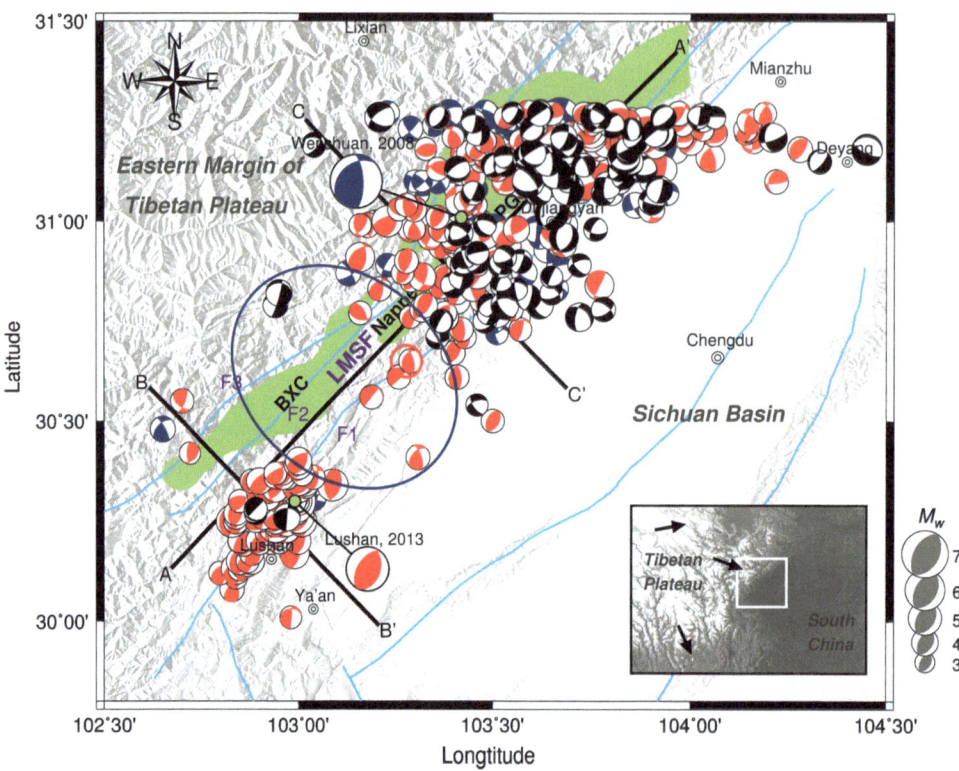

Figure 1
Focal mechanisms (beach balls) of 471 earthquakes determined in this study. The *color* and *size* of beach ball denote faulting type and moment magnitude, respectively. *Black*, *red* and *blue* beach balls are corresponding to normal, thrust and strike-slip faulting types, respectively. The *red circle* indicates 1970 Dayi earthquake (M_s 6.2). The northern and southern segments mentioned in the text are roughly separated by the *blue elliptic*. LMSF, Longmenshan Fault *F1*, Pengxian–Guanxian fault *F2*, Yingxiu–Beichuan fault *F3*, Wenchuan–Maoxian fault. *BXC* Baoxing complex; *PGC* Pengguan complex. BXC and PGC, together with the nappe are after Xu et al. (2008). *Inset* the location of our study region and the *black arrows* indicating large-scale block motion

margin of Tibetan plateau under a dominant NWW–SEE (Xu 2001; Xu et al. 2008) stress field.

The earthquake is essentially resulted from the release of strain energy accumulated by tectonic stress loading. GPS data have found the markedly low convergence velocity in the Longmenshan range (<3 mm/year, after Xu et al. 2008), and thus, the tectonic stress and fluid pressure have to take a long time to accumulate during the interseismic period. Wang et al. (2010) estimated that the largest moment deficit would be on the southwestern segment of the LMSF in eastern Tibet. They further suggested that unruptured segment is capable of producing M_w 7.7 earthquake in the next 50 years after Wenchuan earthquake. Liu et al. (2014) found that the Lushan earthquake released about 1/3 estimated moment

deficit on the southern Longmenshan fault, and the seismic gap had the highest stress increase, most likely to produce the next big earthquake in this region. However, Pei et al. (2014) used a Pg inversion method to find that low velocity is distributed in the upper crust of the seismic gap region. Wang et al. (2015) used P- and S-wave arrival times to reveal that lower crust velocity may be associated with the seismic gap from seismic imaging. They, therefore, argued that a large event was unlikely due to the ductile nature of the material. If this is the case, a ductile creeping may account for the deformation in the region to accommodate the interaction between the Sichuan basin and the Tibetan plateau. Other studies (Bai et al. 2010; Li et al. 2013; Zhao et al. 2012) have also found that the structure in the region is quite complex. To investigate which mechanism is more likely in control of the seismogenic process, and to further understand the deep structure and dynamics in this region, a comprehensive study on the detailed stress distribution and its variation during the co-seismic and post-seismic periods in the region is much more needed.

A number of algorithms have been developed to study stress states in a region. Especially, the focal mechanisms of earthquakes can reveal the stress distribution along the fault in the seismogenic depth (Xu et al. 2010; Yi et al. 2012). Yang et al. (2015) inverted focal mechanism of aftershocks of the Lushan earthquake and found a nearly uniform stress states in that area, while Yi et al. (2012) found that the stress distribution in the aftershock zone of the Wenchuan earthquake is more complex. The stress parameters derived in these cases represent stress status of the rupture process of each earthquake. Hardebeck and Michael (2004) established an algorithm to determine stress parameters for each subarea by fitting source mechanism of all earthquakes in that subarea. They introduced the spatial and temporal damping parameters and developed damped linear inversion technique to produce regional-scale models of stress distribution (Hardebeck and Michael 2006). This technique has been used in many areas to study regional stress fields (e.g. Zhao et al. 2013; Luo et al. 2015).

In this study, we first determined the focal mechanisms of 471 earthquakes with magnitude $M \geq 3$ from January 2008 to July 2014 near the seismic gap, and then applied the damped linear inversion to derive the regional stress field using the focal mechanisms.

2. Data and Method

In this study, we collected broadband seismograms from permanent stations of Sichuan seismic network (triangles in Fig. 2). The earthquake catalogue was downloaded from the China Earthquake Network Center (CENC). The data analysis and focal mechanism inversion were performed using the CPS package (Herrmann 2014). All the waveform records were deconvolved to remove instrument responses to obtain

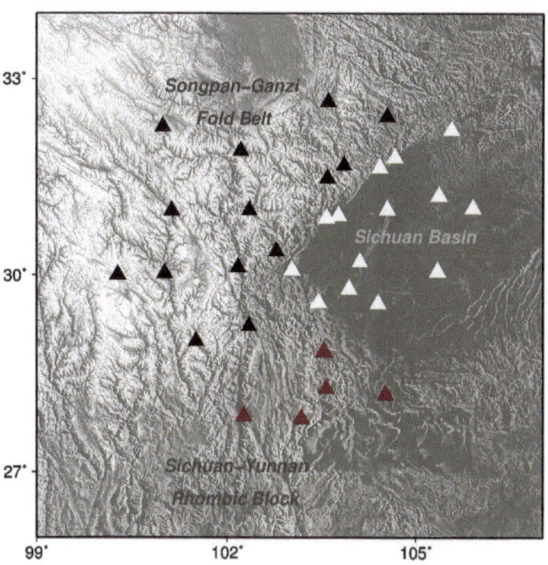

Figure 2
Locations of seismic stations used in this study. Different velocity models are used for ray paths associated with stations in different colors

ground velocity (m/s) and then rotated to vertical, radial and transverse components. For each earthquake, the waveforms with an epicentral distance less than 300 km were preserved, although an inverse distance weighting would be applied to each trace in the inversion. This step was followed by a quality control process to typically delete low signal-to-noise ratio (SNR) waveforms with a bandpass filter of 0.02–0.1 Hz.

Earthquakes studied are distributed mainly along the LMSF, which separates Songpan–Ganzi fold belt and South China block (Fig. 1). Seismic velocities vary dramatically across the LMSF and its adjacent region (Liang and Song 2004; Zhu et al. 2012). To account for the possible effects of heterogeneity on focal mechanism solution, the region covered by seismic stations was divided into three tectonic units: Sichuan basin, eastern and southeastern margin of Tibet plateau (Fig. 2). Three different velocity models were used for Green's functions computation for paths in different tectonic units, with S-wave velocities being averaged from receiver function inversion (Langston 1979; Yang et al. 2015), while P-wave velocities were derived using a constant $V_p/V_s = 1.732$. Our previous study (Yang et al. 2015) had shown the validity of using such a model division.

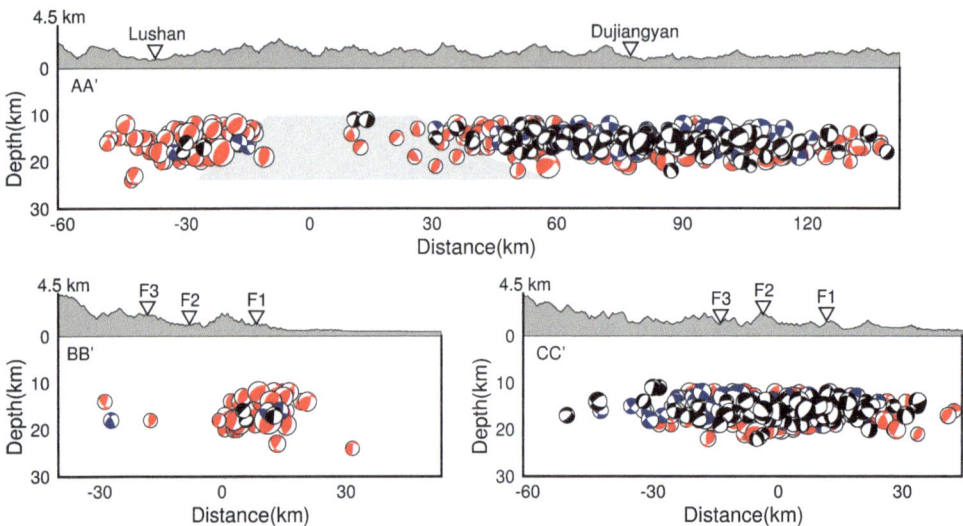

Figure 3
Focal mechanisms in cross-sectional view along the profile *AA'*, *BB'* and *CC'*. The *black*, *red* and *blue* colors of beach ball in three profiles indicate normal, thrust and strike-slip focal mechanisms, respectively

A grid search for the strike, dip, rake angles and source depth was applied to find the best fit between the predicted and observed waveforms in the source inversion. Prior to grid search, the predicted and observed waveforms were cut from 10 s before and 180 s after the initial P-wave arrival. Moreover, these waveforms for a certain event were simultaneously filtered in a band depending on its magnitude (Herrmann et al. 2011). We selected the passband of 0.01–0.06 Hz for the Lushan and Wenchuan mainshock, 0.02–0.08 Hz for $4 \leq M \leq 6$, and 0.02–0.1 Hz for $3 < M < 4$.

2.1. *Focal Mechanism Solutions*

Focal mechanisms of 471 earthquakes with moment magnitude larger than 2.97 were determined (Fig. 1). Faulting types are divided by the plunge of P, B and T axes (Zoback 1992); for a small number of earthquakes whose faulting types cannot be distinguished by this way, we classify them based on rake angles: normal faulting with a rake between $-135°$ and $-45°$, thrust faulting with a rake between $45°$ and $135°$, rest are strike-slip faulting. For a better analysis of the spatial segmentation characteristic of the focal mechanisms, the study region is divided into north and south segments

by the seismic gap (blue elliptic in Fig. 1). From Fig. 1, we can find the normal, thrust and strike-slip mechanisms in the north segment. But in a significant contrast, most of the earthquakes that occurred in southern segment are dominated by thrust faulting.

Figure 3 shows focal mechanisms in cross-sectional view along the profile AA' and the other two profiles BB' and CC' perpendicular to the strike of LMSF. Depth section for profile AA' reflects a chimney-shaped seismic gap, which is consistent with the low-velocity and high Poisson's ratio zone determined by Wang et al. (2015) via seismic tomography. This seems to imply that the compressional stresses due to the collision of eastern margin of Tibetan plateau and Sichuan basin may be at least partially absorbed by ductile flow of inelastic materials in the crust, although it is unclear how much stress energy is released by ductile creeping and how much are accumulated to make next big earthquake.

Another overall feature is that most earthquakes are located in the depth of 12–20 km, with only 2 % events deeper than 20 km (Fig. 4). The distribution of focal depth is similar with many studies of the earthquake relocation in this region (Lü et al. 2008; Zhu et al. 2008; Chen et al. 2009; Han et al. 2014; Fang et al. 2015). The thrust and strike-slip events

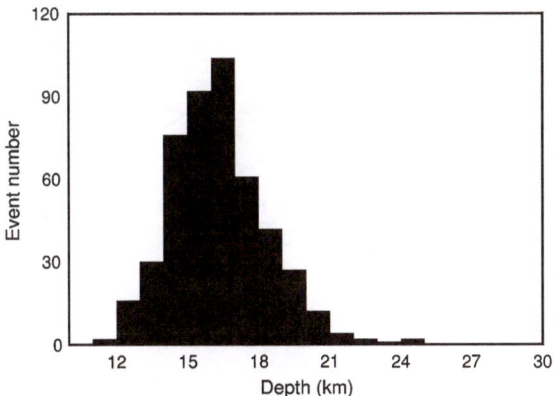

Figure 4
Distributions of source depths determined in this study

north of Dujiangyan. Similar region-dependent complexity is observed for T axes.

Figure 7 shows the statistics of strike, dip, rake, azimuth and plunge of P axis for northern and southern segment, respectively. The strikes are roughly oriented in the NE–SW direction which is consistent with the strike of LMSF. The dip angles vary slightly from north to south. Significant quantities of rake angles in north are between $-30°$ and $-150°$, consistent with the large amount of normal faulting events. Another significant difference between two segments is shown in the azimuths of P axis, as shown in Fig. 6a, while the plunges of P axis are general in the range of $0°–30°$, reflecting nearly horizontal compressions.

were in majority in northern segment, obviously controlled by the Wenchuan mainshock of thrust faulting with moderate dextral strike-slip component (Fig. 5). But it is worth to notice that considerable quantities of normal events occurred in this region, reflecting a complex stress field. The focal mechanisms of most earthquakes in southern segment are similar with Lushan mainshock, showing almost pure thrust faulting.

The principal compression (P axis) and extension (T axis) stress axes of focal mechanism solution represent the equivalent stress field that caused an earthquake. Figure 6 shows the horizontal projection of P and T axes of the 471 focal mechanisms. The P axes in southern segment generally orient NW–SE. However, in the northern segment, P axes generally orient in NWW–SEE and W–E to the south of Dujiangyan. But they are much more complex to the

3. Stress Field Study

Hardebeck and Michael (2006) developed a damped linear inversion technique to produce regional-scale models of stress orientation. Here, we used a 2D inversion to investigate regional stress field near the seismic gap based on the determined focal mechanisms. The study region was divided into subareas with grid spacing of $0.1° \times 0.1°$ and $0.05° \times 0.05°$, respectively. Each earthquake was assigned to the nearest grid point. A damping parameter e was chosen by investigating the trade-off curve of the model length and data variance, near the corner of the curve (Fig. 8), to produce a more accurate and also smoothed result. In this study, the optimal value of e was determined to be 1.1. 2000

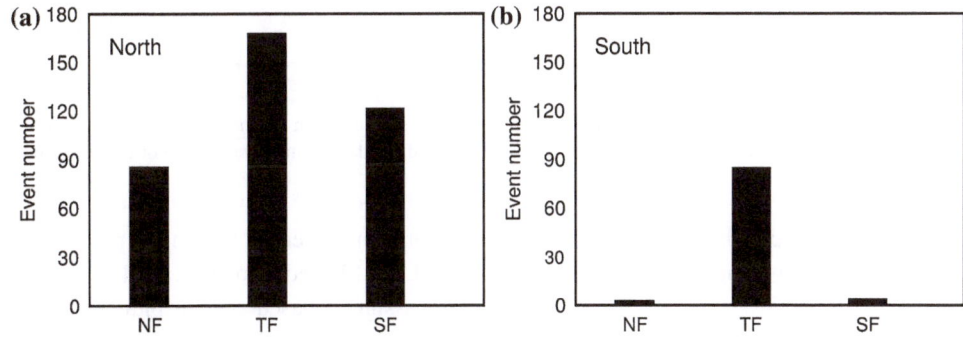

Figure 5
Distributions of faulting types in northern segment (a) and southern segment (b). *NF* normal faulting; *TF* thrust faulting; *SF* strike-slip faulting

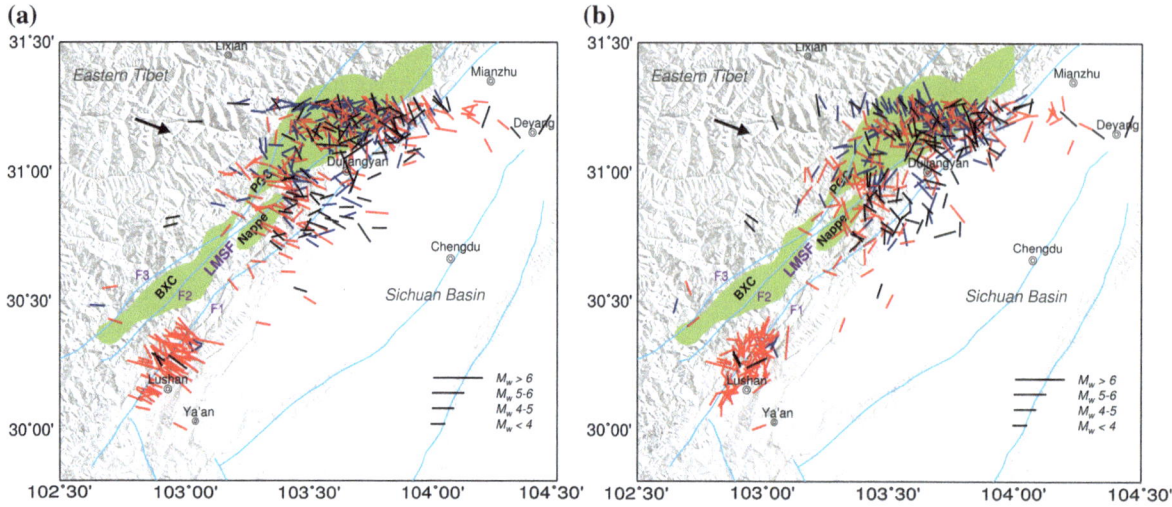

Figure 6
Horizontal projection of principal compression (**a**) and extension (**b**) stress axes of the focal mechanisms. The *length* and *color of bars* indicate the moment magnitude and the faulting type of earthquakes, respectively. *Black*, *red* and *blue* are corresponding to the normal, thrust and strike-slip faulting types, respectively

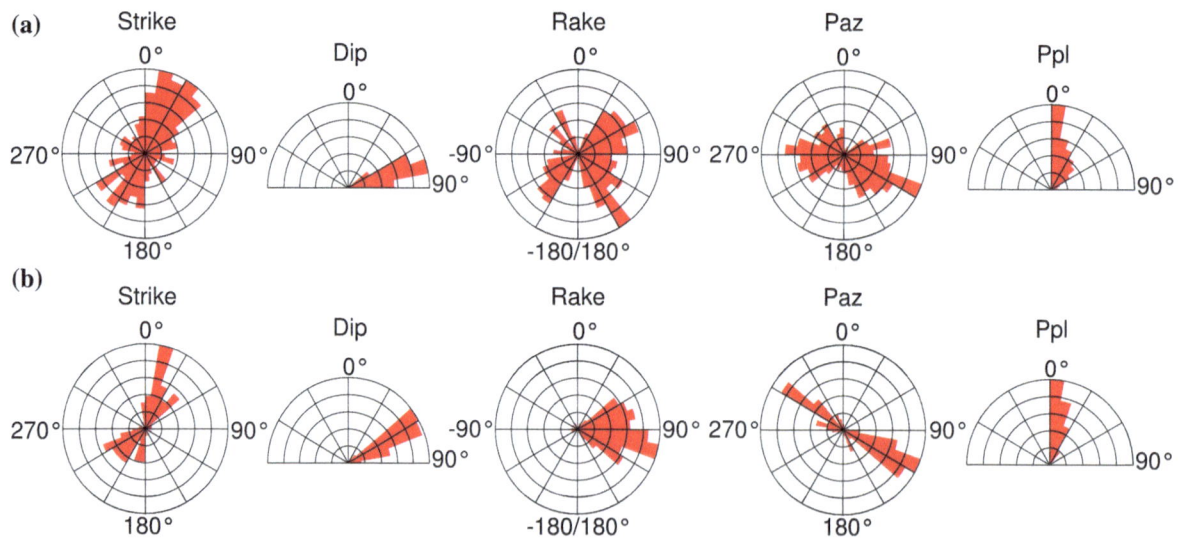

Figure 7
Statistics for the parameters of fault planes and principal compressional axes (P axes) in the northern (**a**) and southern (**b**) segments, respectively. From *left* to *right*, the *5 rose diagrams* represent strike, dip, rake, azimuth and plunge of P axis, respectively

bootstrap resampling for input dataset and 95 % of confidence interval are used (Martínez-Garzón et al. 2014).

The 2D stress maps based on different grid sizes are illustrated in Fig. 9. The maximum principal compression axis ($\sigma 1$) is depicted with different colors for different faulting types. Following Zhao et al. (2013) and Luo et al. (2015), the faulting types are assigned based on the plunge of the maximum and minimum principal compression stress axes: black bar indicating normal faulting with plunge >35° for $\sigma 1$ and plunge <35° for $\sigma 3$; red bar

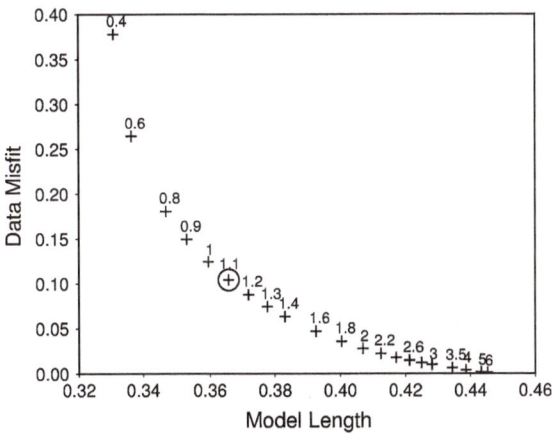

Figure 8
Trade-off between model length and data misfit for the full range of possible values of damping parameter

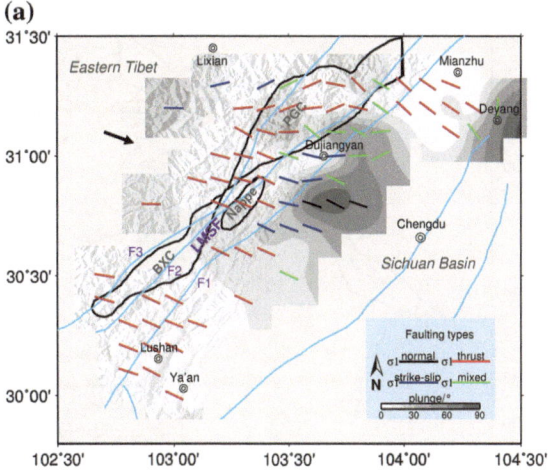

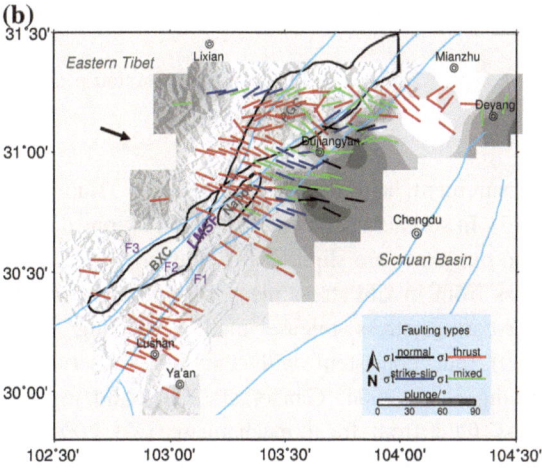

Figure 9
Stress inversion results with a grid size of 0.1° (**a**) and 0.05° (**b**). The maximum principal compressional axes σ1 are plotted on each grid point. The *black, red, blue* and *green bars* for *σ1* indicate normal, thrust, strike-slip and mixed faulting, respectively. The *gray shading* shows the plunges of *σ1* (plunges increase from *shallow* to *deep gray*)

indicating thrust faulting with plunge <35° for σ1 and plunge >55° for σ3; blue bar indicating strike-slip faulting with plunge <35° for both σ1 and σ3; green bar indicating a mixed faulting which cannot be categorized as the three types above. Figure 9a shows a dominant thrust faulting stress regime in southern segment, with nearly horizontal maximum compressions oriented in NWW–SEE. This is significantly different from Luo et al. (2015) whose results show a sharp stress rotation across the F1 although there is no earthquake to the east of the F1 in their studies at all. However, northern segment is characterized as a much more complex stress environment. In general, regions along the F3 are dominated by thrust faulting. Stresses along the F2 and F1 show strong variation from north to south. Large plunges of σ1 are found to the east of the northern segment and they imply a normal environment. The orientations of σ1 show an anti-clockwise rotation around Dujiangyan and a clockwise rotation in the northern end of study region.

The large grid size may produce a smooth stress field in a large scale with a sacrifice on fine details. Figure 9b presents another stress map based on a grid size of 0.05° × 0.05°. The general pattern is similar between the two, but Fig. 9b does show more details than Fig. 9a, especially for the area near the Dujiangyan city. To the west of the Dujiangyan city, there appears to be a seismic gap (Figs. 1, 9a). The orientation patterns of σ1 obviously break into two branches in northern segment, with one dominantly in

NWW–SEE direction and the other one roughly rotating from NEE–SWW to NW–SE. In general, the small gap in the Pengguan complex appears to be an asperity that causes the σ1 to rotate around it. Faulting types and plunges of σ1 show a similar variation pattern with Fig. 9a.

4. Discussions

The stress field can be obtained from different datasets, such as focal mechanism, in situ stress

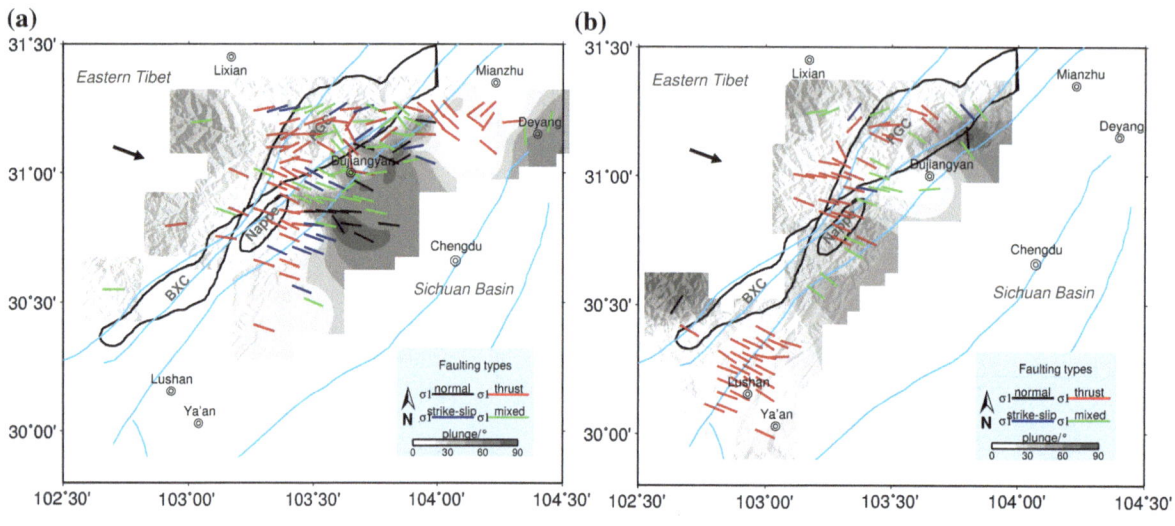

Figure 10
Same as Fig. 9b but for different time periods. **a**, **b** for events before and after January 2009, respectively

measurement, and shear wave splitting (Luo et al. 2015). In southern segment, the σ1 directions derived from our study are slightly different from the crustal stress from in situ stress measurement and borehole over-coring stress release method (Huang et al. 2009), but consistent with the GPS observations (Zhang et al. 2004; Gan et al. 2007) and regional stress field from focal mechanisms (Xu 2001; Xu et al. 2008). In northern segment, both in situ stress measurement and GPS observations are in a constant NWW–SEE direction, coinciding with our results around the Wenchuan mainshock. Seismic anisotropic study from regional shear wave splitting reveals more complicated fast directions in southern segment, with NE–SW in northeastern, NWW–SEE in southern and dominantly W–E in eastern (Gao et al. 2014). Considering that the regional shear wave splitting technique determines the average stress regime in the crust, differences between results from splitting analysis and this study may indicate that the stress regime in the southern segment varies laterally and vertically.

Investigations on the stress regime of different time periods (Fig. 10) and depths (Fig. 11) indicate more complex temporal and spatial variations. From Fig. 10, orientations of σ1 in the areas north of the "asperity" show an obvious change after 2009, especially in the central Pengguan complex, where

the orientations vary from constant NEE–SWW before 2009 to NW–SE after 2009. Another feature is that the earthquake activity is more concentrated on F2 and F3 after 2009 and barely extended into Sichuan basin. The number of normal and strike-slip events dropped significantly after 2009. The seismic distribution shows a slightly southward extension along F1 with NWW–SEE compression direction and relative higher plunges than that before 2009 which is consistent with the regional stress filed.

Stress distribution derived from different depth ranges tells the variation of stress in vertical direction (Fig. 11). Figure 11a displays events shallower than 14 km (b) displays for events in the depth of 14–18 km and (c) displays for events deeper than 18 km. The Pengguan complex is almost aseismic in shallow depth (11–13 km) in its central part. But σ1 show the rotational feature similar with Fig. 9b in its marginal area. In depth (19–24 km), σ1 show the constant NNW–SSE direction north of the "asperity" and NE–SW direction south of the "asperity". This is significantly different from that in shallow depth. Especially for the stress field south of the seismic gap, the σ1 directions are almost parallel to the LMSF, and they are surprisingly different from that of the shallow depth. The plunges of σ1 also show significant variations with depths. Thus, the σ1 variations with depths and areas may imply that the

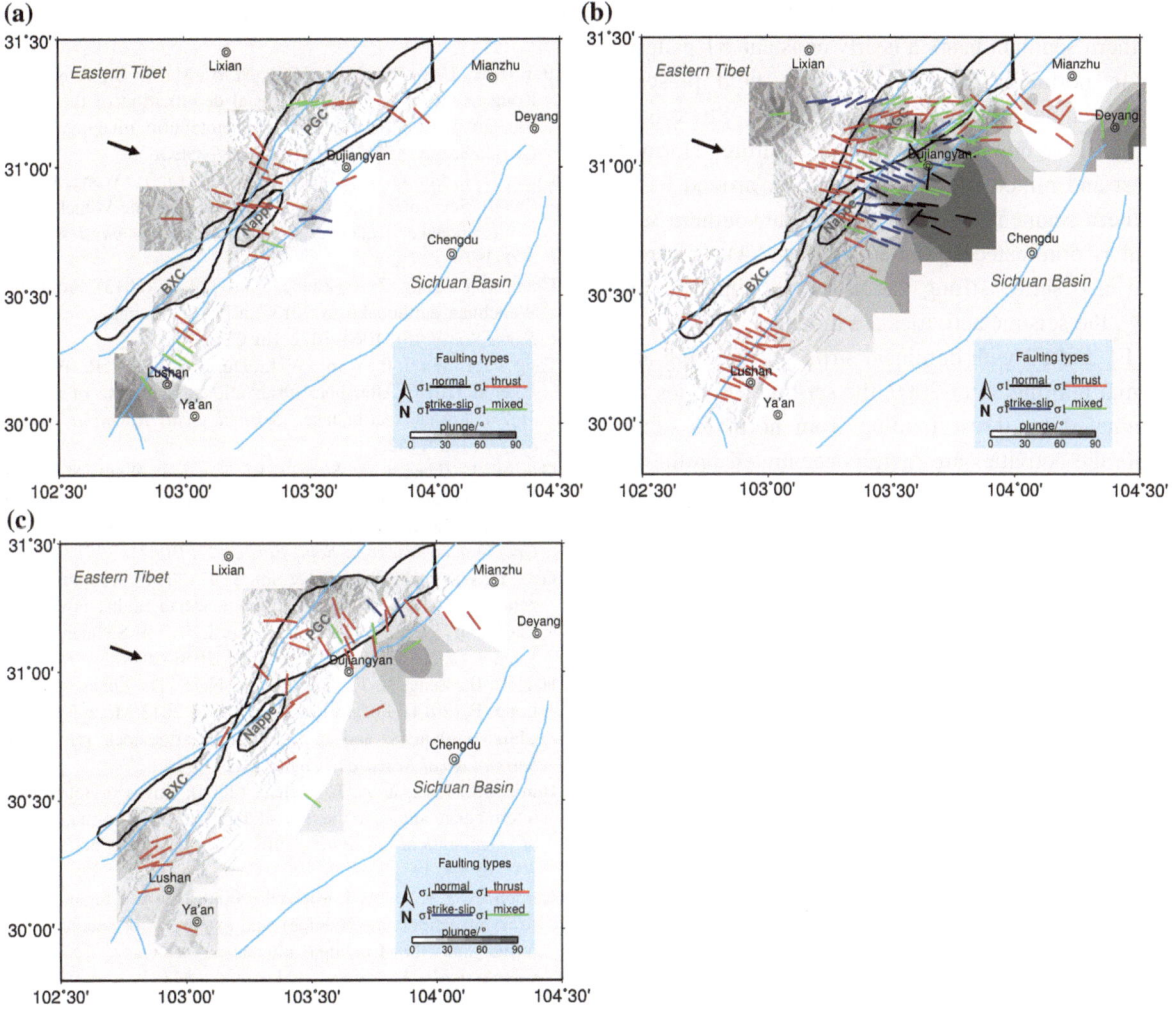

Figure 11
Same as Fig. 9b but derived from events in different depths. **a** based on events shallower than 14 km, **b** based on events in the depth of 14–18 km and **c** based on events deeper than 18 km

movement in depth is decoupled from the movement in shallow depths. This is consistent with the general conception that the materials become successively ductile due to the increase of temperature with depth.

5. Conclusions

Based on the focal mechanisms of 471 earthquakes and the derived stress field around the seismic gap in the southern Longmenshan fault system (LMSF), the following major conclusions may be

made: (1) between the seismically active northern and southern segments, the chimney-shaped seismic gap is collocated with a low-velocity zone determined by tomographic studies. Although there is no way to tell how much stress energy is dissipated by ductile flow and how much is accumulated to form a big earthquake in the future, (2) in terms of stress distribution along the LMSF, the seismic gap appears to be a transition zone between a more complicated northern segment and a much simpler southern segment; in general, the σ1 shows an obvious rotation around the "asperity" located to the west of

Dujiangyan city in northern segment, while the southern segment bears a nearly constant σ1 pattern; (3) in northern segment, the stress field presents significant degree of horizontal variations. Stress field along F3 is dominated by thrust faulting. Normal, thrust and mixed stress regime appear around F1 in northern segment; on the contrary, the southern segment is dominated by thrust faulting; (4) the stress field also shows strong variation with time. Before 2009, the seismic activities are more concentrated on the F1 and F2 with dominant strike-slip faulting and normal faulting; after 2009, the seismic activities are dominated by thrust faulting from north to south, while the activities are more concentrated on the F3 in northern segment and F1 in southern segment; and (5) the σ1 varies in different depths and the variation pattern is totally different from north to south. It may imply the decoupled movement between shallow layer and deep layer.

6. Data and Resources

The seismic data used in this paper are provided by Sichuan seismic network. The earthquake catalogue was obtained from China Earthquake Network Center (CENC) at http://www.csndmc.ac.cn/newweb/index.jsp (last accessed Feb. 2016). The program package CPS for focal mechanism determination was written by Herrmann (2014), and can be obtained upon request. The stress inversion software MSATSI was developed by Martínez-Garzón et al. (2014), and can be obtained upon request. All figures in this manuscript are produced by the generic mapping tools (GMT) of Wessel and Smith (1995).

Acknowledgments

We appreciate Mian Liu and one anonymous reviewer for their constructive comments and suggestions that have helped to improve this paper. This work was partially supported by National Natural Science Foundation of China (41340009), Sichuan Science and Technology Support Plan (2015RZ0032, 2015SZ0224) and the Creative Team Plan of Chengdu University of Technology.

References

Bai, D. H., Unsworth, M. J., Meju, M. A., Ma, X. B., Teng, J. W., Kong, X. R., et al. (2010). Crustal deformation of the eastern Tibetan plateau revealed by magnetotelluric imaging. *Nature Geoscience, 3*, 358–362. doi:10.1038/NGEO830.

Chen, J. H., Liu, Q. Y., Li, S. C., Guo, B., Li, Y., Wang, J., et al. (2009). Seismotectonic study by relocation of the Wenchuan M_S 8.0 earthquake sequence. *China Journal Geophysics, 52*(2), 390–397.

Chen, Y. T., Yang, Z. X., Zhang, Y., & Liu, C. (2013). From 2008 Wenchuan earthquake to 2013 Lushan earthquake. *Sci. China Earth Sci., 43*(6), 1064–1072. **(in Chinese)**.

Fang, L. H., Wu, J. P., Wang, W. L., Du, W. K., Su, J. R., Wang, C. Z., et al. (2015). Aftershock observation and analysis of the 2013 M_S7.0 Lushan earthquake. *Seismological Research Letters, 86*(4), 1135–1142.

Gan, W. J., Zhang, P. Z., Shen, Z. K., Niu, Z. J., Wang, M., Wang, Y. G., et al. (2007). Present-day crustal motion within the Tibetan Plateau inferred from GPS measurements. *Journal of Geophysical Research: Solid Earth, 112*, B08416.

Gao, Y., Wang, Q., Zhao, B., & Shi, Y. T. (2014). A rupture blank zone in middle south part of Longmenshan faults: effect after Lushan M_S 7.0 earthquake of 20, April 2013 in Sichuan, China. *Science China Earth Sciences, 57*(9), 2036–2044.

Han, L. B., Zeng, X. F., Jiang, C. S., Ni, S. D., Zhang, H. J., & Long, F. (2014). Focal mechanism of the 2013 M_W 6.6 Lushan, China earthquake and high-resolution aftershock relocations. *Seismological Research Letters, 85*(1), 8–14.

Hardebeck, J. L., & A. J. Michael (2004). Stress orientations at intermediate angles to the San Andreas fault, California, *Journal of Geophysical Research*, 109, no B11303. doi:10.1029/2004JB003239.

Hardebeck, J. L., & A. J. Michael (2006). Damped regional-scale stress inversions: methodology and examples for southern California and the Coalinga aftershock sequence, *Journal of Geophysical Research*, 111, no. B11310. doi:10.1029/2005JB004144.

Herrmann, R. B. (2014). Computer programs in seismology, Version 3.30, http://www.eas.slu.edu/eqc/eqccps.html (last accessed Dec 2015).

Herrmann, R. B., Malagnini, L., & Munafò, I. (2011). Regional Moment Tensors of the 2009 L'Aquila Earthquake Sequence. *Bulletin of the Seismological Society of America, 101*(3), 975–993.

Huang, R. Q., Wang, Z., Pei, S. P., & Wang, Y. S. (2009). Crustal ductile flow and its contribution to tectonic stress in Southwest China. *Tectonophysics, 473*, 476–489.

Langston, C. A. (1979). Structure under Mount Rainier, Washington, inferred from teleseismic body waves. *J. Geophy. Res., 84*(B9), 4749–4762.

Li, Z. W., Tian, B. F., Liu, S., & Yang, J. S. (2013). Asperity of the 2013 Lushan earthquake in the eastern margin of Tibetan Plateau from seismic tomography and aftershock relocation. *Geophysical Journal International, 195*, 2016–2022. doi:10.1093/gji/ggt370.

Liang, C. T., & Song, X. D. (2004). Tomographic inversion of Pn travel times in China. *Journal Geophysical Research, 109*, B11304. doi:10.1029/2003JB002789.

Liu, M., Luo, G., & Wang, H. (2014). The 2013 Lushan earthquake in China tests hazard assessments. *Seismological Research Letters, 85*(1), 40–43.

Lü, J., Su, J. R., Jin, Y. K., Long, F., Yang, Y. Q., Zhang, Z. W., et al. (2008). Discussion on relocation and seismo-tectonics of the M_S 8.0 Wenchuan earthquake sequences. *Seismology and Geology, 30*(4), 917–925.

Luo, Y., Zhao, L., Zeng, X. F., & Gao, Y. (2015). Focal mechanisms of the Lushan earthquake sequence and spatial variation of the stress field. *Science China Earth Science, 58*(7), 1148–1158. doi:10.1007/s11430-014-5017-y.

Martínez-Garzón, P., Kwiatek, G., Ickrath, M., & Bohnhoff, M. (2014). MSATSI: A MATLAB© package for stress inversion combining solid classic methodology, a new simplified user-handling and a visualization tool. *Seismological Research Letters, 85*(4), 896–904. doi:10.1785/0220130189.

Pei, S. P., H. J. Zhang, J. R. Su, and Z. X. Cui (2014). Ductile gap between the Wenchuan and Lushan earthquakes revealed from the two-dimensional Pg seismic tomography. Sci. Rep. 4, 6489. doi:10.1038/srep06489.

Wang, Z., J. R. Su, C. X. Liu, and X. L. Cai (2015). New insights into the generation of the 2013 Lushan Earthquake (M_s 7.0), China. J. Geophys. Res. 120, 3507–3526, doi:10.1002/2014JB011692.

Wang, H., Liu, M., Shen, X. H., & Liu, J. (2010). Balance of seismic moment in the Songpan-Ganze region, eastern Tibet: implications for the 2008 Geat Wenchuan earthquake. *Tectonophysics, 491*, 154–164.

Wessel, P., and W. H. F. Smith (1995). New version of Generic Mapping Tools released, Eos Trans. AGU 76, p. 329.

Xu, Z. H. (2001). A present-day tectonic stress map for eastern Asia region. *Acta Seismologica Sinica, 23*, 492–501. **(in Chinese)**.

Xu, Y., Herrmann, R. B., & Koper, D. K. (2010). Source parameters of regional small-to-moderate earthquakes in the Yunnan-Sichuan Region of China. *Bulletin of the Seismological Society of America, 100*, 2518–2531. doi:10.1785/0120090195.

Xu, Z. Q., Ji, S. C., Li, H. B., Hou, L. W., Fu, X. F., & Cai, Z. H. (2008a). Uplift of the Longmenshan range and the Wenchuan earthquake. *Episodes, 31*, 291–301.

Xu, X. W., Wen, X. Z., Han, Z. J., Chen, G. H., Li, C. Y., Zheng, W. J., et al. (2013). Lushan M_s 7.0 earthquake: a blind reserve-fault earthquake. *Chinese Science Bulletin, 58*(20), 1887–1893. doi:10.1007/s11434-013-5999-4. **(in Chinese)**.

Xu, J. R., Zhao, Z. X., & Ishikawa, Yuzo. (2008b). Regional characteristics of crustal stress field and tectonic motions and around Chinese mainland. *Chinese Journal of Geophysics, 51*(3), 770–781.

Yang, Y. H., Liang, C. T., & Su, J. R. (2015). Focal mechanism inversion based on regional model inverted from receiver function and its application to the Lushan earthquake sequence. *Chinese Journal of Geophysics, 58*(10), 3583–3600. doi:10.6038/cjg20151013. **(in Chinese)**.

Yi, G.X., Long, F., and Zhang, Z.W. (2012). Spatial and temporal variation of focal mechanisms for aftershocks of the 2008 M_s 80 Wenchuan earthquake. Chin. J. Geophys. 55, no. 4, 1213-1227, doi:10.6038/j.issn.0001-5733.2012.04.017.

Zhang, P. Z., Shen, Z. K., Wang, M., Gan, W. J., Bürgmann, R., Molnar, P., et al. (2004). Continuous deformation of the Tibetan Plateau from global positioning system data. *Geology, 32*(9), 809–812.

Zhao, L., Luo, Y., Liu, T. Y., & Luo, Y. J. (2013). Earthquake Focal Mechanisms in Yunnan and their Inference on the Regional Stress Field. *Bulletin of the Seismological Society of America, 103*(4), 2498–2507. doi:10.1785/0120120309.

Zhao, G. Z., Unsworth, M. J., Zhan, Y., Wang, L. F., Chen, X. B., Jones, A. G., et al. (2012). Crustal structure and rheology of the Longmenshan and Wenchuan Mw 7.9 earthquake epicentral area from magnetotelluric data. *Geology, 40*(12), 1139–1142. doi:10.1130/G33703.1.

Zhu, A. L., Xu, X. W., Diao, G. L., Su, J. R., Feng, X. D., Sun, Q., et al. (2008). Relocation of the M_S 8.0 Wenchuan earthquake sequence in part: preliminary seismotectonic analysis. *Seismology and Geology, 30*(3), 759–767.

Zhu, J. S., Zhao, J. M., Jiang, X. T., Fan, J., & Liang, C. T. (2012). Crustal flow beneath the eastern margin of the Tibetan plateau. *Earthquake Science, 25*, 469–483. doi:10.1007/s11589-012-0871-1.

Zoback, M. L. (1992). First- and second-order patterns of stress in the lithosphere: the world stress map project. *Journal of Geophysical Research, 97*(B8), 11703–11728.

(Received March 1, 2016, revised June 14, 2016, accepted July 20, 2016, Published online August 18, 2016)

Pure Appl. Geophys. 174 (2017), 2269–2278
© 2016 Springer International Publishing
DOI 10.1007/s00024-016-1428-3

Pure and Applied Geophysics

Parametrizing Physics-Based Earthquake Simulations

KASEY W. SCHULTZ,[1] MARK R. YODER,[1] JOHN M. WILSON,[1] ERIC M. HEIEN,[2] MICHAEL K. SACHS,[1]
JOHN B. RUNDLE,[1,2] and DON L. TURCOTTE[2]

Abstract—Utilizing earthquake source parameter scaling relations, we formulate an extensible slip weakening friction law for quasi-static earthquake simulations. This algorithm is based on the method used to generate fault strengths for a recent earthquake simulator comparison study of the California fault system. Here we focus on the application of this algorithm in the Virtual Quake earthquake simulator. As a case study we probe the effects of the friction law's parameters on simulated earthquake rates for the UCERF3 California fault model, and present the resulting conditional probabilities for California earthquake scenarios. The new friction model significantly extends the moment magnitude range over which simulated earthquake rates match observed rates in California, as well as substantially improving the agreement between simulated and observed scaling relations for mean slip and total rupture area.

Key words: Virtual Quake, Virtual California, earthquake simulations.

1. Introduction

The stress drop is a fundamental earthquake source property used for earthquake hazard assessment and is particularly important to numerical earthquake simulations. There are numerous extant numerical earthquake codes that simulate stress interactions between faults at the regional scale. These simulations typically produce synthetic seismic catalogs of many tens of thousands of years. Tullis et al. (2012a, b) recently summarized the properties of four extant regional-scale earthquake simulators with different friction laws. For example, RSQSim uses rate-and-state friction laws (Richards-Dinger and Dieterich 2012), while AllCal used velocity weakening friction (Ward 2000, 2012) and Virtual Quake (then named Virtual California) used a much simpler cellular automata approach that used stress drops that were prescribed for the simulator comparison on a fault-wise basis. This paper will present the method used to generate those stress drops—or fault strengths depending on the particular simulator's implementation—used in the earthquake simulator comparison. However, the main purpose of this paper is to introduce the new slip-weakening friction law used in Virtual Quake simulations.

First we will introduce the Virtual Quake simulator physics and the stress drop model. Then, as a case study, we explore the effect of the new friction law on simulations of the California fault system as provided by Field et al. (2014). Finally we will conclude with conditional earthquake probabilities derived from the Virtual Quake simulation.

2. Virtual Quake

Virtual Quake (VQ) is a boundary element code that uses stress interactions and friction laws to simulate realistically driven fault systems. VQ has been designed to explore the long term statistical behavior of topologically complex fault networks (Rundle 1988a, b; Rundle et al. 2006a, b, c). VQ simulates many tens of thousands of years of earthquakes on any fault network. VQ consists of three major components: a fault model, a set of quasi-static elastic interactions (Green's functions), and a rupture (earthquake) model.

Today, Virtual Quake is a modern, open-source scientific code that simulates earthquakes in a high

[1] Department of Physics, University of California-Davis, One Shields Ave., Davis 95616, Canada. E-mail: kwschultz@ucdavis.edu
[2] Department of Earth and Planetary Sciences, University of California-Davis, Davis, USA.

performance computing environment (Schultz et al. 2015; Sachs et al. 2012; Heien and Sachs 2012). Although the main application of Virtual Quake has been seismic hazard assessment and computing conditional probabilities for large California earthquake scenarios (Rundle et al. 2005; Schultz et al. 2015), VQ has a wide range of applications from computing coseismic gravity changes (Schultz et al. 2014) to studying the effects of asperities on the recurrence times of earthquakes (Yikilmaz et al. 2010).

For completeness, it is worth summarizing the major simplifications of Virtual Quake. VQ models faults in a single-layer elastic half-space, with no viscoelastic layer. The elastic half-space assumes a homogeneous flat earth, and we do not account for rheological properties or elastic discontinuities. Also, Virtual Quake partitions planar fault surfaces into rectangular fault elements instead of filling the curved surfaces more completely with triangular elements. One may expect that by using rectangular elements that leave gaps and overlaps between elements, we may be introducing errors in stress calculations. However, Barall and Tullis (2016) have recently shown that rectangles perform better than triangles. For more detailed descriptions of the Virtual Quake simulator see the following: Schultz et al. (2015, 2016), Sachs et al. (2012) and Heien and Sachs (2012).

2.1. Fault Model

For the earthquake simulator comparison studies in Tullis et al. (2012a, 2012b), all simulators used the UCERF2 California fault model provided by the United States Geological Survey (USGS) (Field et al. 2009). The model for the California fault system used in this study is the UCERF3 model, described in detail in Field et al. (2014). While the model represents the most comprehensive California fault system model to date, the model is not complete and there are many other known faults at smaller scales that have been neglected.

This model, shown in Fig. 1, is comprised of 313 faults corresponding to the known major faults in California. Each fault is partitioned into square elements that are approximately 3 km × 3 km, for a total of 21,833 elements.

2.1.1 Modifications to the UCERF3 Model

It is worth highlighting the modifications we have made to the standard UCERF3 fault model as presented by Field et al. (2014). After inspecting initial test simulations of the UCERF3 fault model with Virtual Quake, we found a systematic overprescription of a 10% aseismic fraction to the slip rates presented in the model. This aseismic fraction acts to reduce the effective stress loading rate on all faults, adding a small monotonic stress decrease with simulation time. With 83% of the 21,833 fault elements having exactly a 10% aseismic fraction, Virtual Quake initially produced earthquake rates that were far below the observed California earthquake rates (Table L12 of Appendix 12 in Field et al. 2014). We are uncertain as to the reasons behind this possible systematic fitting error. The only way to mitigate this effect was to neglect the aseismic fraction for all fault elements with less than 10% aseismic fraction.

Additionally we taper the long term slip rates along the faults. In order to avoid stress concentrations at the bases of our faults, we taper the slip rates to decrease with depth. Also in order to avoid stress concentrations at the tips of faults, we taper the slip rates to decrease to near zero at the ends of the faults. We only horizontally taper the last 12 km toward the end of each fault. Furthermore, we renormalize the slip rates on a fault by fault basis in order to preserve the original moment rates of the system.

2.2. Fault Element Interactions

The simulation physics in Virtual Quake is based on stress transfer between faults via the accumulation of a slip deficit—the amount of slip each fault should move in a given time period given the long term slip rate. The behavior of the system is determined by interactions between elements from the Green's function and the stress release from elements during events.

2.2.1 Stress Green's Functions

Interactions between fault elements depend on the relative position and orientation of each fault

Figure 1
A Google Earth rendering of the UCERF3 California fault model (shown above ground), 3 km × 3 km mesh resolution (Field et al. 2014). Faults colors are long term slip rates from UCERF3, the largest being along the San Andreas fault (*red*)

element, and are calculated using stress Green's functions (Okada 1992) at the start of the simulation. The equations for the change in stress on a fault element due to relative motion of all the other fault elements is given by Rundle et al. (2006a, b).

In Virtual Quake simulations, the stress field is evaluated only at element centers and slip is applied uniformly across an element's surface and along the element's rake angle as defined in the fault model. Under these assumptions, we can reduce equation for the stress on some element (denoted as A) at time t to:

$$\sigma_{ij}^A(t) = \sum_B T_{ij}^{AB} s^B(t) \tag{1}$$

where iteration over element B adds the contributions from all other elements, and $s^B(t)$ is the slip deficit for element B—the distance element B has back slipped from its equilibrium position at time t. The friction laws used in Virtual Quake only require the shear stress along the element rake vector and normal stress perpendicular to the element. Hence we can reduce the tensor T_{ij}^{AB} to T_s^A for shear stresses and T_n^A for normal stresses, leading to the following expressions for shear and normal stresses on some element A:

$$\sigma_s^A(t) = \sum_B T_s^{AB} s^B(t) \quad \text{and} \quad \sigma_n^A(t) = \sum_B T_n^{AB} s^B(t) \tag{2}$$

More details about the stress interaction physics are described in Rundle et al. (2006a) and Sachs et al. (2012).

2.2.2 Static Failure

To determine when a simulated earthquake begins, Virtual Quake uses a combined static-dynamic friction law. The law uses the Coulomb failure function (CFF) (Stein 1999) which is defined for each element A:

$$\text{CFF}^A(t) = \sigma_s^A(t) - \mu_s^A \sigma_n^A(t), \tag{3}$$

where

$$\mu_s = \frac{\Delta\sigma}{\rho g d}, \tag{4}$$

is the element's static coefficient of friction and depends on the stress drop $\Delta\sigma$, the depth d of the fault element, gravity g, and density of the halfspace ρ in order to give a fault strength that is constant with depth (Ward 2000).

During long term stress loading, an element A will slip at time t_f when $\mathrm{CFF}^A(t_f) = 0$; this is called static failure. When one or more elements slip, the simulation begins the rupture propagation phase described below.

2.3. Rupture Model

A simulated earthquake in a Virtual Quake simulation consists of many "sweeps" or iterations over the fault elements. With each sweep, we check the state of each element and add elements to the rupture if they meet our failure criteria. We allow failed elements to slip during each sweep, and then re-evaluate the new stress state for the entire system using Eq. 2.

A simulated earthquake begins when one or more fault elements have statically failed, they are allowed to slip. Through stress interactions, these initial slipped fault elements will change the stress on nearby fault elements. More fault elements are allowed to join the rupture and slip even if they do not reach their failure stresses. Virtual Quake includes this approximation to the stress intensity during rupture to facilitate rupture propagation; we call this dynamic failure. We introduce a free parameter in VQ, called the dynamic triggering factor, to control the extent of rupture propagation. For an element on the same fault as the initial statically failed fault element to join the rupture, the following relation must be satisfied.

$$\frac{\mathrm{CFF}_{\mathrm{init}} - \mathrm{CFF}_{\mathrm{final}}}{\mathrm{CFF}_{\mathrm{init}}} > \eta \qquad (5)$$

In Virtual Quake simulations, the value of the dynamic trigger factor η is the main tuning parameter used to fit observed earthquake rates. In the simulation data presented later in this study, we take $\eta = 0.5$. This requires a 50% stress increase on a nearby element for that element to join the rupture. This value for the parameter is consistent with the Virtual Quake of the UCERF2 fault model in Sachs et al. (2012) and Tullis et al. (2012a, b).

2.3.1 Using Stress Drops to Determine Slip

The initial rupture—the first fault element to slip—during an event is always caused by static failure. During rupture propagation the first element to fail slips back towards the equilibrium position. For any element, denoted by A, the amount of slip during the initial static failure, Δs^A, is related to the stress drop defined for that element in the fault model, $\Delta\sigma^A$, by Sachs et al. (2012)

$$\Delta s^A = \frac{1}{K_L^A}\left(\Delta\sigma^A - \mathrm{CFF}^A\right) \qquad (6)$$

where K_L^A is the element's stiffness or self-stress, defined as

$$K_L^A = T_s^{AA} - \mu_s^A T_n^{AA}. \qquad (7)$$

Once an element has ruptured due to static or dynamic failure it will no longer slip in the event due to these failures. However, ruptured elements may slip further due to the movement of other elements.

During each rupture sweep, ruptured elements may slip further due to movement on other elements and the resulting stress changes. The amount of slip on a failed element is related to the movement of all other elements through the stress Green's functions. In Virtual Quake, we take a simplistic approach to determining these interdependent slip values by calculating the slip on all other elements when a given element moves. However, this is highly inefficient because the slip on a given element will in turn cause slip on all ruptured elements, which will cause slip on all ruptured elements, and so on forever. Instead, the system is described using a set of linear equations—a matrix version of Eq. 6—relating the slip of each element to the stress on all other elements in the final state of a given sweep.

This relationship between element slips during a sweep is determined for each ruptured element A as:

$$\sum\left(T_s^{AB} - \mu_s^B T_n^{AB}\right)s_B = \Delta\sigma^A - \mathrm{CFF}^A. \qquad (8)$$

Since non-ruptured blocks do not change their slip, they are excluded from the system. This system is then solved using Gaussian elimination. Once the slip is calculated for all ruptured elements, a new stress state for the entire system is calculated using Eq. 2. The rupture process is repeated by again checking the static and dynamic failure conditions until no more elements have failed.

3. Stress Drop Algorithm

Tullis et al. (2012b) presents the results from a number of physics-based earthquake simulators—including Virtual Quake, then named Virtual California—modeling the California fault system. Fault locations and topology were taken from the UCERF2 fault model (Field et al. 2009). An additional, fundamental component of this model was the fault "strengths", or the stress drop values defined for each fault depending on each simulator's implementation. In Virtual Quake, we use these values directly to prescribe slip during a simulated earthquake (see $\Delta\sigma$ in Eqs. 6 and 8). In this section we will outline the method used to compute these stress drops and then introduce the improvement to this method that underlies Virtual Quake's new friction law.

It is also worth highlighting that Virtual Quake is a quasi-static, regional stress-accumulation model with a cellular automata approach to rupture dynamics instead of time dependent rate and state dynamics. The model introduced in this section is intended to improve the current simulation physics to mimic the slip weakening feature of a more realistic time and state dependent friction law.

3.1. Stress Drops in Virtual Quake

The stress drops for the fault elements are computed via Leonard (2010) scaling relations—a recent study improving upon Wells and Coppersmith (1994)—along with an analytical solution for the shear stress change per unit slip on a vertical rectangular fault. The simulation is sensitive to these values; large stress drops tend to produce larger earthquakes with higher mean slip and longer periods between events, while smaller stress drops produce more earthquakes with lower mean slip.

One of Virtual Quake's free parameters ΔM globally adjusts the stress drop values. This parameter was used in conjunction with the algorithm described below in the earthquake simulator comparison (Tullis et al. 2012b) to tune the stress drops on a fault-wise basis, fitting recurrence interval data where available. To keep our algorithm scalable, and reproducible by other researchers, we do not fit this tuning parameter on a fault wise basis and instead we choose a global value that produces earthquake rates that best fit observed earthquake rates.

3.1.1 Scaling Relations

We begin by using scaling relations to determine a different characteristic magnitude M_{char} and a characteristic mean slip $\bar{s}_{\mathrm{char}}$ for each element in a fault based on the expected magnitude from the fault's geometry using Leonard (2010) scaling relations,

$$M_{\mathrm{char}} = 4.0 + \log_{10}(A) + \Delta M, \qquad (9)$$

where A is the surface area of the fault in km^2. The definition of moment magnitude M is,

$$M = \frac{2}{3}\log_{10}(\mu A \bar{s}) - 6.0, \qquad (10)$$

where μ is the rigidity (shear modulus) of the elastic half space in Pa, A here is the rupture area in m^2, and $\bar{s}$ is the mean slip in m (Kanamori and Anderson 1975).

We can determine the expected, characteristic mean slip $\bar{s}_{\mathrm{char}}$ from the definition of moment magnitude to be,

$$\bar{s}_{\mathrm{char}} = \frac{10^{\frac{3}{2}(M_{\mathrm{char}}+6.0)}}{\mu A}, \qquad (11)$$

where M^{char} is the expected moment magnitude for an earthquake on the fault with area A if the entire fault ruptures.

We employ the the analytical solution (personal communication Dec. 30, 2015) to equation 1b from Ward (2012) for stress drop per unit slip across a vertical strike slip fault. We then use the characteristic mean slip from Eq. 11 to determine the characteristic stress drop for each fault element,

$$\Delta\sigma = -\frac{2\mu\bar{s}_{\mathrm{char}}}{(1-v)\pi R}\left((1-v)\frac{L}{W}+\frac{W}{L}\right), \qquad (12)$$

where v is the Poisson ratio of the elastic half space, W is the down-dip width of the fault, L is the length of the fault in the direction of slip, and $R = \sqrt{L^2 + W^2}$; all in MKS units.. This is in agreement with the Equation 1 from Leonard (2010)—rewritten from Kanamori and Anderson (1975)—with $\Delta\sigma = \frac{C\mu\bar{s}}{R}$ where C is a geometric constant depending on fault type.

3.1.2 Maximum Stress Drops

Figure 2 shows a histogram of the stress drops for the UCERF3 faults produced by this algorithm. These stress drops are larger than observed stress drops of a few MPa (Scholz 1990), however as the reader will see in the next Section these represent the maximum stress drops and in practice the mean stress drops are lower.

3.2. Dynamic Stress Drops for Frictional Weakening

We do not want to directly plug in the stress drops from the algorithm in Sect. 3.1 into Eqs. 6 and 8. Doing so would prescribe the same characteristic stress drop and slip for every simulated earthquake, and hence would lead to small earthquakes slipping as much as very large earthquakes. Indeed this behavior resulted in an additional tuning parameter for older versions of Virtual Quake, whereby the slip for small earthquakes was reduced by a factor (Sachs

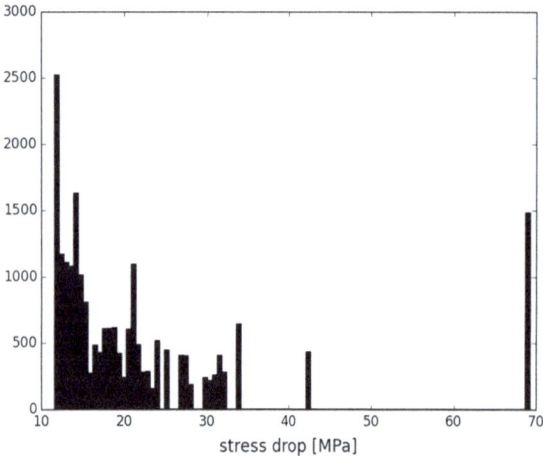

Figure 2
A histogram of the stress drops for each fault element in the UCERF3 model produced by the algorithm described in Sect. 3, using the free parameter $\Delta M = 0.7$

et al. 2012). This method did not take into account any dynamic aspects of fault friction.

We can improve this algorithm by modifying the stress drop during a simulated earthquake to mimic some features of both slip weakening and velocity weakening friction found in laboratory experiments (Byerlee 1970 and Chapter 2 of Scholz 1990). If we modify Eqs. 9 and 11 so that they increase with increasing rupture area, then we can facilitate larger slips once an earthquake has begun and the rupture area is increasing. This amounts to reevaluating the Eqs. 9 and 11 at every iteration of the rupture model using the current rupture area. The new relations then become

$$M^*_{\text{char}} = \begin{cases} 4.0 + \log_{10}(A_r) + \Delta M, & A_r < A_{\text{fault}} \\ 4.0 + \log_{10}(A_{\text{fault}}) + \Delta M, & A_r \geq A_{\text{fault}} \end{cases}$$

(13)

and

$$\bar{s}^*_{\text{char}} = \begin{cases} \dfrac{10^{\frac{3}{2}(M^*_{\text{char}}+6.0)}}{\mu A_r}, & A_r < A_{\text{fault}} \\ \dfrac{10^{\frac{3}{2}(M^*_{\text{char}}+6.0)}}{\mu A_{\text{fault}}}, & A_r \geq A_{\text{fault}} \end{cases}$$

(14)

During the interseismic, stress-loading phase of the simulation we set the fault strengths (Eq. 4) to their maximum values as determined by the algorithm in Sect. 3 using the entire fault area A_{fault} to compute the stress drops. Then once a rupture begins and fault elements continue to join the rupture and slip, we reevaluate the stress drops at every rupture model iteration using the current rupture area A_r in Eqs. 13 and 14. As new fault elements join the rupture, the stress drops for all rupturing elements increases with increasing rupture area up to the maximum stress drop for each fault when the rupture area is equal to the element's fault area. Figure 3 shows how the stress drop would change for a hypothetical single-fault earthquake where fault elements are added to the rupture one by one until the entire fault has slipped.

After the fault elements slip and reduce the stresses sufficiently such that all CFF < 0, the rupture propagation phase is over. At this point all faults are allowed to restore their strengths and stress drops to the interseismic maximum values and the simulator solves for the next static failure. By facilitating more

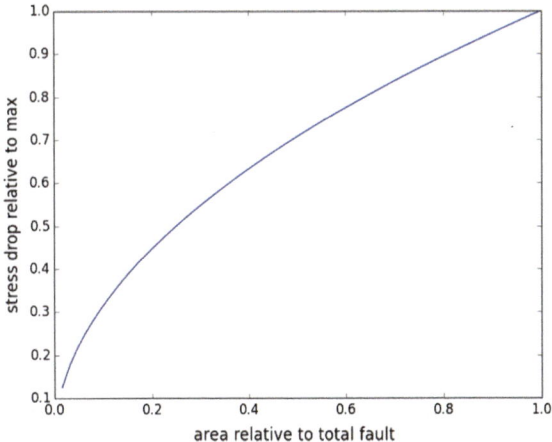

Figure 3
The dynamic stress drop algorithm. The stress drop is allowed to increase with increasing rupture area as the simulated rupture grows by re-evaluating the characteristic slip (Eq. 14) using the current rupture area at every iteration of the rupture model

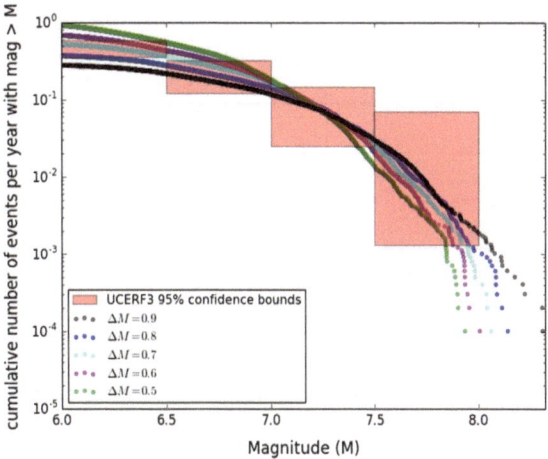

Figure 4
Frequency-magnitude plot for 10,000 year-long simulations of the UCERF3 fault model using the dynamic stress drop model from Sect. 3.2. Here we vary the stress drops by varying ΔM from Eqs. 9 and 13. *Red boxes* denote 95% confidence bounds for observed California earthquakes from 1932 to 2011—Table L12 in Appendix L of Field et al. (2014)

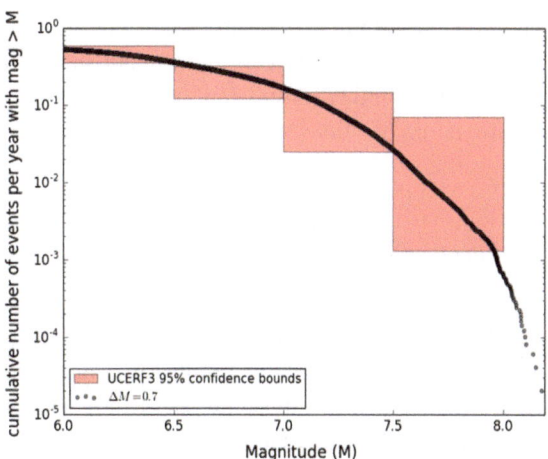

Figure 5
Frequency-magnitude plot for a 50,000 year-long simulation of the UCERF3 fault model with the dynamic stress drop model described in Sect. 3.2, using $\Delta M = 0.7$. Observed earthquake rates in California from from 1932 to 2011 with 95% confidence bounds (Field et al. 2014)

and more slip with increasing rupture area, and by enforcing that fault strength is restored after each earthquake, this dynamic stress drop model provides Virtual Quake with some aspects of slip weakening and velocity weakening friction.

4. Virtual Quake Simulations

Before evaluating the results of the dynamic stress drop model outlined in the previous section, it is best to address the effect of tuning the stress drop parameter ΔM (Eqs. 9 and 13) on the simulated earthquake rates. This parameter along with the dynamic trigger factor η (Eq. 5) are the two free parameters for Virtual Quake, allowing the modeler to tune the simulation to match observed earthquake rates. Figure 4 shows how the simulated earthquake frequency-magnitude relation is affected when we change ΔM. Increasing ΔM prescribes a higher maximum stress drop and hence a higher maximum slip for simulated earthquakes, which in turn produces fewer small earthquakes with $M < 7$.

Figure 5 shows a 50,000 UCERF3 simulation implementing the dynamic stress drop model with the best fitting value of $\Delta M = 0.7$ from Fig. 4. The value of $\Delta M = 0.7$ was chosen as to best align Virtual Quake simulated earthquake rates with the observed

California earthquakes from 1932 to 2011 as reported in Table L12 of Appendix L from Field et al. (2014).

Using this simulation, we can compute the conditional probabilities for California earthquake scenarios using the techniques presented in Schultz et al. (2015) and Rundle et al. (2005). Querying the

Table 1

Conditional probabilities for different magnitude earthquakes in California in the next 1, 5 and 15 years derived from Virtual Quake simulation data

		1 year	5 years	15 years
$N = 25,475$	$M \geq 5$	0.45	0.96	1.00
$N = 23,774$	$M \geq 6$	0.43	0.95	1.00
$N = 7495$	$M \geq 7$	0.15	0.57	0.92

The number of simulated earthquakes used in computing the probabilities are listed on the left. Methods are described in detail in Schultz et al. (2015) and Yoder et al. (2015)

ANSS global composite earthquake catalog using a latitude and longitude rectangular bounds with $114.0°W < \mathrm{lon} < 125.2°W$ and $32.2°N < \mathrm{lat} < 42.0°N$, we obtain the observed time since the last California earthquakes (ANSS 2016). There have been 1.0 years since the last $M \geq 5$ earthquake ($M = 5.7$ on January 28, 2015), 1.4 years since the last $M \geq 6$ earthquake ($M = 6.0$ on August 24, 2014), and 5.8 years since the last $M \geq 7$ earthquake ($M = 7.2$ on April 4, 2010). Using these values, we compute the following conditional probabilities for earthquakes in California in the next 1, 5, and 15 years from today shown in Table 1. For further discussion of Virtual Quake conditional probabilities see Yoder et al. (2015).

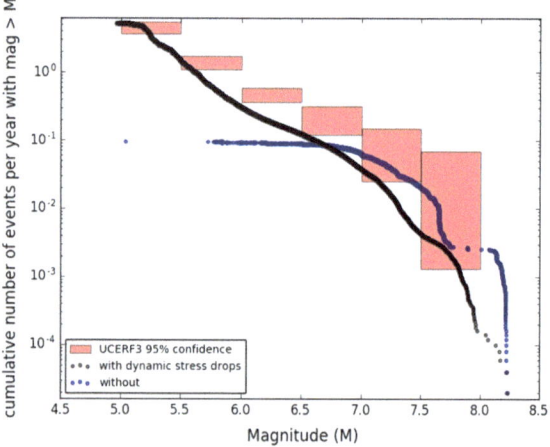

Figure 6
Frequency-magnitude plot for a 50,000 year-long simulation of the UCERF3 fault model with the dynamic stress drop model described in Sect. 3.2, using $\Delta M = 0.9$ (parameters differ from the simulation in Fig. 5), compared to a previous simulation with the same parameters that does not use the dynamic stress drop model. Observed earthquake rates in California from from 1932 to 2011 with 95% confidence bounds (Field et al. 2014)

4.1. Effects of the Dynamic Stress Drop Model

To gauge the effects of the dynamic stress drop model we ran two simulations of the UCERF3 model for 50,000 years, one with the original friction model and the other with the dynamic stress drop model. Figures 6, 7 and 8 show the differences between the simulations.

Figure 6 shows a stark difference in the earthquake rates for magnitudes smaller than 6.5. This is due to the fact that the original model universally prescribed a slip according to the stress drop derived from assuming a full fault rupture. In the new model, smaller earthquakes occur more frequently by permitting smaller ruptures to progress according to the stress state of neighboring areas of the fault. The new model only prescribes a slip that produces a full fault stress drop when the stresses on surrounding areas of the fault are high enough for the full fault to join the rupture. Hence the old model tends to prescribe more

characteristic earthquakes whereas the new model allows ruptures to grow in a manner more consistent with slip weakening friction, with slip growing as the rupture grows. This effect is also present for larger magnitude earthquakes, with the new model producing large earthquakes with a broader range of magnitudes instead of being clustered in more characteristic events.

The dynamic stress drop model also improves the rupture area and mean slip scaling relations. Figure 7 confirms that the mean slip was indeed too large for most magnitudes using the previous friction model. Both Figs. 7 and 8 show that the fit to the observed empirical scaling relations in Leonard (2010) are greatly improved.

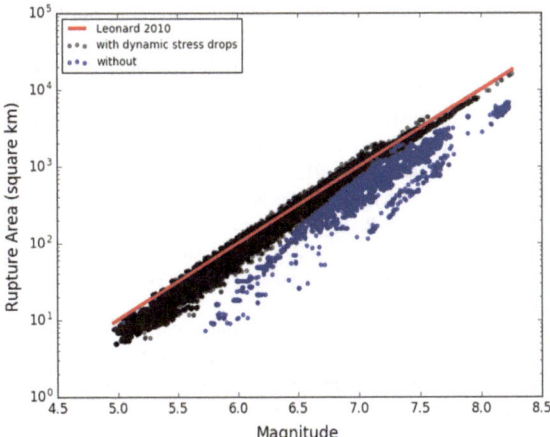

Figure 7

Mean slip-magnitude plot for the simulations shown in previous figure, 50,000 years of a UCERF3 fault model simulation with and without the dynamic stress drop model. The observed scaling relation from Leonard (2010) shown as the *red line*

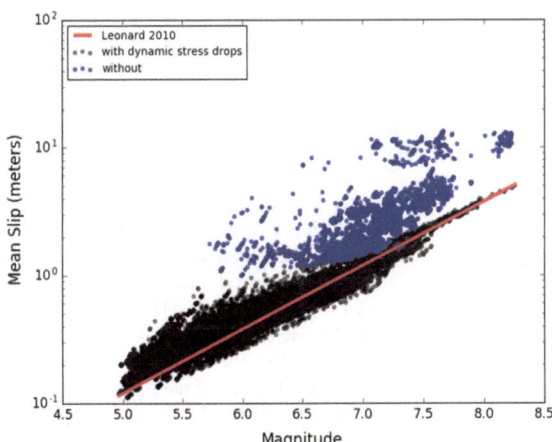

Figure 8

Rupture area-magnitude plot or the simulations shown in previous figure, 50,000 years of a UCERF3 fault model simulation with and without the dynamic stress drop model. The observed scaling relation from Leonard (2010) shown as the *red line*

5. Conclusions and Discussion

To improve the friction laws in Virtual Quake, we employ a stress drop algorithm analogous to slip-weakening friction; the larger the earthquake, the lower the resistance to slip. This algorithm is based on a method that was used to generate fault strengths in a recent study comparing multiple modern earthquake simulators (Tullis et al. 2012b). This new friction model significantly extends the moment magnitude range over which simulated earthquake rates match observed rates in California, nearly doubling the magnitude range of simulated earthquakes. Additionally implementing the dynamic stress drop friction model substantially improved the agreement between simulated and observed scaling relations for mean slip and total rupture area as reported by Leonard (2010). This algorithm is fully extensible, can be employed in Virtual Quake simulations of any fault model, and can be more robustly employed by tuning the free parameter to fit each fault's recurrence interval data where available.

Acknowledgements

We would like to thank Prof. Steven N. Ward for his initial version of this stress drop algorithm, and for his advice guiding the development of this study. This research was supported by National Aeronautics and Space Administration (NASA) Earth and Space Science fellowship number NNX11AL92H. Virtual Quake is hosted by the NSF-supported Computational Infrastructure for Geodynamics (CIG). Virtual Quake is open source scientific software, is available for download and comes with a user manual (Schultz et al. 2016). URL:http://geodynamics.org/cig/software/vq/.

REFERENCES

ANSS. 2016. UC Berkeley Seismological Laboratory. http://www.quake.geo.berkeley.edu/anss/catalog-search.html. Accessed 25 January 2016.

Barall, M., & Tullis, T. E. (2016). The performance of triangular fault elements in earthquake simulators. *Seismological Research Letters*, 87(1), 164–170. doi:10.1785/0220150163.

Byerlee, J. (1970). The mechanics of stick-slip. *Tectonophysics*, 9(5), 475–486.

Field, E. H., Dawson, T. E., Felzer, K. R., Frankel, A. D., Gupta, V., Jordan, T. H., et al. (2009). Uniform california earthquake rupture forecast, version 2 (ucerf 2). *Bulletin of the Seismological Society of America*, 99(4), 2053–2107.

Field, E. H., Arrowsmith, R. J., Biasi, G. P., Bird, P., Dawson, T. E., Felzer, K. R., et al. (2014). Uniform california earthquake rupture forecast, version 3 (ucerf3)-the time-independent model. *Bulletin of the Seismological Society of America*, 104(3), 1122–1180.

Heien, E., & Sachs, M. (2012). Understanding long-term earthquake behavior through simulation. *Computing in Science Engineering*, 14(5), 10–20.

Kanamori, H., & Anderson, D. L. (1975). Theoretical basis of some empirical relations in seismology. *Bulletin of the Seismological Society of America*, *65*(5), 1073–1095.

Leonard, M. (2010). Earthquake fault scaling: Self-consistent relating of rupture length, width, average displacement, and moment release. *Bulletin of the Seismological Society of America*, *100*(5A), 1971–1988. doi:10.1785/0120090189.

Okada, Y. (1992). Internal deformation due to shear and tensile faults in a half-space. *Bulletin of the Seismological Society of America*, *82*(2), 1018–1040.

Richards-Dinger, K., & Dieterich, J. H. (2012). Rsqsim earthquake simulator. *Seismological Research Letters*, *83*(6), 983–990.

Rundle, J. B. (1988a). A physical model for earthquakes: 1. Fluctuations and interactions. Journal of Geophysical Research: Solid. *Earth*, *93*(B6), 6237–6254.

Rundle, J. B. (1988b). A physical model for earthquakes: 2. Application to southern California. Journal of Geophysical Research: Solid. *Earth*, *93*(B6), 6255–6274.

Rundle, J. B., Rundle, P. B., Donnellan, A., Turcotte, D. L., Shcherbakov, R., Li, P., et al. (2005). A simulation-based approach to forecasting the next great san francisco earthquake. *Proceedings of the National Academy of Sciences of the United States of America*, *102*(43), 15363–15367.

Rundle, J. B., Rundle, P. B., Donnellan, A., Li, P., Klein, W., Morein, G., et al. (2006a). Stress transfer in earthquakes, hazard estimation and ensemble forecasting: Inferences from numerical simulations. *Tectonophysics*, *413*(12), 109–125.

Rundle, P., Rundle, J., Tiampo, K., Donnellan, A., & Turcotte, D. (2006b). Virtual california: Fault model, frictional parameters, applications. In *Computational Earthquake Physics: Simulations, Analysis and Infrastructure, Part I, Pageoph Topical Volumes* (pp. 1819–1846). Birkhäuser, Basel.

Rundle, P., Rundle, J., Tiampo, K., Donnellan, A., & Turcotte, D. (2006c). Virtual california: Fault model, frictional parameters, applications. *Pure and Applied Geophysics*, *163*(9), 1819–1846.

Sachs, M. K., Heien, E. M., Turcotte, D. L., Yikilmaz, M. B., Rundle, J. B., & Kellogg, L. (2012). Virtual California Earthquake Simulator. *Seismological Research Letters*, *83*(6), 973–978.

Scholz, C. H. (1990). *The Mechanics of Earthquakes and Faulting*. Cambridge: Cambridge University Press.

Schultz, K. W., Sachs, M. K., Heien, E. M., Rundle, J. B., Turcotte, D. L., & Donnellan, A. (2014). Simulating gravity changes in topologically realistic driven earthquake fault systems: First results. *Pure and Applied Geophysics* (in press). doi:10.1007/s00024-014-0926-4.

Schultz, K. W., Sachs, M. K., Yoder, M. R., Rundle, J. B., Turcotte, D. L., Heien, E. M., et al. (2015). Virtual quake: Statistics, coseismic deformations and gravity changes for driven earthquake fault systems. *International Association of Geodesy Symposia*, *1–9*, doi:10.1007/1345_2015_134.

Schultz, K. W., Heien, E. M., Sachs, M. K., Wilson, J. M., Yoder, M. R., Rundle, J. B., & Turcotte, D. L. (2016). *Virtual Quake User Manual, Version 2.1.2*. Davis, California: Computational Infrastructure for Geodynamics. https://geodynamics.org/cig/software/vq/vq_manual_2.1.2.pdf.

Stein, R. S. (1999). The role of stress transfer in earthquake occurrence. *Nature*, *402*(6762), 605–609. doi:10.1038/45144.

Tullis, T. E., Richards-Dinger, K., Barall, M., Dieterich, J. H., Field, E. H., Heien, E. M., et al. (2012a). A Comparison among Observations and Earthquake Simulator Results for the allcal2 California Fault Model. *Seismological Research Letters*, *83*(6), 994–1006.

Tullis, T. E., Richards-Dinger, K., Barall, M., Dieterich, J. H., Field, E. H., Heien, E. M., et al. (2012b). Generic earthquake simulator. *Seismological Research Letters*, *83*(6), 959–963.

Ward, S. N. (2000). San francisco bay area earthquake simulations: A step toward a standard physical earthquake model. *Bulletin of the Seismological Society of America*, *90*(2), 370–386. doi:10.1785/0119990026.

Ward, S. N. (2012). Allcal earthquake simulator. *Seismological Research Letters*, *83*(6), 964–972. doi:10.1785/0220120056.

Wells, D. L., & Coppersmith, K. J. (1994). New empirical relationships among magnitude, rupture length, rupture width, rupture area, and surface displacement. *Bulletin of the Seismological Society of America*, *84*(4), 974–1002.

Yikilmaz, M. B., Turcotte, D. L., Yakovlev, G., Rundle, J. B., & Kellogg, L. H. (2010). Virtual california earthquake simulations: simple models and their application to an observed sequence of earthquakes. *Geophysical Journal International*, *180*(2), 734–742. doi:10.1111/j.1365-246X.2009.04435.x.

Yoder, M. R., Schultz, K. W., Heien, E. M., Rundle, J. B., Turcotte, D. L., Parker, J. W., et al. (2015). The Virtual Quake earthquake simulator: a simulation-based forecast of the El Mayor-Cucapah region and evidence of predictability in simulated earthquake sequences. *Geophysical Journal International*, *203*(3), 1587–1604. doi:10.1093/gji/ggv320.

(Received February 2, 2016, revised October 21, 2016, accepted November 9, 2016, Published online November 22, 2016)

Pure Appl. Geophys. 174 (2017), 2279–2293
© 2016 Springer International Publishing
DOI 10.1007/s00024-016-1385-x

Spatial Evaluation and Verification of Earthquake Simulators

JOHN MAX WILSON,[1] MARK R. YODER,[1] JOHN B. RUNDLE,[1,2] DONALD L. TURCOTTE,[1] and KASEY W. SCHULTZ[1]

Abstract—In this paper, we address the problem of verifying earthquake simulators with observed data. Earthquake simulators are a class of computational simulations which attempt to mirror the topological complexity of fault systems on which earthquakes occur. In addition, the physics of friction and elastic interactions between fault elements are included in these simulations. Simulation parameters are adjusted so that natural earthquake sequences are matched in their scaling properties. Physically based earthquake simulators can generate many thousands of years of simulated seismicity, allowing for a robust capture of the statistical properties of large, damaging earthquakes that have long recurrence time scales. Verification of simulations against current observed earthquake seismicity is necessary, and following past simulator and forecast model verification methods, we approach the challenges in spatial forecast verification to simulators; namely, that simulator outputs are confined to the modeled faults, while observed earthquake epicenters often occur off of known faults. We present two methods for addressing this discrepancy: a simplistic approach whereby observed earthquakes are shifted to the nearest fault element and a smoothing method based on the power laws of the epidemic-type aftershock (ETAS) model, which distributes the seismicity of each simulated earthquake over the entire test region at a decaying rate with epicentral distance. To test these methods, a receiver operating characteristic plot was produced by comparing the rate maps to observed $m > 6.0$ earthquakes in California since 1980. We found that the nearest-neighbor mapping produced poor forecasts, while the ETAS power-law method produced rate maps that agreed reasonably well with observations.

Key words: Earthquake simulators, ETAS, Earthquake forecasting, RELM.

1. Introduction

Of the various natural hazards, earthquakes pose a unique threat because they occur with little or no warning and shaking lasts no more than a few seconds to a few minutes. In contrast to natural hazards such as hurricanes and tropical storms (which can be tracked by monitoring meteorological data), tectonic stress and damage accumulation at depths where earthquakes occur cannot, at this time, be observed directly. Consequently, earthquake hazard management and mitigation efforts are typically long-term ($\Delta t \sim 30$ years) forecasts, often based on Poisson statistics of recurrence intervals (which may not be well defined) (Field 2007b), and typically focus on structural engineering, building codes, economic loss models, and disaster response (Glasscoe et al. 2014).

Recent events such as the 2014 $m = 6$ South Napa (California, USA), 2010 $m = 8.8$ Maule (Chile), 2010 $m = 7$ Haiti, the 2015 $m = 7.84$ Gorkha (Nepal) and the 2011 $m = 9$ Tohoku-oki (Japan) earthquakes remind us that on the one hand, efforts in this area have produced successes: investments in seismic hazard science and enforcement of construction codes save lives and mitigate fiscal losses. On the other hand, sobering statistics of fatalities and financial losses from these events, the Tohoku-oki event in particular (Kajitani et al. 2013), indicate that even in the best case scenario—for a nation well hardened against seismic hazard—there is much work to be done. In order to protect lives and property, it is necessary to significantly improve our understanding of the recurrence intervals and expected magnitudes (in short, the probabilities) of large earthquakes in populated areas.

For the purposes of planning, seismic mitigation in building construction, and the setting of insurance rates, probabilities of large and damaging earthquakes must be computed. For these reasons, earthquake forecasting methods have been developed using a variety of approaches. Here, we distinguish between an earthquake prediction and an earthquake

---
[1] Department of Physics, University of California, Davis, CA 95616-8677, USA. E-mail: jhnwilson@ucdavis.edu
[2] Department of Earth and Planetary Sciences, University of California, Davis, CA 95616-8605, USA.

forecast. We define a prediction to be a deterministic statement that can be verified by a single observation. A forecast, on the other hand, we define to be a statement of probability that requires multiple observations to establish a confidence level. To the best of our knowledge, every developed country in the world has a forecast that is used for planning, mitigation, and insurance. Most of these are primarily based on the computation of earthquake rates of occurrence, combined with an assumption of Poisson statistics for the time dependence (Anagnos and Kiremidjian 1988). Note that Poisson statistics characterize processes in which events are uncorrelated. In addition, Poisson statistics have no dependence on the detailed time history of prior events, since the only parameter entering the statistical model is the rate of event occurrence. For this reason, Poisson statistics are said to characterize processes having no "memory."

In the absence of direct measurements of precursory observables such as tectonic stress and damage accumulation at depth along fault surfaces, the majority of contemporary seismic hazard assessment is based on calculating the expected recurrence intervals between large (i.e., $m > 7$) earthquakes for a given region or on a particular fault segment. These calculations are based on observed seismicity, historical accounts, and paleo-seismic studies of ancient earthquakes (Rundle et al. 2011; Field et al. 2009, 2014; Nanjo et al. 2006; Nanjo 2010; Petersen et al. 2014). Of these data, historical accounts are restricted in time and almost certainly incomplete. Paleoseismic catalogs are extremely limited by the various challenges of measuring events, and so too are incomplete and otherwise problematic sources for high resolution data (Parsons 2008). The best currently available data, based on seismic observations, provide times, locations, and magnitudes of recent earthquakes. Unfortunately, these catalogs are only complete for the past few decades. GPS data (http://earthquake.usgs.gov/monitoring/gps/) provide very accurate measurements of surface deformation with excellent temporal resolution, but spatial resolution is limited, and again, data are only available from the early 1990s forward.

In contrast, the recurrence intervals for damaging earthquakes can easily span hundreds or thousands of years. In California, USA for example, according to the advanced national seismic systems (ANSS) search engine (http://www.anss.org), only six earthquakes with $m > 7$ have occurred between 1915 and the time of writing this manuscript (2015): Kern County (July 21, 1952, $m = 7.3$), Landers (June 28, 1992, $m = 7.3$), Hector Mine (October 16, 1999, $m = 7.1$), El Mayor-Cucapah (April 10, 2010, $m = 7.2$), and those in Santa Barbara County (November 4, 1927, $m = 7.3$) and Cape Mendocino (January 22, 1923 and November 8, 1980, both $m = 7.2$). Data to estimate the expected recurrence interval for earthquakes are insufficient unless about ten recurrences are observed (Ward and Goes 1993).

One approach to mitigating this problem of limited data that continue to show improving promise is the use of physics-based earthquake simulators. Simulators can (relatively) quickly calculate earthquake histories over tens of thousands or even millions of years. Given these long simulated catalogs of earthquake sequences, rates and probabilities of large, hazardous earthquakes can be calculated with improved statistical significance and data can be analyzed to isolate specific rupture sources (Yoder et al. 2015b; Schultz et al. 2015; Sachs et al. 2012a). In general, earthquake simulators facilitate studies that are difficult or impossible to perform, with confidence, based on currently available observed seismicity.

Tullis et al. (2012) conducted a detailed comparison of four leading earthquake simulators. In addition to a detailed discussion of the specific physics, stress propagation, and computational frameworks employed by each simulator, the authors compared recurrence interval statistics and various scaling behaviors for these simulators based on extended catalogs provided by each simulator team. Building on this aim to better understand the roles that earthquake simulators will play in 21st century seismic hazard assessment, in this paper we compare the consistency of the simulator outputs to the historical record. Specifically, we develop a method of comparing simulated catalogs to observed seismicity rather than attempting to explicitly determine which simulator produces the most realistic catalog based on its agreement with the past 30 years of observed earthquake activity.

Unfortunately, the only spatial information typically included in observational catalogs is the point-like hypocenter location. To facilitate direct comparison between simulated catalogs (in which all earthquakes occur on well-defined faults) and observed earthquakes (in which epicenters typically occur near but often not precisely on known faults), the effect of large earthquake rupture spatial extent must be reintroduced. To this end, we introduce two mapping algorithms. In the first, we use a simple nearest-neighbor type method to map observed earthquake epicenters to specific locations on model fault segments. In the second, we develop a method based on Epidemic-type aftershock (ETAS) and aftershock scaling (Yoder et al. 2015a) to distribute simulated seismicity continuously over the entire study region at a decaying rate with respect to epicentral distance.

We use a receiver operating characteristic (ROC) metric (Molchan 1997; Rundle et al. 2011; Jolliffe and Stephenson 2003) to test these methods against the locations of observed $m > 6.0$ earthquakes since 1980. While the nearest-neighbor mapping produces very little gain over a random forecast, the ETAS method generates rate maps that successfully identify regions of high seismic activity.

2. Background

Within the context of this manuscript, the "earthquake simulators" studied are numerical models that include:

1. Realistic and (relatively) detailed fault geometries.
2. A mechanism of stress-loading the fault system through back-slip that is consistent with observed slip rates and so that the time-averaged earthquake rate is similar to the observed rate.
3. Interactions between faults mediated by elastic or viscoelastic stress transfer.
4. Friction laws motivated by laboratory or field observations.

Within the constraints of this definition, earthquake simulators can vary significantly in the physics, stress propagation, and computational models they employ. Given the complexity of earthquake systems and the large computational scope of the problem (Rundle et al. 2003), in general, it is not necessarily clear how to "correctly," (and practically) emulate earthquake sequences on large scales such as the entirety of the state of California. That is to say, it is not obvious what physics must be simulated explicitly using partial differential equations (PDEs)—or some other rigorous but computationally expensive approach—and which behaviors can be parameterized. In practice, simulators employ various approximations, simplified geometries, and other "short-cuts" to make the problem computationally tractable. The question then becomes: what are the salient physics and geometries necessary to accurately calculate earthquake probabilities in California?

In order to address this questions, Tullis et al. (2012) conducted a comparison of four leading earthquake simulators: AllCal (Ward 2012), Virtual Quake (known at the time as Virtual California) (Sachs et al. 2012c), RSQSim (Richards-Dinger and Dieterich 2012), and ViscoSim (Pollitz 2012). In addition to a detailed discussion of the specific physics, stress propagation, and computational frameworks employed by each simulator, simulator teams provided an extended (simulated) catalog of earthquakes corresponding to thirty thousand years of seismic activity. To facilitate meaningful comparison between the output catalogs, each simulator used "allcal2," a fault model based on the UCERF2 geometry and slip rates (Field et al. 2009). allcal2 is available for download at http://scec.usc.edu/research/eqsims/documentation.html. Additionally, the output catalogs were of a standard format, providing simulated event hypocenter location, magnitude, time, and rupture extent. Given these catalogs, Tullis et al. (2012) compared various statistical and scaling behaviors between the simulators, including frequency-magnitude (Gutenberg-Richter or GR), rupture length, and average slip scaling relations as well as the probability distributions of recurrence intervals for large earthquakes. Slight differences were observed in simulator performance: Generally, AllCal and RSQSim produced similar frequency-magnitude relations which matched the expected GR power law, while Virtual California produced fewer small events and ViscoSim produced more. It was speculated that these differences resulted from how each simulator treated the weakening of a fault during rupture. Due to RSQSim's modeling of rate and state friction, it was the only simulator to demonstrate aftershocks following a large simulated event. Despite these differences, Tullis et al. (2012) concluded that the simulators' results were in general agreement with each other and observations.

In this paper, picking up where Tullis et al. (2012) left off, we compare the spatial distributions of these simulated earthquake catalogs not to one another, but to the rates and locations of real, observed earthquakes. In order to quantify the agreement between simulated and observed catalogs, we implement a testing framework based on the Regional Earthquake Likelihood Model (RELM) study (Schorlemmer and Gerstenberger 2007; Schorlemmer et al. 2007; Lee et al. 2011). In the RELM exercise, a region surrounding the state of California (defined by Table 1) was divided into $0.1° \times 0.1°$ spatial bins. Forecasting teams then assigned a probability that a large earthquake would occur in each bin. Each forecast was then evaluated by comparing the number of real earthquakes that occurred in high probability forecast bins.

In order to extend methodology similar to that of the RELM study to the outputs of physics-based numerical earthquake simulators, methods for mapping simulated on-fault seismicity to off-fault regions where real earthquakes are observed must first be developed. Such approaches are not unique; we wish to show that there exists some method by which the spatial distributions of synthetic and observed earthquakes become reasonably well correlated. We present two methods, one in which observed seismicity is mapped to the nearest faults, and one in which simulated seismicity is smeared over the full test region via a power law derived from ETAS. We then verify the resulting forecasts by employing statistical metrics borrowed largely from the meteorological sciences (Jolliffe and Stephenson 2003; Rundle et al. 2011) and consistent with standards established as part of the Collaboratory for the Study of Earthquake Predictability (CSEP) (Zechar et al. 2010).

3. Methods

3.1. Modeling Off-Fault Seismicity

The RELM study divided California into $0.1° \times 0.1°$ bins and scored forecasts based only on the earthquake epicenters. The epicentral location of an earthquake is often the only spatial information provided in earthquake catalogs, including the ANSS catalog (used here). While this is sufficient for such statistical analyses as frequency-magnitude relations, it is inadequate for the spatial verification of earthquake simulators, and ignores the finite rupture extent of large, $m > 6$ earthquakes.

We begin our comparison by pulling epicentral location data from the thirty thousand year simulator catalogs used in the Tullis et al. (2012) comparison study. In both of the following methods, normalized, binned spatial distributions were created from these long-term catalogs, which are assumed to be the time-averaged expected distributions for each simulator.

3.1.1 Event-Shifting

An initial, straight forward, solution to this problem is to simply map observed earthquakes to their nearest fault element in the allcal2 fault model. This approach is based on a fractal, branching model of fault systems (Sachs et al. 2012b) and the assumption

Table 1

Vertex latitude, longitude polygon defining the test region around the state of California for both this paper and the RELM study (Schorlemmer and Gerstenberger 2007)

Vertices of testing region	
Latitude	Longitude
43.0	−125.2
43.0	−119.0
39.4	−119.0
35.7	−114.0
34.3	−113.1
32.9	−113.5
32.2	−113.6
31.7	−114.5
31.5	−117.1
31.9	−117.9
32.8	−118.4
33.7	−121.0
34.2	−121.6
37.7	−123.8
40.2	−125.4
40.5	−125.4

that (1) events occurring near a major fault are likely occurring on a branch of that fault, and (2) the salient, large-scale behavior of a fault system can be approximated by mapping moment release events on these branches back to the fault.

In assigning the observed earthquakes to locations along the allcal2 fault model, most of the RELM test region was empty of simulated or observed earthquakes. We therefore restricted the testing region for the ROC skill score (see Sect. 3.2) to the spatial bins which contained modeled faults. No further redistribution of seismic activity was performed; the bins containing simulated and observed epicenters alone were compared. A normalized spatial histogram in

log scale for each simulator is shown in Fig. 1. New, shifted locations for observed earthquakes are shown in Fig. 2, compared with their original locations.

3.1.2 Power-Law Smoothing

In order to better capture the spatial extent of simulated earthquakes and allow large simulated events to better forecast nearby seismicity, we apply a smoothing function derived from the epidemic-type aftershock (ETAS) model (Ogata 1989) to each simulated event. This distributes the "influence" of each earthquake continuously across all of space, decaying with distance from the epicenter, and we are

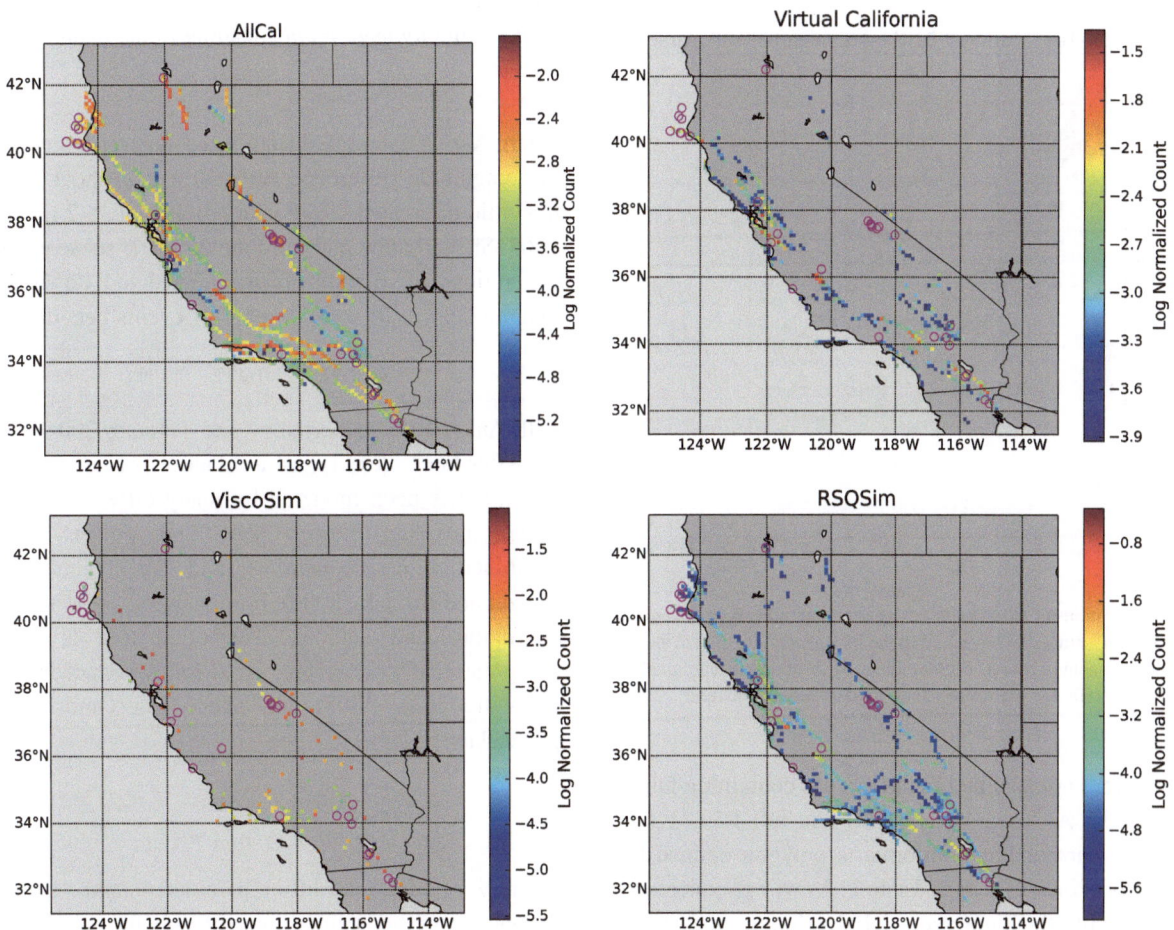

Figure 1

Log spatial histogram for AllCal (*upper left*), Virtual California (*upper right*), ViscoSim (*lower left*), and RSQSim (*lower right*). Maps produced by counting normalized number of simulated event epicenters in each 0.1° × 0.1° bin. *Color scale* is different for each to show detail. Observed event epicenters, shifted to nearest fault model element, shown as *magenta circles*

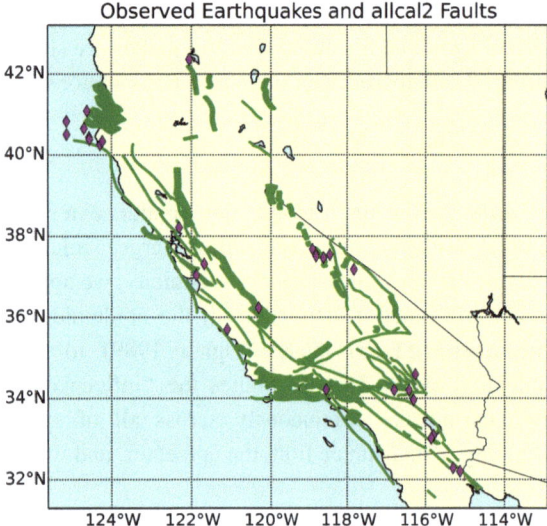

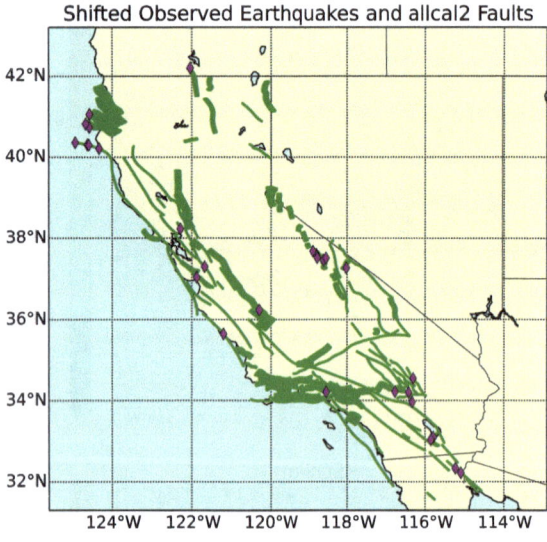

Figure 2

Top Locations of observed $m \geq 6.0$ earthquake epicenters (*magenta diamonds*) since 1980 overlaid on the locations of the allcal2 model fault elements (*green*). *Bottom* Same, with observed earthquakes shifted to the nearest fault model element location

magnitude in combination with scaling behaviors and a model based on the earthquake's finite rupture length (and area). The result is a time-independent map of aftershock rate densities that covers the entire test region.

In this model, Gutenberg-Richter scaling (Gutenberg and Richter 1954), describing the number N of earthquakes with magnitude greater than some catalog-completion magnitude m_c in a given region and time span:

$$N(>m_c) = 10^{a-bm_c}, \qquad (1)$$

where a represents a total seismicity rate and b defines the magnitude scaling between large and small quakes. In the context of aftershocks associated with a mainshock of magnitude m^*, we can rewrite total seismicity as $a = bm^*$, leading to the expression

$$N(>m_c)_{\text{Omori}} = 10^{b(m^* - \Delta m^* - m_c)}, \qquad (2)$$

where Δm^* is a modified Båth's constant representing the magnitude difference between a mainshock and it's child aftershocks, and is approximately 1.2 (Båth 1965; Shcherbakov and Turcotte 2004). If subsequent generations of aftershocks are considered down to magnitude m_c, the total number of events becomes

$$N(>m_c)_{\text{Omori}} = 10^{b(m^* - \Delta m - m_c)} \qquad (3)$$

with $\Delta m \approx 1.0$ (Shcherbakov and Turcotte 2004)

Aftershock density as a function of epicentral distance has been modeled by many different functions, including power laws and power laws combined with exponentials, and other functions. (Felzer and Brodsky 2006). For this work, we desire a continuous and smooth distribution that is also computationally tractable. We follow precedence (Turcotte et al. 2007) and choose an Omori-like power law

$$N' = \frac{dN}{dr} = \frac{1}{\chi(r_0 + r)^q}, \qquad (4)$$

where χ, r_0, and q are fitting parameters, and r is the distance from the "center," presumably the epicenter, of the event. The scaling exponent q is determined observationally (see Felzer and Brodsky 2006; Yoder et al. 2015a, and references therein). The parameters χ and r_0 (and of course, the maximum or "initial"

therefore not confined to the regions containing faults (Fig. 3). As such, we consider the full RELM test region surrounding California, a polygon defined by the coordinates listed in Table 1. As we are concerned only with the time-averaged distribution of earthquakes, we employ only the spatial component of an ETAS variant developed by Yoder et al. (2015a) in which the spatial distribution of an earthquake's aftershocks are determined from the earthquake's

linear density $N_{\max} = 1/\chi r_0^q)$ are determined by imposing near and far-field constraints.

Following Yoder et al. (2015a) and Shcherbakov et al. (2004, 2006), we first impose the normalization condition

$$\int_0^\infty N' \mathrm{d}r = N Omori, \qquad (5)$$

which leads to the expression

$$\chi = \frac{r_0^{1-q}}{N Omori(q-1)}. \qquad (6)$$

Yoder et al. (2015a) introduced a near-field constraint in which an earthquake rupture is modeled as a self-similar series of aftershocks, packed together, effectively "filling" the mainshock rupture area. Combining GR scaling, the number Nas of aftershock earthquakes filling such a rupture with fractal dimension $1 < D < 2$ is

$$\log(Nas) = \left(\frac{2}{2+D} \log(1+D/2) \right. \\ \left. + \frac{2D}{2+D} \left(\frac{m^* - \Delta m - m_c}{2} \right) \right). \qquad (7)$$

Combining Eqs. 4 and 7, setting $r = 0$, and dividing by half the rupture length, the maximum linear density of aftershocks is

$$N'(r=0) = \frac{2}{L_r \chi r_0^q}. \qquad (8)$$

Equations 6–8 can be combined to find an expression for r_0:

$$r_0 = \frac{L_r}{2} \frac{N Omori(q-1)}{Nas}, \qquad (9)$$

and in practice, the scaling relation for rupture length L_r can be substituted as necessary,

$$\log(L_r) = \frac{m}{2} - \Delta\lambda. \qquad (10)$$

For our analysis, we consider only events of magnitude greater than 6.0 ($m_c = 6.0$). $\Delta\lambda$ is 1.75 (Kagan 2002; Yoder et al. 2011), and q is taken to be 1.5 (Ogata and Zhuang 2006).

In order to align these seismicity halos with the underlying fault structure, an elliptical transformation is applied to the circular Omori distribution. While the strike angle could be obtained directly from the UCERF2 fault model, we wish to develop a method applicable to event catalogs for which the underlying fault structure is not known. Yoder et al. (2015a) performed a linear Chi-squared analysis on events occurring within five rupture lengths of the epicenter. We improve upon this method by conducting principal component analysis (PCA) (Jolliffe 2014) of the same set of events. A covariance matrix is computed for the locations of the events. For the eigenvector $\vec{v}_1$ associated with the largest principle eigenvalue λ_1 of the covariance matrix, the fault strike is given by $\frac{\pi}{2} - \arctan(v_{1y}/v_{1x})$. For the aspect ratio of the ellipse, we defer to Kagan (2002) and set the semiaxis ratio to 2.0.

r_{ij} is the aftershock rate from earthquake i assigned to bin j of the RELM test region. To find rate r_j for the entire simulation catalog, we sum together the contribution from each simulated earthquake:

$$r_j = \sum_i r_{ij} \qquad (11)$$

r_j is a measure of the probability of occurrence of an earthquake in bin j.

3.2. Retrospective Forecast Verification and the ROC Area Skill Test

We ultimately desire to compare the produced rate maps to the ANSS catalog of observed $m \geq 6$ earthquakes in the test region since 1980. Figure 2 shows the locations of these epicenters as well as the locations of the allcal2 fault elements. Because earthquakes are binary events (they either occur or they do not), a method must be used to translate a map of rates (drawn from a continuum) to a yes/no binary prediction. To this end, we apply the ROC test of binary forecasts (Molchan 1997; Rundle et al. 2011; Jolliffe and Stephenson 2003). We sort bins r_j for a particular simulator by rate. This rank ordering results in a list of rate values r_k, where k indexes the kth highest rate bin. It should be noted that, as bins are compared only within a single simulator, each simulator's rate map need not be normalized to the same total seismicity rate; only the relative rates between bins for a single simulator affect bin ordering.

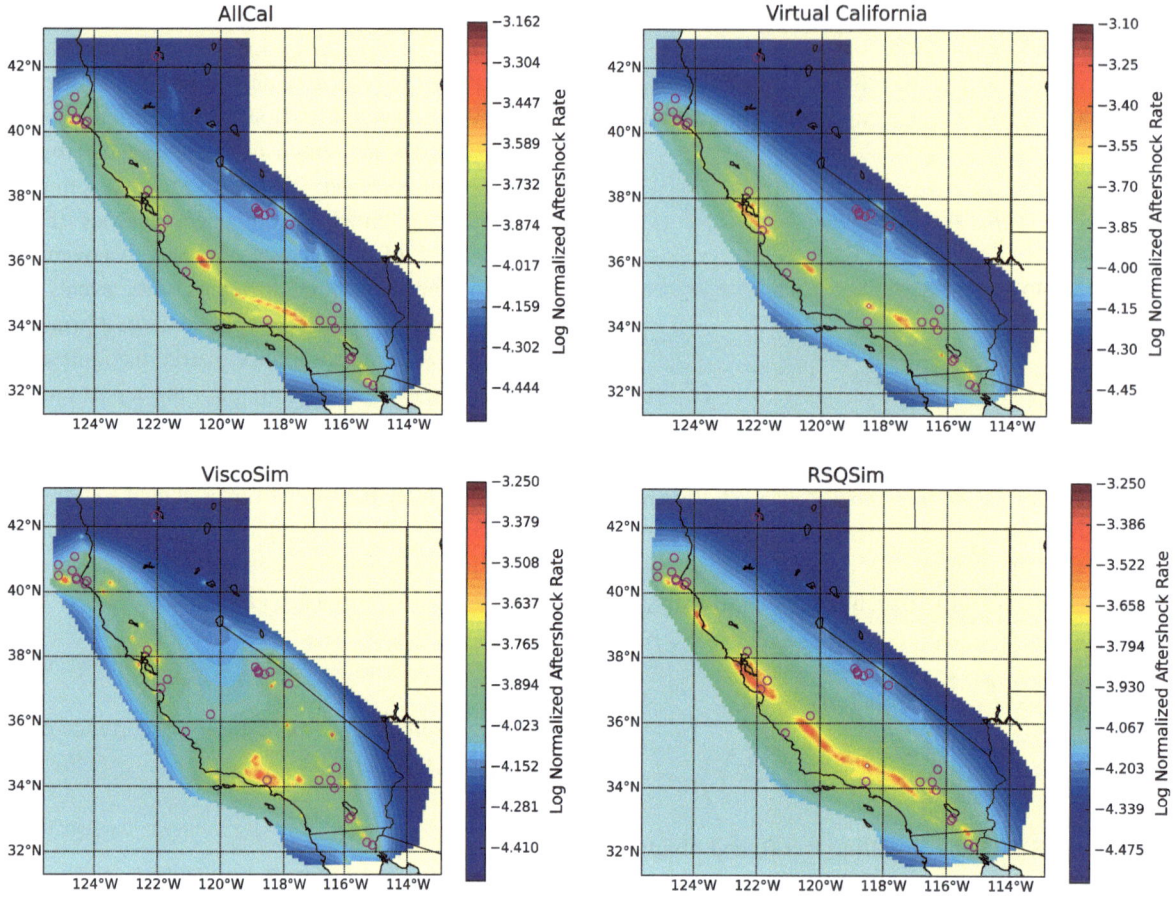

Figure 3
Time-independent log rate maps for AllCal (*upper left*), Virtual California (*upper right*), ViscoSim (*lower left*), and RSQSim (*lower right*). Maps produced by applying a spatial ETAS field for each simulated event. *Color scale* is different for each to show detail. Observed event epicenters shown as *magenta circles*

To partition the list of rates into "yes" and "no" forecasts, we define a rate threshold T such that all bins for which $r_k > T$ are considered to be forecasts of "yes," while those where $r_k < T$ are considered as forecasts of "no." The relationship between the forecast bins and observations can be grouped into four contingencies: $\mathbb{A}$-True Positives, $\mathbb{B}$-False Positives, $\mathbb{C}$-False Negatives, and $\mathbb{D}$-True Negatives (Table 2).

We define the Hit Rate as follows: $H = \mathbb{A}/(\mathbb{A} + \mathbb{C})$. Similarly, False Alarm Rate is defined as follows: $F = \mathbb{B}/(\mathbb{B} + \mathbb{D})$. The total number of bins is $N = \mathbb{A} + \mathbb{B} + \mathbb{C} + \mathbb{D}$.

We initially set T so that only $r_1 > T$, resulting in only a single bin considered as a "yes" forecast. We then find the values $\mathbb{A}$, $\mathbb{B}$, $\mathbb{C}$, and $\mathbb{D}$ by comparing the

Table 2

ROC contingencies

	Observed	
	Yes	No
Forecast		
Yes	$\mathbb{A}$	$\mathbb{B}$
No	$\mathbb{C}$	$\mathbb{D}$

forecast against observed earthquakes, and Hit Rate and False Alarm Rate are calculated and recorded. We then decrease T such that $r_2 > T$ to include the two highest-rate bins, repeat the calculations, reduce T to include three bins, etc, until T is reduced such that $T < r_N$ and all bins are considered to be "yes" forecasts.

As an example of the technique, consider a hypothetical test region of 50 spatial bins and 5 earthquakes all occupying different bins. In our first pass, our threshold T is set so that only the highest-ranked bin is a "yes" forecast, and we find that an earthquake is found in that bin. $\mathbb{A}$ is the number of successful positive-forecast bins, so $\mathbb{A} = 1$. There are no bins that were positive forecasts and had no earthquakes, so $\mathbb{B} = 0$. $\mathbb{C}$ counts the number of bins containing earthquakes that were negative forecasts: $\mathbb{C} = 4$. $\mathbb{D}$ counts the remaining bins that were negative forecasts and contain no earthquakes: $\mathbb{D} = 45$. The hit rate would be $H = 1/(1 + 4) = 1/5$, and false alarm rate would be $F = 0/(0 + 45) = 0$. Imagine that after we lower the threshold to include the next highest-ranked bin as a positive forecast, we find that no additional earthquakes fall there. Therefore, $\mathbb{A} = 1$,

$\mathbb{B} = 1, \mathbb{C} = 4, \mathbb{D} = 44$. Hit Rate is again $H = 1/5$, and now False Alarm Rate is 1/45. We would repeat this process until all 50 bins are considered positive forecasts, with $H = 1$ and $F = 1$.

We then create a plot of $H(T)$ against $F(T)$ for each value of T. The area under this curve is related to the quality of the forecast; a forecast in which all earthquakes fell into the highest-ranked bins (a perfect forecast) would have a vertical rise in Hits initially, and a horizontal increase of False Alarms at the end, yielding an area of 1. A forecast in which all earthquakes fell into the lowest-ranked bins would have a horizontal increase followed by a vertical climb, resulting in an area of 0. A line of slope 1 (with area under the curve of 0.5) represents the theoretical results of a uniformly random rate map. In order to express the quality of the forecast relative to this random baseline,

Table 3

ANSS catalog of m > 6.0 earthquakes for the test region from 1980 to 2015

Observed earthquakes used in ROC analysis

Year	Month	Day	Hour	Minute	Latitude	Longitude	Magnitude
1980	5	25	16	33	37.59	−118.83	6.1
1980	5	25	16	49	37.68	−118.90	6.0
1980	5	25	19	44	37.55	−118.81	6.1
1980	5	27	14	50	37.50	−118.81	6.2
1980	6	9	3	28	32.20	−115.12	6.3
1980	11	8	10	27	41.08	−124.62	7.2
1983	5	2	23	42	36.23	−120.31	6.7
1984	4	24	21	15	37.31	−121.68	6.2
1984	9	10	3	14	40.50	−125.13	6.6
1984	11	23	18	8	37.46	−118.61	6.1
1986	7	21	14	42	37.54	−118.44	6.4
1987	11	24	1	54	33.09	−115.79	6.2
1987	11	24	13	15	33.02	−115.85	6.6
1989	10	18	0	4	37.04	−121.88	7.0
1991	8	17	19	29	40.25	−124.29	6.0
1992	4	23	4	50	33.96	−116.32	6.1
1992	4	25	18	6	40.34	−124.23	6.69
1992	4	26	7	41	40.43	−124.57	6.45
1992	4	26	11	18	40.38	−124.56	6.57
1992	6	28	11	57	34.2	−116.44	7.3
1992	6	28	15	5	34.20	−116.83	6.3
1993	5	17	23	20	37.18	−117.83	6.1
1993	9	21	5	45	42.36	−122.06	6.0
1994	1	17	12	30	34.21	−118.54	6.7
1999	10	16	9	46	34.60	−116.27	7.1
2003	12	22	19	15	35.70	−121.10	6.5
2010	1	10	0	27	40.65	−124.69	6.5
2010	4	4	22	40	32.29	−115.30	7.2
2014	3	10	5	18	40.83	−125.13	6.8
2014	8	24	10	20	38.22	−122.31	6.02

we subtract 0.5 from the area under the ROC curves to yield the Area Skill Score. Positive values of this score represent forecasts that were more successful than a random forecast, negative scores for those less successful than random. For our analysis, all $m \geq 6$ observed earthquakes in the ANSS catalog for the test region since 1980 were used (Table 3).

To estimate the uncertainty in the ROC analysis, we use a standard bootstrap method (Chernick 2011; Davison and Hinkley 1997). Taking the bootstrap approach to the thirty observed earthquakes, we create 1000 sample earthquake lists by sampling earthquakes with replacement. From the catalog of thirty observed earthquakes, a new random set of thirty earthquakes is selected, allowing events to be selected more than once. For example, at a probability of $\left(\frac{1}{30}\right)^{30}$, a bootstrap catalog could potentially be a set of thirty copies of the same earthquake. We then compute the ROC curve and area skill score for each of the thousand bootstrap catalogs (Fig. 5).

Additionally, we compare the ROC score for observed earthquakes against those of 1000 random synthetic catalogs drawn from the ETAS rate field (Fig. 6). Thirty epicenter locations are randomly selected using the rate field as a spatial probability distribution. The ROC score of the ETAS field is calculated against each of these catalogs of random

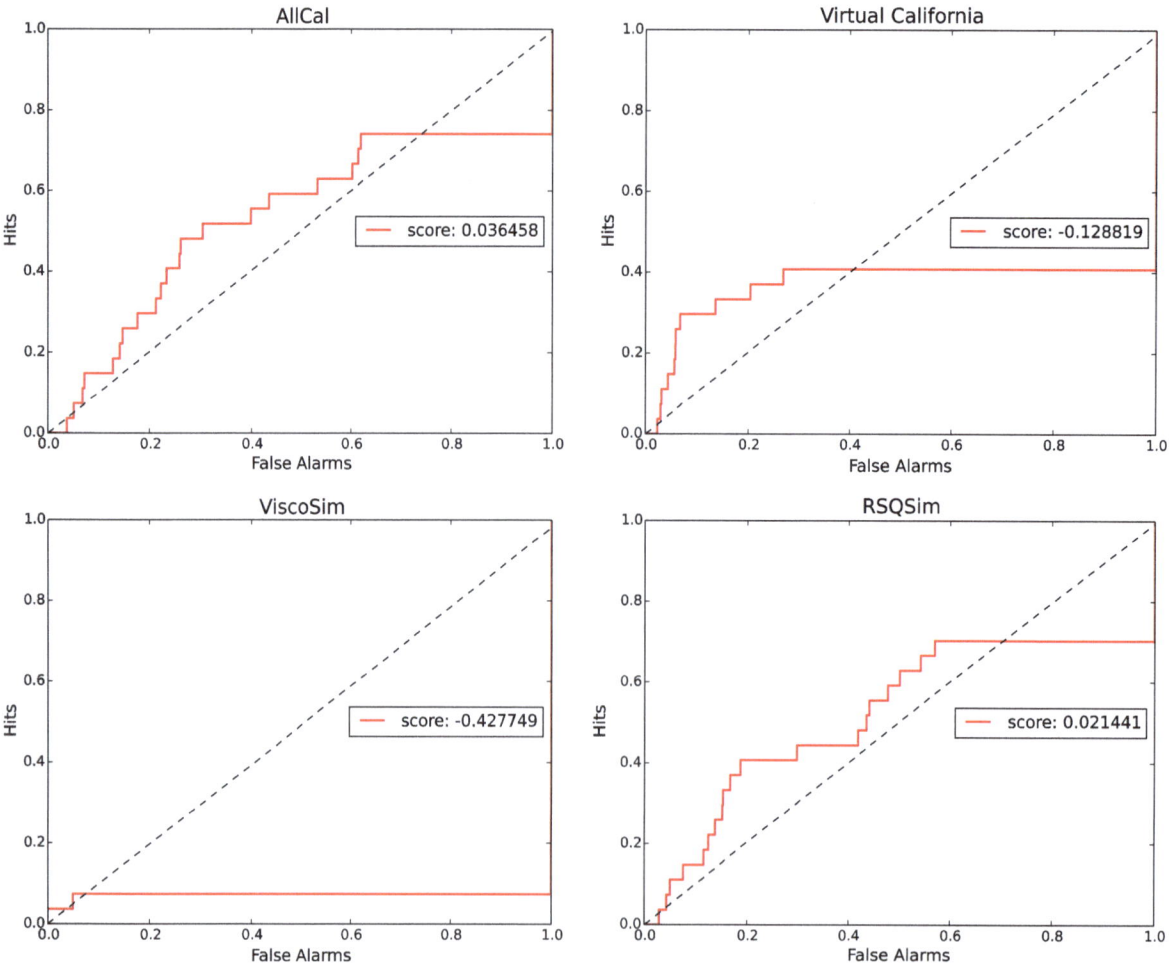

Figure 4
ROC plots for each simulator for event-shifted observed earthquake catalog. Score given as area under curve and above the diagonal, with a perfect score of 0.5. Higher scores indicate better forecasts

locations in the same manner as for the observed catalog. We calculate the central 95 % for the random catalogs' ROC scores.

4. Discussion

The nearest neighbor algorithm produced results of little use for verification. Although this method facilitates direct comparisons of observed and synthetic (simulator-based) catalogs, its practical value is limited; after observed earthquake epicenters were shifted to the nearest modeled fault, they frequently remained outside of the exact spatial bins populated by simulated

events. As a result, analyses produced only marginal gains over random selection (see Fig. 4). This is apparently due to event clustering that takes place in the simulators to different degrees. In order for this method to yield useful results for verification against a limited observational catalog, either simulated or observed seismicity would need to be further redistributed, though in doing so, we could bypass the process of relocating observed events altogether.

The ETAS power-law smoothing method produces continuous maps which are reasonably well correlated with observed earthquakes (Fig. 5). The initial rise in ROC curves for all simulators indicates successful correlation between high-rate regions of

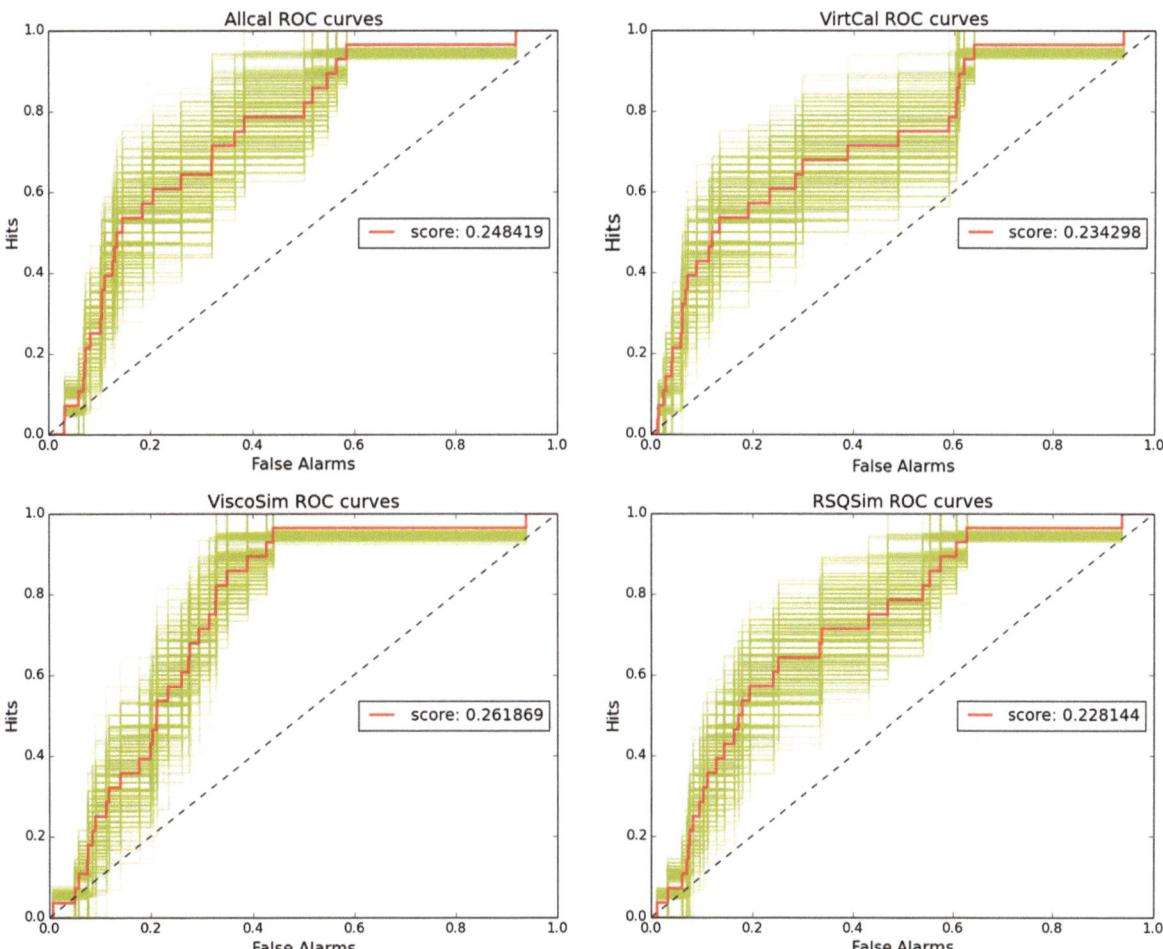

Figure 5
ROC plots for each simulator for observed earthquake catalog (*red*) and bootstrap catalogs (*yellow*). Score given as area under curve and above the diagonal, with a perfect score of 0.5. Higher scores indicate better forecasts. Spread in the bootstrap ROC curves indicates the uncertainty in the observed catalog's score

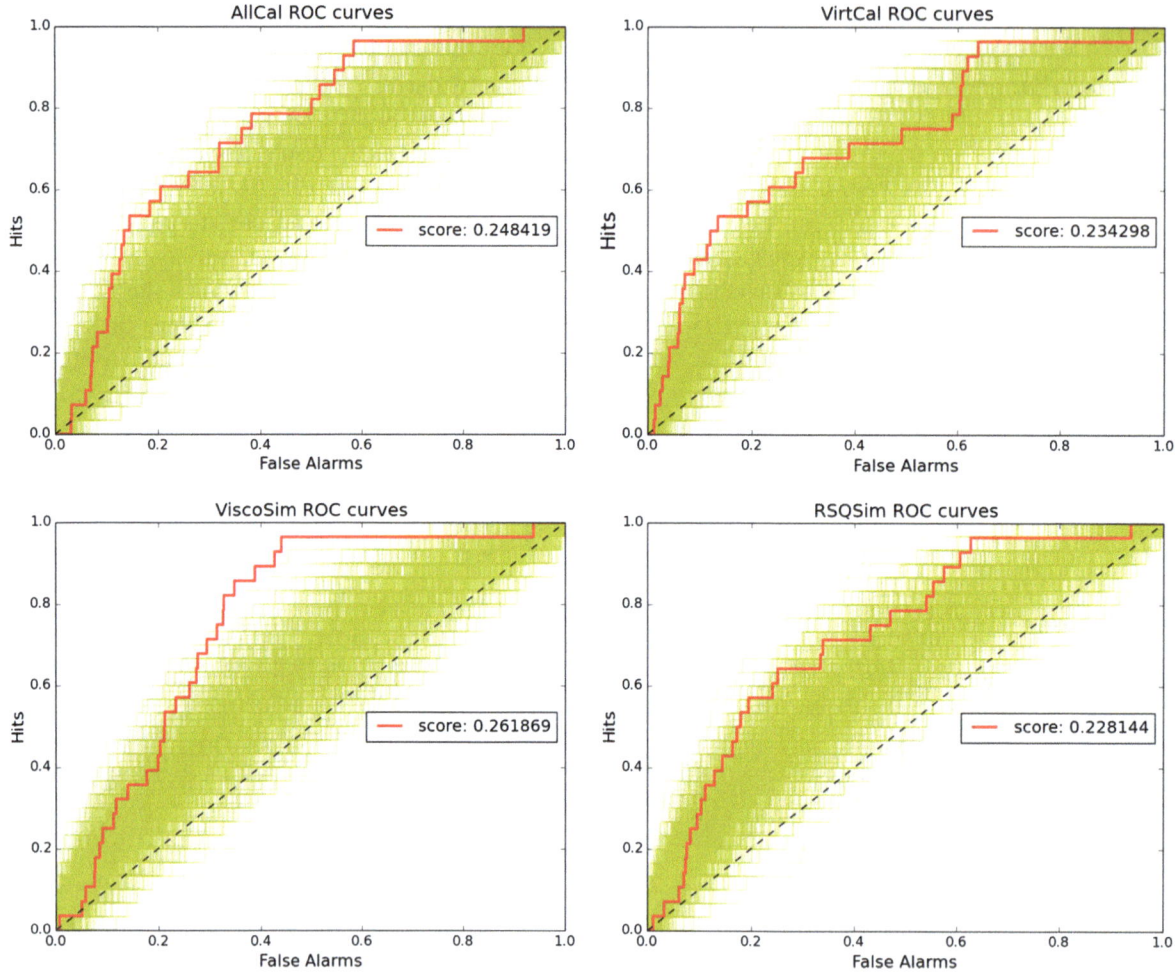

Figure 6
ROC plots of random catalogs generated from ETAS map (*yellow*) and ROC plot for observed catalog (*red*)

the map and observed earthquake locations. The horizontal portion at the end of each curve is expected; this corresponds to the regions in which no seismicity occurred nor was forecast. These regions make a large contribution to the area under each curve. Changing the area of such regions included in the test polygon would modify how much of this "padding" is present in the ROC curves, changing the percent difference between simulator results, but not changing the simulators' relative ranking. The less steep portion in the middle of the AllCal, Virtual California, and RSQSim plots correspond to regions where the ETAS rate map held moderately high values, but where observed seismicity was sparse—for example, the California Central Valley (Fig. 6).

The comparison of the ROC scores for catalogs of event locations randomly generated from the ETAS rate map distribution to the ROC score for the observed earthquakes yields information indicating possible improvements to the model. We find the observed score lies close to or outside the upper 2σ of the distribution of random catalog scores for each simulator. AllCal and ViscoSim both lie outside, while Virtual California and RSQSim both lie just inside of 2σ (Fig. 7). This, promisingly, indicates that the highest-rate locations on the rate maps are correlated with locations of real earthquake occurrence, but the large gap between average random catalog score and observed score indicates that the simulator catalogs were overly smoothed. That

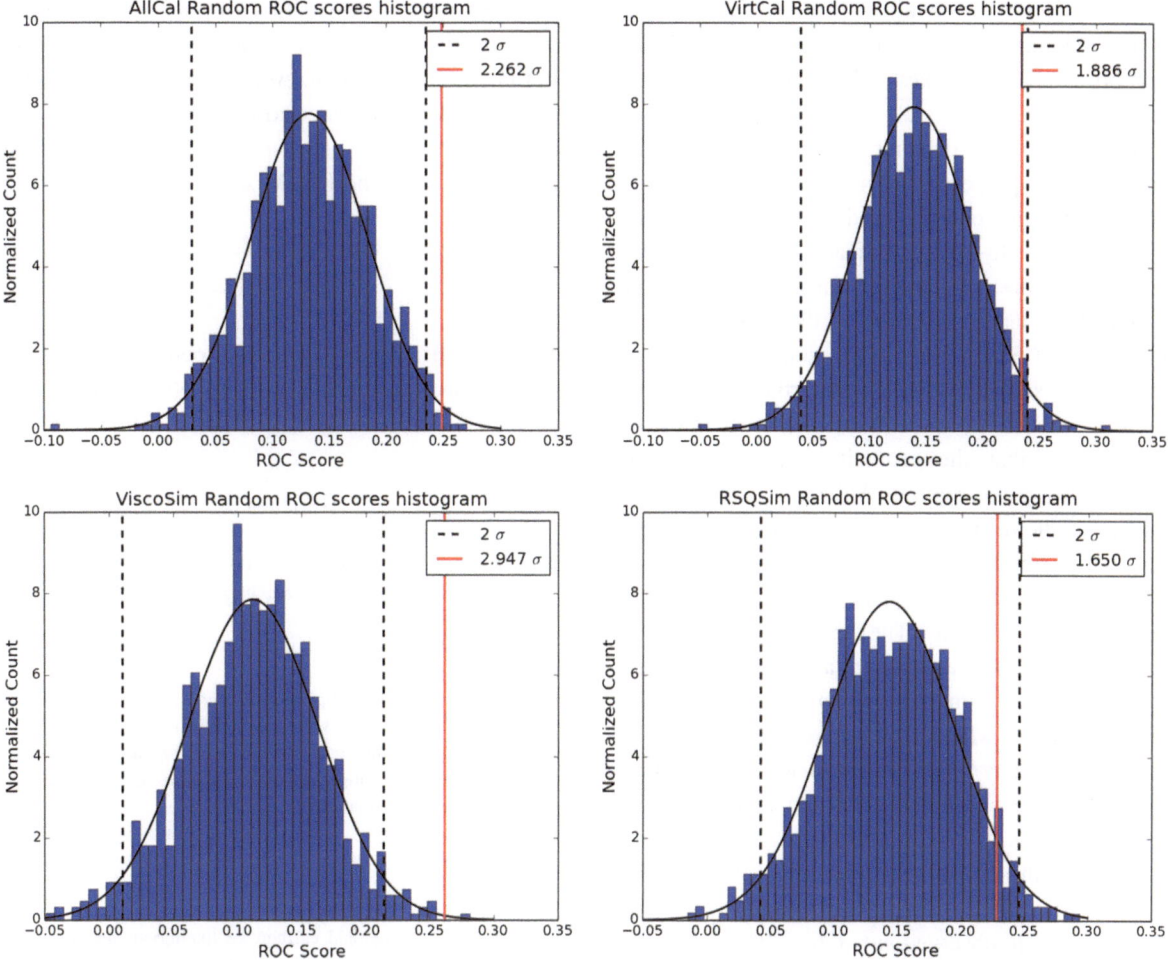

Figure 7
Histograms of ROC scores of random catalogs generated from ETAS rate map. ROC score for observed earthquakes marked by *red line*. The score for the observed catalog performs near the 2σ mark for each simulator, indicating that observed earthquakes fall in high rate bins from the ETAS method

is, much of the "probability" was distributed too evenly.

The exact ROC scores for each simulator, as well as the values of their rate maps, are sensitive to the parameters of the ETAS distribution used. The limitations in this method discussed above could be improved by including a magnitude dependence in the elliptical aspect ratio of the aftershock distributions. Increasing the aspect ratio for very large simulated events would reduce the width of the high rate value halo surrounding the San Andreas fault. An additional parameter for

further consideration is the power law exponent q, the value of which could be refined with further aftershock observations.

5. Conclusion

Numerical earthquake simulators provide a means to study phenomenon that have natural time scales much longer than the recorded history of complete seismic data. In order to ensure accurate forecasts and hazard assessment, earthquake simulators must be verified against observation. Tullis et al. (2012) studied the agreement between four such simulators with observed earthquakes and each other for several statistical tests.

Hazard analysis depends heavily on spatial distribution; we propose applying spatial verification methods similar to those used in the RELM study (Field 2007a) to the earthquake simulators. One challenge, addressed here, is that simulators output earthquakes located precisely on the fault model (in this case, UCERF2 Field et al. 2009). In order to compare these outputs with real earthquakes, which often occur off of known faults, a method must be used to redistribute the seismicity of the two data sets to the same spatial regions. Although there does not exist a unique method for achieving this, we present two such methods. The first is a simplistic approach, whereby observed earthquakes are shifted to the nearest fault element. The second method is a smoothing of the simulator catalog via power law decay borrowed from the epidemic-type aftershock (ETAS) model, which continuously distributes the catalog seismicity over the entire test region. This method produces continuous earthquake rate density maps that cover the entire test region.

To test these methods, the receiver operating characteristic (ROC) plot was employed, comparing the rate maps to observed $m > 6.0$ earthquakes since 1980. We found that the nearest-neighbor mapping produced poor forecasts, due to the nearest spatial bin for most earthquakes still lying outside of the high-rate bins for the simulators at the spatial resolution explored $(0.1° \times 0.1°)$. The ETAS power-law smoothing method produced rate maps that agreed with observations. These ETAS results were further analyzed by comparing them against catalogs of earthquake epicenters randomly generated from the ETAS rate map. The observed ROC scores for all simulators outperformed both uniformly distributed random catalogs as well as random catalogs derived from the rate maps themselves. These maps are therefore successful in causing observed and simulated earthquake spatial distributions to be reasonably well correlated, though the discrepancy between the mean ROC area skill score of random catalogs and that of the observed catalog beg further consideration. Future studies on the sensitivity of the random scores

to the ETAS rate distribution elliptical aspect ratio as well as the power law exponent q could further optimize this technique. Additionally, these distributions could be compared to existing estimates of the spatial distribution of seismicity in the test region, such as those of the RELM study.

Acknowledgments

JMW and JBR would like to acknowledge support for this research from NASA Grant NNX12A22G and SCEC/USC Grant USC32774854-NSF FFT.

REFERENCES

Anagnos, T., & Kiremidjian, A. S. (1988). A review of earthquake occurrence models for seismic hazard analysis. *Probabilistic Engineering Mechanics*, 3(1), 3–11.

Båth, M. (1965). Lateral inhomogeneities of the upper mantle. *Tectonophysics*, 2(6), 483–514.

Chernick, M. R. (2011). *Bootstrap methods: A guide for practitioners and researchers* (vol. 619). Hoboken, New jersey: Wiley.

Davison, A. C., & Hinkley, D. (1997). *Bootstrap methods and their applications*. Cambridge: Cambridge Series in Statistical and Probabilistic Mathematics.

Felzer, K. R., & Brodsky, E. E. (2006). Decay of aftershock density with distance indicates triggering by dynamic stress. *Nature*, 441(7094), 735–738.

Field, E. H. (2007a). Overview of the working group for the development of regional earthquake likelihood models (relm). *Seismological Research Letters*, 78(1), 7–16.

Field, E. H. (2007b). A summary of previous working groups on california earthquake probabilities. *Bulletin of the Seismological Society of America*, 97(4), 1033–1053.

Field, E. H., Arrowsmith, R. J., Biasi, G. P., Bird, P., Dawson, T. E., Felzer, K. R., et al. (2014). Uniform california earthquake rupture forecast, version 3 (ucerf3)the time-independent model. *Bulletin of the Seismological Society of America*, 104(3), 1122–1180.

Field, E. H., Dawson, T. E., Felzer, K. R., Frankel, A. D., Gupta, V., Jordan, T. H., et al. (2009). Uniform california earthquake rupture forecast, version 2 (ucerf 2). *Bulletin of the Seismological Society of America*, 99(4), 2053–2107.

Glasscoe, M., Rosinski, A., Vaughan, D., & Morentz, J. (2014). Disaster response and decision support in partnership with the california earthquake clearinghouse. In *AGU Fall Meeting Abstracts* (vol. 1, p. 07).

Gutenberg, B. & Richter, C. (1954). *Seismicity of the earth and associated phenomena*. Princeton, New Jersey: Princeton University Press.

Jolliffe, I. (2014). *Principal Component Analysis*. Wiley StatsRef: Statistics Reference Online.

Jolliffe, I. T., & Stephenson, D. B. (2003). *Forecast verification: a practitioner's guide in atmospheric science*. Chichester, West Sussex, England: Wiley

Kagan, Y. Y. (2002). Aftershock zone scaling. *Bulletin of the Seismological Society of America, 92*(2), 641–655.

Kajitani, Y., Chang, S. E., & Tatano, H. (2013). Economic impacts of the 2011 tohoku-oki earthquake and tsunami. *Earthquake Spectra, 29*(s1), S457–S478.

Lee, T.-T., Turcotte, D. L., Holliday, J. R., Sachs, M. K., Rundle, J. B., Chen, C.-C., et al. (2011). Results of the regional earthquake likelihood (relm) test of earthquake forecasts in california. *Proceedings of the National Academy of Sciences, 108*(40), 16533–16538.

Molchan, G. M. (1997). Earthquake prediction as a decision-making problem. *Pure and Applied Geophysics, 149*(1), 233–247.

Nanjo, K., Holliday, J., Chen, C.-C., Rundle, J., & Turcotte, D. (2006). Application of a modified pattern informatics method to forecasting the locations of future large earthquakes in the central japan. *Tectonophysics, 424*(3), 351–366.

Nanjo, K. Z. (2010). Earthquake forecast models for italy based on the ri algorithm. *Annals of Geophysics, 53*(3), 117–127.

Ogata, Y. (1989). Statistical model for standard seismicity and detection of anomalies by residual analysis. *Tectonophysics, 169*(1), 159–174.

Ogata, Y., & Zhuang, J. (2006). Space-time etas models and an improved extension. *Tectonophysics, 413*(1), 13–23.

Parsons, T. (2008). Appendix c: Monte carlo method for determining earthquake recurrence parameters from short paleoseismic catalogs: Example calculations for california. *US Geological Survey Open File Report, 1437-C*, 32.

Petersen, M. D., Moschetti, M. P., Powers, P. M., Mueller, C. S., Haller, K. M., Frankel, A. D., et al. (2014). *Documentation for the 2014 update of the united states national seismic hazard maps*. Technical report, US Geological Survey.

Pollitz, F. F. (2012). Viscosim earthquake simulator. *Seismological Research Letters, 83*(6), 979–982.

Richards-Dinger, K., & Dieterich, J. H. (2012). Rsqsim earthquake simulator. *Seismological Research Letters, 83*(6), 983–990.

Rundle, J. B., Holliday, J. R., Yoder, M., Sachs, M. K., Donnellan, A., Turcotte, D. L., et al. (2011). Earthquake precursors: activation or quiescence? *Geophysical Journal International, 187*(1), 225–236.

Rundle, J. B., Turcotte D. L., Shcherbakov R., Klein W., & Sammis C. (2003). Statistical physics approach to understanding the multiscale dynamics of earthquake fault systems. *Reviews of Geophysics, 41*, 1019. doi:10.1029/2003RG000135.

Sachs, M., Turcotte, D. L., Holliday, J. R., & Rundle, J. (2012a). Forecasting earthquakes: The relm test. *Computing in Science & Engineering, 14*(5), 43–48.

Sachs, M., Yoder, M., Turcotte, D., Rundle, J., & Malamud, B. (2012b). Black swans, power laws, and dragon-kings: Earthquakes, volcanic eruptions, landslides, wildfires, floods, and soc models. *The European Physical Journal-Special Topics, 205*(1), 167–182.

Sachs, M. K., Heien, E. M., Turcotte, D. L., Yikilmaz, M. B., Rundle, J. B., & Kellogg, L. H. (2012c). Virtual california

earthquake simulator. *Seismological Research Letters, 83*(6), 973–978.

Schorlemmer, D., Gerstenberger, M., Wiemer, S., Jackson, D., & Rhoades, D. (2007). Earthquake likelihood model testing. *Seismological Research Letters, 78*(1), 17–29.

Schorlemmer, D., & Gerstenberger, M. C. (2007). Relm testing center. *Seismological Research Letters, 78*(1), 30–36.

Schultz, K. W., Sachs, M. K., Heien, E. M., Yoder, M. R., Rundle, J. B., Turcotte, D. L., & Donnellan, A. (2015). Virtual quake: Statistics, co-seismic deformations and gravity changes for driven earthquake fault systems. In *International Association of Geodesy Symposia* (pp. 1–9). doi:10.1007/1345_2015_134.

Shcherbakov, R., & Turcotte, D. L. (2004). A modified form of båth's law. *Bulletin of the Seismological Society of America, 94*(5), 1968–1975.

Shcherbakov, R., Turcotte D. L., & Rundle J. B. (2004). A generalized Omori's law for earthquake aftershock decay. *Geophysical Research Letters, 31*, L11613. doi:10.1029/2004GL019808.

Shcherbakov, R., Turcotte, D. L., & Rundle, J. B. (2006). Scaling properties of the parkfield aftershock sequence. *Bulletin of the Seismological Society of America, 96*(4B), S376–S384.

Tullis, T. E., Richards-Dinger, K., Barall, M., Dieterich, J. H., Field, E. H., Heien, E. M., et al. (2012). A comparison among observations and earthquake simulator results for the allcal2 california fault model. *Seismological Research Letters, 83*(6), 994–1006.

Turcotte, D. L., Holliday J. R., & Rundle J. B. (2007). BASS, an alternative to ETAS. *Geophysical Research Letters, 34*, L12303. doi:10.1029/2007GL029696.

Ward, S. N. (2012). Allcal earthquake simulator. *Seismological Research Letters, 83*(6), 964–972.

Ward, S. N., & Goes, S. D. (1993). How regularly do earthquakes recur? a synthetic seismicity model for the san andreas fault. *Geophysical Research Letters, 20*(19), 2131–2134.

Yoder, M. R., Rundle, J. B., & Glasscoe, M. T. (2015a). Near-field ETAS constraints and applications to seismic hazard assessment. *Pure and Applied Geophysics, 172*(8), 2277–2293.

Yoder, M. R., Schultz, K. W., Heien, E. M., Rundle, J. B., Turcotte, D. L., Parker, J. W., & Donnellan, A. (2015b). The Virtual Quake earthquake simulator: a simulation-based forecast of the El Mayor-Cucapah region and evidence of predictability in simulated earthquake sequences. *Geophysical Journal International, 203*(3), 1587–1604.

Yoder, M. R., Turcotte, D. L., & Rundle, J. (2011). *Record-breaking earthquake precursors*. PhD thesis.

Zechar, J. D., Schorlemmer, D., Liukis, M., Yu, J., Euchner, F., Maechling, P. J., et al. (2010). The collaboratory for the study of earthquake predictability perspective on computational earthquake science. *Concurrency and Computation: Practice and Experience, 22*(12), 1836–1847.

(Received September 25, 2015, revised August 8, 2016, accepted August 23, 2016, Published online September 7, 2016)

Pure Appl. Geophys. 174 (2017), 2295–2310
© 2016 Springer International Publishing
DOI 10.1007/s00024-016-1403-z

Radar Determination of Fault Slip and Location in Partially Decorrelated Images

JAY PARKER,[1] MARGARET GLASSCOE,[1] ANDREA DONNELLAN,[1] TIMOTHY STOUGH,[1] MARLON PIERCE,[2] and JUN WANG[2]

Abstract—Faced with the challenge of thousands of frames of radar interferometric images, automated feature extraction promises to spur data understanding and highlight geophysically active land regions for further study. We have developed techniques for automatically determining surface fault slip and location using deformation images from the NASA Uninhabited Aerial Vehicle Synthetic Aperture Radar (UAVSAR), which is similar to satellite-based SAR but has more mission flexibility and higher resolution (pixels are approximately 7 m). This radar interferometry provides a highly sensitive method, clearly indicating faults slipping at levels of 10 mm or less. But interferometric images are subject to decorrelation between revisit times, creating spots of bad data in the image. Our method begins with freely available data products from the UAVSAR mission, chiefly unwrapped interferograms, coherence images, and flight metadata. The computer vision techniques we use assume no data gaps or holes; so a preliminary step detects and removes spots of bad data and fills these holes by interpolation and blurring. Detected and partially validated surface fractures from earthquake main shocks, aftershocks, and aseismic-induced slip are shown for faults in California, including El Mayor-Cucapah (M7.2, 2010), the Ocotillo aftershock (M5.7, 2010), and South Napa (M6.0, 2014). Aseismic slip is detected on the San Andreas Fault from the El Mayor-Cucapah earthquake, in regions of highly patterned partial decorrelation. Validation is performed by comparing slip estimates from two interferograms with published ground truth measurements.

Key words: Radar, interferometry, fault slip, computer vision, Canny algorithm.

1. Introduction

While earthquake mainshocks often break the surface in sliding events with offsets of several meters, the smaller surface fracture events that are often created in the vicinity of large earthquakes (including silent slip events) are also of importance.

They are important to society, commonly requiring road resurfacing; more extreme damage to infrastructure is also well known, such as the 1963 fault-triggered failure of the Baldwin Hills reservoir, whose flood resulted in the destruction of hundreds of homes in Los Angeles (Hudson and Scott 1965). Northridge earthquake (M6.9, 1994) triggered co-motion on the nearby Mission Hills fault (Johnson et al. 1996). The Landers (Fialko 2004) and Hector Mine (Fialko et al. 2002) earthquakes triggered deformation on nearby faults, detected by satellite repeat-pass InSAR. Rymer et al. (2011) recount eight prior triggered earthquake events detected in the Salton Trough, California, found by field investigations. They document dozens of field verifications of triggered slip detected in interferometric phase jumps in UAVSAR repeat-pass images (uavsar.jpl.nasa.gov) In this work "fault" refers to semi-permanent, deeply rooted crustal structures that support slip, "surface fracture" to a detectable lineation displaying slip on the surface, often associated with a neighboring earthquake event. There can also be localized shear, detectable in the interferogram but not by field investigation and typically associated with subsurface fault slip; for convenience such localized shear (near-surface fracture) will be lumped with "surface fracture," when the two are indistinguishable to the computer vision algorithm. Note that surface fractures near a mainshock may be due to immediate (elastic) or delayed (afterslip, relaxation) processes; and these fractures may or may not be associated with local seismicity.

Examples in this work rely on UAVSAR data, although they can be extended to satellite InSAR as well (repeat-pass interferometry, or RPI). UAVSAR is an airborne system, and the RPI data relies on the same principles as satellite InSAR, but with differing strengths and weaknesses. Advantages of UAVSAR

[1] Jet Propulsion Laboratory, California Institute of Technology, 4800 Oak Grove Drive, Pasadena, CA 91109, USA. E-mail: jay.w.parker@jpl.nasa.gov
[2] Indiana University, 2709 East 10th Street, Bloomington, IN 47408, USA.

include a public data policy, high resolution (roughly 7 m pixel size in released ground-range geolocated phase maps, that reflect surface deformation), and flight-path flexibility (direction and visit time vary according to experiment design). Also, UAVSAR data quality is not affected by ionospheric disturbance. Weaknesses are that the width of an imaged swath is limited to roughly 20 km (typical satellite swaths are 70–350 km); also RPI ground-range phase images are affected by limitations of the aircraft: distortions due to unplanned aircraft motions can be but partially compensated in data processing. The wide range of incidence angle in UAVSAR makes the data somewhat more difficult to interpret: the near edge of a swath has a zenith incidence angle of about 20°; the far edge angle is about 70°. Both are affected by radar propagation delay due to spatial variability of atmospheric water vapor, and by decorrelation that results in a steady decrease in useful image pixels as repeat-pass time increases.

Data images from the UAVSAR program are freely available from two sources, http://uavsar.nasa.gov and http://geo-gateway.org. The latter has been developed to create web service-based tools that rely on the UAVSAR data. For example thumbnail images of the unwrapped ground range data are overlain with Google Maps, with a variety of features to enable navigation to data strips of interest (Parker et al. 2015; Wang et al. 2012). Also, the tools support line plots of line-of-sight deformation values on user-specified lines. As of this writing there are 1250 available RPI data images, concentrated in California (where the aircraft is based) but including locations as far as Japan and Iceland.

This large number of high-resolution images presents a need for feature extraction, automated by computer vision algorithms. Since unwrapped phase images correspond to the component of deformation (between the repeat visits) in the line of sight of the instrument, detectable boundaries corresponding to phase discontinuities correspond (much of the time) to surface expression of fault slip. Note we do not find that InSAR can see the surface fault itself, as the feature size is smaller than the imaging pixel scale; but it is often easy to observe fault-induced discontinuities in phase in the unwrapped image, because one side of a fault has moved (relatively) toward the instrument, and the other side has moved away. Image processing detects a coherent motion of many pixels defining regions on both sides of the fault.

Surface fracture detection is a highly favorable feature to explore early in developing feature detection. Its pattern is nearly orthogonal to typical instances of image distortion due to atmospheric water vapor (which tends to vary slowly across an image) and uncompensated aircraft motion. Features that may be confused with faults, such as the edge of a vegetated area (a farm may be wetter in one pass than in another one months later, affecting the radar phase and mimicking a fault slip at its edge), are commonly easy to identify by comparing with optical images and maps.

Phase gradient approaches have been used to find surface fractures following the Landers (Price and Sandwell 1998) and Hector Mine (Sandwell et al. 2000) earthquakes in the Mojave Desert, and near the 1995 Nuweiba earthquake in the Gulf of Aqaba (Red Sea) (Baer et al. 2001). A phase gradient product is produced by the GMTSAR package (http://topex.ucsd.edu/gmtsar/). Their approach computes a phase gradient from the real and imaginary part of the complex interferogram, avoiding reliance on phase unwrapping, which is problematic in complex scenes. To use the computer vision algorithm directly, this work operates on phase gradients created directly from an unwrapped phase product of UAVSAR, and therefore is limited to interferogram portions where this phase unwrapping is reliably produced. Both approaches use a combination of Gaussian smoothing with a numerical derivative filter, resulting in similar combined filter response functions.

The computer vision subject corresponding to detection of surface fractures in RPI images is edge detection. Some prior well-developed work in this field (Canny 1986) has been implemented in open-source software. To make our work easy to share, we apply the opencv Python library, which is easy to obtain and integrate into free Python development environments.

Because there is scientific and emergency response value in detecting triggered or silent fault slip, the method described here is valuable in its own right; estimation of fault slip amplitude and mechanism (often requiring additional information) can be performed in a subsequent stage of processing, not described here.

2. UAVSAR Interferograms

UAVSAR geometry and data interpretation are addressed by Donnellan et al. (2014). In the context of surface fracture detection, it is important to note the following:

The detected slip on a fault is based on the deformation difference across a determined boundary. This difference is taken from the components of deformation in the direction of the aircraft (Fig. 1); from a pixel location on the ground, the aircraft has an elevation angle and an azimuth (angle clockwise from north). Therefore, the line-of-sight estimate of the slip amplitude is an underestimate of the true slip; and the result of this underestimation can be practically zero, in the case where both sides of the fault are moving perpendicular to the aircraft (Fig. 1). Therefore in a landscape of interest any fault slipping in an unfavorable direction is unlikely to result in a detected edge (although very large slip would be detected).

As noted, for UAVSAR the near edge of a swath has a zenith angle of 20°: nearly vertical. This implies that strike-slip fault motion near this swath edge will be difficult to detect. Similarly at the far edge of the swath, strike-slip faults striking at favorable angles will be easy to detect, but some normal and thrust faults may have undetectable slip. These method defects can be overcome with RPI passes at multiple viewing angles, such as the combined data sets in Donnellan et al. (2015); but this is not always possible. Focusing here on slip detection and partial information about slip, examples are drawn from single interferograms.

3. Fracture Detection

While the key to determining fault slip location is the computer vision Canny method, considerable preprocessing and postprocessing is performed to improve the data product. These steps are essential because the Canny method is not, by itself, suited to images with missing data; it supports no means of masking poor data or judicious treatment of data gaps. Interferograms invariably have holes, many at the single pixel level and hence invisible to the eye without magnification, others covering irregular patches spanning thousands of pixels. Without preprocessing, the Canny method will draw a

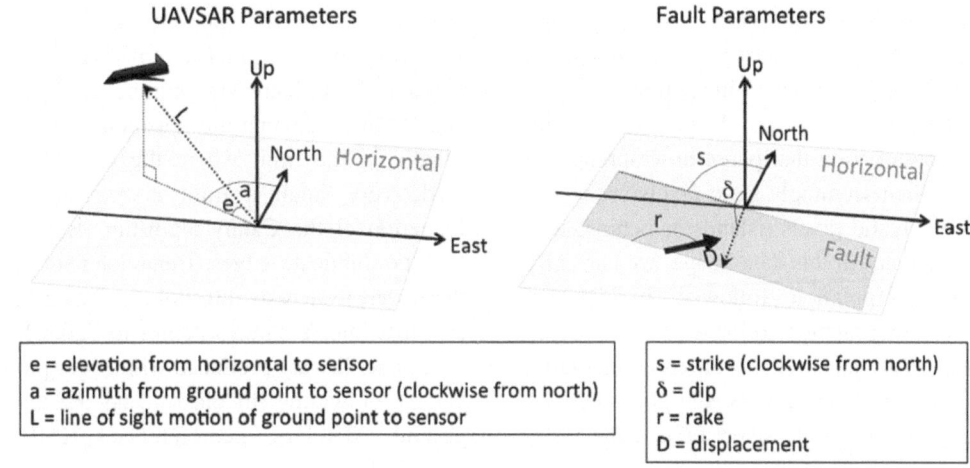

Figure 1

Line-of-sight geometry figure (from Donnellan et al. 2015). Parameters and conventions for UAVSAR observations and fault parameters. Line-of-sight changes are of the ground point relative to the instrument. A positive value means that the point on the ground moved toward the instrument on the aircraft, but the line length shortens. Strike is defined such that the fault always dips to the right when moving along the strike. Rake is defined by motion of hanging wall (*upper block*) relative to the footwall (*lower block*). Rake convention is such that 180° = right lateral, −90° = normal, 0° = left lateral, 90° = thrust

detected edge about every pixel and region of missing data, adding excessive clutter to the final image of surface fractures.

The following steps are used to process the images to produce detections of surface fractures shown below.

Initial data consists of a ground-range georeferenced unwrapped interferogram image file (ending with unw.grd) and the corresponding coherence image file (ending with cor.grd), along with matching metadata (file ending with.ann), and (purely for visual reference) a compressed markup image (ending with unw.kmz), all routinely produced by the UAVSAR project. The following steps prepare a region for automatic edge detection, illustrated for a very small patch of deformed desert in Fig. 2 [the large-scale pattern of deformation is due to the El Mayor-Cucapah (EMC) event]. To reduce processing time the user may (and typically will) provide a georeferenced polygon, produced by Google Earth using the polygon tool, typically drawn by the user to include features observed in the compressed markup image.

The data is reduced by averaging 3×3 blocks of product pixels, and by eliminating product pixels that do not meet quality control criteria (low local coherence, or excessive variance in the phase). Optionally, the product image (converted to mm of line of sight deformation) may be flattened by removing a smooth function; this has been found helpful when coseismic interferograms include a high-gradient region near the main rupture.

Additional data pixels are flagged as invalid, based on the observation that phase unwrapping can be unreliable when seemingly good pixels are nearly surrounded by invalid pixels (islands or peninsulas).

To treat the holes in this data image, we currently use bilinear interpolation followed by iterated smoothing operating on the unreliable pixels. We find this working rapidly and well for all cases including single pixels, decorrelated regions, and pixels beyond the bounding region used to select part of the interferogram for study.

A mild smoothing step is applied: a truncated 2-D Gaussian filter kernel, with width (sigma) about product pixels, is convolved with the image to reduce noise without substantially interfering with the broader Canny processing kernel.

Since the Canny method is designed to operate on 8-bit grayscale values (0–255) we scale the preprocessed data to this scale to maximize contrast. The threshold is set by the user (in millimeter, then scaled accordingly), to a value calibrated such that the user specifies a threshold corresponding to a deformation discontinuity in mm.

The Canny method initially detects possible edges by identifying high gradients, rapidly approximated by a convolution kernel in each dimension truncating the first derivative of a Gaussian hump, large enough to reduce any background noise (5×5 pixels, in work presented here). This is followed by modest non-linear enhancements: suppress candidate edges that are not at the gradient maximum; reduce multiple nearby edges to one; and apply hysteresis to favor edges marking smaller discontinuities that are connected to edges marking larger discontinuities. This hysteresis corresponds to the behavior of many faults, that slip is reduced near the ends of a slip event.

The Canny method is applied, with a 5×5 processing kernel, and hysteresis threshold 75 % of the user-specified edge threshold, and a mild smoothing step: a truncated 2-D Gaussian filter kernel, with width (sigma) about the scale of a product pixel, is convolved with the image to reduce noise without interfering with the Canny processing kernel. Note in Fig. 2 how the effects of a low-correlation highway are separated from a slipping fault.

Generally more pixels are marked as the threshold is reduced, but increasing numbers are single dots or clumps of pixels that are not indicative of fault slip. Finally in the limit where the chosen threshold is made very small a limit is reached, probably a reflection of the Canny algorithm eliminating what may be duplicate edges. Behavior with and without the lower, hysteresis threshold is shown in Fig. 3 for an EMC interferogram (item 2 in Table 1).

Image files are produced with edges plotted with a variety of widths (in black), and transparent background. Matching georeferenced KML files are produced as well: the edge map may immediately be read by Google Earth or similar applications, and made into overlays with the interferogram (for helpful checking and presentation) and background geography (indicating where possible fractures intersect critical infrastructure such as highways and dams).

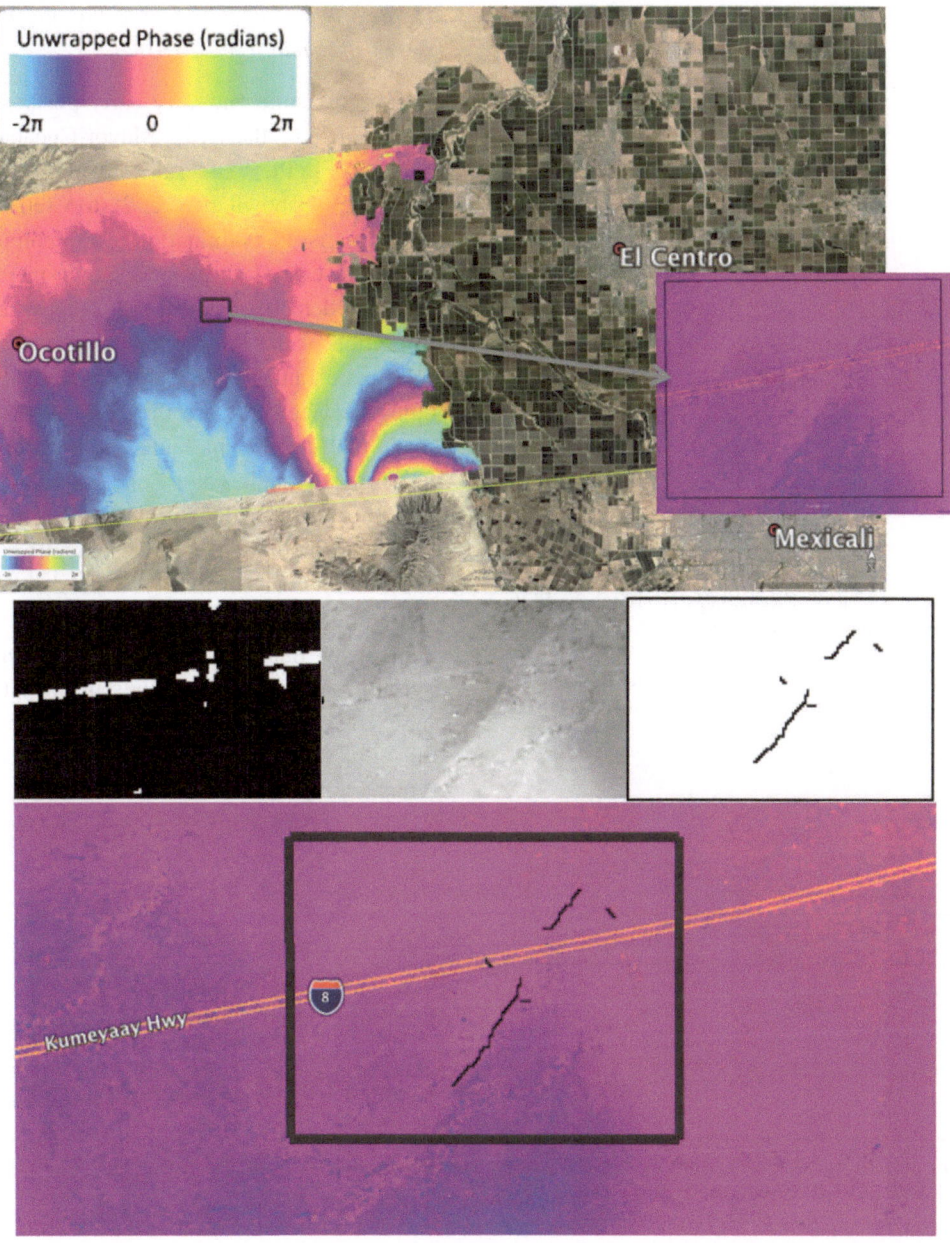

Figure 2
Computer vision steps applied to UAVSAR unwrapped interferogram. Google Earth image of repeat-pass interferogram representing the EMC event (plus 9 days). Zoomed image where unnamed fault intersects four-lane highway. Map of pixel quality (*white* are considered unusable), highlighting highway. *Grayscale patched image*, using bilinear interpolation and iterative void-pixel smoothing to reassign values to low-quality pixels. Image of detected edge pixels using 35-mm threshold (assumed to represent 35-mm slip event surface fracture). Note slip is not detected at highway crossing

4. Results

We show maps of detected edges that may indicate fault slip for three regions, the Yuha Desert close to the California -Baja California Norte boundary, the Bombay Beach segment of the San Andreas Fault, and the northern part of the South Napa Earthquake rupture. Geographic locations of these study areas are outlined in Fig. 4. For each interferogram the observation dates and other flight parameters are listed in Table 1.

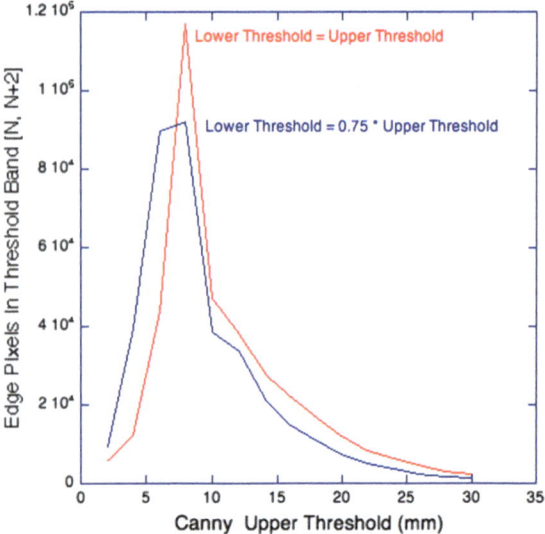

Figure 3
Edges vs threshold. Count of detected edge pixels for the EMC Coseismic case (see Fig. 5a, below), in each 2 mm range of thresholds. The peak at 8–10 mm appears to be due to reaching the speckle noise floor, where further detection of surface fractures is overwhelmed by Canny production of spurious edges from background noise

4.1. El Mayor-Cucapah Coseismic and Early Postseismic, Line 26501

The coseismic and very early postseismic deformation from the El Mayor-Cucapah (EMC) Earthquake in the southernmost California UAVSAR flight line is shown in Fig. 2 (top left) and Fig. 5 (top). This is the culmination of the computer vision analysis discussed above and shown in Fig. 2. Fault slip along detected phase jumps (edges) is shown for the portion of the interferogram west of the agricultural areas south of the Salton Sea, in the Imperial Valley. Key features are the West and East Laguna Salada faults, the Yuha fault, and a small area in the Northern Centinella Fault Zone.

Rymer et al. (2011) provide tabulated locations and values of slip detected at waypoints in fault-local field reconnaissance shortly after the EMC event. Figure 5 illustrates the locations of the waypoints covered by this interferogram, together with the overlain map of discontinuities in the interferogram phase (detected as edges in the computer vision

Table 1

Repeat-pass interferograms used in this work, with portions displayed in Figs. 2, 5, 6, 7, 8, 9, 10, 11

Figure	Line	Bearing	First visit	Second visit	Interval (days)
Figures 2, 5[1]	26501	265	21-Oct-2009	12-Apr-2010	173
Figure 6[2]	26501	265	21-Oct-2009	28-Sep-2010	342
Figure 7[3]	26514	265	21-Sep-2009	21-Oct-2010	395
Figure 8[4]	26501	265	13-Apr-2010	11-Feb-2014	1400
Figures 9, 10[5]	05512	55	29-May-2014	29-Aug-2014	92
Figure 11[6]	05512	55	29-Aug-2014	15-Oct-2015	412

[1] Coseismic El Mayor-Cucapah (M7.2, April 4, 2010), flight line closest to California—Baja California Norte border. *Data Product:* SanAnd_26501_09083-010_10027-001_0173d_s01_L090HH_01

[2] Coseismic and early postseismic El Mayor-Cucapah (M7.2, April 4, 2010), flight line closest to California—Baja California Norte border. *Data Product:* SanAnd_26501_09083-010_10071-001_0342d_s01_L090HH_01

[3] Coseismic El Mayor-Cucapah (M7.2, April 4, 2010), view includes southernmost San Andreas Fault at Bombay Beach. *Data Product:* SanAnd_26514_09074-102_10074-000_0395d_s01_L090HH_01

[4] Postseismic El Mayor-Cucapah (M7.2, April 4, 2010), flight line closest to California—Baja California Norte border. Long time span: about 3.8 years. *Data Product:* SanAnd_26501_10028-000_14017-011_1400d_s01_L090HH_01

[5] Coseismic South Napa Earthquake (M6.0, August 24 2014), northern portion of rupture. *Data Product:* SanAnd_05512_14068-002_14128-005_0092d_s01_L090HH_01_bh1

[6] Postseismic South Napa Earthquake (M6.0, August 24 2014), northern portion of rupture. *Data Product:* SanAnd_05512_14128-005_15152-005_0412d_s01_L090HH_01

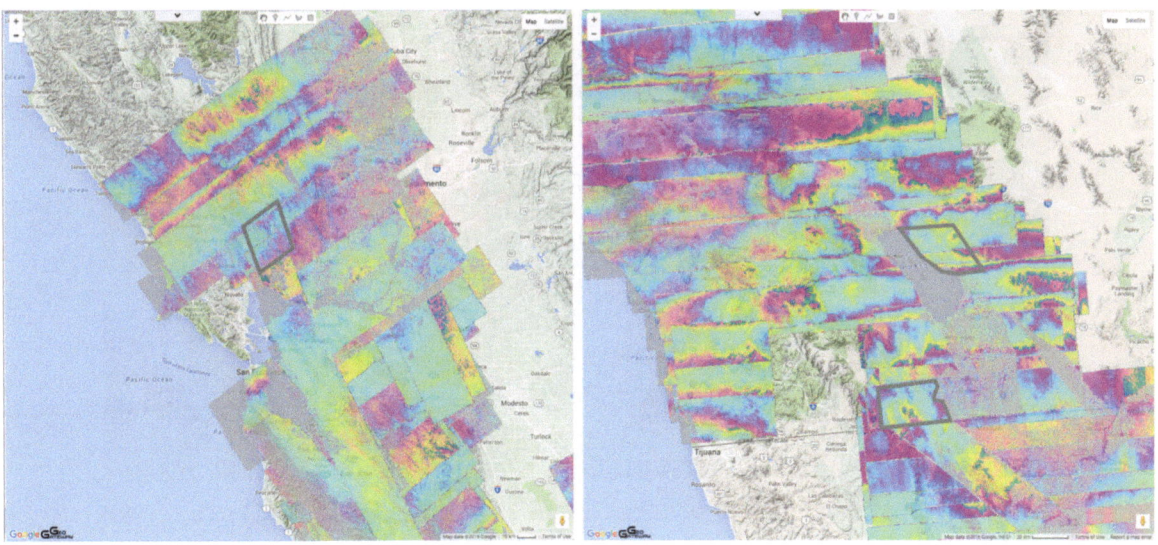

Figure 4
Geographical locations of UAVSAR interferogram portions selected for computer vision fault slip determination. *Left gray diamond* marks study area for South Napa Earthquake-related examination. San Francisco Bay is just south of center. *Right upper polygon* marks study area for induced slip along Bombay Beach segment of San Andreas Fault. Lower polygon marks area for study of three interferograms indicating slip near the EMC Earthquake near the California–Mexico border. Images are from http://geo-gateway.org, which displays colored strips for all interferograms produced by the UAVSAR project and user-supplied KML objects such as polygons (interferograms may be overlapped by later interferograms). *Scale* appears at bottom, and can also be deduced from roughly 20-km wide UAVSAR swaths

algorithm) suggesting surface fractures. The way-point slip values projected into the radar line-of-site direction are compared with the inferred radar-based slip for the four faults found to slip in association with the EMC event, West Laguna Salada Fault, East Laguna Salada Fault, Yuha Fault, and the Northern Centinella Fault Zone. Both maps of slip location correspond well (within roughly 100 m), and projected slip amounts are clearly correlated, but with substantial scatter. This may be due to limitations in the method, but also due to measuring different things (a surface fracture may have different slip than the differential deformation of the regions adjacent to the fracture, as measured by UAVSAR) and measuring at different times in an active, early postseismic period.

4.2. El Mayor-Cucapah Coseismic and Moderate Postseismic, Line 26501

The interferogram presented in Fig. 6 combines the coseismic deformation of the EMC earthquake with the main part of the postseismic development, including the first 177 days after the earthquake. Black lines indicate fault slip according to the computer vision method described above; thick lines exceed a 50-mm threshold or meet the Canny lower threshold requirement (75 % of the main threshold, and adjacent and roughly aligned with main threshold edges), thin lines exceed 35 mm or meet the Canny lower threshold requirement. Dominant fault slip areas are near the southern edge (bottom) of the plot, adjacent with the EMC rupture lying to the south (not shown), and trace the East and West Laguna Salada Faults and the Yuha Fault. A network of thin and broken lines north of the Laguna Salada area reveals a pattern of near-perpendicular cross faulting. At left center is a northwest-trending faint line with many cross-direction faint edges, extending nearly to Coyote Wells. This unmapped fault is in the position and direction to be a transitional fault between the well-known Laguna Salada faults to the south and the major Elsinore fault whose southern end is marked by a nearly east–west marked line of fault slip north of Ocotillo. It appears to be a buried fault, not generating a strong computer vision detection, and is associated with the June 15 2010 M5.7 aftershock at this location.

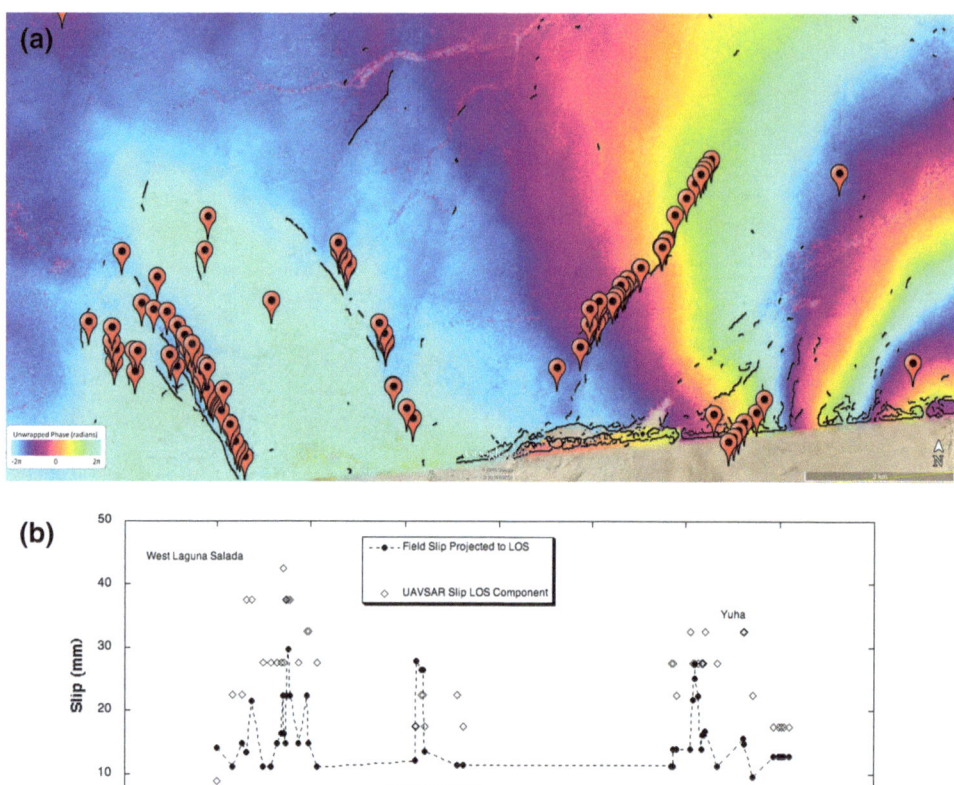

Figure 5

Validation: **a** EMC, Rymer et al. waypoints displayed with 15-mm threshold phase edges (representing surface ruptures). **b** Slip projected into radar line of sight, Rymer et al. waypoints and computer vision detected edges

4.3. San Andreas Triggered Slip from EMC, Line 26514

Figure 7 shows the interferogram portion adjacent to the Salton Sea (appearing blank at the southwestern part of the figure). The feature of primary interest is slip on the San Andreas Fault, over a span of visit times consistent with the EMC event, whose rupture terminates roughly 100 km to the south. Rymer et al. (2011) report creepmeter stepwise motion of 33 mm on this fault segment, coincident with the EMC event (Ferrum station). In places the estimate of the line of sight component of slip exceeds 30 mm. Some the other background pattern of dots and short segments

appears in an area of water seepage between the California Aqueduct and the Salton Sea.

4.4. El Mayor-Cucapah Postseismic to 2014, Line 26501

Figure 8 is from the same flight line as Figs. 5 and 6, and shows an interferogram with an exceptionally long visit time, 1400 days, all in the postseismic time period. The distribution of pixels of unacceptable quality (shown in white on BW map at upper left) has many more and larger white patches than in the previous case, an indication of

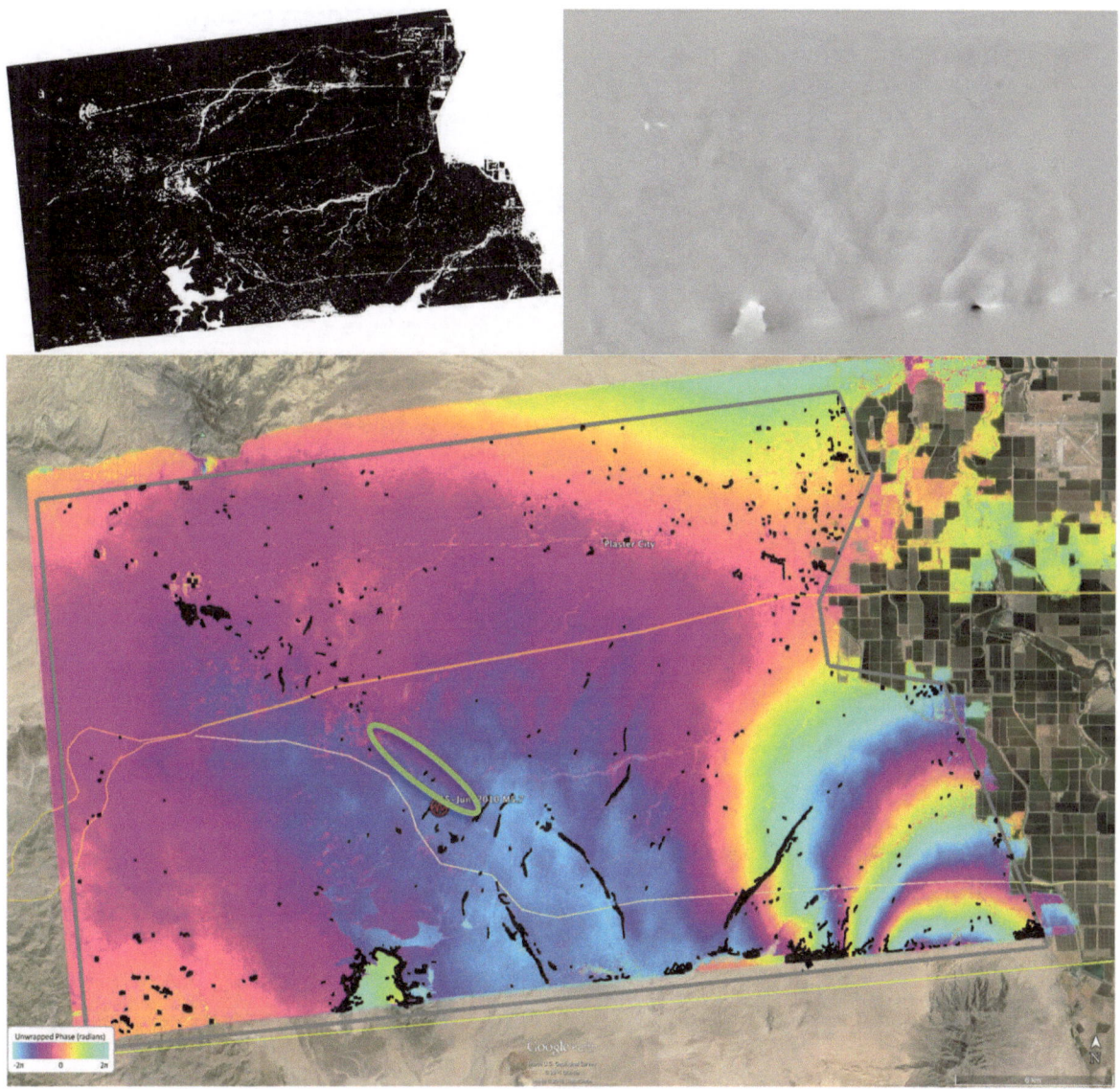

Figure 6

Arrangement for this and subsequent images: *Upper left* map of pixels of useful quality (*black*) and bad pixels (*white*) which are not used in subsequent analysis. *Upper right gray scale* image of interferometric phase in region of interest, after bad pixel regions are patched using a combination of interpolation and repeated void-only smoothing. *Below* unwrapped phase interferogram, single color cycle is 24 cm (4 pi). *Heavy line* is detected edges with line of sight jumps roughly 25 mm or greater. *Faint lines* are detected edges with LOS jumps roughly 15 mm or greater. The trace of the emerging buried fault is indicated by an olive ellipse, here and in Fig. 8. Interferogram representing coseismic and several months' postseismic deformation from EMC earthquake. Yuha fault postseismic slip is evident at lower right, trending northeast. Buried transition fault between Laguna Salada and Elsinore faults is at left center, trending northwest

how coherence suffers over times of several years. This figure also shows motion on the Yuha fault at lower right, but very little motion on the Laguna Salada faults near the south edge: these are evidently not involved in the postseismic process. However, the transitional fault at left center and the perpendicular fault at its southern end show marked slip. Deformation profiles show the transitional fault continued to slip after the M5.7 aftershock at this location. This buried transition fault between Laguna Salada and Elsinore faults is at left center, trending northwest.

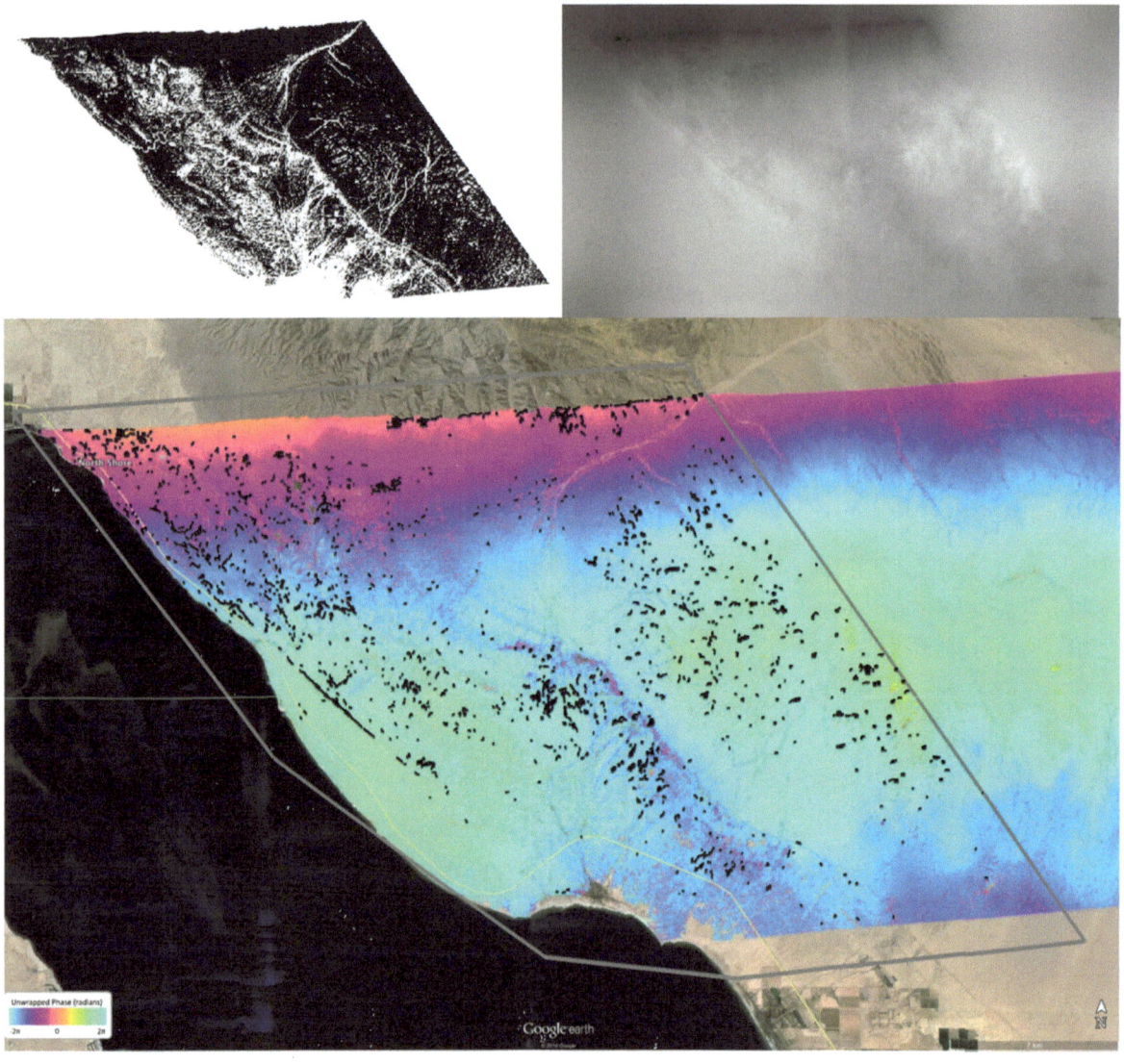

Figure 7
Interferogram representing coseismic EMC deformation on southernmost portion of San Andreas, the Bombay Beach segment. The evident N–S color gradient is consistent with a visit-to-visit change in mean atmospheric water vapor, whose influence on the phase image varies with the large range of UAVSAR zenith angles across the image (arrangement according to Fig. 6 caption)

4.5. South Napa Coseismic, Line 05512

Figure 9 shows the interferogram in the area around Napa, California, bracketing the time of the South Napa earthquake. In the coseismic interferogram, the blank feature to the west is not a body of water (as it resembles), but rather a range of wooded hills, where patchy radar incoherence defeated efforts

to unwrap the phase and create a continuous deformation map.

The chief feature in Fig. 9 is the coseismic rupture, just south of the blank central area. It displays a smooth path over most of its length, but breaks into several apparent surface rupture paths directly west of Napa. Rough north-trending traces

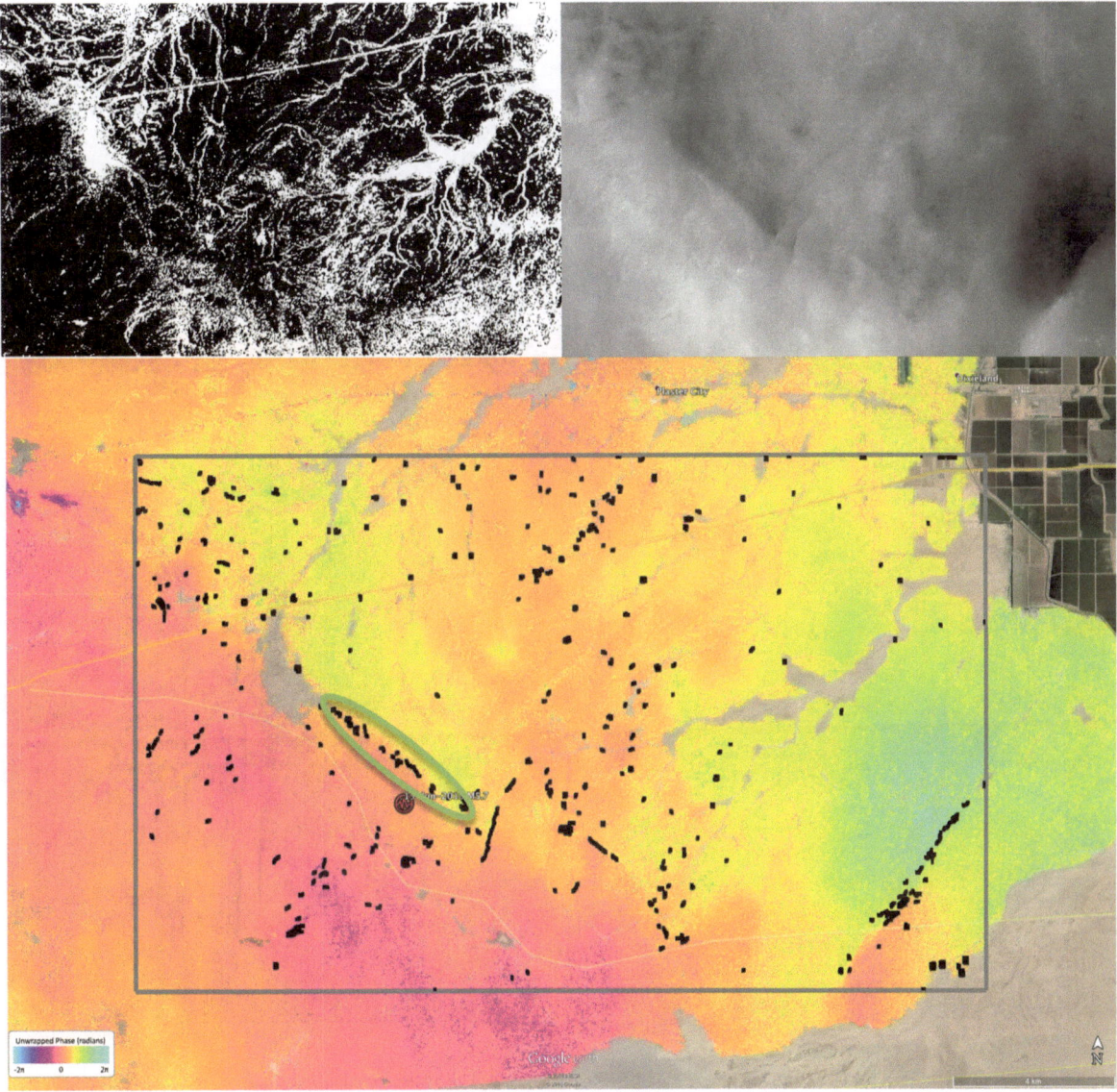

Figure 8
Interferogram representing EMC postseismic deformation (more than 3 years). Yuha fault postseismic slip is evident at lower right, trending northeast. Buried transition fault between Laguna Salada and Elsinore faults is at left center, trending northwest, indicated by light green thick line (arrangement according to Fig. 6 caption)

east of the main rupture may be associated with the secondary fault that damaged a runway at the Napa County Airport farther to the south. The main rupture trace shows continuing postseismic slip, as has been documented in Bray et al. (2014).

The UAVSAR estimate of the projected slip compares reasonably with the fault local measurements of slip reported in Bray et al. (2014). Figure 10

plots the radar-inferred line-of-sight component of slip at a set of waypoints together with the fault-local measured slip projected into the radar line of sight. Both types of measurements show a maximum slip at approximately latitude 38.29, with a rapid decline in slip towards the north. The southern margin shows differences, which are probably explained by the rapid postseismic surface slip in this area documented

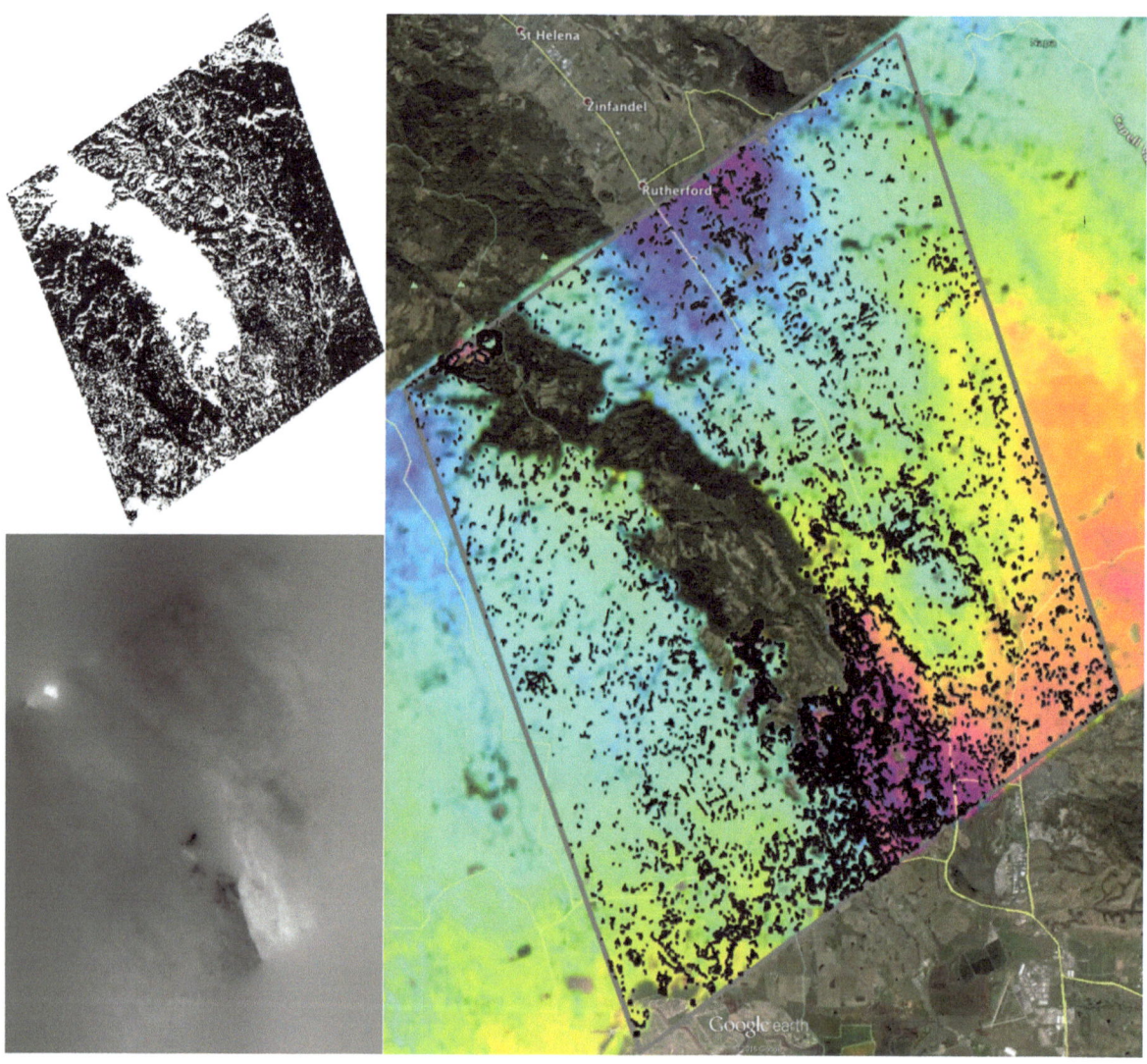

Figure 9
Interferogram representing coseismic deformation from South Napa earthquake. This phase map has been corrected courtesy Brian Hawkins, UAVSAR project, using expected far-field low deformation to choose the unwrapping discontinuity across the fault, which spans the interferogram south to north. Because the South Napa rupture crosses this entire scene, standard phase unwrapping tools could not determine the offset across the rupture, requiring additional information (the elastic behavior in the farther part of the image) (arrangement according to Fig. 6 caption, but rotated)

by Lienkaemper et al. (2016): the UAVSAR visit is 4–5 days after the field measurements at these sites.

4.6. South Napa Postseismic, Line 05512

Figure 11 shows the interferogram for the chief part of the postseismic deformation after the South Napa Earthquake. The area that was a large blank feature in Fig. 9 is successfully imaged here

(Fig. 11), during the postseismic period. The low-coherence region of Fig. 9 filled in with a consistent phase map. The key feature detected here is the map of continuing deformation along the main rupture after the earthquake. The many clutter dots and short lines of Fig. 9 are not seen in the postseismic deformation map of Fig. 11, suggesting actual deformation gradients at these sites that result from immediate earthquake effects such as differential

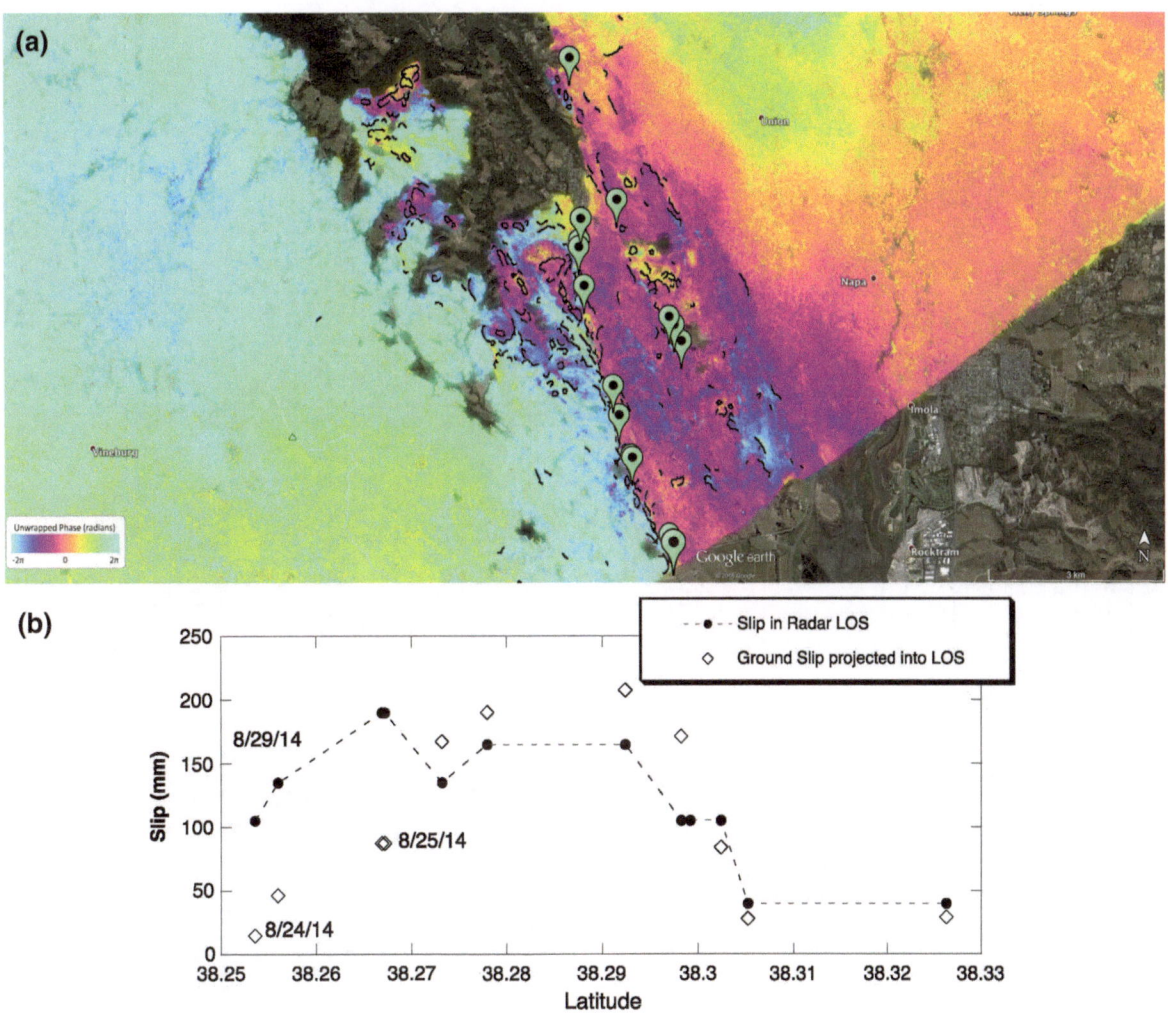

Figure 10

a Napa region, coseismic interferogram with GEER report waypoints and >35 mm detected edges. **b** Comparison of radar line-of-sight slip with projected ground truth measurements at waypoints. Discrepancy at left three points is consistent with known postseismic slip occurring in the first few days (dates of UAVSAR estimated slip and representative fault-local measurements are indicated)

settling, poroelastic effects, or local responses to the changing stress field.

5. Discussion

Comparison of field measurements and computer vision detection of dislocations in UAVSAR phase-based interferograms show high correspondence of locations corresponding to fault surface slip. Estimation of surface slip along these faults shows a degree of agreement with substantial discrepancy.

Dislocations and localized shear zones identified by InSAR sample a different aspect of a slipping fault than surface dislocations measured by field instruments than narrow shear zones measured by InSAR. Rather than a caliper or human-scale measurement, the work here uses a noise-reducing operator combining many pixels; represent relative displacement of patches of the earth's surface, covering areas of roughly 50 m by 100 m on each side. Therefore, for example when there is a slip deficit within roughly 100 m of the surface of a slipping fault, our methods would detect greater slip (reflecting slip at depth)

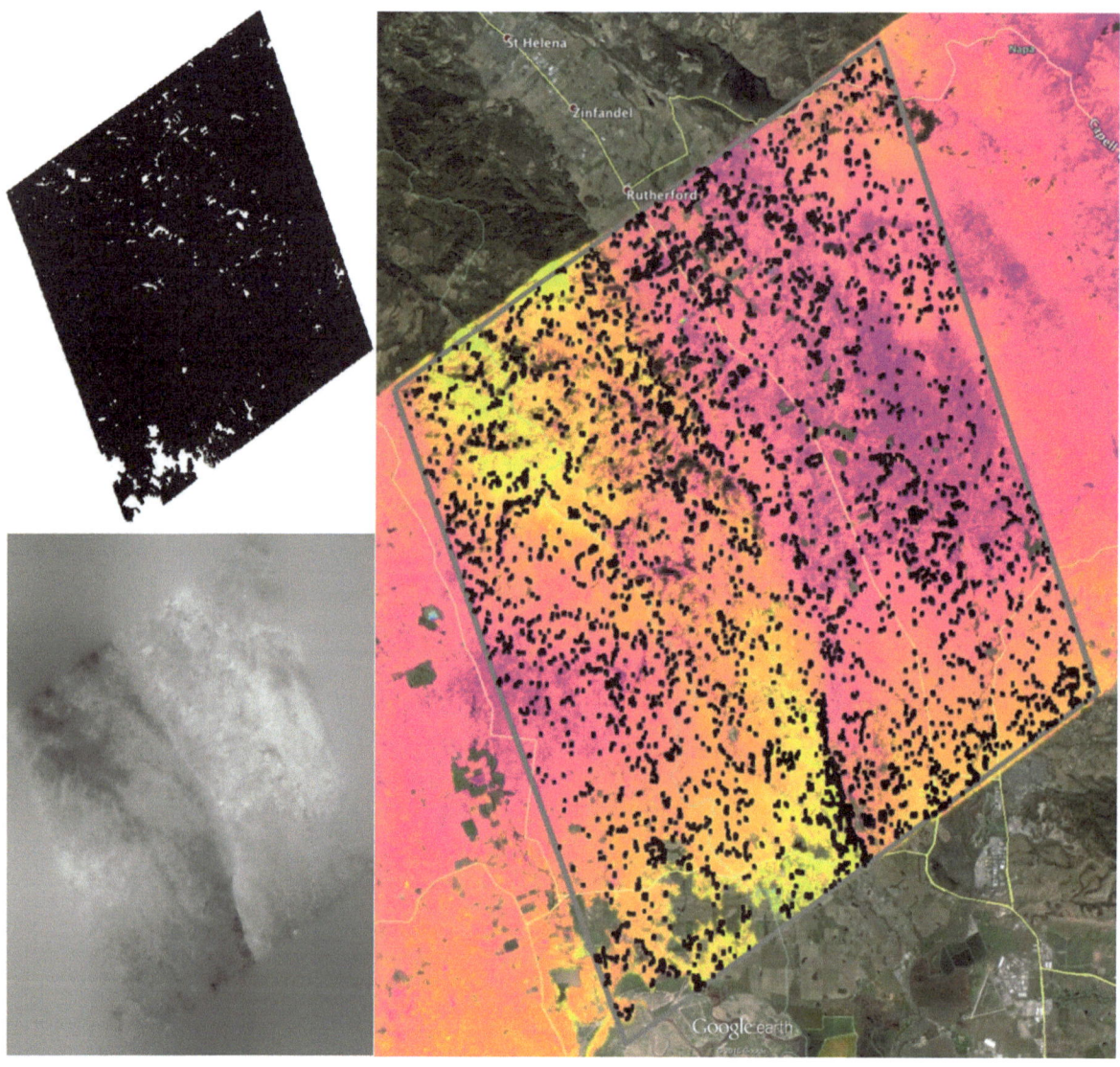

Figure 11
Although this repeat time is over 1 year, the bad pixel map (*upper left*) has very few white pixels. Note the empty space in Fig. 9 is now filled in: phase unwrapping is successful in the postseismic phase. The main rupture shows postseismic slip in the feature at bottom center (arrangement according to Fig. 6 caption, but rotated)

than that measured by a field geologist at the surface dislocation. Interferograms from a single look direction (as we treat here) cannot distinguish the mechanism (rake) of the rupture, but can be compared with field-examined ruptures of known or measured mechanism.

The detected edges in the EMC Coseismic and early Postseismic (Figs. 5, 6), and the triggered San Andreas Fault segment (Fig. 7) correspond closely

with the surveyed faults in Rymer et al. (2011), indicating that detections with thresholds as low as 15 mm, forming continuous lines or curves with length exceeding roughly 1 km represent fault ruptures that reach the surface or nearly so. We interpret the South Napa Coseismic and Postseismic images similarly, concluding that while the main rupture fault is present in both images, there are many short and looping structures evident in the coseismic image in particular that should not be regarded as faulting. They could be due to differential subsidence from shaking or rapid changes in the stress environment, irregular folding, or soil moisture patterns, perhaps due to poroelastic squeezing of water from some volumes into adjacent ones. Whatever the cause, the pattern of roughly 50 cm phase edges is most intense near the rupture fault (and, it appears, where the fault bends near Browns Valley Road), and the pattern is not evident in the Postseismic image. These details suggest this textured deformation is caused by the main shock rather than other possible sources.

The M5.7 June 15 2004 aftershock near Ocotillo is examined in Rymer et al. (2011). They observe a postseismic radar phase change along a NW trend, but do not observe slip at matching locations. They conclude this is probably slip at depth, producing "… shear too diffuse to map at the surface" by field measurements. UAVSAR interferograms show this feature is a shearing zone spread over roughly 500 m, and therefore probably too broad to detect as surface fractures. However, images (Fig. 6) show secondary fractures, some with roughly 10- to 15-mm slip along faults nearly perpendicular to this color gradient feature, and extending ~ 1 km as consistent detected fractures. The evidence of Figs. 6 and 8 show this subsurface slip continues to grow after October 2010.

6. Conclusions

Synthetic aperture interferometry provides a useful level of reconnaissance of silent slip, triggered surface fracture, and active fault morphology, one of particular value in uninstrumented areas after an earthquake. Airborne interferometry illustrated by the NASA UAVSAR program provides data images of deformation that carry information about emerging fault networks at high (~ 10 mm) sensitivity and high (~ 20 m) resolution. Employing computer vision techniques including infilling, flattening, and Canny edge detection provides an automated way to extract maps and values of surface slip, including slip beyond one radar phase cycle as long as phase unwrapping can be reliably employed.

The extracted slipping fault networks from the EMC event (viewed in far southern California, USA)

and the South Napa earthquake near California's San Francisco Bay and the state capitol, Sacramento, are demonstrated to carry comparable information to in situ measurements. Therefore, this remote sensing view of surface fractures inferred from jumps in radar phase can provide useful information when there are no in situ measurements, and can provide a key element of a unified synoptic view when presented together with fault-local measurements.

Determination of fault slip and emerging fault networks is valuable for science and also for hazard detection. The postseismic EMC interferogram shows apparent subsurface slip on a northwest trending fault, which is not previously mapped, but forms a link between the Laguna Salada and Elsinore faults, illuminating the faulting structure to the southwest of the North America-Pacific plate boundary. Hazard detection is illustrated by the radar-produced images of previously unknown faults in the Yuha desert due to the EMC event, which cross several roads and highways that required repair. The Napa earthquake experience illustrates the need for rapid assessment of critical infrastructure, where investigators used earthquake satellite images (utilizing human inspection of the phase images) and fault-local reconnaissance to find surface fractures corresponding to a damaged airport runway and a natural gas pipeline that was found to be undamaged but required inspection (Bray et al. 2014). Automated and rapid interferogram analysis for surface fractures is leading to geographic data products that will be immediately compared with highways, bridges, pipelines, dams, and other critical infrastructure, leading to informed priorities for inspection and repairs after an earthquake.

Acknowledgments

This work was carried out at the Jet Propulsion Laboratory, California Institute of Technology under contract with NASA. The work was funded by NASA Earth and Space Science program, NASA's EarthScope Geodetic Imaging UAVSAR program, Advanced Information Systems Technology (AIST) program for QuakeSim work, and the ACCESS program for GeoGateway development. We thank the UAVSAR team and in particular Yunling Lou, Brian Hawkins, Naiara Pinto, and Yang Zheng for collection and processing of the UAVSAR data.

REFERENCES

Baer, G., Shamir, G., Sandwell, D., & Bock, Y. (2001). Crustal deformation during 6 years spanning the Mw = 7.2 1995 Nuweiba earthquake analyzed by interferometric synthetic aperture radar. *Israel Journal of Earth Sciences, 50*, 9–22.

Bray, J., Cohen-Waeber, J., Dawson, T., Kishida, T., & Sitar, N. (2014). Geotechnical engineering reconnaissance of the August 24, 2014 M 6 South Napa earthquake. Geotechnical Extreme Events Reconnaissance (GEER) Association Report Number GEER, 37.

Canny, J. (1986). A computational approach to edge detection. *IEEE Transactions on Pattern Analysis and Machine Intelligence, 6*, 679–698.

Donnellan, A., Grant Ludwig, L., Parker, J. W., Rundle, J. B., Wang, J., Pierce, M., et al. (2015). Potential for a large earthquake near Los Angeles inferred from the 2014 La Habra earthquake. *Earth and Space Science, 2*, 378–385.

Donnellan, A., Parker, J., Hensley, S., Pierce, M., Wang, J., & Rundle, J. (2014). UAVSAR observations of triggered slip on the Imperial, Superstition Hills, and East Elmore Ranch Faults associated with the 2010 M 7.2 El Mayor-Cucapah earthquake. *Geochemistry, Geophysics, Geosystems, 15*, 815–829.

Fialko, Y. (2004). Probing the mechanical properties of seismically active crust with space geodesy: study of the coseismic deformation due to the 1992 Mw7. 3 Landers (southern California) earthquake. *Journal of Geophysical Research: Solid Earth (1978–2012), 109*(B3), 1978–2012.

Fialko, Y., Sandwell, D., Agnew, D., Simons, M., Shearer, P., & Minster, B. (2002). Deformation on nearby faults induced by the 1999 Hector Mine earthquake. *Science, 297*, 1858–1862.

Johnson, A. M., Fleming, R. W., Cruikshank, K. M., & Packard, R. F. (1996). Coactive fault of the Northridge earthquake—Granada Hills Area, California. Open File Report 96-523, US Department of the Interior, US Geological Survey.

Hudson, D. E., & Scott, R. F. (1965). Fault motions at the Baldwin Hills reservoir site. *Bulletin of the Seismological Society of America, 55*, 165–180.

Lienkaemper, J. J., DeLong, S. B., Domrose, C. J., & Rosa, C. M. (2016) Afterslip behavior following the M6.0, 2014 South Napa earthquake with implications for afterslip forecasting on other seismogenic faults. Seismological Research Letters, *87*(11). doi:10.1785/0220150262.

Parker, J., Donnellan, A., Glasscoe, M., Fox, G., Wang, J., Pierce, M., et al. (2015). Advantages to Geoscience and disaster response from QuakeSim implementation of interferometric radar maps in a GIS database system. *Pure and Applied Geophysics, 172*, 2295–2304.

Price, E. J., & Sandwell, D. T. (1998). Small-scale deformations associated with the 1992 Landers, California, earthquake mapped by synthetic aperture radar interferometry phase gradients. *Journal Geophysical Research, 103*, 27001–27016.

Rymer, M. J., Treiman, J. A., Kendrick, K. J., Lienkaemper, J. J., Weldon, R. J., Bilham, R., Wei, M., Fielding, E. J., Hernandez, J. L., Olson, B. P., & Irvine, P. J. (2011). Triggered surface slips in southern California associated with the 2010 El Mayor-Cucapah, Baja California, Mexico, earthquake (No. 2010-1333). US Geological Survey.

Sandwell, D. T., Sichoix, L., Agnew, D., Bock, Y., & Minster, J. B. (2000). Near real-time radar interferometry of the Mw 7.1 Hector Mine Earthquake. *Geophysical Reseach Letters, 27*, 3101–3104.

Wang, J., Pierce, M., Ma, Y. M., Fox, G. C., Donnellan, A., Parker, J. W., et al. (2012). Using service-based geographical information system to support earthquake research and disaster response. *IEEE Computing in Science and Engineering, 14*, 21–30. doi:10.1109/MCSE.2012.592012.

(Received April 11, 2016, revised August 29, 2016, accepted September 16, 2016, Published online September 26, 2016)

Pure Appl. Geophys. 174 (2017), 2311–2330
© 2016 Springer International Publishing
DOI 10.1007/s00024-016-1326-8

Pure and Applied Geophysics

Detecting Significant Stress Drop Variations in Large Micro-Earthquake Datasets: A Comparison Between a Convergent Step-Over in the San Andreas Fault and the Ventura Thrust Fault System, Southern California

T. H. W. Goebel,[1] E. Hauksson,[2] A. Plesch,[3] and J. H. Shaw[3]

Abstract—A key parameter in engineering seismology and earthquake physics is seismic stress drop, which describes the relative amount of high-frequency energy radiation at the source. To identify regions with potentially significant stress drop variations, we perform a comparative analysis of source parameters in the greater San Gorgonio Pass (SGP) and Ventura basin (VB) in southern California. The identification of physical stress drop variations is complicated by large data scatter as a result of attenuation, limited recording bandwidth and imprecise modeling assumptions. In light of the inherently high uncertainties in single stress drop measurements, we follow the strategy of stacking large numbers of source spectra thereby enhancing the resolution of our method. We analyze more than 6000 high-quality waveforms between 2000 and 2014, and compute seismic moments, corner frequencies and stress drops. Significant variations in stress drop estimates exist within the SGP area. Moreover, the SGP also exhibits systematically higher stress drops than VB and shows more scatter. We demonstrate that the higher scatter in SGP is not a generic artifact of our method but an expression of differences in underlying source processes. Our results suggest that higher differential stresses, which can be deduced from larger focal depth and more thrust faulting, may only be of secondary importance for stress drop variations. Instead, the general degree of stress field heterogeneity and strain localization may influence stress drops more strongly, so that more localized faulting and homogeneous stress fields favor lower stress drops. In addition, higher loading rates, for example, across the VB potentially result in stress drop reduction whereas slow loading rates on local fault segments within the SGP region result in anomalously high stress drop estimates. Our results show that crustal and fault properties systematically influence earthquake stress drops of small and large events and should be considered for seismic hazard assessment.

Key words: Stress field heterogeneity, source parameter inversion, spatial stress drop variation, asperity strength, slip rates.

1. Key Points

- significant spatial variation in stress drops beyond measurement uncertainties
- stress drops show no systematic increase with higher differential stresses
- stress drops are higher in regions with larger stress field heterogeneity
- stress drops are approximately correlated with tectonic loading rates

2. Introduction

A robust identification of possible spatial variations in stress drop is essential to advance our understanding of earthquake physics and scaling relations. Self-similar earthquake scaling is observed at many scales, implying that stress drops remain constant across a wide range of earthquake magnitudes and that fault slip increases systematically with rupture area (e.g., AKI 1967; PRIETO *et al.* 2004; WALTER *et al.* 2006; SHEARER 2009; KWIATEK *et al.* 2011; GOODFELLOW and YOUNG 2014). However, other studies highlight a possible deviation from self-similarity at regional and global scales indicating that stress drops may change with earthquake magnitude and for different regions (e.g. KANAMORI *et al.* 1993; HARRINGTON and BRODSKY 2009; LIN *et al.* 2012; OTH 2013).

Here, we use the seismic record of ∼6000 small earthquakes in southern California between 2000 and 2014 to identify possible spatial variations in stress drops and search for correlated parameters that may reveal underlying, physical mechanisms. The static

¹ Earth and Planetary Sciences, University of California, Santa Cruz, California, USA. E-mail: thw.goebel@gmail.com

² Seismological Laboratory, California Institute of Technology, Pasadena, California, USA.

³ Earth and Planetary Sciences, Harvard University, Cambridge, Massachusetts, USA.

stress drop during an earthquake is defined as the ratio of earthquake slip to rupture dimension (e.g., AKI 1967; KANAMORI and ANDERSON 1975; SHEARER 2009).

Static stress drop can rarely be measured directly and is commonly inferred from the seismic moment and shape of the source spectra (e.g., BRUNE 1970). Unlike the seismic moment, which is computed from long-period ground motions, stress drop estimates are sensitive to high-frequency energy radiation which is generally more difficult to measure. As a consequence, more sources of uncertainty arise, related to limited recording bandwidths, low-quality seismic records and variable attenuation structure PRIETO et al. (2007); SHEARER et al. (2006); ABERCROMBIE (2013, (2015). High-frequency attenuation is especially problematic for small events, which can cause an artificial break-down in self-similarity ABERCROMBIE (1995). In addition, apparent variations in stress drop estimates may be the result of changing rupture velocity as rupture velocity and stress drop can both influence the amount of high-frequency energy radiation ATKINSON AND BERESNEV (1997); BERESNEV (2009). All of these factors potentially contribute to the generally observed large scatter in stress drops which are partially an expression of measurement and modeling uncertainty but may also be evidence for physical differences in rupture processes.

There are many observations of systematically varying stress drop estimates beyond expectations of measurement uncertainty. For example, earthquakes on the Parkfield segment of the San Andreas fault were suggested to exhibit similar source pulse widths for a variety of event magnitudes so that stress drop varied between 0.18 and 63 MPa HARRINGTON and BRODSKY (2009). High stress drops for on-fault events at Parkfield were also suggested by NADEAU and JOHNSON (1998), although this result was questioned by later studies which suggested stress drop variations in Parkfield to be comparable to other areas SAMMIS and RICE (2001); ALLMANN and SHEARER (2007). In southern California, a comprehensive study of P-wave spectra from over 60,000 earthquakes found no correlation between stress drop and distance from major faults SHEARER et al. (2006), while a study of global earthquakes with $M > 5$ revealed higher stress drops for intraplate compared

to plate boundary events ALLMANN and SHEARER (2009).

At present, the main physical parameters that drive stress drop variations remain unresolved. In southern California, higher-than-average stress drops were identified in some regions containing a relatively high fraction of normal-faulting events whereas the mainly reverse-faulting aftershocks of the Northridge earthquake have lower-than-average stress drops SHEARER et al. (2006). In contrast, a global study found higher-than-average stress drops for strike-slip events ALLMANN and SHEARER (2009) while a significant correlation with faulting type was absent in a recent high-resolution study GOEBEL et al. (2015). Spatial and temporal heterogeneity in stress drops may also be a result of variations in seismic coupling and transient slip processes before main shocks, for example, expressed by differences in foreshock and aftershock source spectra in southern California CHEN and SHEARER (2013). Furthermore, stress drops are lower in regions with high heat flow OTH (2013) but may increase with depth SHEARER et al. (2006); YANG and HAUKSSON (2011); OTH (2013); HAUKSSON (2014). Stress drops have also been observed to vary as a function of recurrence intervals and loading rates in the laboratory and nature (e.g., KANAMORI et al. 1993; HE et al. 2003; GOEBEL et al. 2015). Slower loading rates and longer healing periods within interseismic periods lead to an increase in asperity strengths and stress drops BEELER et al. (2001); MCLASKEY et al. (2012).

Here, we try to unravel measurement uncertainty and potential physical differences in earthquake source processes by analyzing source spectra of a large population of micro-earthquakes ranging from $M_L = 0.3–3.5$. Micro-earthquakes or microseismicity here refers to events that are commonly too small to be felt. Analyzing a large number of small-magnitude earthquakes enables us to better resolve statistically significant variations in stress drops even if scatter is high. A similar approach was used recently by GOEBEL et al. (2015) for a more localized region close to San Gorgonio Pass and is extended here to a wider region including the western and eastern Transverse Ranges in southern California. Building on earlier studies, we now also characterize seismicity statistics using Gutenberg-Richter b-value and fractal analysis as well as stress heterogeneity within the study areas. In the following, stress heterogeneity refers to heterogeneity in both principle stress orientations and types of faulting. Our study takes advantage of the newly available broadband waveform records of microseismicity that became available at a large scale after ~2000. These records have proved to be pivotal for a more reliable determination of spectral shapes at high frequencies (see Sect. 4.1 and Fig. 6). Note that the stress drop estimates reported here may not directly measure static stress drop as defined by the ratio of slip to rupture dimension. Instead the seismically inferred stress drop estimates primarily provide insight into how energetic earthquakes are at high frequencies.

3. Data

3.1. Seismicity Data and Tectonic Setting

We analyze stress drop variations, seismicity statistics, stress orientations and types of faulting within the greater San Gorgonio Pass (SGP) and Ventura basin (VB) areas. These areas are located in the eastern and western Transverse Ranges, respectively, and are characterized by complex active fault systems that accommodate components of Pacific and North American Plate motions. Faulting in the Ventura area is dominated by east-west trending surface emergent and blind-thrust faults that uplift mountains on the north and south sides of the basin. The basin is comprised of up to 10 km of Cenozoic sediments. In contrast, deformation in the San Gorgonio Pass region is driven by the San Andreas fault, with slip partitioned onto several strike-slip fault splays with thrust faults localized by restraining bends on these systems PLESCH et al. (2007). Many of these faults are unfavorably oriented with respect to the relative direction of plate motion (Fig. 1).

The SGP area exhibits high geometric complexity and loading rate variations within the San Andreas fault system. It is occasionally referred to as 'structural knot', not least due to its potential in restricting the propagation of large magnitude ruptures through this area (e.g., MATTI and MORTON 1993; LANGENHEIM et al. 2005; DAIR and COOKE 2009; COOKE and DAIR 2011).

This potential limitation on maximum earthquake magnitude has large implications for the seismic hazard in near-by urban areas like Los Angeles and San Bernardino (e.g. MAGISTRALE and SANDERS 1996; GRAVES et al. 2008). The deformation across the geologically complex VB is accommodated by series of thrust faults. For this region, the seismic hazard is mainly driven by rapid contraction rates of about 10 mm/year MARSHALL et al. (2013); BURGETTE et al. (2015) and the potential of a large rupture involving a series of linked thrust faults, which could result in a M_w 7.7–8.1 earthquake ROCKWELL (2011); MARSHALL et al. (2013); HUBBARD et al. (2014).

This study examines earthquake records in SGP and VB between 2000 and 2014 which largely excludes the Northridge aftershock sequence. We deliberately selected smaller scale study areas to isolate the seismicity in Ventura and SGP from seismicity on the San Jacinto fault as well as from the Northridge, Landers or Joshua tree aftershock sequences. We estimate earthquake stress drops for ~1000 events in VB with magnitudes between ~1.3 and ~3.2 and for ~5000 events in SGP with magnitude between ~0.5 and ~3.5. Few events exist outside of these magnitude ranges including the largest magnitude events, i.e. the M_w 4.8 Isla Vista event in 2013 and the M_w 4.9, Yucaipa event in 2005 (Fig. 1b,c) with no significant influence on average stress drop estimates.

The SGP and VB regions present an ideal natural laboratory to examine the extent and scale of resolvable, physical stress drop variations. Such physical differences may be driven by some of the specific crustal and faulting conditions in SGP and VB such as: (1) Regional-scale tectonics, i.e., dominant compressional system in VB vs. a mixed strike-slip and compressional system in SGP, allowing us to study the connection between large-scale tectonics and average stress drop variations. (2) The large range of focal depth in VB and SGP which include some of the deepest seismicity in Southern California (>22 km). Differences in focal depths provide a test-bed for examining potential stress drop variations with increasing differential stresses at greater depth (where differential stress is defined as scalar difference between largest and least compressive principle stress axis, i.e., $\sigma_1 - \sigma_3$). (3) Different regional seismicity characteristics related to magnitude

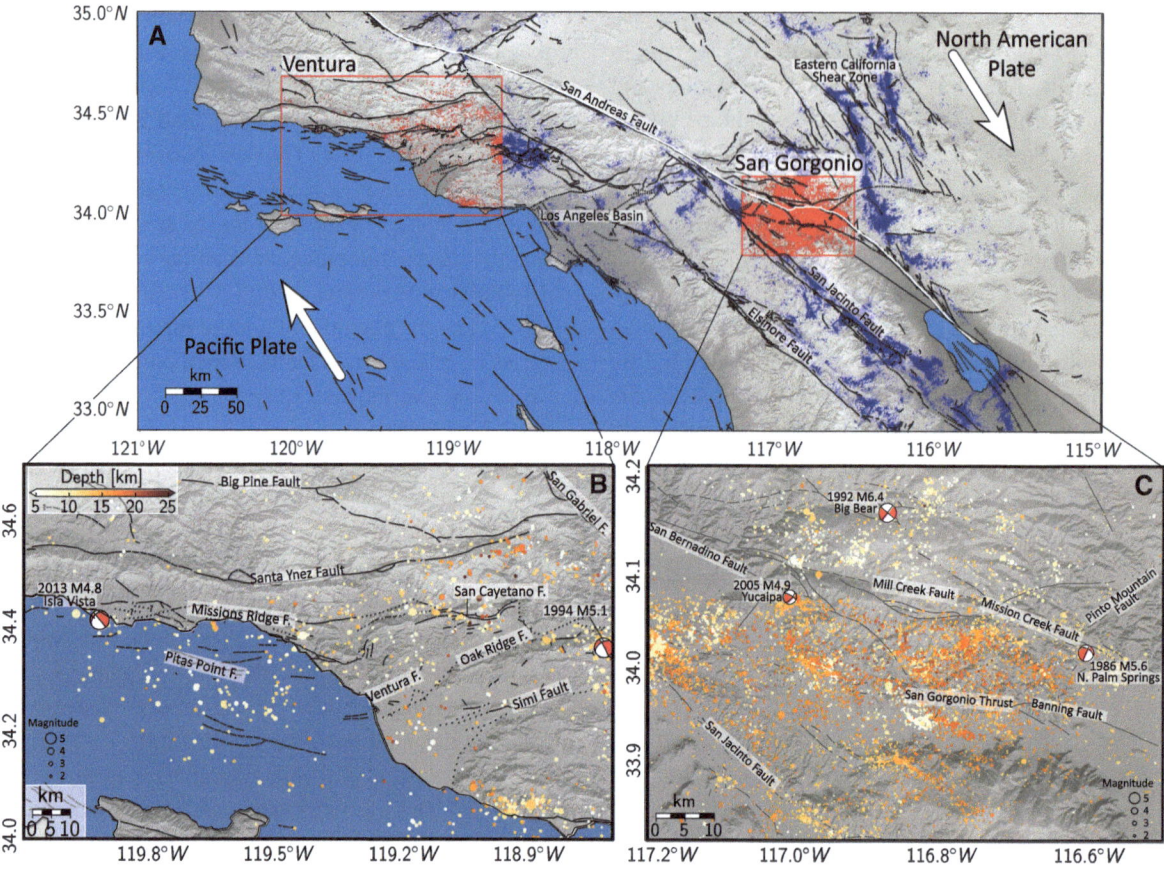

Figure 1
Seismicity and major faults in southern California (*A*). Study regions (SGP: [117.2W, 33.8N] to [116.5W, 34.2N] and VB [120.1W, 34.0N] to [118.7W, 34.7N]) and corresponding seismicity are highlighted in *red, white arrow* show approximate direction of relative plate motion. Seismicity colored with depth and scaled with magnitude in the greater VB (*B*) and SGP(*C*) regions. Fault traces are highlighted by *black lines*

distributions and spatial clustering. Investigating these clustering statistics enables us to assess potential joint variations in seismicity statistics and stress drop. (4) Different fault properties that result from variations in lithology between SGP and VB.

3.2. Seismicity statistics

We quantify seismicity statistics including both the spatial distribution of hypocenters and magnitude distributions in VB and SGP. The latter showed little variations within the study areas. Based on a best-fitting power-law model using the minimum Kolmogorov–Smirnov distance between power-law-fit and observed data Clauset *et al.* (2009), we determined *b*-values close to unity in both regions (i.e. $b = 1.0 \pm 0.04$ in VB and $b = 1.1 \pm 0.02$ in SGP) with magnitudes of

completeness, $M_c = 2.0$ in VB and $M_c = 1.5$ in SGP (Fig. 2a).

The spatial hypocenter distribution, on the other hand, showed significant variations between VB and SGP. To quantify this distribution, we computed the pair correlation function, $C(r)$, at all scales and for all event pairs, N in VB and SGP with separation distance, s, less than r so that $C(r) = N(s < r)/N_{tot}^2$ (e.g., Wyss *et al.* 2004; Goebel *et al.* 2013a). After log transformation, this function is approximately linear between ~ 1 and ~ 30 km, indicating that the interevent distances are fractal within this range of distances. The slope of the correlation function is the fractal dimension, D_2, which provides insight into the degree of spatial clustering. Random spatial earthquake distributions in 3-D are connected to $D_2 \sim 3$. For seismicity distributions that are concentrated along major fault zones, the fractal

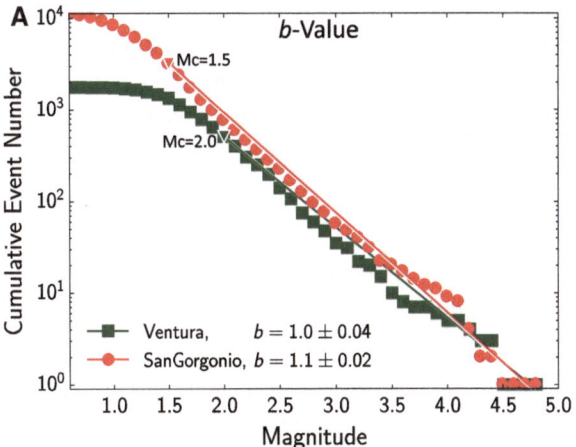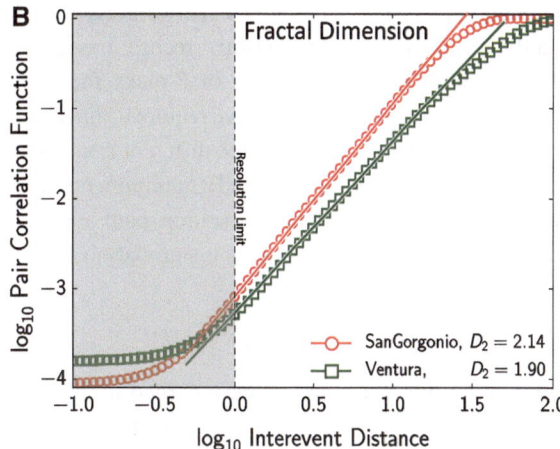

Figure 2

Comparison between *b*-values (**a**) and fractal dimensions of seismicity hypocenters (**b**) in SGP and VB. The G + R *b*-values do not deviate significantly from unity in both regions whereas the differences in fractal dimensions are statistically significant (see text for details). The lower limit of the fractal range of the correlation function (resolution limit in **b**) is influenced by the average horizontal hypocenter uncertainty, which is on the order of 1 km in SGP and VB

dimension decreases down to values of $D_2 \sim 2$ for predominant planar seismicity. Seismicity localization within fault zones or close to a point in space can result in D_2 values below 2.

We observed significant variations in fractal dimensions between VB with $D_2 = 1.90$ and SGP with $D_2 = 2.14$. This difference is an expression of the more distributed hypocenter locations in SGP and more localized seismicity along major faults in VB (Fig. 2b). We tested the statistical significance and influence of potential sampling bias on these results. To this aim, we randomly drew 1000 hypocenters from both distributions and resampled 1000 times resulting in similar values of $D_2 = 2.1 \pm 0.02$ in SGP and of $D_2 = 1.9 \pm 0.05$ in VB. We then employed a student *t* test and a two sided Kolmogorov–Smirnov test of the interevent distance distributions. Both tests confirm that the differences in spatial hypocenter distributions are significant at the 99 % level.

4. Determining Stress Drops From Stacked Source Spectra

We determine stress drops from corner frequencies and seismic moment of source terms in the log-frequency domain using an algorithm developed by PRIETO *et al.* (2004) and SHEARER *et al.* (2006). Using

the recorded waveforms of all earthquakes in SGP and VB across the Caltech/USGS Southern California Seismic Network (SCSN), we separate source, path and station contributions by iteratively stacking over common receivers, paths and events. Our method allows us to analyze large data sets in a uniform fashion so that potential methodological biases present in Green's functions deconvolutions using earthquake pairs can be minimized. The stress drop estimates involve four key analysis steps: (1) Separating recorded spectra into source, path and site spectra; (2) Calibrating relative moment estimates to absolute seismic moments using local magnitudes; (3) Correcting the high-frequency portion of the source spectra for attenuation using robust, regional empirical Green's function estimates; (4) Fitting the corrected source spectra to obtain source parameters for each event. Each of these four steps is described in the following sections.

4.1. Separation Into Source, Path and Site-Response Spectra

We start by transforming the portion of the recorded waveforms within a 0.5 s time window before and 1.28 s after P-arrival into the frequency domain. This comparably short time window provides reliable results for small-magnitude earthquakes and short S-P times, which dominate our dataset. We perform an initial quality assessment of the seismic record based on signal-to-noise ratio and number of P-picks for each event. For the spectral inversion, we require a signal-to-noise ratio of $SNR > 5$ within three different frequency bands (5–10, 10–15, 15–20 Hz) and a minimum number of 5 P-picks. The convolution of source, path and site contributions can be expressed as a summation in the log-frequency domain:

$$d_{ij}(f) = e_i(f) + t_{ij}(f) + s_j(f), \qquad (1)$$

where d_{ij} is the logarithm of the recorded amplitude spectrum, e_i and s_j are the event and station terms and t_{ij} is the travel time term between the ith event and jth station. This system of equations can be solved iteratively by estimating event, station and path terms as the average of the misfit to the observed spectra minus the other terms (e.g., ANDREWS 1986; WARREN and SHEARER 2000; SHEARER et al. 2006; YANG et al. 2009). For robustness, we suppressed outliers by assigning L1 norm weights to spectra with large misfit residuals.

The path terms, t_{ij}, in Eq. 1 were discretized at each iteration by binning at 1 s intervals according to the corresponding P-wave travel times. The stacked path terms capture the average, large-scale effects of geometric spreading and attenuation along the ray path. The observed, systematic spectral amplitude decrease with distance is in approximate agreement with expectation from a simple attenuation model and $Q = 550$ (Fig. 3c, d).

At this point, our method does not include take-off angle dependent differences in recorded spectra arising from radiation pattern and directivity effects, which are a potential source of uncertainty within the source spectra estimates (e.g., KANEKO and SHEARER 2014). However, these differences are reduced by stacking spectra from many stations and thus averaging over the focal sphere. The robustness of the spectral inversion method for large data sets was previously verified by a comparison with synthetic data ALLMANN and SHEARER (2007).

4.2. Calibration to Absolute Seismic Moment

Relative seismic moments, Ω_0, were determined for individual events and source spectra from corresponding low-frequency spectral level which were averaged over the first three data points above 1 Hz. We then calibrate the relative moments using the catalog magnitudes, assuming that the low-frequency amplitudes are proportional to moment, and that the catalog magnitude is equal to moment magnitude at $M_L = 3$.

4.3. Empirical Green's Function and High-Frequency Correction

Before determining stress drops for individual events, we compute an average Empirical Green's Function (EGF) from the stacked source spectra within 0.2 magnitude bins to correct for high-frequency attenuation. The EGF was determined by fitting a constant stress drop, Brune-type spectral model (see following section) between 2–20 Hz to the magnitude-binned spectra. The EGF is then computed from the average misfit between theoretical and observed spectral shapes. The corrected, magnitude-binned spectra can be described by a Brune-type model, i.e., they show constant values at long periods and f^{-2} fall-off at short periods above the corner frequency, f_c (Fig. 3a). Furthermore, correcting the mag-binned spectra for differences in seismic moments by shifting along a f^{-3}-line results in a data-collapse, indicating self-similar behavior and constant stress-drop at a large scale (see inset in Fig. 3). Our best-fitting constant stress-drop model with reasonable fit to the mag-binned spectra has a stress drop of 6.1 MPa in SGP and 1.0 MPa in VB. At the scale of the entire study regions, the stacked source spectra in the SGP and VB region show constant stress drops and self-similar source scaling; however, at smaller scales we also observe spatially varying stress drops which will be investigated below.

In the following, a regional EGF is used to correct average source terms of individual events to search for possible smaller scale variations in stress drops. The EGFs in Fig. 3 differ substantially between SGP and VB, indicating differences in attenuation structure between the basin-dominated Ventura region and SGP. To investigate how these differences may effect stress drop estimates, we test spatially varying EGF's for different sub-regions in SGP and VB as discussed in detail in Sect. 4.2.

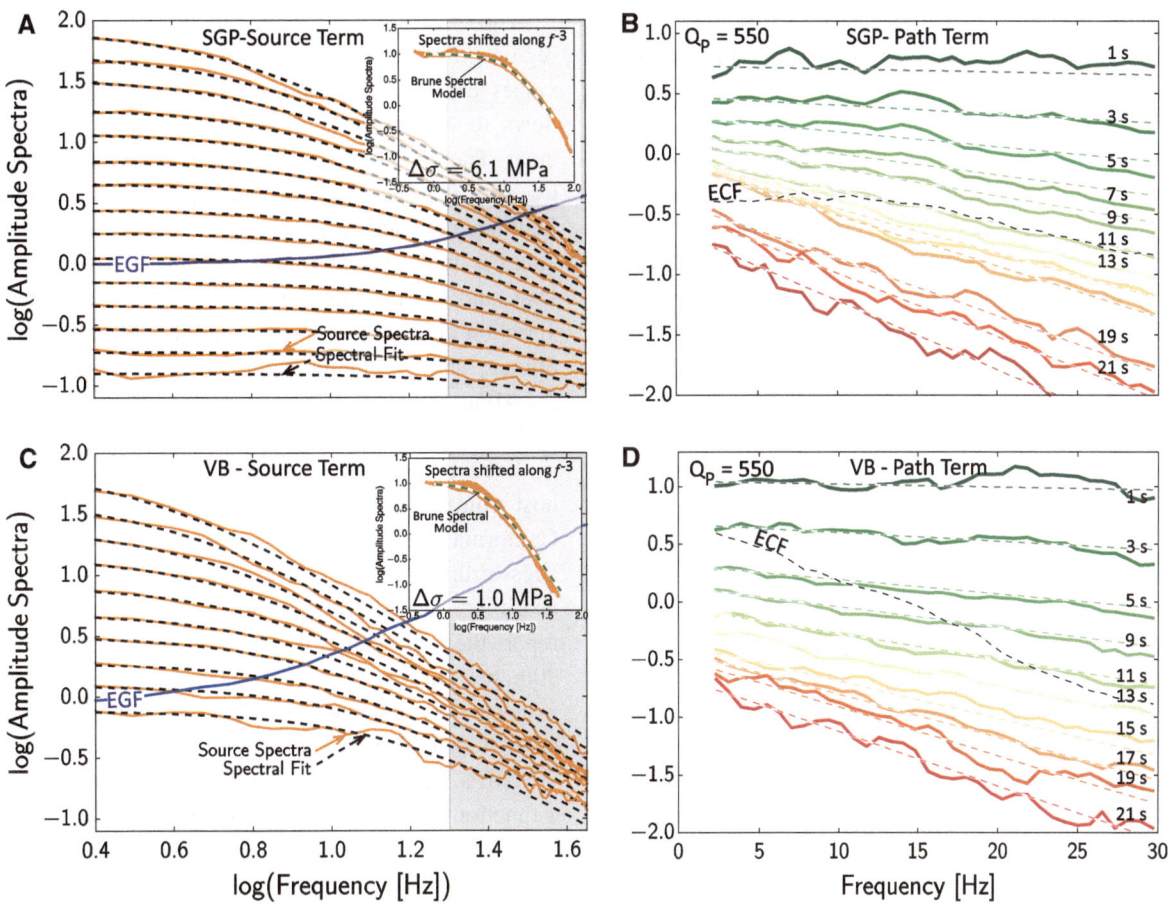

Figure 3

Source and path terms of amplitude spectra in SGP and VB. **a, c** Source spectra (*orange*) stacked in 0.2 magnitude bins are displayed together with the empirical Green's Function (*blue curve*) and Brune-type spectral fits. *White areas* highlight the frequency range for spectral fitting. The inset shows the regional stress drop estimate after shifting the magnitude-binned spectra along f^{-3}. **b, d**: Path terms stacked over 2 s travel time bins (*colored curves*) are shown on the right, together with a standard attenuation model for Southern California with $Qp = 550$ (*colored dashed lines*). Similarly, to the EGF, we used an empirical correction function (ECF) to remove common terms from all path terms (*black dashed curves*)

4.4. Spectral model and stress drop estimates

To determine corner frequencies and stress drops, we first correct source spectra using the regional EGF, and then fit individual event spectra with a Brune-type f^{-2} model BRUNE (1970). The model for displacement spectra, $u(f)$, has the following form:

$$u(f) = \frac{\Omega_0}{1 + (f/f_c)^2} \qquad (2)$$

where Ω_0 describe the low-frequency plateau and f_c the corner frequency.

For a circular, isotropic rupture and constant rupture velocity, the stress drop ($\Delta\sigma$) and P-wave corner frequency can be related by ESHELBY (1957); MADARIAGA (1976):

$$\Delta\sigma = M_0\left(\frac{f_c}{0.42\beta}\right)^3 \qquad (3)$$

where M_0 is the seismic moment, β is the shear wave velocity at the source and the rupture velocity is $0.9 \times \beta$. Different models, such as those of BRUNE (1970), SATO and HIRASAWA (1973), MADARIAGA

(1976), and KANEKO and SHEARER (2014), will yield differences in absolute stress drop values that vary by up to a factor of five, even when the same rupture velocity is assumed. Consequently, stress drop variations should be compared between studies using the same source model. It should be noted that seismically inferred stress drop estimates are only a measure of the physically released stress if the above mentioned model assumptions (such as circular crack, constant rupture velocity, isotropic rupture) are an appropriate description of earthquake ruptures.

5. Results

We report on results from a comparative analysis of stress drop variations in VB and SGP. Our analysis showed that average stress drop estimates show no systematic variation as a function of earthquake magnitude, but stress drop estimates for individual events vary strongly between 0.1 and 100 MPa. This scatter may not solely be due to measurement uncertainty but rather indicate physical variations in stress drop estimates. We test whether statistical significant stress drop variations are resolvable in large (>100 s earthquake) sample sizes. We start by showing smoothed spatial representations of stress drop estimates and binned depth variations. We then examine the robustness of stress drop variations by analyzing corner frequency and seismic moment estimates as well as spectral shapes of stacked source terms.

5.1. Spatial Variation in Average Stress Drop Estimates

The smoothed spatial variations in individual event stress drops are large in SGP ranging from 1 to more than 20 MPa, compared to VB where stress drop estimates are between 0.5 and 2 MPa throughout the study area (Fig. 4). The smoothed maps were computed from individual event stress drops and by using regional EGFs in VB and SGP to correct for high-frequency attenuation prior to fitting a Brune-type model to the source spectra. Stress drop values were then smoothed using the log-normal mean of the 200 closest events.

The smoothed stress drop map in SGP reveals two regions of significantly elevated stress drops with values above 20 MPa with decreasing stress drops down to ~2 MPa southeast and northwest of these areas. The main region is located between the surface traces of the San Gorgonio thrust and Mission Creek fault and coincides with a 'structural knot' in the San Andreas fault system (e.g. LANGENHEIM et al. 2005; DAIR and COOKE 2009). The VB region shows less variation about the average value of 1 MPa with the 5th and 95th percentile occupying values of 0.3 and 3.4 MPa. Slightly elevated average stress drops of ~2 MPa can be observed in the Malibu coastal area.

Even though our study regions in VB and SGP host some of the deepest seismicity in southern California, we observe no systematic variations of stress drops with depth. We computed average stress drops in 2 and 4 km depth-bins in SGP and VB and report the log-mean stress drop as well as 10th and 90th percentiles (Fig. 5). In SGP, stress drop estimates are largely constant within the upper 8 km and show a sharp increase at ~9 km from ~3 MPa to ~6 MPa. In the VB, stress drops increase slightly as a function of depth from 0.5 to 1 MPa. This increase may partially be due to increasing rupture velocities with depth which can be tested using a 1D shear wave velocity model for Southern California and by assuming that variations in shear wave velocities are a good proxy for rupture velocity variations. The apparent increase in stress drop with depth is reduced to $\Delta\sigma = 0.7$ MPa at 2 km and $\Delta\sigma = 0.9$ MPa at 18 km depth. Thus, we conclude that, although there is an abrupt increase in stress drop below 10 km in SGP, there is no generic relationship between stress drop and focal depth.

We also examine the role of shear wave velocity variations as a proxy for rupture velocity changes at each hypocenter using the 3D CVHM velocity model for Southern California SHAW et al. (2015). The 3D model includes a more gradual increase in seismic velocities with depth in VB compared to SGP. However, the strongest changes in velocities are concentrated in the upper 5 km above the seismically most active depths. Using the 3D velocities as input in Eq. , we find that average stress drop increase to 1.12 MPa in VB and decrease down to 4.7 MPa in SGP. For velocity changes to account for the entire

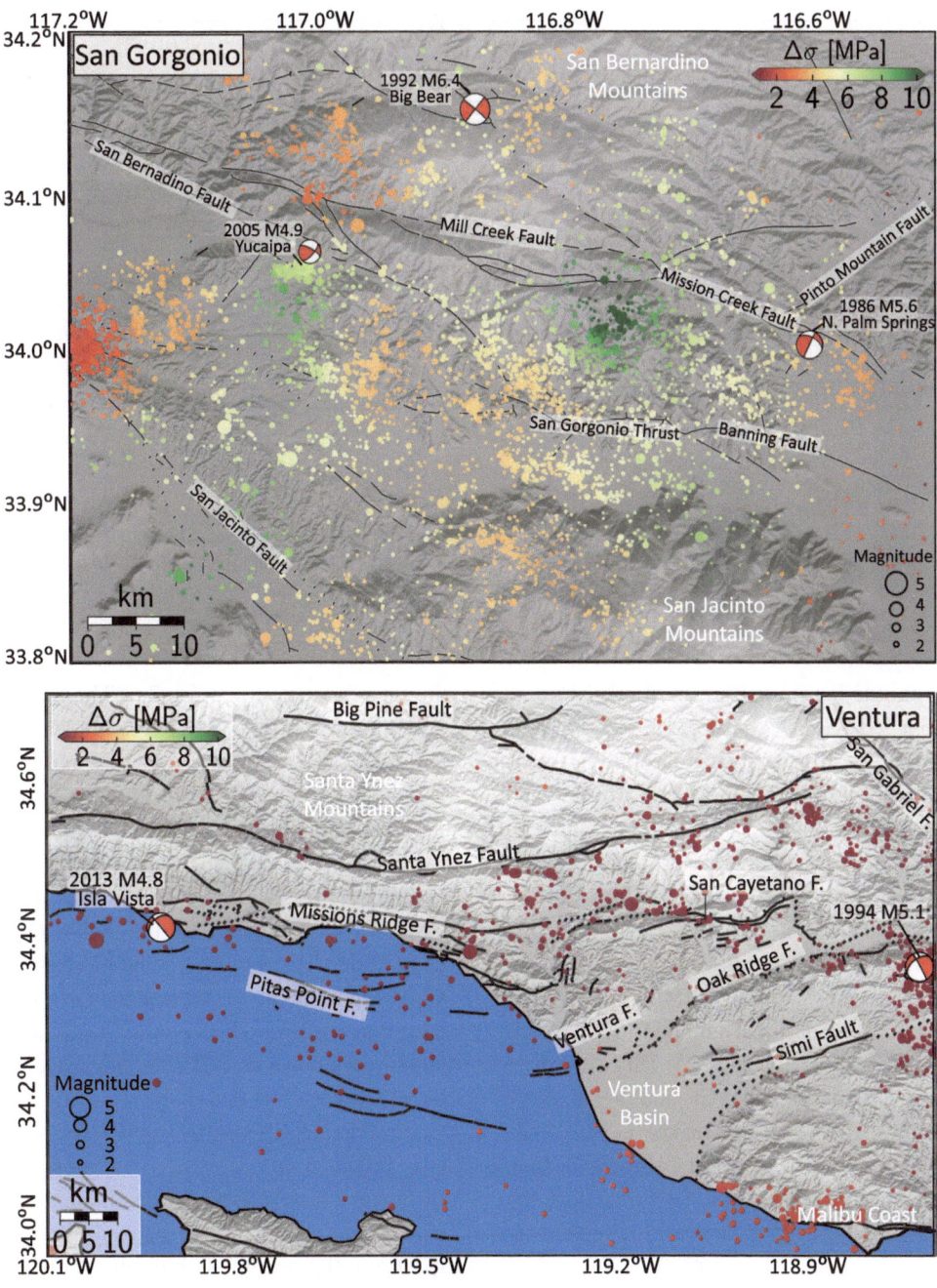

Figure 4
Smoothed spatial variation in stress drop estimate using the log-normal mean of the 200 closest events in SGP and VB. Stress drop estimates vary substantially in SGP between 1 and 20 MPa whereas stress drops in VB are mostly close to 1 MPa (see legend in upper right for stress drop values corresponding to *marker colors*)

difference in stress drops between the two study regions an average difference in velocities of 1.5 km/s would be required across all depths which is generally not observed.

In addition to spatial variations of stress drop within the study areas, there is a significant difference in average stress drop between the regions. The average stress drop in SGP ($\Delta\sigma_{mean}$ = 4.8 MPa),

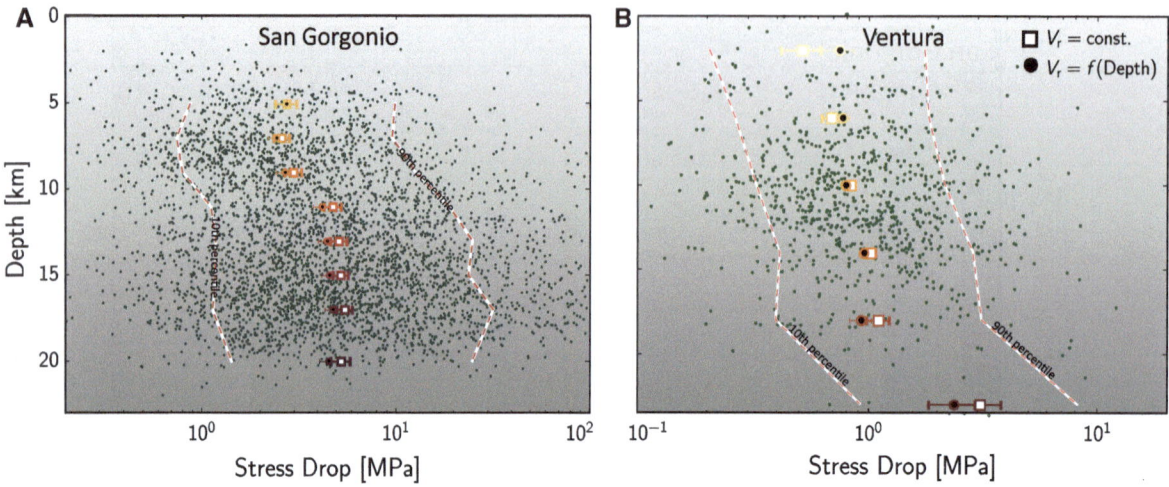

Figure 5
Stress drop variations as a function of depth in San Gorgonio (**a**) and Ventura (**b**). Stress drops increase abruptly below 10 km in SGP and increase marginally with depth in VB. *Colored squares* illustrate the log-mean stress drop in 2 km bins, *colored circles* are log-mean stress drops after correcting for increase in rupture velocity (V_r) with depth. *Green dots* are stress drop estimates for individual earthquakes and *red dashed lines* the corresponding 10th and 90th percentile

computed from the log-normal mean of all values, is higher than in VB ($\Delta\sigma_{mean} = 0.9$ MPa, Fig. 6). This difference in stress drop estimates is highly significant based on a two-sided Kolmogorov–Smirnov and student's *t* test (see GOEBEL *et al.* 2015 for details) so that the hypothesis that both stress drop catalogs are sampling the same population can be rejected at the 99 % level.

Moreover, stress drop estimates in SGP show much scatter between 0.1 and 100 MPa whereas the scatter is about one order of magnitude lower in VB ranging from 0.1 to 10 MPa (Fig. 6). This difference in stress drop distributions in SGP and VB is more pronounced for estimates based on the newly available broadband stations in Southern California, which better resolve high stress drops in SGP and reduce the scatter in VB, indicating a decrease in epistemic uncertainty.

We test the robustness of resolved stress drop differences between VB and SGP as well as within a high and low stress drop region in SGP (see insets in Fig. 7). To this aim, we separately invert the low and high stress drop regions in SGP for source, path and site terms and compute local EGFs to account for possible lateral variations in attenuation structure. The results confirm that stress drop estimates are significantly different, with variations that go beyond

the uncertainties in f_c (Fig. 7a). The uncertainty in f_c is determined from the 90 % confidence interval of the misfit between modeled and observed spectra using a grid-search algorithm for which more pronounced minima in the misfit function result in narrower confidence intervals. Based on these 90 % confidence intervals, we expect the uncertainty in stress drop estimates to be about one order of magnitude. This rough estimate of stress drop uncertainty approximately corresponds to the observed scatter in VB (Fig. 7b); however, does not explain the larger variations in SGP nor the systematic differences between VB and SGP.

Systematic differences in stress drop estimates are also reflected in differences in spectral shapes of earthquake source terms. To facilitate the comparison between source terms, we stacked the corresponding spectra in 0.2 magnitude bins and shifted the stacked spectra along a f^{-3} to correct for differences in seismic moment (Fig. 8). Such a shift leads to a collapse of spectra if stress drops are constant. Differences in stress drops are observable by relative differences in frequency content, corresponding to possible differences in high-frequency fall-off and corner frequencies. The spectra from high and low stress drop regions in SGP show approximately self-similar scaling within each region but emphasize that

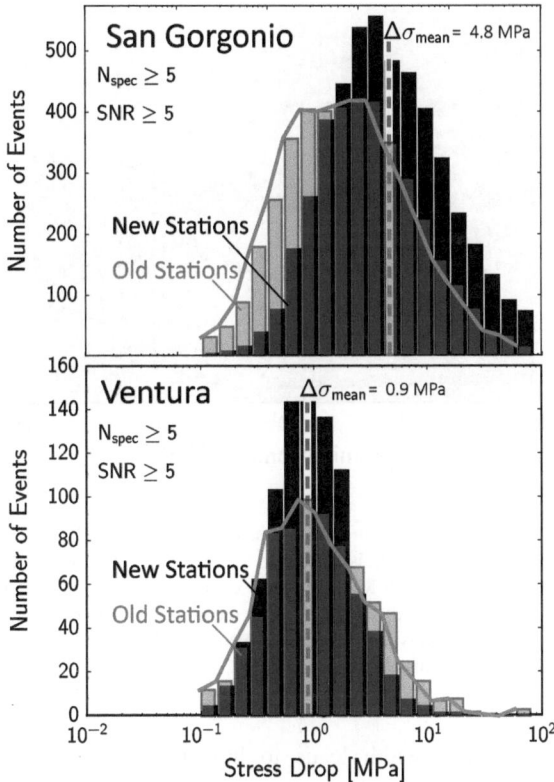

Figure 6
Histogram of stress drops in SGP (*top*) and VB (*bottom*). The histograms clearly illustrate the higher average stress drop (*vertical dashed lines*) and larger scatter in SGP. For comparison, we also plotted stress drop estimates using previous instrumentation (*gray vertical bars* and *curves*). Computing stress drops from new broadband station records led to a decrease in scatter and a better resolution of high stress drop events especially in SGP

the difference in stress drops between these regions is rooted in significantly different corner frequencies (Fig. 8). Moreover, the differences in source spectra between VB and SGP demonstrate that corner frequencies in SGP are systematically higher than in VB for similar size earthquakes.

In the following section, we explore possible biases in stress drop estimates related to lateral variations in attenuation structure and seismic moment computations.

5.2. Possible Source of Bias in Stress Drop Estimates Leading to Systematic Differences

While the statistically significant variations in large stress drop populations are intriguing and may indicate systematic, physical differences in source

processes, there are also a number of sources for systematic bias. Sources of uncertainty include seismic moment and EGF estimates as well as uncertainties in spectral fits and corner frequencies.

Since seismic moments are computed by linear regression of long-period spectral amplitudes Ω_0 and local magnitude M_L, differences in the regression between VB and SGP can lead to systematic biases in stress drop estimates. We test this by simply using the same regression relationship for VB and SGP. This approach leads to a poorer fit of the data in VB while at the same time lowering the average stress drop in VB even further to ~ 0.7 MPa. Thus, the systematically lower stress drops in VB are not a result of a bias in seismic moment estimate.

Another source of systematic variation are lateral differences in attenuation structure that are not accounted for in the regional EGF and travel time-binned path terms. This is especially problematic in regions such as the greater Ventura area where many ray paths travel through the deep sedimentary basin. These basin effects may lead to significantly attenuated high-frequency energy and lower corner frequencies.

We perform two tests to evaluate the influence of basin effects and attenuation structure: (1) We created a sub-set of data in the VB study area that includes only stations and earthquakes north of Ventura basin and repeat the stress drop calculation. (2) We perform the stress drop analysis using spatially varying EGFs for subsets of data, comprised of the 200 nearest neighbor events. Neither of these tests result in a reduction of stress drop difference between SGP and VB (Table 1).

The last source of bias considered here are systematic uncertainties in f_c estimates. We determined the uncertainty in corner frequencies using the 90 % confidence interval of the spectral fit of source terms. Propagating f_c uncertainty into stress drop estimates results in an expected variation of about one order of magnitude which cannot readily explain the systematic differences observed here. However, f_c uncertainty also depends on the fitting bandwidth and absolute values of f_c relative to the Nyqvist frequency. For a reliable estimation of f_c, our spectral fitting method requires a notable curvature within the fitting bandwidth of 1–20 Hz. Conceivably, this

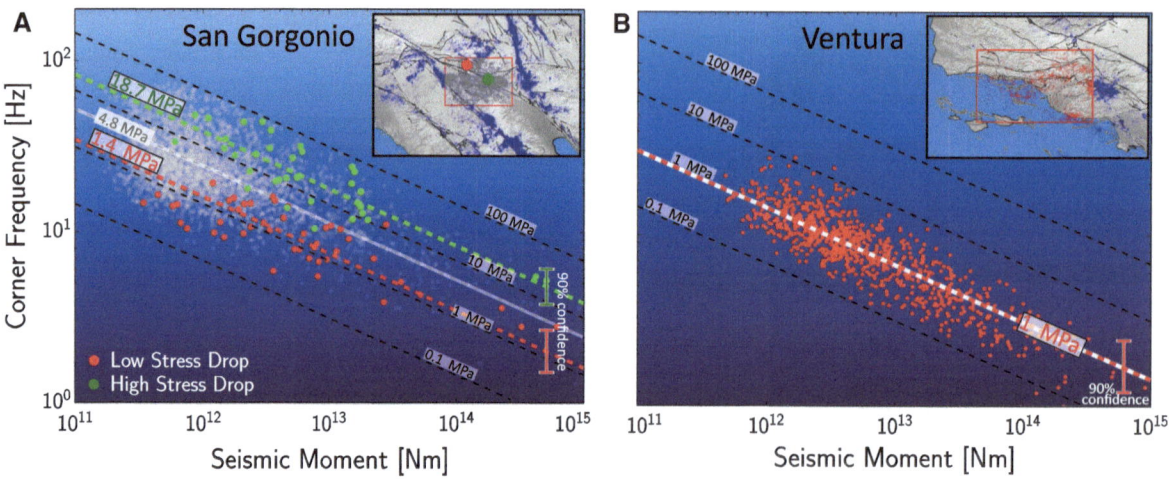

Figure 7
Corner frequency and seismic moment determined from source spectra of individual earthquakes in SGP (*left*) and VB (*right*). Constant stress drops are illustrated by *dashed lines* (assuming constant rupture velocity). While the VB region is connected to lower stress drop estimates (∼1 MPa) and less scatter the SGP region is characterized by larger average stress drop (∼4.8 MPa) and more scatter. The scatter in SGP is partially related to stress drop variations in geographically distinct areas illustrated by *red* and *green markers* in the inset in **a**. The *green* and *red error bars* at the bottom right of each plot represent the average 90 % confidence interval of all corner frequencies. The corresponding stress drop uncertainty is approximately one order of magnitude

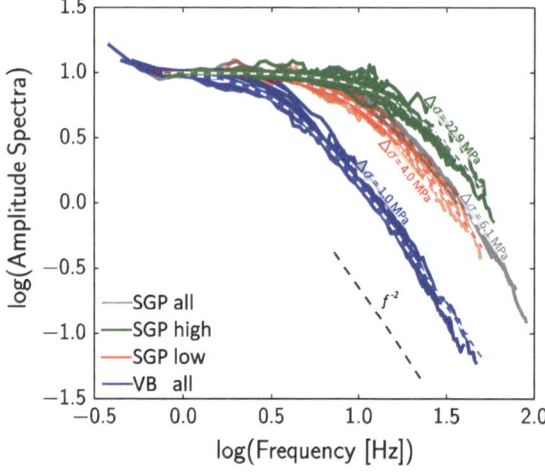

Figure 8
Magnitude-binned, stacked source spectra after correcting for differences in M_0 for the greater VB (*blue*) and SGP (*gray*) as well as for the low (*red*) and high (*green*) stress drop regions in SGP (see insets in Fig. 7 for locations). The source spectra are stacked in 0.2 magnitude bins and shifted along f^{-3}. Self-similar scaling leads to spectral collapse after shifting. Note that the quality of spectral collapse is also influenced by the number of events which are used to compute average spectra, leading to apparently better results for larger study regions

limited bandwidth leads to higher uncertainty for high f_c values which correspond to less spectral curvature below 20 Hz. This effect may in part explain the larger scatter of stress drop values in SGP but is likely not responsible for the observed systematic stress drop differences.

5.3. Stress Field Heterogeneity in the Western and Eastern Transverse Ranges

To obtain a more detailed insight into the stress field heterogeneity across the Transverse ranges, we computed stress orientations and faulting regimes in SGP and VB. The stress field estimates allow for a correlation of stress drop with dominant type of faulting as well as with the overall variability in stress orientations.

To invert for stress fields in VB and SGP, we use high-quality focal mechanisms data sets of individual earthquakes. Focal mechanisms were computed under considerations of uncertainties related to velocity structure, take-off angle and polarity information

Table 1

Log-mean stress drop estimates for regionally varying empirical green functions (EGFs)

Region	Regional eGF	Spatially varying eGF
San Gorgonio	4.8 (1.0, 24.8)	5.1 (0.7, 26.4)
Ventura	0.9 (0.3, 2.5)	0.8 (0.2, 4.2)
North of Ventura Basin	0.7 (0.2, 1.5)	0.5 (0.1, 1.6)

Stress drops are in MPa and values in parentheses are 10th and 90th percentile

using the HASH algorithm HARDEBECK and SHEARER (2002). We then determine regional stress field variations using a damped inversion methodology, implemented in the SATSI algorithm HARDEBECK and MICHAEL (2006); MARTÍNEZ-GARZÓN *et al.* (2014). The inversion uses sets of neighboring earthquake focal mechanisms to find the stress orientations that best fit the data. The damped stress inversion avoids over-fitting of the data, and is thereby less sensitive to artifacts while at the same time reliably resolving stress variations required to fit the data HARDEBECK and MICHAEL (2006); YANG and HAUKSSON (2013).

The computed stress field in both VB and SGP is dominated by north-south compression but shows significantly more heterogeneity in SGP (Fig. 9). Stress heterogeneity is expressed in the variability of faulting regimes in SGP, ranging from normal to strike-slip and intermediate faulting regimes. More-over, focal mechanism solutions for individual events (not shown) vary substantially and indicate strong thrust components toward the east of SGP. This stress field heterogeneity is most clearly observed in the variations of P-axes orientations. In VB, the P-axis trend is predominantly horizontal and the plunge varies between $\sim 5°$ E and $\sim 15°$ W of north. In SGP, on the other hand, the P-plunge varies between horizontal and vertical between the normal, interme-diate and strike-slip faulting regimes and the P-trend varies between $\sim 10°$ E and $\sim 15°$ W of north (see polar plots in Fig. 9).

The resolved stress fields in VB and SGP are in agreement with the larger scale tectonic setting in the study areas with VB being dominated by a system of east-west trending thrust faults whereas SGP hosts an interconnected system of strike-slip, thrust and nor-mal faults resulting in the observed variation in average faulting types.

If stress drop variations were connected to absolute level of differential stress, we would expect higher stress drop in the thrust-fault dominated VB, since thrust faulting is commonly associated with higher differential stresses (e.g. SIBSON 1974). This is clearly not observed. Moreover, we examined a possible correlation between spatially varying *b*-value estimates and stress drops in SGP using a 200-event nearest neighbor sampling approach. The SGP region showed little spatial variation in *b*-value, occupying values between 0.9 and 1.15. Moreover, these variations showed no significant correlations to changes in stress drops further suggesting that changes in average differential stresses are of secondary importance for changes in stress drop.

6. Discussion

6.1. Uncertainty and Physical Variations in Stress Drop Estimates

It has been subject of ongoing debate whether earthquake source parameter scale in a self-similar way for different magnitudes and study regions, or whether stress drops vary systematically as a function of fault properties and crustal conditions. Our detailed investigation of stress drop variability revealed systematic lateral differences in stress drop estimates with distinctly lower stress drops in VB than SGP. The spatial separation of low and high stress drop region in SGP as well as significant differences between SGP and VB, suggests that stress drop variations are in part driven by physical variations in source processes.

Spatial variations in stress drop estimates are at some level connected to the inherent heterogeneity in crustal attenuation structure and limitations of

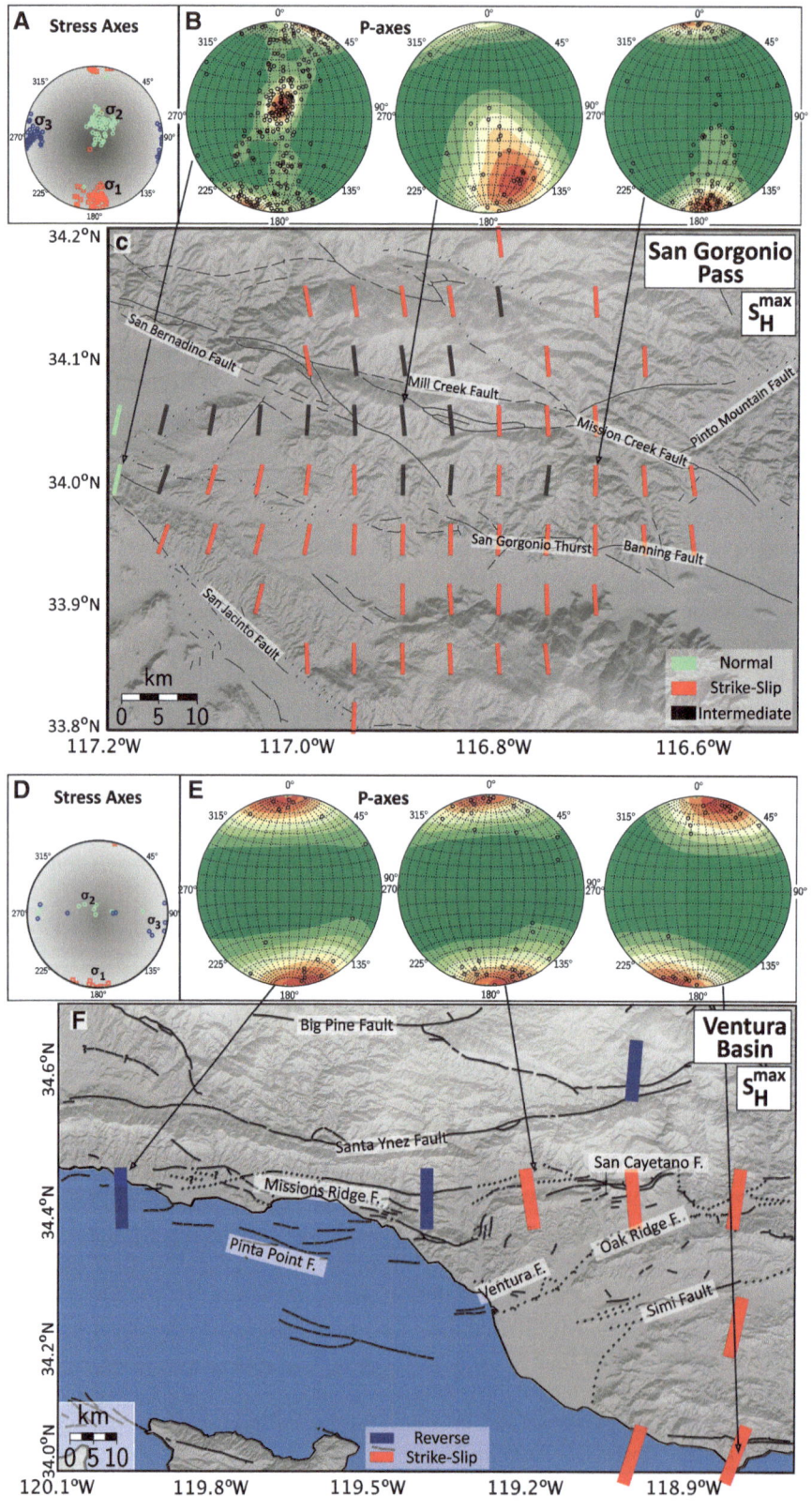

◄
Figure 9
Stress field heterogeneity in SGP (*A–C*) and VB (*D–E*) as well as direction of S_H^{max} and corresponding faulting type (*C, F*). Principal stress orientations ($\sigma_1, \sigma_2, \sigma_3$) are indicated by *red, green* and *blue circles* in insets A and D. *Red colors* in B and E indicate plunge and azimuth with the highest density of P-axes orientation for three different subregions in SGP and VB. Both the principal stress and average P-axes orientations indicate more stress heterogeneity in SGP than in VB. The average types of faulting (see legend in *C*) in SGP includes normal, strike-slip and intermediate whereas VB is dominated by reverse and strike-slip mechanisms. Note that the spatial resolution of S_H^{max} in VB is limited because of the lower seismic activity

seismic instrumentation which both complicate a simple detection of physical stress drop variations. Limited recording bandwidth and sampling rates are especially problematic during the estimate of high corner frequencies for small magnitude events like in the present study. Here, we attempted to mitigate these issues using large data sets and averaging over many stations as well as stacking results over many events. This approach seems successful in resolving significant differences in stress drop estimates between VB and SGP as well as within the structural knot of the San Andreas fault at San Gorgonio Pass.

6.2. *What are the Primary Controlling Parameters For Stress Drop Variations?*

In the following, we examine five plausible factors (i.e., changes in differential stress level, loading rates, pore pressures, lithologic variations, and stress field heterogeneity) that may lead to stress drop variations and discuss in how far they may be responsible for the here observed variations in stress drops:

1. Stress drops may be affected by relatively higher differential stresses. Proxies for such stress variations are lower *b*-value SCHORLEMMER *et al.* (2005); GOEBEL *et al.* (2013b), increased thrust faulting (e.g. SIBSON 1974) and larger focal depth. None of these three parameters is significantly correlated with stress drops so that differential stress is likely of little importance for stress drop variations.

2. We explored the role of fault healing and asperity strength connected to differences in fault loading rates. In frictional sliding experiments, fault

healing and asperity-strength scale with the logarithm of hold times between slip events, resulting in a percentage increase in friction per decade increase in hold times (e.g. MARONE 1998; BEELER *et al.* 2001). Similarly, stress drop of subsequent slip events, $\Delta\tau_s$ can be expressed as a function of normal stress, σ_n, frictional parameters, a_f and b_f and the logarithm of loading velocity, V_1, relative to some reference velocity, V_0 (e.g., GU and WONG 1991; HE *et al.* 2003; RUBIN and AMPUERO 2005):

$$\Delta\tau_s = \sigma_n(b_f - a_f)\ln(V_0/V_1) \qquad (4)$$

At room temperatures and low normal stresses the resulting changes in stress drop are expected to be small, likely below what can be resolved with our method. However, at seismogenic conditions which cannot easily be reproduced in the laboratory and include high temperatures and confining pressures, healing rates and the effect of loading rate may be amplified (e.g., MARONE 1998). Such a systematic influence of tectonic loading rates on stress drops along the San Andreas fault was indeed observed in a recent study GOEBEL *et al.* (2015). Similarly, rapid contraction rates are observed across the Ventura basin based on GPS measurements and geologic reconstructions occupying values between 6 and 10 mm/year DONNELLAN *et al.* (1993); MARSHALL *et al.* (2013). If faults are connected at depth and slip rapidly as suggested by YEATS (1983) and HUBBARD *et al.* (2014), they may be key agents that accommodate rapid contraction over geologic time scales. Such a scenario would also explain the lower stress drops connected to weaker asperities on more rapidly loaded faults in VB compared to SGP. Moreover, recurrence intervals of major earthquakes within the SGP fault system are likely much longer than in VB (e.g. YULE and SIEH 2003; SCHARER *et al.* 2007), suggesting a likely coupling between slower loading rates, longer recurrence intervals and elevated stress drops.

3. Increased pore-fluid pressures as a result of deep sedimentary units that are under large compressive stresses may further lower stress drops in the Ventura area. Such an influence of higher pore pressures on stress drop estimates has been documented during fluid-injection in Basel, Switzerland GOERTZ-ALLMANN *et al.* (2011). The influence of fluid pressure on stress drop is

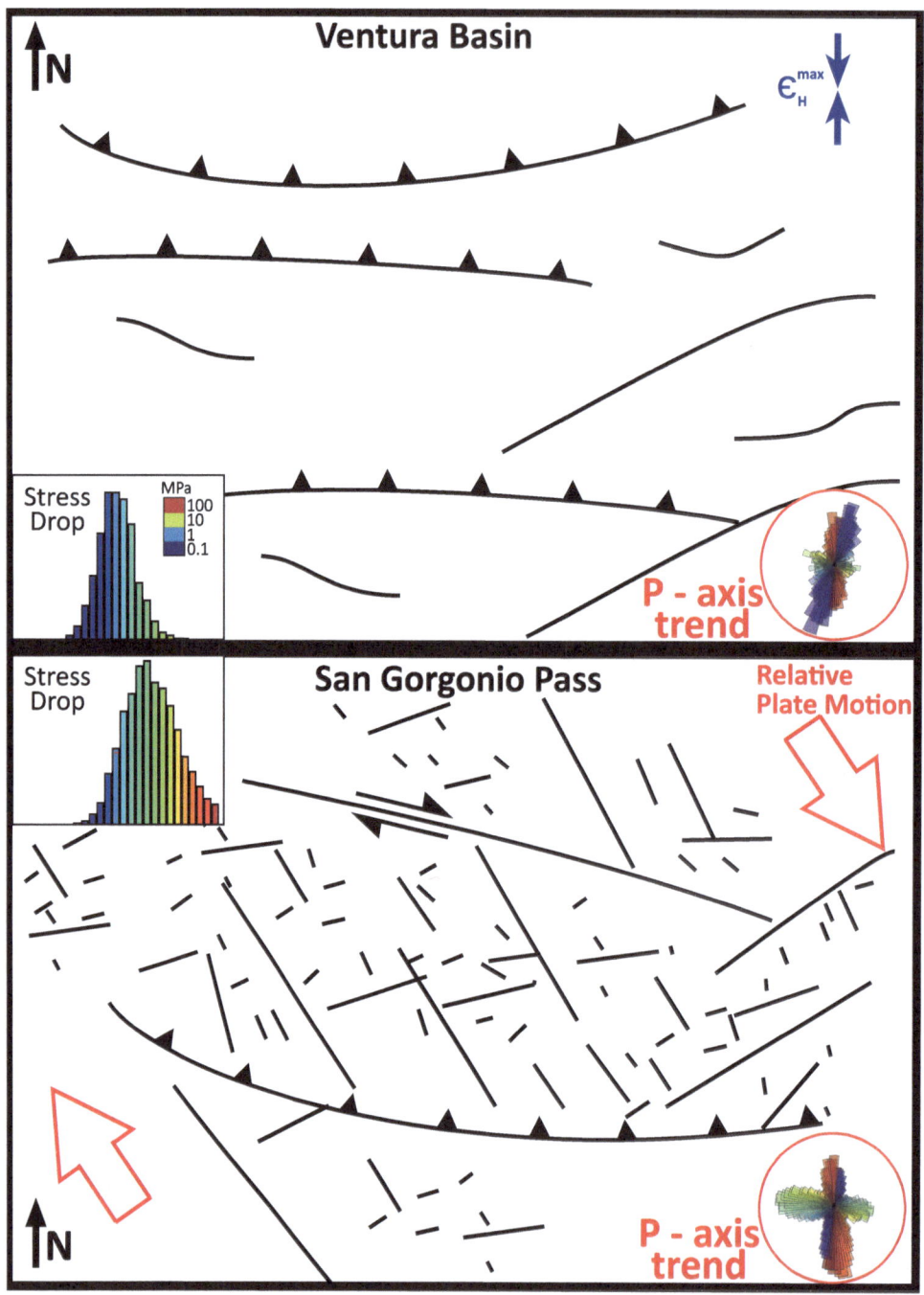

Figure 10
Schematic representations of joint variations in inferred fault geometry and stress field heterogeneity in SGP and VB. The greater VB (*top*) is characterized by rapid N-S contraction rates across the basin, generally more homogeneous stresses and more localized seismicity along major fault traces [average direction of maximum horizontal strain is indicated by *blue arrows*, after San Andreas strain removal from MARSHALL *et al.* (2013)]. The SGP region (*bottom*), on the other hand, is characterized by more distributed seismicity, heterogenous slip rates which are lowest in the area of highest stress drop and a strongly heterogeneous stress field resulting in normal, strike-slip and thrust-type focal mechanisms

difficult to resolve because of the comparably low seismic event density within sedimentary basins and lack of measurements to resolve pressure variations at seismogenic depth.

4. Systematic difference in lithology and crustal composition may influence stress drop variations. Variations in rock composition and connected changes in brittle–ductile transition temperatures may influence the comparably rapid decrease in stress drop estimates north of the San Andreas fault in the SGP region GOEBEL et al. (2015). In this region, the dominant rock composition changes rather abruptly across the Mission creek segment from more brittle feldspar-dominated to more ductile quartzitic rocks. This change in lithology is also supported by changes in seismic velocities within the area.

5. In addition to the geologic differences between VB and SGP, we showed that the overall stress field heterogeneity, including variations in stress orientations and average types of faulting, is significantly higher in SGP than VB. The stress heterogeneity in SGP may be closely linked to fault geometric complexity of the compressional step-over in the San Andreas fault system (e.g., CATCHINGS et al. 2009; CARENA et al. 2004; DAIR and COOKE 2009). Based on the fractal analysis of hypocenters, we found that SGP hosts more isolated rupture planes which can rarely be associated with major fault zones, whereas the fractal dimension of earthquake hypocenter distributions in VB indicates a more localized fault network. This is in agreement with a more heterogeneous stress field in SGP. Thus, the larger geometric and stress field heterogeneity together with much off-fault seismicity may contribute to the higher than average stress drops in SGP (Fig. 10). This implies that at present much of the seismicity in SGP does not reflect the expected long-term deformation along the major faults in the region but occurs in form of heterogeneous faulting throughout the seismogenic crust.

In summary, our results suggest that loading rate, stress field heterogeneity and lithologic differences may be the primary parameters that control the observed stress drop variations in SGP and VB. More specifically, larger variations in stress orientations and types of faulting as well as more distributed faulting promote higher than average stress drops in SGP in addition to possible differences in tectonic loading rates and lithology. In VB, on the other hand, the comparably simpler stress field and alignment of expected, average fault slip vectors and principle stress orientations lead to lower stress drops and more localized faulting.

Our stress drop estimates indicate scale-dependent self-similarity expressed in large stress drop variations at the scale of the entire study area in SGP. These large variations are likely a result of spatially varying stress drops around the average value of ~ 4.8 MPa with several sub-regions showing significant deviations from the average value. The scale at which significant stress drop variations occur is related to the scale of crustal heterogeneity, e.g., controlled by the wavelengths of stress field heterogeneity and loading rate changes.

If the described physical mechanisms related to fault slip and stress field heterogeneity are controlling rupture processes of small earthquakes, these conditions are expected to also influence large earthquake ruptures. Namely, if the amount of high-frequency source energy is controlled by crustal conditions and loading rates within a specific area the same will hold true for both small and large earthquakes, leading to high stress drops of small and large earthquakes, for example, in SGP.

7. Conclusion

We analyzed source spectra and stress drops of more than 6000 micro-earthquakes in SGP and VB and resolved statistically significant variations in earthquake stress drop. Our results suggest that these stress drop variations are scale-dependent, i.e. at the scale of the study areas (~ 100 km), we observed low stress drop of ~ 1 MPa in VB whereas the SGP area showed elevated stress drops (~ 5 MPa) and larger variations. At a smaller scale of ~ 5–10 km the variations in average stress drop estimates in SGP between 2 to 20 MPa are mainly connected to differences in fault loading rates and lithologic differences.

A comparison of seismicity statistics and stresses between the study regions revealed joined variations in spatial earthquake distribution (fractal dimension), stress drops and stress field heterogeneity, so that more localized faulting and homogeneous stress fields seem to favor lower stress drops. Thus stress drop variations may be driven by the overall stress heterogeneity (i.e., variations in stress orientations and type of faulting), which is best resolvable at the scale of 50–100 km, as well as by changes in loading rates and lithology which are best resolvable at the ~ 10 km scale.

Moreover, the dependence of stress drop on crustal conditions in specific areas implies that stress drops of both small and larger magnitude earthquakes should be affected similarly, resulting in generally more high-frequency seismic energy radiation from these areas. For example, we expect that the structural knot in SGP produces high stress drops in both small and large earthquake ruptures. Such large ruptures and high stress drops are of prime interest in engineering seismology and should be considered in ground motion predictions.

Acknowledgments

The initial manuscript benefitted from comments by Xiaowei Chen and Grzegorz Kwiatek. We would like to thank Michele Cooke for her detailed review. T. Goebel and E. Hauksson were supported by NEHRP/USGS grant G15AP00095 and the Southern California Earthquake Center (SCEC) under contribution number 12017 and 14033. SCEC is funded by NSF Cooperative Agreement EAR-0529922 and USGS Cooperative Agreement 07HQAG0008. We have used waveforms and parametric data from the Caltech/USGS Southern California Seismic Network (SCSN); DOI: 10.7914/SN/CI; stored at the Southern California Earthquake Center. DOI: 10.7909/C3WD3xH1.

REFERENCES

ABERCROMBIE, R. E. (1995), *Earthquake source scaling relationships from −1 using seismograms recorded at 2.5 km depth*, J. Geophys. Res., *100*(B12), 24,015–36.

ABERCROMBIE, R. E. (2013), *Comparison of direct and coda wave stress drop measurements for the Wells, Nevada, earthquake sequence*, J. Geophy. Res., *118*, doi:10.1029/2012JB009638.

ABERCROMBIE, R. E. (2015), *Investigating uncertainties in empirical green's function analysis of earthquake source parameters*, J. Geophy. Res., *120*, doi:10.1002/2015JB011984.

AKI, K. (1967), *Scaling law of seismic spectrum*, J. Geophys. Res., *72*, 1217–1231.

ALLMANN, B. P., and P. M. SHEARER (2007), *Spatial and temporal stress drop variations in small earthquakes near Parkfield, California*, J. Geophs. Res., *112*(B4), B04,305.

ALLMANN, B. P., and P. M. SHEARER (2009), *Global variations of stress drop for moderate to large earthquakes*, J. Geophys. Res., *114*(B1), B01,310.

ANDREWS, D. J. (1986), *Objective determination of source parameters and similarity of earthquakes of different size*, in Earthquake Source Mechanics, pp. 259–267, doi:10.1029/GM037p0259.

ATKINSON, G. M., and I. BERESNEV (1997), *Don't call it stress drop*, Seismological Research Letters, *68*(1), 3–4.

BEELER, N. M., S. H. HICKMAN, and T.-f. WONG (2001), *Earthquake stress drop and laboratory-inferred interseismic strength recovery*, J. Geophys. Res., *106*(B12), 30,701–30,713.

BERESNEV, I. A. (2009), *The reality of the scaling law of earthquake-source spectra?*, J. Seismol., *13*(4), 433–436.

BRUNE, J. N. (1970), *Tectonic stress and the spectra of seismic shear waves from earthquakes*, J. Geophys. Res., *75*(26), 4997–5009.

BURGETTE, R. J., K. M. JOHNSON, and W. C. HAMMOND (2015), Observations of vertical deformation across the western transverse ranges and constraints on ventura area fault slip rates, 2015 SCEC Annual Meeting Abstracts, p. 201.

CARENA, S., J. SUPPE, and H. KAO (2004), *Lack of continuity of the San Andreas fault in Southern California: Three-dimensional fault models and earthquake scenarios*, J. Geophys. Res., *109*(B4).

CATCHINGS, R., M. RYMER, M. GOLDMAN, and G. GANDHOK (2009), *San Andreas fault geometry at Desert Hot Springs, California, and its effects on earthquake hazards and groundwater*, Bull. Seismol. Soc. Am., *99*(4), 2190–2207.

CHEN, X., and P. M. SHEARER (2013), *California foreshock sequences suggest aseismic triggering process*, Geophys. Res. Letts., *40*(11), 2602–2607.

CLAUSET, A., C. R. SHALIZI, and M. E. J. NEWMANN (2009), *Power-law distributions in empirical data*, SIAM review, *51*(4), 661–703.

COOKE, M. L., and L. C. DAIR (2011), *Simulating the recent evolution of the southern big bend of the San Andreas fault, Southern California*, J. Geophys. Res., *116*(B4).

DAIR, L., and M. L. COOKE (2009), *San Andreas fault geometry through the San Gorgonio Pass, California*, Geology, *37*(2), 119–122.

DONNELLAN, A., B. H. HAGER, and R. W. KING (1993), *Discrepancy between geological and geodetic deformation rates in the Ventura Basin*, Nature, *366*(6453), 333–336.

ESHELBY, J. D. (1957), *The determination of the elastic field of an ellipsoidal inclusion, and related problems*, Proceedings of the Royal Society of London. Series A. Mathematical and Physical Sciences, *241*(1226), 376–396.

GOEBEL, T. H. W., C. G. SAMMIS, T. W. BECKER, G. DRESEN, and D. SCHORLEMMER (2013a), *A comparison of seismicity characteristics and fault structure in stick-slip experiments and nature*, Pure Appl. Geophys., doi:10.1007/s00024-013-0713-7.

GOEBEL, T. H. W., D. SCHORLEMMER, T. W. BECKER, G. DRESEN, and C. G. SAMMIS (2013b), *Acoustic emissions document stress changes over many seismic cycles in stick-slip experiments*, Geophys. Res. Letts., *40*, doi:10.1002/grl.50507.

GOEBEL, T. H. W., E. HAUKSSON, J.-P. AMPUERO, and P. M. SHEARER (2015), *Stress drop heterogeneity within tectonically complex regions: A case study of San Gorgonio Pass, southern California*, Geophys. J. Int., *202*(1), 514–528, DOI 10.1093/gji/ggv160.

GOERTZ-ALLMANN, B. P., A. GOERTZ, and S. WIEMER (2011), *Stress drop variations of induced earthquakes at the basel geothermal site*, Geophys. Res. Letts., *38*(9).

GOODFELLOW, S., and R. YOUNG (2014), *A laboratory acoustic emission experiment under in situ conditions*, Geophysical Research Letters, *41*(10), 3422–3430.

GRAVES, R. W., B. T. AAGAARD, K. W. HUDNUT, L. M. STAR, J. P. STEWART, and T. H. JORDAN (2008), *Broadband simulations for M_w 7.8 southern San Andreas earthquakes: Ground motion sensitivity to rupture speed*, Geophys. Res. Lett., *35*(22).

GU, Y., and T.-f. WONG (1991), *Effects of loading velocity, stiffness, and inertia on the dynamics of a single degree of freedom spring-slider system*, J. Geophys. Res., 96(B13), 21,677–21,691.

HARDEBECK, J. L., and A. J. MICHAEL (2006), *Damped regional-scale stress inversions: Methodology and examples for southern California and the Coalinga aftershock sequence*, Journal of Geophysical Research: Solid Earth (1978–2012), *111*(B11).

HARDEBECK, J. L., and P. M. SHEARER (2002), *A new method for determining first-motion focal mechanisms*, Bulletin of the Seismological Society of America, *92*(6), 2264–2276.

HARRINGTON, R. M., and E. E. BRODSKY (2009), *Source duration scales with magnitude differently for earthquakes on the San Andreas Fault and on secondary faults in Parkfield, California*, Bull. Seismol. Soc. Am., *99*(4), 2323–2334.

HAUKSSON, E. (2014), *Average stress drops of southern California earthquakes in the context of crustal geophysics: Implications for fault zone healing*, Pure Appl. Geophys., pp. 1–12, doi:10.1007/s00024-014-0934-4.

HE, C., T.-f. WONG, and N. M. BEELER (2003), *Scaling of stress drop with recurrence interval and loading velocity for laboratory-derived fault strength relations*, J. Geophys. Res., *108*(B1), doi:10.1029/2002JB001890.

HUBBARD, J., J. H. SHAW, J. DOLAN, T. L. PRATT, L. MCAULIFFE, and T. K. ROCKWELL (2014), *Structure and seismic hazard of the Ventura Avenue Anticline and Ventura Fault, California: Prospect for large, multisegment ruptures in the western Transverse Ranges*, Bull. Seismol. Soc. Am., doi:10.1785/0120130125.

KANAMORI, H., and D. L. ANDERSON (1975), *Theoretical basis of some empirical relations in seismology*, Bull. Seismol. Soc. Am., *65*, 1073–1095.

KANAMORI, H., J. MORI, E. HAUKSSON, T. H. HEATON, L. K. HUTTON, and L. M. JONES (1993), *Determination of earthquake energy release and m_l using terrascope*, Bull. Seismol. Soc. Am., *83*(2), 330–346.

KANEKO, Y., and P. M. SHEARER (2014), *Seismic source spectra and estimated stress drop derived from cohesive-zone models of circular subshear rupture*, Geophys. J. Int., doi:10.1093/gji/ggu030, (in press).

KWIATEK, G., K. PLENKERS, and G. DRESEN (2011), *Source parameters of picoseismicity recorded at Mponeng deep gold mine, South Africa: implications for scaling relations*, Bull. Seism. Soc. Am., *101*(6), 2592–2608.

LANGENHEIM, V. E., R. C. JACHENS, J. C. MATTI, E. HAUKSSON, D. M. MORTON, and A. CHRISTENSEN (2005), *Geophysical evidence for wedging in the San Gorgonio Pass structural knot, southern San Andreas fault zone, southern California*, Geological Society of America Bulletin, *117*(11-12), 1554–1572.

LIN, Y.-Y., K.-F. MA, and V. OYE (2012), *Observation and scaling of microearthquakes from the Taiwan Chelungpu-fault borehole seismometers*, Geophys. J. Int., *190*(1), 665–676.

MADARIAGA, R. (1976), *Dynamics of an expanding circular fault*, Bull. Seismol. Soc. Am., *66*(3), 639–666.

MAGISTRALE, H., and C. SANDERS (1996), *Evidence from precise earthquake hypocenters for segmentation of the San Andreas fault in San Gorgonio Pass*, J. Geophys. Res., *101*(B2), 3031–3044.

MARONE, C. (1998), *Laboratory-derived friction laws and their application to seismic faulting*, Annu. Rev. Earth Planet. Sci., *26*, 643–696.

MARSHALL, S. T., G. J. FUNNING, and S. E. OWEN (2013), *Fault slip rates and interseismic deformation in the western transverse ranges, california*, Journal of Geophysical Research: Solid Earth, *118*(8), 4511–4534.

MARTÍNEZ-GARZÓN, P., G. KWIATEK, M. ICKRATH, and M. BOHNHOFF (2014), *MSATSI: A MATLAB package for stress inversion combining solid classic methodology, a new simplified user-handling, and a visualization tool*, Seismological Research Letters, *85*(4), 896–904.

MATTI, J. C., and D. M. MORTON (1993), *Paleogeographic evolution of the San Andreas fault in southern California: A reconstruction based on a new cross-fault correlation*, Geological Society of America Memoirs, *178*, 107–160.

MCLASKEY, G. C., A. M. THOMAS, S. D. GLASER, and R. M. NADEAU (2012), *Fault healing promotes high-frequency earthquakes in laboratory experiments and on natural faults*, Nature, *491*(7422), 101–104.

NADEAU, R. M., and L. R. JOHNSON (1998), *Seismological studies at parkfield VI: Moment release rates and estimates of source parameters for small repeating earthquakes*, Bull. Seismol. Soc. Am., *88*(3), 790–814.

OTH, A. (2013), *On the characteristics of earthquake stress release variations in Japan*, Earth and Planetary Science Letters, *377*, 132–141.

PLESCH, A., et al. (2007), *Community fault model (CFM) for southern California*, Bulletin of the Seismological Society of America, *97*(6), 1793–1802.

PRIETO, G. A., P. M. SHEARER, F. L. VERNON, and D. KILB (2004), *Earthquake source scaling and self-similarity estimation from stacking P and S spectra*, J. Geophys. Res., *109*(B8).

PRIETO, G. A., D. J. THOMSON, F. L. VERNON, P. M. SHEARER, and R. L. PARKER (2007), *Confidence intervals for earthquake source parameters*, Geophys. J. Int., *168*(3), 1227–1234.

ROCKWELL, T. K. (2011), Large co-seismic uplift of coastal terraces across the Ventura Avenue anticline: Implications for the size of earthquakes and the potential for tsunami generation, 2011 Annual Meeting Abstracts, 21, (Plenary talk).

RUBIN, A. M., and J.-P. AMPUERO (2005), *Earthquake nucleation on (aging) rate and state faults*, J. Geophys. Res., *110*(B11), doi:10.1029/2005JB003686.

SAMMIS, C. G., and J. R. RICE (2001), *Repeating earthquakes as low-stress-drop events at a border between locked and creeping fault patches*, Bull. Seismol. Soc. Am., *91*(3), 532–537.

SATO, T., and T. HIRASAWA (1973), *Body wave spectra from propagating shear cracks*, J. Phys. Earth, *21*(4), 415–431.

SCHARER, K. M., R. J. WELDON, T. E. FUMAL, and G. P. BIASI (2007), *Paleoearthquakes on the southern San Andreas fault, Wrightwood, California, 3000 to 1500 BC: A new method for evaluating paleoseismic evidence and earthquake horizons*, Bull. Seismol. Soc. Am., *97*(4), 1054–1093.

SCHORLEMMER, D., S. WIEMER, and M. WYSS (2005), *Variations in earthquake-size distribution across different stress regimes*, Nature, *437*, 539–542, DOI 10.1038/nature04094.

SHAW, J. H., et al. (2015), *Unified structural representation of the southern California crust and upper mantle*, Earth and Planetary Science Letters, *415*, 1–15.

SHEARER, P. M. (2009), Introduction to seismology, Cambridge University Press.

SHEARER, P. M., G. A. PRIETO, and E. HAUKSSON (2006), *Comprehensive analysis of earthquake source spectra in southern California*, J. Geophs. Res., *111*(B6), B06,303.

SIBSON, R. H. (1974), *Frictional constraints on thrust, wrench and normal faults*, Nature, *249*, 542–544.

WALTER, W. R., K. MAYEDA, R. GOK, and A. HOFSTETTER (2006), The scaling of seismic energy with moment: Simple models compared with observations, Earthquakes: Radiated energy and the physics of faulting, pp. 25–41.

WARREN, L. M., and P. M. SHEARER (2000), *Investigating the frequency dependence of mantle Q by stacking P and PP spectra*, J. Geophys. Res., *105*(B11), 25,391-25.

WYSS, M., C. G. SAMMIS, R. M. NADEAU, and S. WIEMER (2004), *Fractal dimension and b-value on creeping and locked patches of the San Andreas fault near Parkfield, California*, Bull. Seismol. Soc. Am., *94*, 410–421.

YANG, W., and E. HAUKSSON (2011), *Evidence for vertical partitioning of strike-slip and compressional tectonics from seismicity, focal mechanisms, and stress drops in the east Los Angeles basin area, California*, Bull. Seismol. Soc. Am., *101*(3), 964–974.

YANG, W., and E. HAUKSSON (2013), *The tectonic crustal stress field and style of faulting along the Pacific North America Plate boundary in Southern California*, Geophys. J. Int., *194*(1), 100–117.

YANG, W., Z. PENG, and Y. BEN-ZION (2009), *Variations of strain-drops of aftershocks of the 1999 İzmit and Düzce earthquakes around the Karadere-Düzce branch of the North Anatolian Fault*, Geophys. J. Int., *177*(1), 235–246.

YEATS, R. S. (1983), *Large-scale Quaternary detachments in Ventura Basin, southern California*, J. Geophys. Res., *88*(B1), 569–583.

YULE, D., and K. SIEH (2003), *Complexities of the San Andreas fault near San Gorgonio Pass: Implications for large earthquakes*, J. Geophys. Res., *108*, doi:10.1029/2001JB000451.

(Received February 2, 2016, revised May 29, 2016, accepted May 30, 2016, Published online June 13, 2016)

Pure Appl. Geophys. 174 (2017), 2331–2349
© 2016 Springer International Publishing
DOI 10.1007/s00024-016-1417-6

Real-Time Earthquake Intensity Estimation Using Streaming Data Analysis of Social and Physical Sensors

Yelena Kropivnitskaya,[1] Kristy F. Tiampo,[1,2] Jinhui Qin,[3] and Michael A. Bauer[4]

Abstract—Earthquake intensity is one of the key components of the decision-making process for disaster response and emergency services. Accurate and rapid intensity calculations can help to reduce total loss and the number of casualties after an earthquake. Modern intensity assessment procedures handle a variety of information sources, which can be divided into two main categories. The first type of data is that derived from physical sensors, such as seismographs and accelerometers, while the second type consists of data obtained from social sensors, such as witness observations of the consequences of the earthquake itself. Estimation approaches using additional data sources or that combine sources from both data types tend to increase intensity uncertainty due to human factors and inadequate procedures for temporal and spatial estimation, resulting in precision errors in both time and space. Here we present a processing approach for the real-time analysis of streams of data from both source types. The physical sensor data is acquired from the U.S. Geological Survey (USGS) seismic network in California and the social sensor data is based on Twitter user observations. First, empirical relationships between tweet rate and observed Modified Mercalli Intensity (MMI) are developed using data from the M6.0 South Napa, CAF earthquake that occurred on August 24, 2014. Second, the streams of both data types are analyzed together in simulated real-time to produce one intensity map. The second implementation is based on IBM Info-Sphere Streams, a cloud platform for real-time analytics of big data. To handle large processing workloads for data from various sources, it is deployed and run on a cloud-based cluster of virtual machines. We compare the quality and evolution of intensity maps from different data sources over 10-min time intervals immediately following the earthquake. Results from the joint analysis shows that it provides more complete coverage, with better accuracy and higher resolution over a larger area than either data source alone.

Key words: Stream computing, high performance computing, parallel computing, physical sensors, socialsensors, hazard estimators.

1. Introduction

Earthquakes are a natural phenomenon that regularly produces significant damage and loss of life. According to the USGS (2015a), every year several million earthquakes occur worldwide. Most are not dangerous due to either their small magnitude or remote location. But others can cause significant economic loss and casualties. For example, the M7.8 Nepal earthquake that occurred on April 25, 2015 killed more than 9000 people, injured more than 23,000 people and destroyed 436,344 houses (NDRRP 2015). Even in well-studied regions with modern building codes such as California, estimated rates of potentially dangerous seismic activity in the north San Francisco Bay area have changed with time. While during the second half of the 19th century, occurrence rates were estimated at one $M \geq 6.0$ earthquake every four years, after the 1906 earthquake, the seismic activity rates significantly decreased until the M6.9 Loma Prieta earthquake in 1989. Today, scientists expect larger and more frequent earthquakes on the basis of increasing regional stresses (USGS 2015c).

Despite the fact that the joint efforts of the scientific and engineering communities have significantly reduced the impact of earthquakes as a result of practical actions and procedures that help to minimize losses after an earthquake, their causes, properties and impacts on human society is still an important topic of scientific research. While engineers are able to build earthquake-resistant buildings, bridges and other infrastructure elements and emergency response services continuously improve population preparedness and risk mitigation, the scientific community continues its studies and simulations of the earthquake source, including the

---

[1] Department of Earth Sciences, Western University, London, Canada. E-mail: ykropivn@uwo.ca; kristy.tiampo@colorado.edu

[2] CIRES, University of Colorado, Boulder CO, USA.

[3] SHARCNET, Western University, London, Canada. E-mail: jqin5@uwo.ca

[4] Department of Computer Science, Western University, London, Canada. E-mail: bauer@uwo.ca

size and extent of the resulting ground motion. Today, continuous analysis of massive volumes of real-time seismic data can be streamed and processed at high speeds and low latency. This technological advantage can be incorporated into post-disaster emergency response systems.

One of the most important measures of damage in emergency response systems is the earthquake intensity. Earthquake intensity quantifies the severity of ground shaking at a given distance from the epicenter and provides a direct measure of the likely damage. Intensity maps are the most common method for spatial representation of intensity levels across a given region. Public and private organizations use intensity maps for both disaster planning and post-earthquake response (Wald et al. 2006a). The MMI scale is the most common intensity scale used today. MMI is a twelve step scale, numbered from I to XII. The numbers represent shaking levels from slight shaking to total destruction (Wood and Neumann 1931; Richter 1958).

There are two general approaches for the estimation of intensity levels. The first approach is designated instrumental intensity level and is based on a regression of kinematic parameters recorded at individual seismic stations. During an earthquake, energy travels in the form of waves from the epicenter and causes ground movement. The two most common measures of ground motion are peak ground acceleration (PGA) and peak ground velocity (PGV), measured in the East–West (EW) direction, North–South (NS) and vertical (UP) directions. In this case, the intensity level is most accurate for locations adjacent to the seismic stations and less accurate at those locations where the ground shaking is obtained through interpolation (Wald et al. 2006b).

The second approach for intensity calculations is based on the shared information from people who experience an earthquake. The current state-of-art is the "Did You Feel It?" program for Community Internet Intensity Maps (CIIM) developed by the USGS. Here, individuals fill in a simple online questionnaire to share information about their earthquake experiences and observations. A CIIM processes the completed questionnaires and assigns average intensity level to each ZIP code. The CIIM is updated in time as additional data are received (Wald et al. 2006a).

The fact that people share their experience and observations in social networks can be extended to improve the CIIM results. In particular, any related observations posted online that can be linked to their geographical location can be incorporated into CIIM analyses. The micro-blogging service Twitter is one potentially significant data source, as it has 255 million active users around the world and connects them to the Internet through both their phones and computers (Campagne et al. 2012). In addition, one of the most important features of Twitter messages (tweets) that makes them useful for improving earthquake response is their real-time nature (Sakaki et al. 2010). Sakaki et al. (2010) first proposed using Twitter users as earthquake sensors and designated them 'social sensors'. To date, several studies have demonstrated the application of Twitter data in post-disaster response systems (Earle et al. 2010, 2011). Earle et al. (2010, 2011) demonstrated how instrument-based event detection and estimation of earthquake location and magnitude could be supplemented by Twitter data. Sakaki et al. (2010) constructed earthquake monitoring, detection and early warning system in Japan based on tweet data. That system sends emails to registered users based on seismic events detected with 96 % probability, using a special tweets classifier based on the tweet keywords, context and the number of words. System notification is delivered much faster when compared to the warnings issued by the Japan Meteorological Agency. Crooks et al. (2012) analyzed the spatial and temporal characteristics of Twitter activity during the M5.8 earthquake that occurred on the east coast of the United States, August 23, 2011 and concluded that Twitter data can complement other sources of data and enhance situational awareness among people. Burks et al. (2014) created a regression model using tweets in the 10 min following an earthquake and integrated it with earthquake characteristics such as moment magnitude, source-to-site distance and shear-wave velocity in the top thirty meters. The main contribution of that work was the demonstration of Tweeter potential as a near real-time complementary data source for earthquake intensity estimation.

In this paper, we take another step forward and present a streaming implementation of an approach for earthquake intensity estimation that integrates data from both social and physical sensors. Results are shown for the M6.0 South Napa, CA earthquake that occurred on August 24, 2014 at 3:20 a.m. local time. The North San Francisco Bay Area portion of the San Andreas Fault system is a complex network of primarily right-lateral strike-slip faults accommodating motion between the North American and Pacific plates. This network of faults is approximately 80 km wide and trends north-northwest in the area of the West Napa Fault. The West Napa Fault transfers slip between a group of related faults (the Contra Costa Shear Zone) which has a maximum slip rate of 1 mm per year.

The epicenter of the South Napa earthquake was located to the south of the city of Napa and to the northwest of American Canyon on the West Napa Fault (Fig. 1). Fifteen thousand people experienced severe shaking, 106,000 people felt very strong shaking, 176,000 felt strong shaking, and 738,000 felt moderate shaking. The duration of shaking lasted from 10 to 20 s, depending on location. One person was killed, approximately 200 people were injured and over $400 million in damage occurred as a result of this event (USGS 2014).

Here, we assume that the rate of tweets that directly reference earthquake shaking is correlated with the intensity of shaking. The goal is to calculate the number of tweets per minute that directly identify each particular earthquake intensity level, based on a set of terms typically used in MMI calculations (Wald et al. 2006a).

2. Selection and Validation of Predictive Relationship Between MMI and Tweets Rate

To develop an empirical relationship between seismic intensity and tweets rate we used two data sets. The first database is provided by the Northern California Earthquake Data Center (NCEDC) and consists of all aggregated reported intensity estimations from instrumental ground motion recordings and represents the most accurate intensity levels after the earthquake. Each value reflects the average estimation of the ground shaking experienced by the public or an assessment of damage level. Figure 2 shows the MMI map with main cities in the region, epicenter of earthquake and seismic stations in the region (NCEDC 2014).

The second dataset is the archive of Twitter records from August 24, 2014 (downloaded from https://archive.org/details/twitterstream). Tweets related to the earthquake (positive tweets) are identified using keywords listed in Table 1. There are several well-known challenges associated with using Twitter data for analytical purposes. One of them is the precise geolocation detection of tweets. Some tweets are geotagged, meaning that they contain the current user location indicator at the time of tweeting. However, the geotagging feature is rarely used by users. For instance, Graham et al. (2014) observed only 0.7 % geotagged tweets among 19.6 million tweets. For tweets containing specific cities, the percentage of geotagging was between 2 and 5 % (Severo et al. 2015). The location of a Twitter user can also be obtained from a field in the user account description. Of user accounts with tweets containing some location information, 7.5 % contained latitude and longitude values, 57 % included a named location, 20.4 % referenced information that helped to identify a country, while 15.1 % provided humorous or nonspatial information (Takhteyev et al. 2012). In this study, we began with geotagged tweets and for tweets with a location assigned in the user profile we assigned that as the location of the current tweet. Finally, for tweets with no georeference parameters we used a text-based geolocation algorithm. That algorithm analyzes the tweets' content and assigns location coordinates according to a coordinates list of major California cities, towns, their shortcuts and keywords. For example, we assigned "SF Bay Area", "SF" and "Bay Area" to San Francisco city coordinates and "South Napa", "SW Napa" and "napa" as city of Napa coordinates.

The Twitter dataset used here was gathered on August 24, 2014, with a total volume of 1.8 gigabytes. The streaming approach provides low latency high-volume access to tweets. The stream of positive tweets on August 24, 2014 is shown in Fig. 3. We limited our analysis to the 10-min period following the earthquake. There were 747 positive tweets

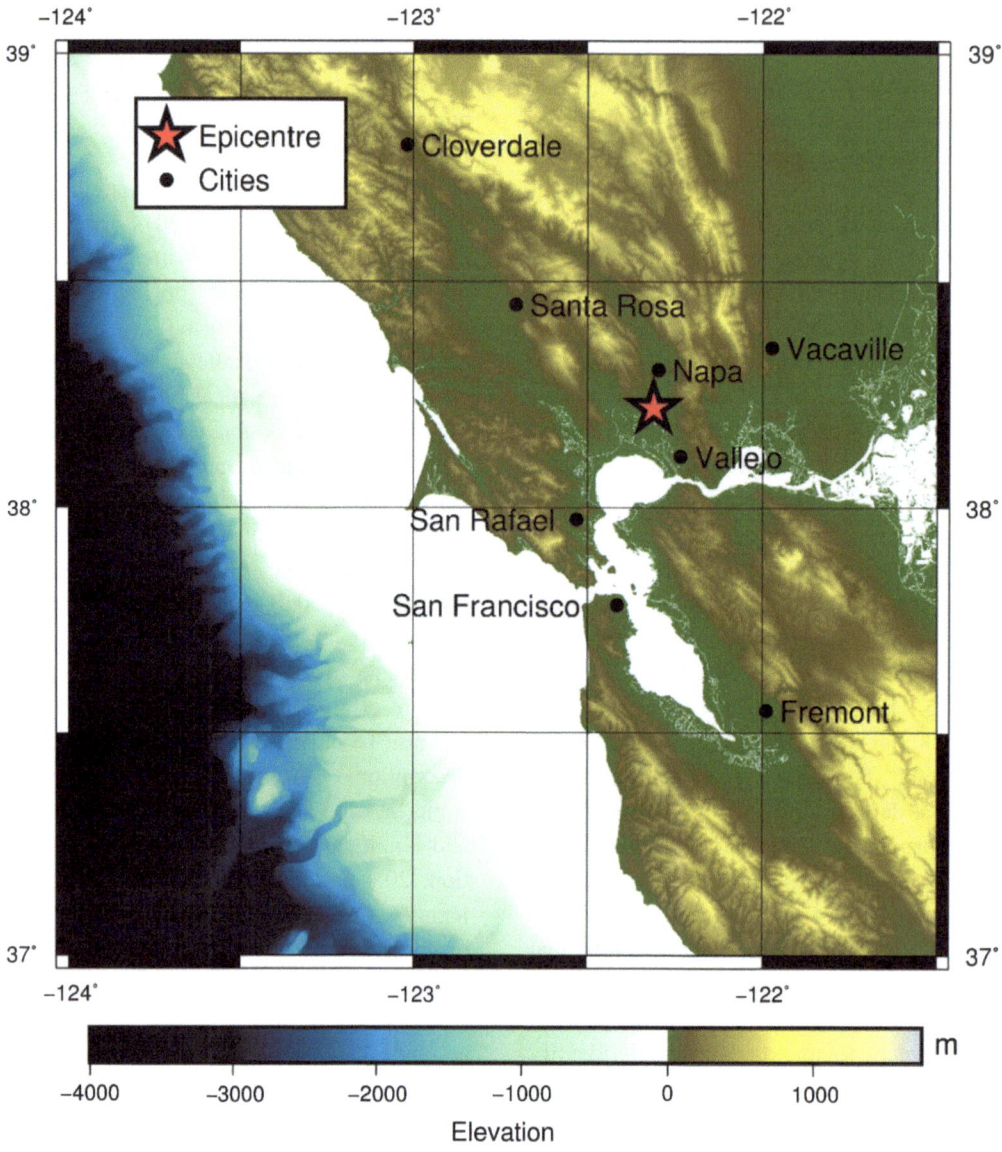

Figure 1
Topography map of the region [based on ETOPO1 dataset (Amante and Eakins 2009)]

during that time interval: 348 were geotagged or contained a user location, 399 were assigned a geolocation based on the above algorithm. Figure 4 shows a map of the spatial distribution of the number of tweets in this analysis. Figure 5 presents a map of population density for northern California. We anticipated that the number of tweets would correlate with population density or, more specifically, the amount of twitter users should correlate with population density. However, comparison of Figs. 4 and 5

suggests that population density in this particular case is not as closely correlated with the spatial pattern of the number of tweets in the first ten minutes after an event. There are three likely explanations for this phenomenon. The first reason is related to the local time of the event. The earthquake occurred at 03:20:44 Pacific Daylight Time, when the majority of the population was likely asleep and not actively using cell phones or laptops to the extent that they would during the daytime. The second reason is

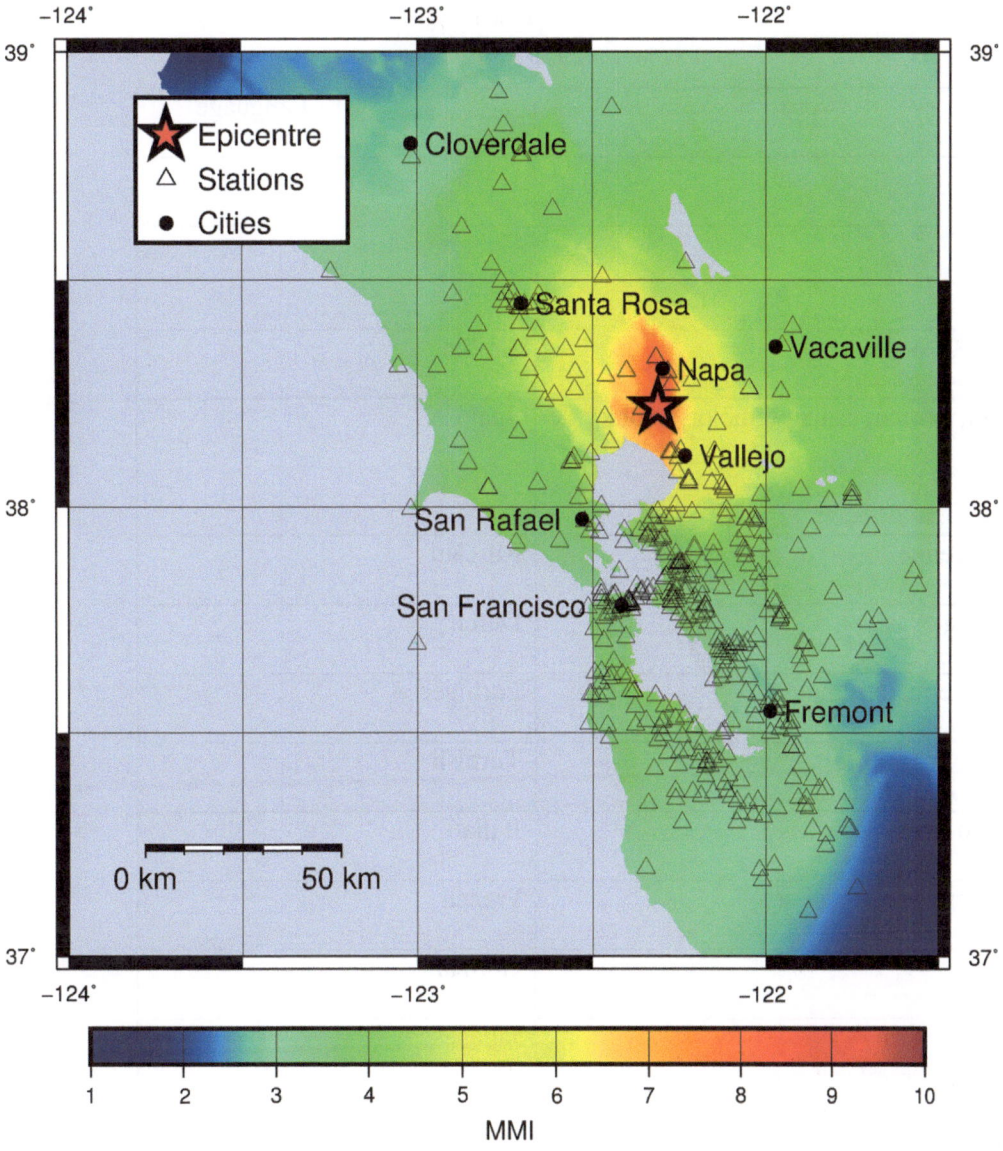

Figure 2
USGS intensity map [created from NCEDC data (2014)]

associated with the time interval of interest. As observed in Fig. 4, people tweeted more immediately after the earthquake in those regions close to the epicenter and with high intensity levels. In those areas with high population density that are further from the event we did not observe high tweet numbers during the first ten minutes. The third reason may be associated with the geolocation technique. The majority of non-geotagged tweets mention the word "Napa" and their location had been assigned to the Napa city coordinates in our analysis. However, at some time after earthquake occurrence, many observers already know what happened and may mention Napa Valley in their tweets, even if they were not actually in the city of Napa. Clearly, with these challenges in current geolocation algorithms there is a significant level of uncertainty in an analysis using data only from social sensors.

Table 1

Keywords used for positive tweets filtering

Word	Language
Σεισμός	Greek
地震	Chinese
أرضية هزة ,زلزال	Arabic
terremoto, seísmo, sismo, temblor, temblor de tierra	Spanish
Землетрясение	Russian
Aardbeving	Dutch
tremor de terra, terremoto	Portugeese
Deprem	Turkish
Terremoto	Italian
tremblement de terre, séisme	French
Erdbeben	German
地震	Japanese
אדמה רעידת	Hebrew
지진	Korean
Jordbävning	Swedish
earthquake, quake	English

To employ tweet rate as an additional earthquake sensor, i.e., to predict MMI of an actual earthquake, we generated a tweet-frequency time series using a sliding time window of 1 min over time steps of 5 s, normalized to the number of tweets per minute. A logarithmic transformation is used here to reduce the

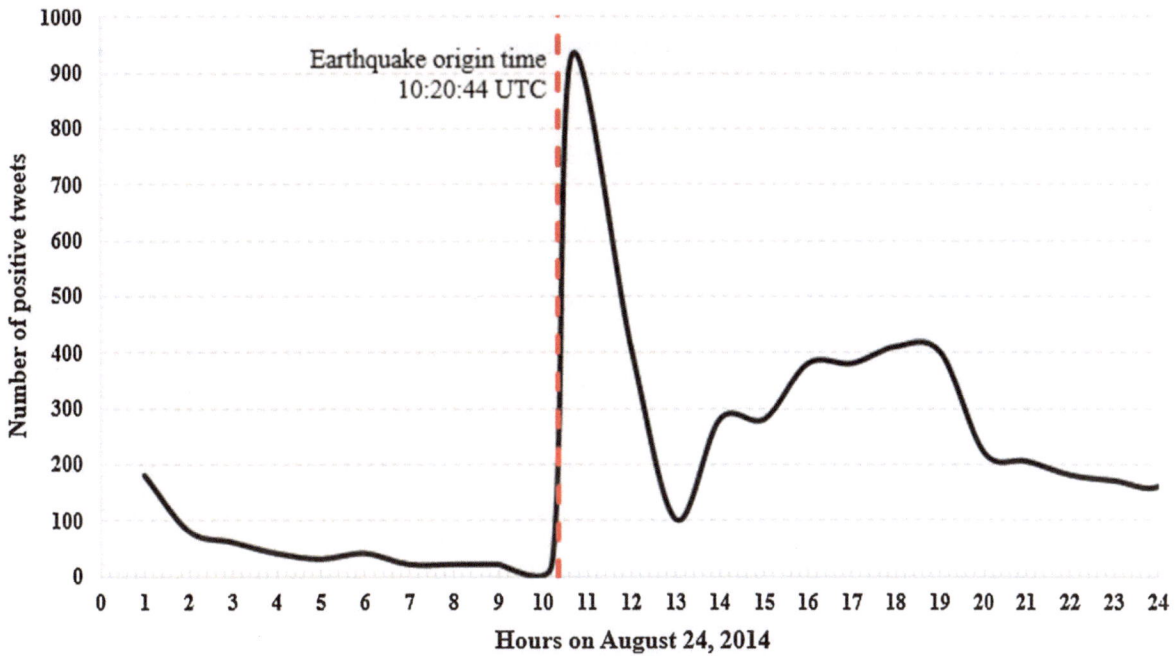

Figure 3
Number of positive tweets in the Napa region on the day of the earthquake

positive skewness in the data. Figure 6 presents the relationship between MMI values and the logarithmic rate of positive tweets obtained in this study. For each intensity point corresponding to the same latitude and longitude location from the first data set, we automatically assigned a value of observed tweet rate, calculated as tweets per minute at the tenth minute after the earthquake. The selection of a 10-min time interval is based on the analysis that showed that the number of positive tweets is constantly increasing over a longer period of time and reaches its maximum value approximately 11 min after the earthquake. In addition, because we are interested in emergency applications of intensity maps, here we consider only a 10-min interval following the earthquake. We regress MMI against the logarithmic mean of the number of tweets per minute to obtain predictive equations within a legitimate range of values for each model (see Table 2). We regress the average ground motion values for specified MMI levels to approximately follow the appropriate trend instead of producing a relationship overly influenced by the greater statistical volume of data at lower intensities.

We applied a least squares solution with 95 % confidence bounds for each model. The three-segment model demonstrates the best fit to the data with the highest coefficient of determination (R-squared) of 0.53 and the lowest root mean square (RMS) error of 0.65 MMI units. However, because R-squared and RMS error cannot determine whether the model estimates and predictions are biased, we also assessed the residual plots. The residuals between predicted and observed data shown in Fig. 7 for each model demonstrate normality in every case.

Several important observations can be made from Fig. 6. First, both two and three-segment models show the positive tweets rate increases slowly up to MMI less than V. The three-segment model also has a greater slope as MMI increases to VIII. At low MMI levels, shaking is light and often goes unnoticed (USGS 2015b). Starting at MMI V it becomes moderate and is felt by nearly everyone, explaining the increase in slope after MMI V. At severe levels of shaking (greater than VIII), even specially designed structures are slightly damaged, potentially affecting communication infrastructure and decreasing the

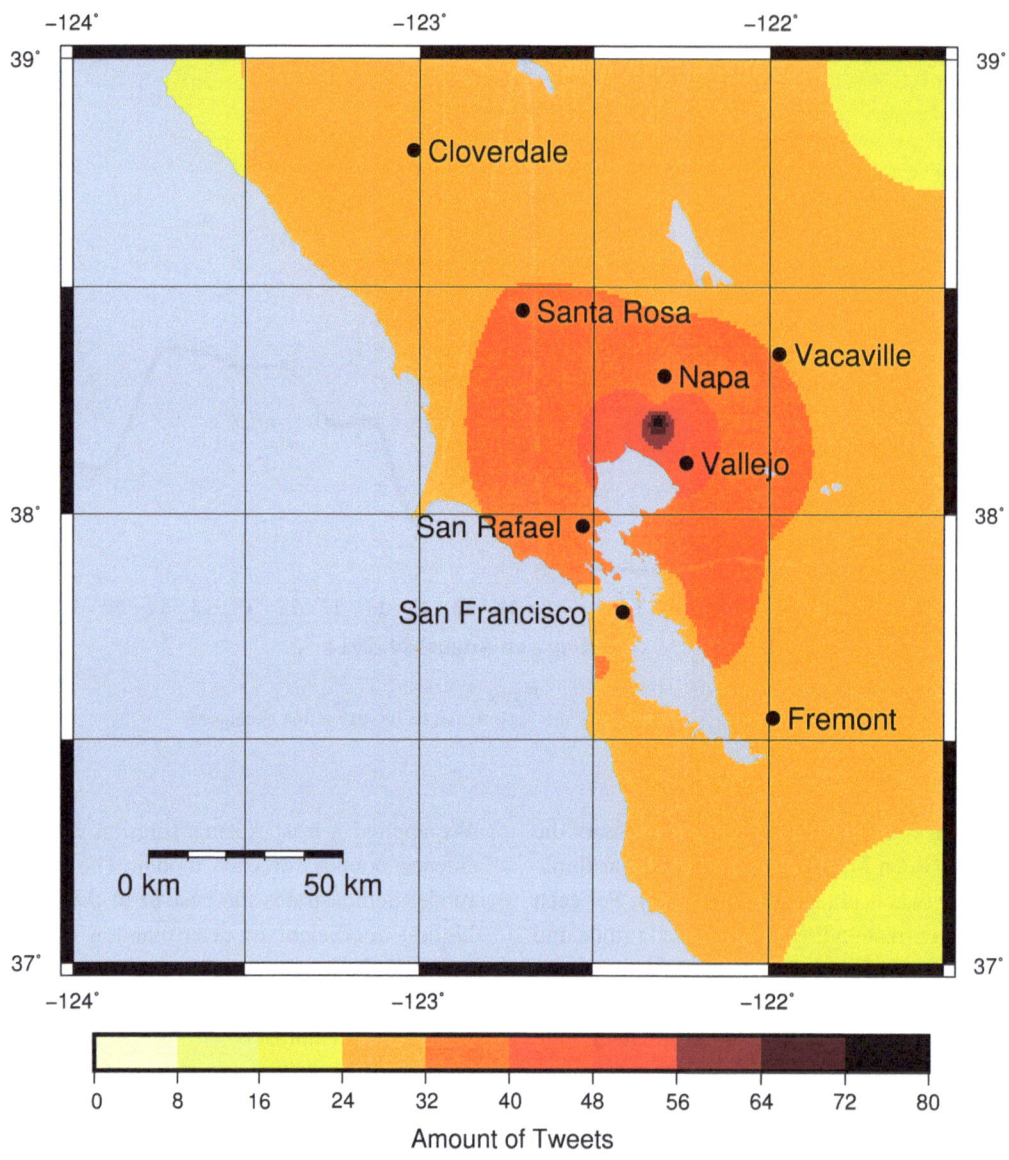

Figure 4
Number of positive tweets 10 min after the earthquake

number of devices with internet access and decreasing the rate of positive tweets.

Twitter data is a simple proxy for MMI level estimation. However, due to the high level of over and underestimation resulting from the use of Twitter data alone, we propose that it should be implemented jointly with instrumental intensity. In that case, two completely different data sources are analyzed in real-time using a streaming environment and methodology.

3. Streaming Methodology and Environment

Both input data types, whether from seismic stations or from Twitter, have a time-dependent nature and can be classified as data streams. A data stream is a sequence of tuples (data packets) received at a sequence of positive real-time intervals. In this case, it is not possible to process the arriving data as a traditional database. Traditional databases are not designed to be used for continuous data loading and

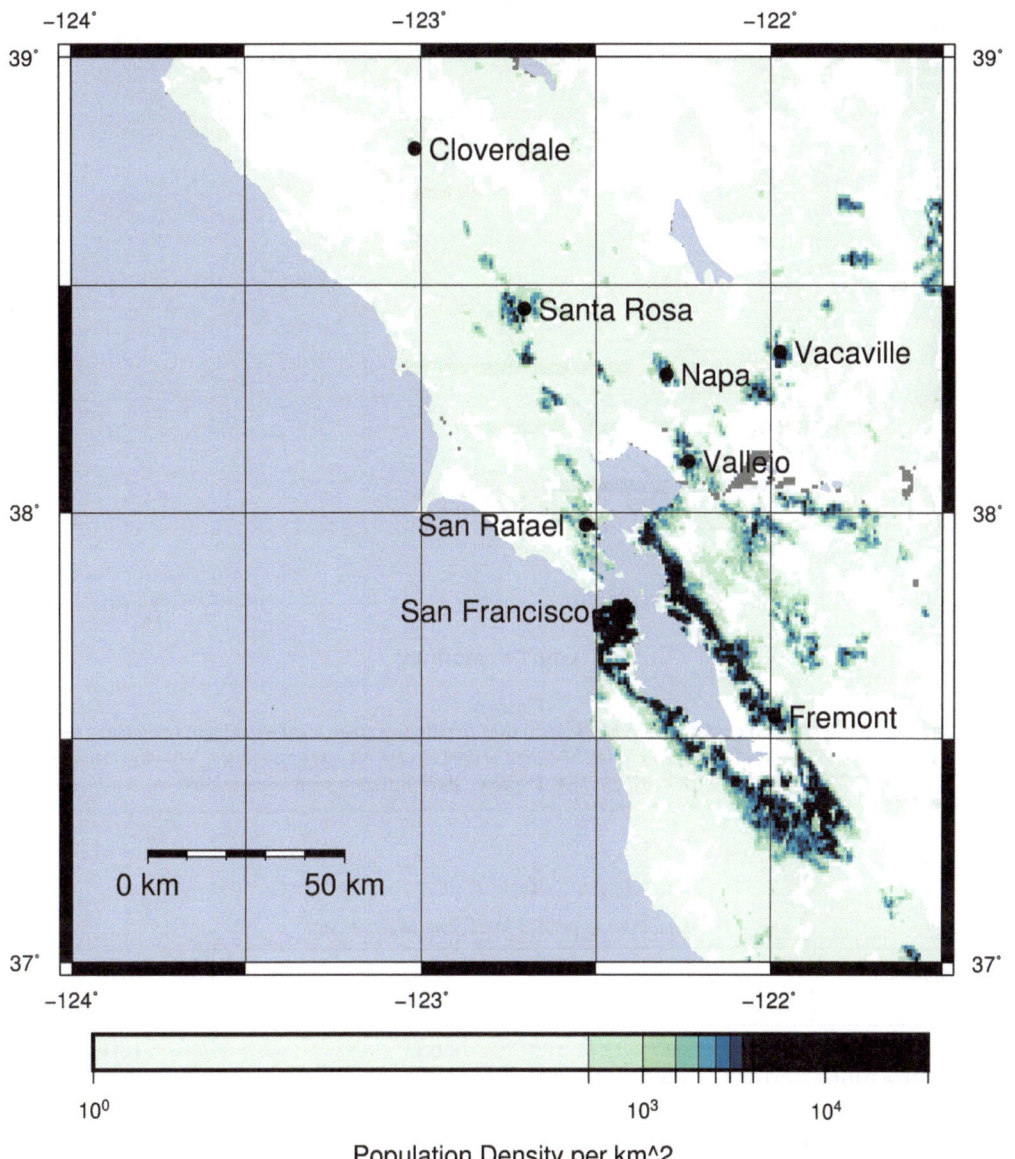

Figure 5
Population density in the region [data from GPWv3 (CIESIN 2005)]

continuous queries (Terry et al. 1992). A Twitter data stream represents a massive volume of data with an average of 5700 tweets per second and requires streaming approaches to be processed with low latency.

IBM created InfoSphere Streams (Streams), a product which provides a runtime platform, a data-centric programming model and a Stream Processing Language (SPL) specifically for complex streaming data analysis. Although there are other software packages for streaming data applications, we chose Streams for this application because it is flexible in accommodating different data sources. In Streams, an application is scalable for deployment on a larger HPC cluster to meet the application needs, and it is easy to implement with the support from various specialized toolkits.

For a Streams application, each processing procedure generally is implemented as an operator, and these operators subsequently are connected to form

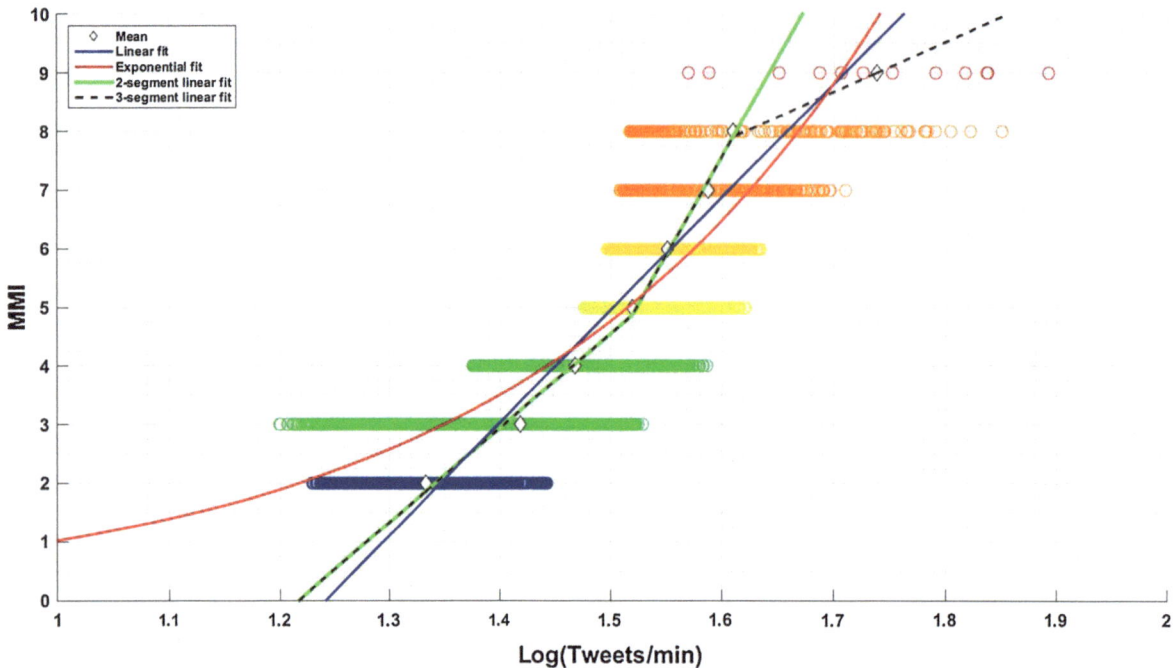

Figure 6
Combined tweet rate dataset (i.e., *colored circles* at each MMI level) used to derive average log(Tweets/min) (*diamonds*) for each MMI level (II—*blue circles*, III—*light green circles*, IV—*green circles*, V—*light yellow circles*, VI—*yellow circles*, VII—*light orange circles*, VIII—*orange circles*, IX—*red circles*). The *lines* show different regression results

Table 2

Equations to predict MMI from tweets rate

Model Name	Equation	Coefficient of determination, R^2	RMS error	Valid range of values Ntweets/min
Linear	MMI = 11.73 × log(Ntweets/min) − 13.47	0.32	0.79	[14.08; 148.4]
Exponential	MMI = 0.0168 × exp(3.67 × log(Ntweets/min))	0.26	0.82	(0; 61.7]
Two-segment linear	MMI = 16.13 × log(Ntweets/min) − 19.65, log(Ntweets/min) < 1.52 MMI = 33.75 × log(Ntweets/min) − 46.42, 1.52 < log(Ntweets/min) < 1.61	0.49	0.69	[16.53; 53.8]
Three-segment linear	MMI = 16.13 × log(Ntweets/min) − 19.65, log(Ntweets/min) < 1.52 MMI = 33.75 × log(Ntweets/min) − 46.42, 1.52 < log(Ntweets/min) < 1.61 MMI = 8.55 × log(Ntweets/min) − 5.86, 1.61 < log(Ntweets/min) < 1.74	0.53	0.65	[16.53; 122.7]

processing pipelines. The processing performance is the critical component of the streaming data application in terms of meeting certain real-time or near real-time requirements important to this work. To promptly and efficiently handle processing workload by making use of the computational resources of a HPC cluster, Streams provides toolkit operators that can easily split an operator's workload into multiple data streams. In the meantime, a processing pipeline also can be duplicated into multiple pipelines to

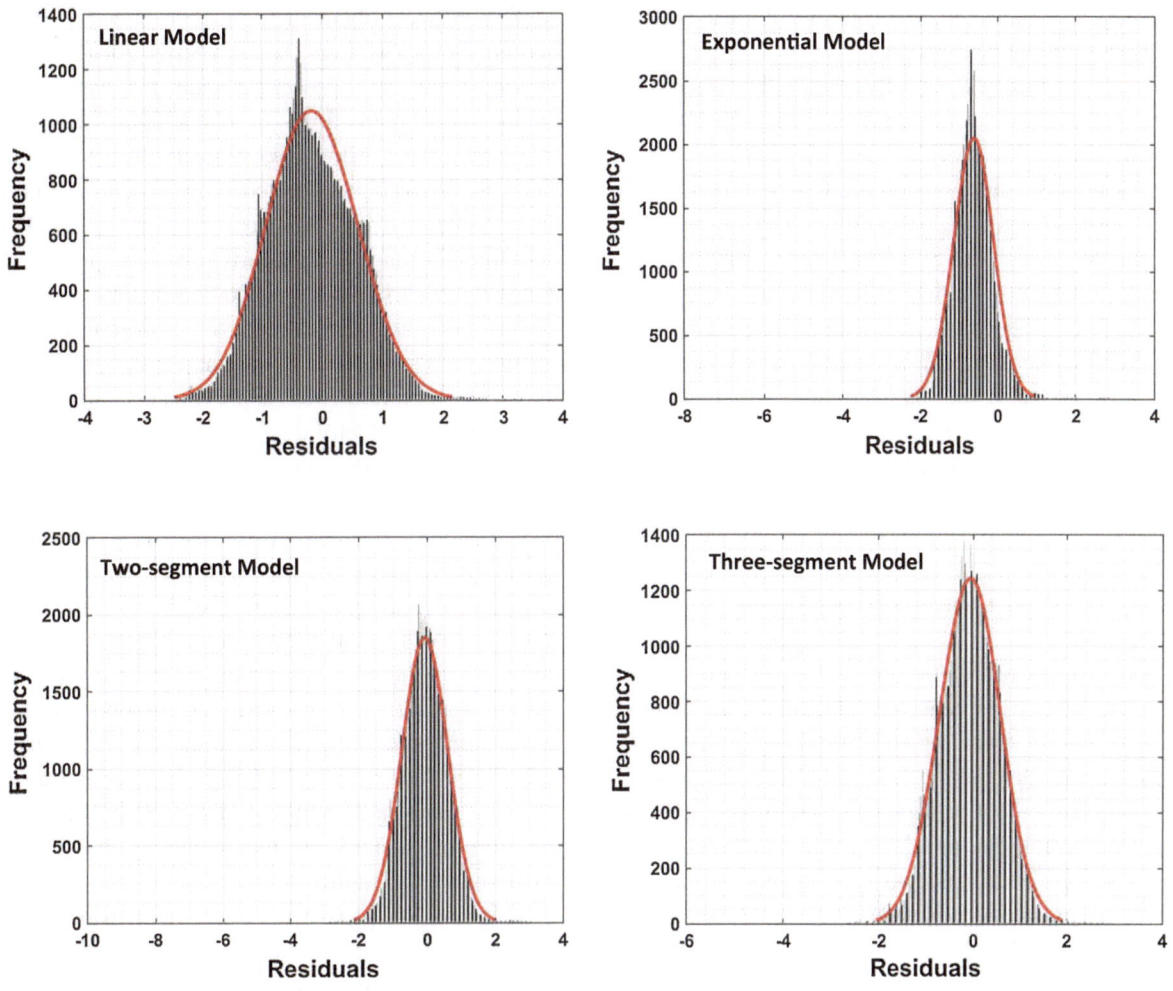

Figure 7
Residuals between predicted and observed data for each model

digest those data streams, which makes an application scalable. The Streams runtime platform makes it easy to deploy those pipelines on a large-scale HPC cluster without the need to manually manage allocation, synchronization or communication among the operators. The Streams runtime model consists of distributed processes. Single or multiple operators form a processing element (PE). The Streams compiler and runtime services determine where best to deploy those PEs, either on a single machine or across a cluster to meet the resource requirements and application performance needs (IBM 2014).

Streams come with a standard toolkit for application development. For example, the toolkit provides different types of data source adapters that can be used to monitor different data sources and pull data from multiple sources at the same time. Various data sources could be a local or remote file directory, a TCP or UDP port, or any web URI. Because data from different sources can be of different types and arrive at a different pace, the toolkit also provides utility operators for synchronizing and/or merging multiple data streams. There also are several specialized toolkits available to speed-up the development work for various applications. For example, here we process Twitter data which is in JSON format (detailed in Sect. 4), and a specialized JSON toolkit for Streams is used to parse and filter out the information needed from those tweet messages. Detailed implementation of this work is presented in the next section.

The experimental environment in this work consists of a cluster of four virtual machines (VMs), each configured with 8-core 2.4 GHz CPU, 16 GB RAM. InfoSphere Streams Version 3.2 has been configured and installed on the VM cluster. The cluster was available through the Southern Ontario Smart Computing Innovation Platform (SOSCIP) cloud, the first research-dedicated analytics cloud in Canada. The SOSCIP cloud also provides access to a broad range of software tools for application development and data analytics, which can be combined with user-specific software configurations to create customized VMs to meet project demands.

4. Implementation and Results

Two separate applications were developed for implementation of intensity level mapping in Streams. The first is for streaming and processing of data from physical sensors and the second from social sensors. Subsequently, both applications were linked into one application for joint processing from both sources.

The pipelined implementation to analyze datasets from seismic stations is shown on Fig. 8 (details are provided in Algorithm 1, Appendix A). Here, the input data are strong motion records from seismic network of USGS/NSMP (USGS National Strong-Motion Project) obtained from the Center for Engineering Strong Motion Data (CESMD 2014), which provides raw and processed data for earthquake engineering applications in cooperation with the USGS and the California Geological Survey (CGS). The USGS/NSMP network contains 33 stations in the area of interest (mapped as triangles on Fig. 9). The duration of strong shaking during the earthquake was generally 10–15 s or less, recorded in time steps of 20 ms. The minimum duration of the strong-motion seismic time series was approximately 20 s while the maximum duration was approximately 80 s. The strong-motion velocity time series are processed and reviewed by USGS/NSMP (noise filtering, baseline or sensor offset corrections performed). However, the corresponding raw, unprocessed data also is available and could be processed here in a preprocessing step using the streaming paradigm. To simulate the real-time nature of the data processing as it would occur in an actual earthquake, we arranged the data input so that the start of each time series corresponded to the actual time recording began after the earthquake occurrence. The initiation of each time series was at different times, although they did overlap. The total duration of shaking in the streaming input was approximately 4 min.

The seismic data is in SMC format, which uses ASCII character codes and consists of text headers, integer headers, real headers, and time series coordinates and values. The header includes information

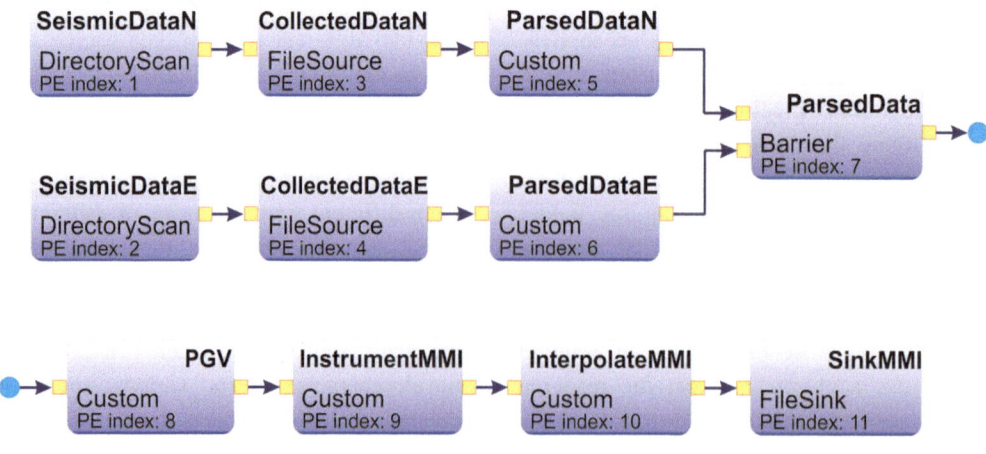

Figure 8
Pipelined application graph for intensity mapping based on seismic data (captured from Streams Studio)

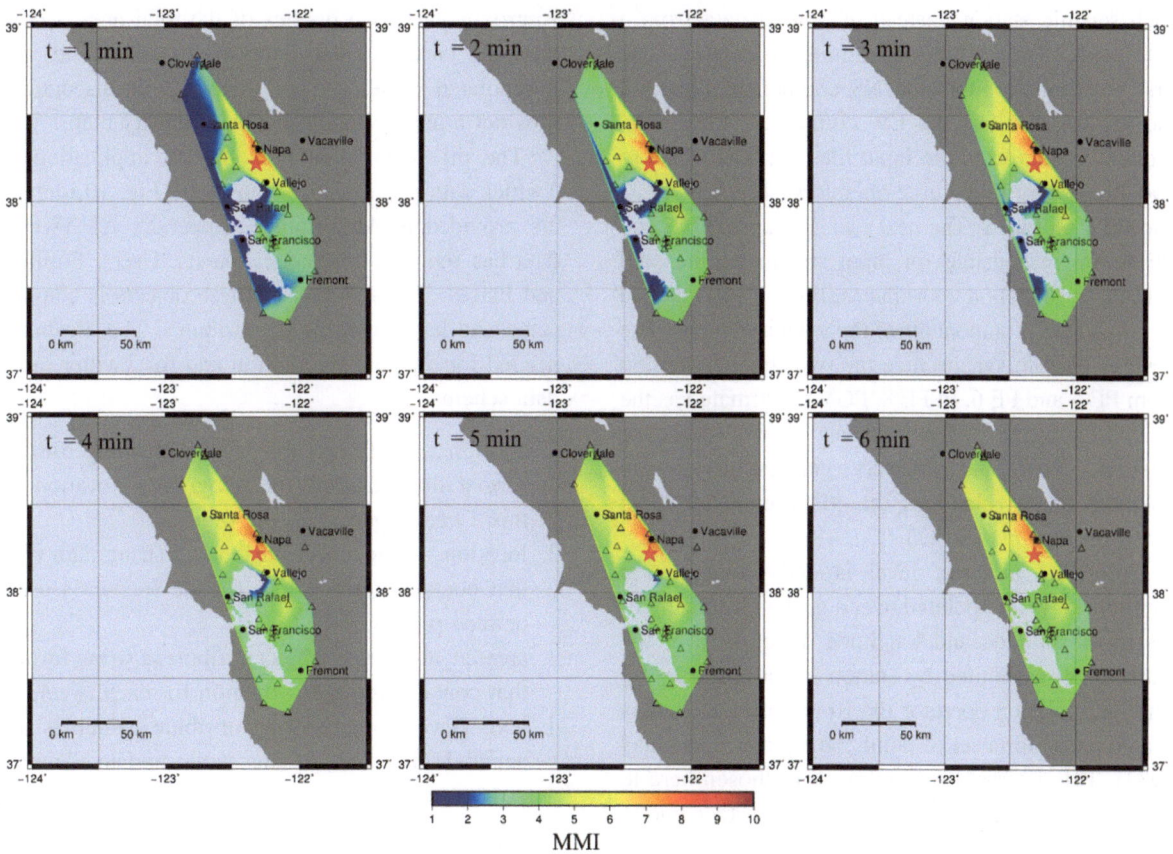

Figure 9
MMI after physical sensors data processing estimated at 1 min intervals after the Napa Valley earthquake. *Red star* represents the epicenter location. *Triangles* represent the location of seismic stations

about the earthquake and the recording physical sensor. Each file also contains either a single time series of acceleration, velocity or displacement or a set of response spectra or Fourier amplitude spectra of corrected acceleration calculated from a single time series. Analog strong-motion records contain traces corresponding to three orthogonal components of motion that are located in three separate SMC format files to represent the record (USGS 2010). Here, the horizontal components (EW, NS) of velocity time series are used in the standard empirical relationship for PGV and MMI. From Atkinson and Kaka (2007), the relationship between PGV and observed MMI can be represented by Eqs. (1) and (2), as follows:

$$MMI = 4.37 + 1.32 \times (logPGV), \quad logPGV \leq 0.48,$$
(1)

$$MMI = 3.54 + 3.03 \times (logPGV), \quad logPGV \geq 0.48.$$
(2)

In our case, these files have the following naming convention:

$$\{StID.NComp.NP.-\_v\}.smc$$

where StID is the station identification and Ncomp is the component name, which has one of three values: HNN is the north direction component, HNE is the east direction component, HNZ is the vertical component of record.

Two DirectoryScan operators control the streaming of the input data source into the horizontal components of the velocity time series, shown as PE 1 and PE 2 on Fig. 8. In real situations, the implementation of these Streams source adaptors could be deployed to directly monitor the data collected at each seismic station, instead of pulling data from a data center for real-time streaming processing. The first operator monitors the NS component files and the second monitors the EW component files. Subsequently, data from the input files is transformed to two streams by FileSource operators, shown as PE 3 and PE 4 on Fig. 8. The next two operators, PE 5 and PE 6, are responsible for input stream parsing and extraction of station coordinates, time of records and north and east components of the velocity values. The PE 7 is used to synchronize the two streams coming from PE 5 and PE 6. In PE 8, PGV is estimated as the maximum horizontal component over a 5 s time interval. Next, the intensity calculation custom operator PE 9 calculates the MMI intensity level based on Eqs. (1) and (2).

After the intensity level is derived at station locations, it is interpolated over a grid with spacing of 3 s in the latitude and longitude direction over the entire area, as limited by station endpoints (Fig. 9). The interpolation operator PE 10 performs Delaunay triangulation for a set of points on a plane (Delaunay 1934). This interpolation approach is chosen here to decrease the roughness of interpolated surface due to the very small number of data points. Finally, the sink operator PE 11 writes the results to the output file in csv format. The output file is refreshed as more data arrives, which can be plugged into a visualization tool to dynamically display MMIs in the area on-the-fly.

Figure 9 shows the output from Streams as the evolution of intensity estimates based on the seismic data, starting 1 min after the Napa Valley earthquake and at 1 min time intervals. However, as expected from the limited amount of data points available from the seismic stations (33 stations), the accuracy is low.

Approximately, one quarter of the total area is covered by instrumental intensity results and the interpolated values are not smooth, with discontinuities not typically found in final intensity maps.

The pipelined implementation of application of Twitter data processing is shown on Fig. 10 (details are provided in Algorithm 2, Appendix A). Twitter data has four main objects: Tweets, Users, Entities, and Places. The anatomy of these objects is complicated and has a number of attributes. The attributes needed for our implementation and for Twitter input data schema are:

1. coordinates—Tweet object attribute in geoJSON format that contains the geographic location of this Tweet reported by the user;
2. location—User object attribute, a string data type that contains the location for the account's user-defined profile;
3. created_at—Tweet object attribute, a string format that contains the UTC creation for each tweet;
4. text—Tweet object string attribute, which is the actual UTF-8 text of the status update (Twitter 2015).

Each tweet in the archive has a creation time attribute with one second accuracy, allowing for the creation of a time series with one second time steps. This time series was fed into Streams, starting at the time of earthquake occurrence, again to simulate real-time processing of the information.

Input Twitter data coming from the source is monitored by DirectoryScan operator PE 1 (Fig. 10) and initially read in JavaScript Object Notation (JSON) format PE 2, an open standard format used to transmit data objects consisting of attribute–value

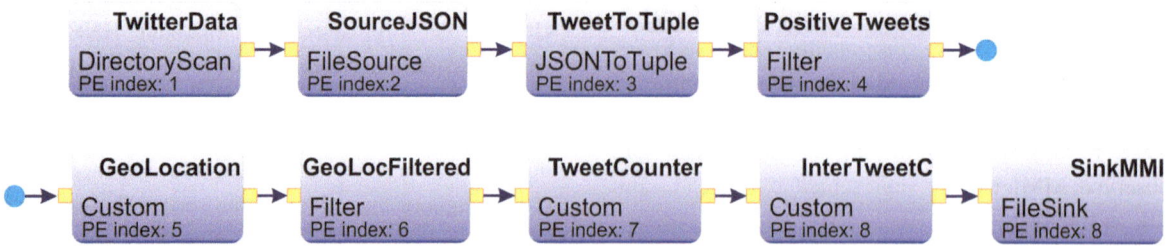

Figure 10
Pipelined application graph of intensity mapping based on Twitter data (captured from Streams Studio)

pairs. The data is transformed from JSON to defined Streams tuples in PE 3. The streaming tuples are filtered for positive tweets when passing through operator PE 4. Not all of the positive tweets have a coordinates attribute. For those positive tweets without a specified location, the location parsing operator PE 5 assigns coordinates according to the geolocation algorithm presented in Sect. 2. In this case, the tweet coordinates obtained are approximate. Subsequently, these positive tweets are filtered PE 6 again to exclude tweets from other regions. The logarithmic number of positive tweets per minute is calculated every 5 s PE 7, which is achieved in Streams by applying a "sliding time window" to data streams when passing through an operator, i.e., the time window is in size of 1 min and the sliding step is 5 s. In step PE 8, the logarithmic number of tweets at every location is interpolated over a spatial grid of 10 min in the latitude and longitude direction for the entire area, using the gridding algorithm with continuous splines in tension of Smith and Wessel (1990). Finally, sink operator PE 9 outputs the results in csv format, which can be plugged into a visualization tool to display.

Figure 11 shows the results obtained from Streams as the logarithmic number of tweets per minute starting from 1 min after the earthquake at 1 min intervals. For instrumental intensity (Fig. 9),

the maximum MMI for the entire area occurs at the seventh minute. Figure 11 shows that the logarithmic number of tweets per minute increased by a factor of four over the 10 min time interval of interest. Also, unlike instrumental intensity, the values are more completely distributed in space (comparing Figs. 9 and 11).

The pipelined implementation of joint processing is shown on Fig. 12 (details are provided in Algorithm 3, Appendix A). Both data sets and all the operators from the above two applications explained above are employed in this analysis. However, because that data is arriving continuously from both sources at the same time, calculations occur at smaller spatial and temporal discretization. Again, the logarithm of the number of tweets per minute are assumed to be correlated with MMI. The correlation analysis PE 17 was performed using the three-segment linear prediction equation (Table 2), the predictive model that resulted from the study presented in Sect. 2. MMI was estimated at every location in the joint streaming analysis of PE 17. In step PE 18, the resulting values of MMI are interpolated on a grid spacing of three seconds in the latitude and longitude directions (Smith and Wessel 1990).

Results from the joint intensity calculation for this case study are shown in Fig. 13. The maps represent the MMI level starting from 1 min after earthquake

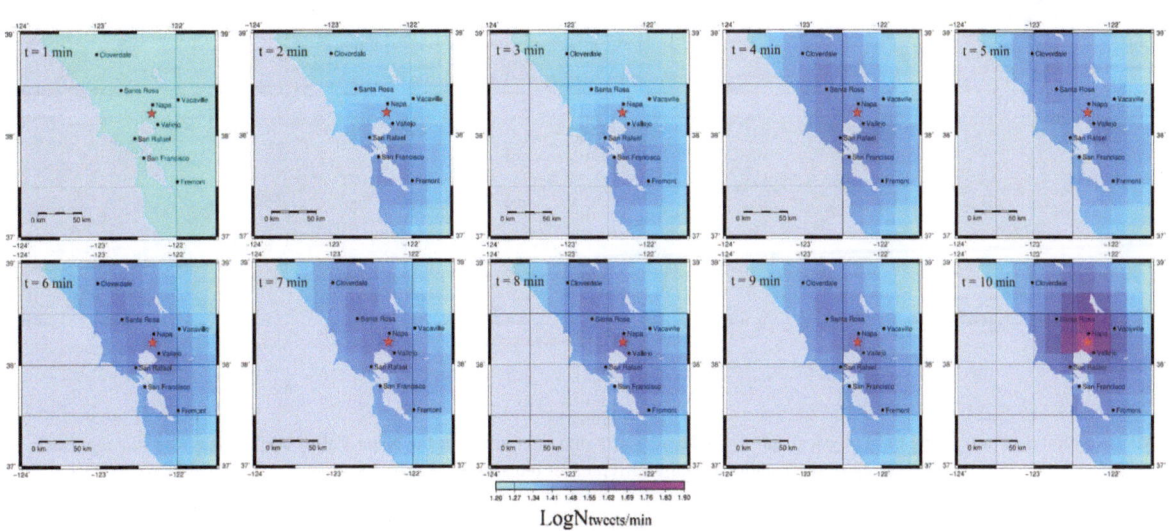

$\mathrm{LogN}_{\text{tweets/min}}$

Figure 11
Logarithmical number of tweets at 1 min intervals after the Napa Valley earthquake. The *red star* represents the epicenter location

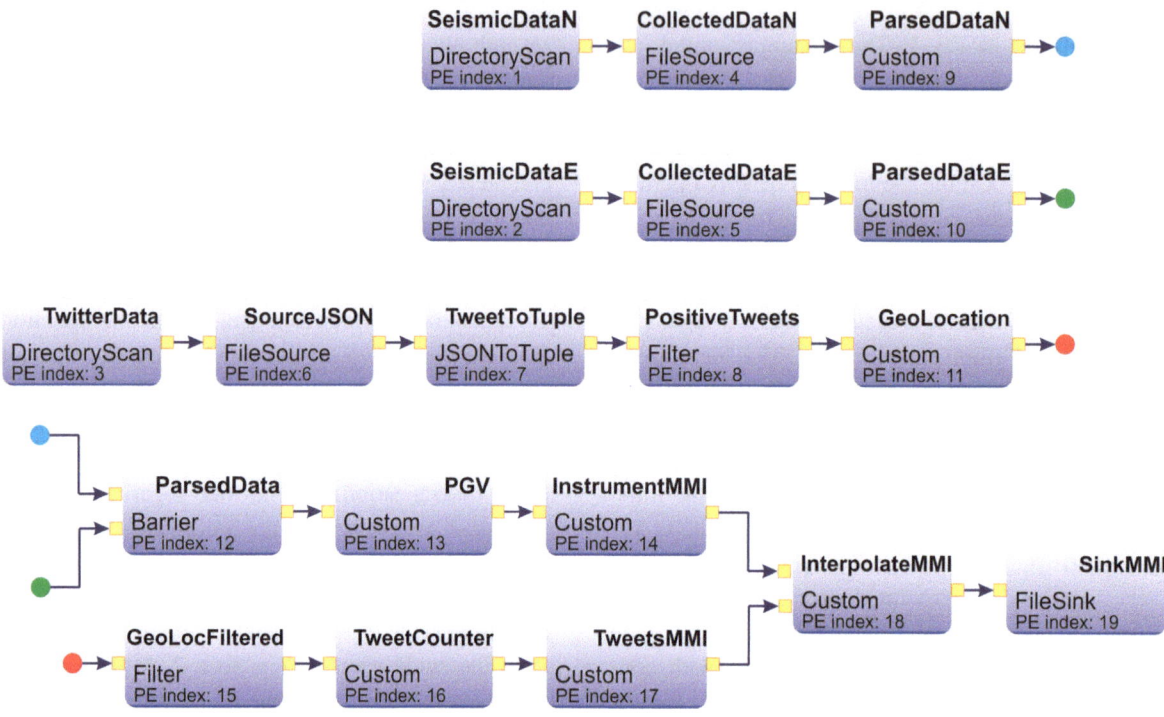

Figure 12
Pipelined application graph of intensity mapping based on joint data (captured from Streams Studio)

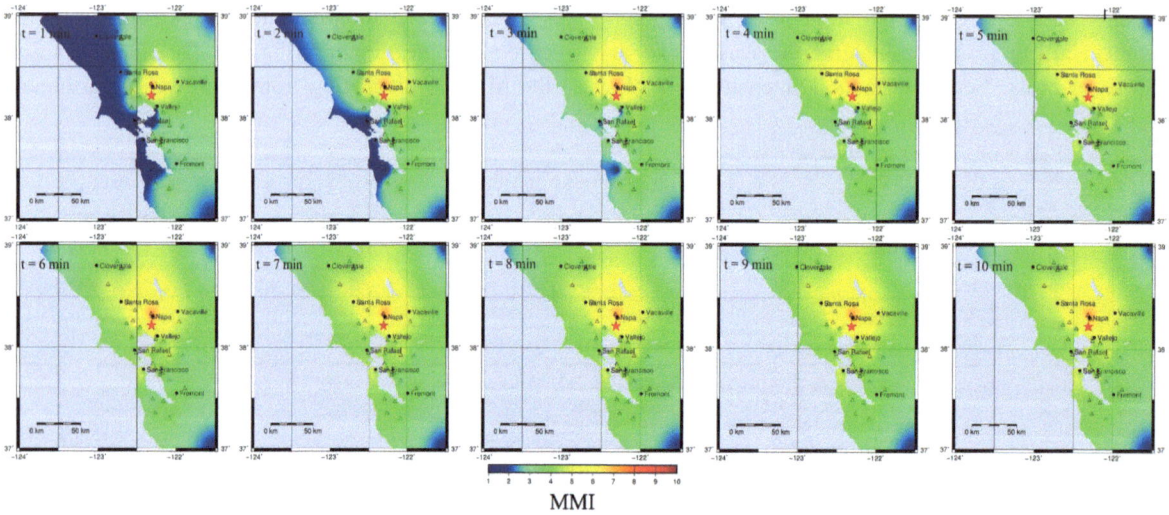

MMI

Figure 13
MMI after joint data processing from physical and social sensors at 1 min intervals after the Napa Valley earthquake. *Red star* represents the epicenter location. *Triangles* represent the seismic station locations

up to 10 min, at 1 min intervals. To quantify the improvement between maps generated based on data just from physical sensors (Fig. 9) and those obtained from joint processing from physical and social sensors (Fig. 13), we calculated the RMS error between those maps obtained at the final minute for physical sensors data (6 min after the earthquake) and the USGS instrumental intensity map provided in Fig. 2. Note that the RMS error calculated here is different from that used to select the best model (Table 2). The RMS error for the maps using only instrumental intensity is 2.8 MMI units. In contrast, for the RMS for the jointly processed intensity map is 0.58 MMI units. The lower RMS and the improvement in resolution for the joint processing case is a result of the increased number of data points used on the final interpolation step compared to those of the maps shown in Fig. 9.

To measure performance characteristics we used the maximum values of the InfoSphere Streams built-in metric for resource utilization monitoring for every host node on the cluster (see Table 3). CPU in milliseconds (ms) represents the time that was used by all PEs on a node. The memory consumption metric shows the amount used by the all PEs on a node, in kilobytes (kB). Load average is a common metric on Linux systems that measures the average CPU load over a period of time. A higher metric represents a system that is increasingly overloaded. InfoSphere Streams multiplies the raw load average from Linux by 100, and normalizes it by the number of processors on each cluster in that instant. In our case, all four hosts were loaded to a level that was less than moderate, providing evidence that the processing workload involved in this particular case could be handled by fewer computational resources for the same performance level. The purpose of this Streams implementation and the execution experiment here is to demonstrate its viability and potential real-time performance In addition, it allows data to be processed continuously on-the-fly while it being collected from multiple sources, which is different from the traditional data processing implementation.

Table 3

Performance metrics over 10 min time interval

Node name on the cluster	CPU (ms)	Memory consumption (kB)	Load average
Cluster-parent	9542	1,682,004	4
Cluster-child1	10,880	2,654,176	1
Cluster-child2	9530	1,197,148	26
Cluster-child3	7800	1,196,550	1

5. Conclusion

This work presents the successful application and potential for emergency management purposes using the analytics cloud. The streaming concept is applied using the IBM InfoSphere Streams product using multiple data sources from an actual earthquake to demonstrate its application in real-time hazard mapping.

Joint streaming processing using tweets and seismic records that were recorded within the 10 min following the Napa Valley earthquake (2014) was used to estimate MMI at particular sites in California. We demonstrate that the logarithmic number of tweets can be used as a proxy for shaking intensity and can be used as a supplementary data source, in conjunction with existing networks of physical sensors, such as seismic stations, to improve intensity estimates in real-time. Results from the joint analysis showed that it provides more complete coverage, with better accuracy and higher resolution over a larger area than either data source alone. Future studies will examine the extent to which the twitter regression relation is applicable in other seismic areas, both in California and worldwide and how potential errors in geolocation affect the accuracy in space and time.

In many areas, the importance of this additional data source could be very significant, due to the complete or partial lack of traditional data sources as a result of the high cost of their installation and ongoing operation. This work demonstrates that Twitter data could be used as independent data source of MMI estimation. In particular, it has significant potential for regions that may not have an extensive seismic network. Finally, implementation of this approach is not event or location dependent, so this approach could be applied to other regions and types of hazard. For example, it could be employed in areas where access is restricted due to flood inundation. Incorporation of social sensor data with traditional data sources using advanced computational processing methods can provide more complete and accurate coverage for rapid hazard response.

Acknowledgments

The research of KFT and YK was made possible by a MITACS Accelerate grant and an NSERC Discovery

Grant and is the result of collaboration between the Western University Computational Laboratory for Fault System Modeling, Analysis, and Data Assimilation and the consortium of Canadian academic institutions, a high performance computing network SHARCNET. Figures were created using GMT plotting software (Smith and Wessel 1990).

Appendix A

Input : Files FE, FN in SMC format with the time-series of PGV horizontal (east and north) components at each station
Output: File J encoding map in xyz format, containing MMI estimation in the region

```
1    FLE ← DirectoryScan(FE);// To monitor new data files FE
2    FLN ← DirectoryScan(FN);// To monitor new data files FN
3    FLE_R ← FileSource(FLE);// To transform data files FLE into stream FLE_R
4    FLN_R ← FileSource(FLN);// To transform data files FLN into stream FLN_R
5    PE ← Custom(FLE_R);      // To parse stream FLE_R into stream PE
6    PN ← Custom(FLN_R);      // To parse stream FLN_R into stream PN
7    P ← Barrier(PE,PN);      // To synchronize streams PE and PN
8    PGV ← Custom(P);         // To calculate PGV stream
9    MMI ← Custom(PGV);       // To calculate MMI stream
10   MMI_I ← Custom(MMI);     // To interpolate MMI stream
11   FileSink(MMI_I);         // To output results from the MMI_I stream into
     file
```

Algorithm 1: Production of instrumental MMI map

Input : File F contains tweets dataset in JSON format
Output: File J encoding map in xyz format, containing tweets rate estimation in the region

```
1    FL ← DirectoryScan(F); // To monitor new data files F
2    FLC ← FileSource(FL); // To transform data files FL into stream FLC
3    JT ← JSONToTuple(FLC); // To parse stream FLC into stream JT
4    PT ← Filter(JT);        // To filter positive tweets into stream PT
5    GeoPT ← Custom(PT);     // To parse the location for positive tweets
6    GeoPTF ← Filter(GeoPT); // To filter geotagged positive tweets
7    TwN ← Custom(GeoPTF);   // To calculate tweets rate
8    TwNI ← Custom(TwN);     // To interpolate tweets rate
9    FileSink(TwIN);         //To output results from TwIN stream into file
```

Algorithm 2: Production of tweets rate map

Input : Files FE, FN in SMC format with the time-series of PGV horizontal (east and north) components at each station, file F contains tweets dataset in JSON format
Output: File J encoding map in xyz format, containing combined MMI estimation in the region

```
1    FLE ← DirectoryScan(FE); // To monitor new data files FE
2    FLN ← DirectoryScan(FN); // To monitor new data files FN
3    FL ← DirectoryScan(F);   // To monitor new data files F
4    FLE_R ← FileSource(FLE);// To transform data files FLE into stream FLE_R
5    FLN_R ← FileSource(FLN);// To transform data files FLN into stream FLN_R
6    FLC ← FileSource(FL); // To transform data files FL into stream FLC
7    JT ← JSONToTuple(FLC); // To parse stream FLC into stream JT
8    PT ← Filter(JT); // To filter positive tweets into stream PT
9    PE ← Custom(FLE_R); // To parse stream FLE_R into stream PE
10   PN ← Custom(FLN_R); // To parse stream FLN_R into stream PN
11   GeoPT ← Custom(PT); // To parse the location for positive tweets
12   P ← Barrier(PE,PN); // To synchronize streams PE and PN
13   PGV ← Custom(P); // To calculate PGV stream
14   MMI ← Custom(PGV); // To calculate MMI stream based on PGV
15   GeoPTF ← Filter(GeoPT); // To filter geotagged positive tweets
16   TwN ← Custom(GeoPTF); // To calculate tweets rate
17   TwMMI ← Custom(TwN); // To calculate MMI stream based on TwN stream
18   MMIC ← Custom(TwMMI,MMI); // To calculate joint MMI stream
19   FileSink(MMIC); //To output results from MMIC stream into file
```

Algorithm 3: Production of combined MMI map

REFERENCES

Amante, C., & Eakins, B. W. (2009). ETOPO1 1 arc-minute global relief model: procedures, data sources and analysis. NOAA Technical Memorandum NESDIS NGDC-24. *National Geophysical Data Center, NOAA*. doi:10.7289/V5C8276M. Accessed 10/10/2015.

Atkinson, G., & Kaka, S. (2007). Relationships between felt intensity and instrumental ground motions for earthquakes in the central United States and California. *Bulletin of the Seismological Society of America, 97*, 497–510.

Burks, L., Miller, M., & Zadeh, R. (2014) Rapid estimate of ground shaking intensity by combining simple earthquake characteristics with tweets. s.l. Tenth U.S. national conference on earthquake engineering frontiers of earthquake engineering.

Campagne, J., Dux, J., Guyot, P., & Julien, D. (2012) Twitter reaches half a billion accounts—More than 140 million in the U.S. http://semiocast.com/en/publications/2012_07_30_Twitter_reaches_half_a_billion_accounts_140m_in_the_US.

Center for International Earth Science Information Network—CIESIN—Columbia University, and Centro Internacional de Agricultura Tropical—CIAT. (2005) Gridded Population of the World, Version 3 (GPWv3): Population Density Grid. Palisades, NY: NASA Socioeconomic Data and Applications Center (SEDAC). doi:10.7927/H4XK8CG2. Accessed 10/10/2015.

Crooks, A., Croitoru, A., Stefanidis, A., & Radzikowski, J. (2012). Earthquake: Twitter as a distributed sensor system. *Transactions in GIS, 17*(1), 124–147.

Delaunay, B. (1934). Sur la sphère vide. A la mémoire de Georges Voronoï. *Bulletin de l'Académie des Sciences de l'URSS, Classe des sciences mathématiques et naturelles, 6*, 793–800.

Earle, P., Bowden, D., & Guy, M. (2011). Twitter earthquake detection: earthquake monitoring in a social world. *Annals of Geophysics, 54*(6), 708–715.

Earle, P., et al. (2010). OMG earthquake! Can Twitter improve earthquake response? *Seismological Research Letters, 81*(2), 246–251.

Graham, M., Hale, S. A., & Gaffney, D. (2014). Where in the world are you? Geolocation and language identification in Twitter. *The Professional Geographer, 66*(4), 568–578. doi:10.1080/00330124.2014.907699.

IBM. (2014). IBM Knowledge Center. http://www-01.ibm.com/support/knowledgecenter/. Accessed 2015.

Nepal Disaster Risk Reduction Portal (2015) Incident report of earthquake 2015. www.drrportal.gov.np. Accessed 12 Oct 2015.

Northern California Earthquake Data Center. (2014). UC Berkeley Seismological Laboratory. *Dataset,*. doi:10.7932/NCEDC.

Richter, C. (1958). *Elementary seismology*. San Francisco: Freeman.

Sakaki, T., Okazaki, M., & Matsuo, Y. (2010) Earthquake shakes Twitter users: real-time event detection by social sensors. World Wide Web Conference (WWW), Raleigh, NC.

Severo, M., Giraud, T., & Pecout, H. (2015). Twitter data for urban policy making: an analysis on four European cities. In C. Levallois (Ed.), *Handbook of Twitter for research*. Écully: EMLYON.

Smith, W. H., & Wessel, P. (1990). Gridding with continuous curvature splines in tension. *Geophysics, 55*, 293–305.

Takhteyev, Y., Wellman, B., & Gruzd, A. (2012). Geography of Twitter networks. *Social Networks, 34*(1), 73–81.

Terry, D., Goldberg, D., Nichols, D., & Andoki, B. (1992). Continuous queries over append-only databases. SIGMOD, pp. 321–330.

The Center for Engineering Strong Motion Data. (2014) CESMD Internet data report. http://www.strongmotioncenter.org/cgi-bin/CESMD/archive.pl. Accessed 10 Sept 2014.

Twitter. (2015). The Twitter platform documentation. https://dev.twitter.com/overview/documentation. Accessed 2015.

United States Geological Survey (2010). National Strong Motion Project. [Online]. http://escweb.wr.usgs.gov/nsmp-data/smcfmt.html. Accessed 2015.

United States Geological Survey (2014) M6.0—6 km NW of American Canyon, California. http://earthquake.usgs.gov/earthquakes/eventpage/nc72282711#general_summary. Accessed 10 Sept 2014.

United States Geological Survey. (2015a). Earthquake facts and statistics. http://earthquake.usgs.gov/earthquakes/eqarchives/year/eqstats.php. Accessed 15 Oct 2015.

United States Geological Survey. (2015b) The modified Mercalli intensity scale. http://earthquake.usgs.gov/learn/topics/mercalli.php. Accessed 15 Oct 2015.

United States Geological Survey. (2015c) The San Andreas and other bay area faults. http://earthquake.usgs.gov/regional/nca/virtualtour/bayarea.php. Accessed 15 Oct 2015.

Wald, D., Quitoriano, V., & Dewey, J. (2006a) USGS "Did you feel it?" community internet intensity maps: macroseismic data collection via the internet. Geneva, Switzerland, First European Conference on Earthquake Engineering and Seismology.

Wald, D., Worden, B., Quitoriano, V., & Pankow, K. (2006b). *ShakeMap manual: technical manual, users guide, and software guide*. Boulder: United States Geological Survey.

Wood, H., & Neumann, F. (1931). Modified Mercalli intensity scale of 1931. *Seismological Society of America Bulletin, 21*(4), 277–283.

(Received February 29, 2016, revised September 26, 2016, accepted October 14, 2016, Published online October 21, 2016)

Pure Appl. Geophys. 174 (2017), 2351–2370
© 2016 Springer International Publishing
DOI 10.1007/s00024-016-1372-2

The Dependency of Probabilistic Tsunami Hazard Assessment on Magnitude Limits of Seismic Sources in the South China Sea and Adjoining Basins

Hongwei Li,[1] Ye Yuan,[1,2] Zhiguo Xu,[1,2] Zongchen Wang,[1] Juncheng Wang,[1] Peitao Wang,[1,2] Yi Gao,[1,2] Jingming Hou,[1,2] and Di Shan[1]

Abstract—The South China Sea (SCS) and its adjacent small basins including Sulu Sea and Celebes Sea are commonly identified as tsunami-prone region by its historical records on seismicity and tsunamis. However, quantification of tsunami hazard in the SCS region remained an intractable issue due to highly complex tectonic setting and multiple seismic sources within and surrounding this area. Probabilistic Tsunami Hazard Assessment (PTHA) is performed in the present study to evaluate tsunami hazard in the SCS region based on a brief review on seismological and tsunami records. 5 regional and local potential tsunami sources are tentatively identified, and earthquake catalogs are generated using Monte Carlo simulation following the Tapered Gutenberg–Richter relationship for each zone. Considering a lack of consensus on magnitude upper bound on each seismic source, as well as its critical role in PTHA, the major concern of the present study is to define the upper and lower limits of tsunami hazard in the SCS region comprehensively by adopting different corner magnitudes that could be derived by multiple principles and approaches, including TGR regression of historical catalog, fault-length scaling, tectonic and seismic moment balance, and repetition of historical largest event. The results show that tsunami hazard in the SCS and adjoining basins is subject to large variations when adopting different corner magnitudes, with the upper bounds 2–6 times of the lower. The probabilistic tsunami hazard maps for specified return periods reveal much higher threat from Cotabato Trench and Sulawesi Trench in the Celebes Sea, whereas tsunami hazard received by the coasts of the SCS and Sulu Sea is relatively moderate, yet non-negligible. By combining empirical method with numerical study of historical tsunami events, the present PTHA results are tentatively validated. The correspondence lends confidence to our study. Considering the proximity of major sources to population-laden cities around the SCS region, the tsunami hazard and risk should be further highlighted in the future.

Key words: Tsunami hazard, corner magnitude, Monte Carlo simulation, return period, South China Sea region.

1. Introduction

The South China Sea (SCS) and its adjoining basins including Sulu Sea and Celebes Sea, hereafter referred as the SCS region for short are identified as potential tsunamigenic source region due to high seismicity of major seismic sources within or surrounding the area. According to historical tsunami catalog of the National Geophysical Data Center (NGDC, available at: http://www.ngdc.noaa.gov/hazard/tsudb.shtml), several devastating tsunamis have been recorded in this region. The latest deadly tsunami was generated by 1976 Mindanao M_w 8.1 earthquake. The waves swept the Moro Gulf of the Celebes Sea and resulted in over 8000 dead or missing. In the wake of 2004 Sumatra tsunami and 2011 Tohoku tsunami, the Pacific Tsunami Warning and Mitigation System (PTWS), through its Medium Term Strategy, called for the establishment of sub-regional tsunami warning and mitigation system in the SCS region to ensure the appropriate warning service. As a basis for the implementation of the tsunami warning system in the SCS region, priority should be given to tsunami hazard and risk identification within and surrounding this area.

A series of studies have been conducted to clarify tsunami hazard in the SCS region from various aspects during the past decade. Deterministic tsunami hazard assessment involving hydrodynamic modeling of tsunami propagation from worst scenarios has been extensively performed in the SCS and adjoining basins, particularly the Manila Trench along the west Luzon Island (Ca and Xuyen 2008; Huang et al. 2009; Megawati et al. 2009; Ruangrassamee and Saelem 2009; Wu et al. 2009; Dao et al. 2009). It is

---
[1] National Marine Environmental Forecasting Center (NMEFC), No. 8 Dahuisi Rd., Haidian District, Beijing 100081, China. E-mail: yuanye@nmefc.gov.cn
[2] Key Laboratory of Research On Marine Hazards Forecasting, State Oceanic Administration, No. 8 Dahuisi Rd., Beijing 100081, China.

supposed that the upper-bound magnitude for Manila Trench could be M_w 9.0–9.3 inferred from long-term plate convergence rate and full fault rupture. Numerical modeling indicates maximum wave amplitude of 3–15 m along the coasts of Philippine, China, Vietnam, Brunei and Malaysia. By carefully examining the largest known earthquake events in the SCS region, Okal et al. (2011) reproduced 14 worst-case scenarios by numerical modeling, and showed the high levels of tsunami hazard existed among the South China Sea, Sulu and Celebes Sea. On the other hand, recent geodetic observations indicated the state of aseismic creeping along the Manila subduction zone, which suggested extremely low possibility of hosting a giant earthquake (Hsu et al. 2012; Wang and Bilek 2014). This conclusion was partially confirmed by the fact that only few large earthquakes with magnitude larger than 7.0 occurred in this area, and majority of seismic moment was released by small and medium-size earthquakes during the past century. Besides, Suppasri et al. (2012) further estimate tsunami risk in the Indian Ocean and the South China Sea by taking coastal population into consideration. The result shows that the population-laden countries like China, the Philippines and Indonesia have higher tsunami risk.

The methodology of Probabilistic Tsunami Hazard Assessment (PTHA) is derived from probabilistic seismic hazard analysis (Cornell 1968), and now have been applied to United States (Geist and Parsons 2006; González et al. 2009), Japan (Annaka et al. 2007; Fukutani et al. 2014; Suppasri et al. 2014; Goda et al. 2015; Goda and Song 2015), Australia (Burbidge et al. 2008), New Zealand (Power et al. 2007; Lane et al. 2013), the Indian Ocean (Horspool et al. 2014; Heidarzadeh and Kijko 2010), the Mediterranean (Sørensen et al. 2012; Lorito et al. 2015), and the Caribbean Sea (Parsons and Geist 2008). The merit of PTHA is capable of providing the likelihood that tsunami wave amplitude at a particular location will exceed a given level within a certain period of time, which is useful in the field of coastal engineering design. To date, only few papers on PTHA have been published in the SCS region. PTHA study in the Southeast Asia is first conducted by Thio et al. (2012). They find that peak wave amplitudes with return period of 475-year could reach up to 6–10 m along the west coast of Luzon Island, which is mainly contributed by Manila Trench, while peak values within the Celebes Sea are much smaller and only range from 1–4 m. However, this finding is inconsistent with the facts that historic seismicity and associated tsunamis are more active in the Celebes Sea. Liu et al. (2007) take Manila Trench as potential tsunami source in the SCS basin and present the probabilistic forecasts of tsunami hazard along the Chinese coast.

There are various sources of uncertainties associated with PTHA, among which the magnitude upper-limit that defines magnitude–frequency relationship for each seismic source is the controlling factor of the PTHA results (Rong et al. 2014). Generally, estimation of magnitude limit is subject to large variations due to limited length of instrumental records, lack of historic large events, poorly defined plate coupling and arbitrary fault segmentation. Considering the complexity of tectonic setting in the SCS region, previous PTHA studies merely presented details on the selection of seismic source parameters that are critical to PTHA, and only focused on specific subduction zone or basin. In the present study, we take the whole SCS region as study area. In combination with review on seismological and tsunami records within the area (Sect. 2), a PTHA study is performed to address common concerns whether and how likely a basin-wide tsunami could be generated and pose threat to the SCS neighboring countries and, more importantly, to highlight the dependency of probabilistic results on magnitude upper bound by adopting different corner magnitudes that could be derived by multiple principles and approaches. It should be noted that submarine landslide-generated tsunamis are not considered here. This study further calls for the need of multiple-disciplinary study on tectonics, plate kinematics, and tsunami hazard in the SCS region.

2. Historical Earthquakes and Tsunamis

The SCS region defined here contains three independent semi-enclosed basins including South China Sea, Sulu Sea and Celebes Sea (Fig. 1). Each

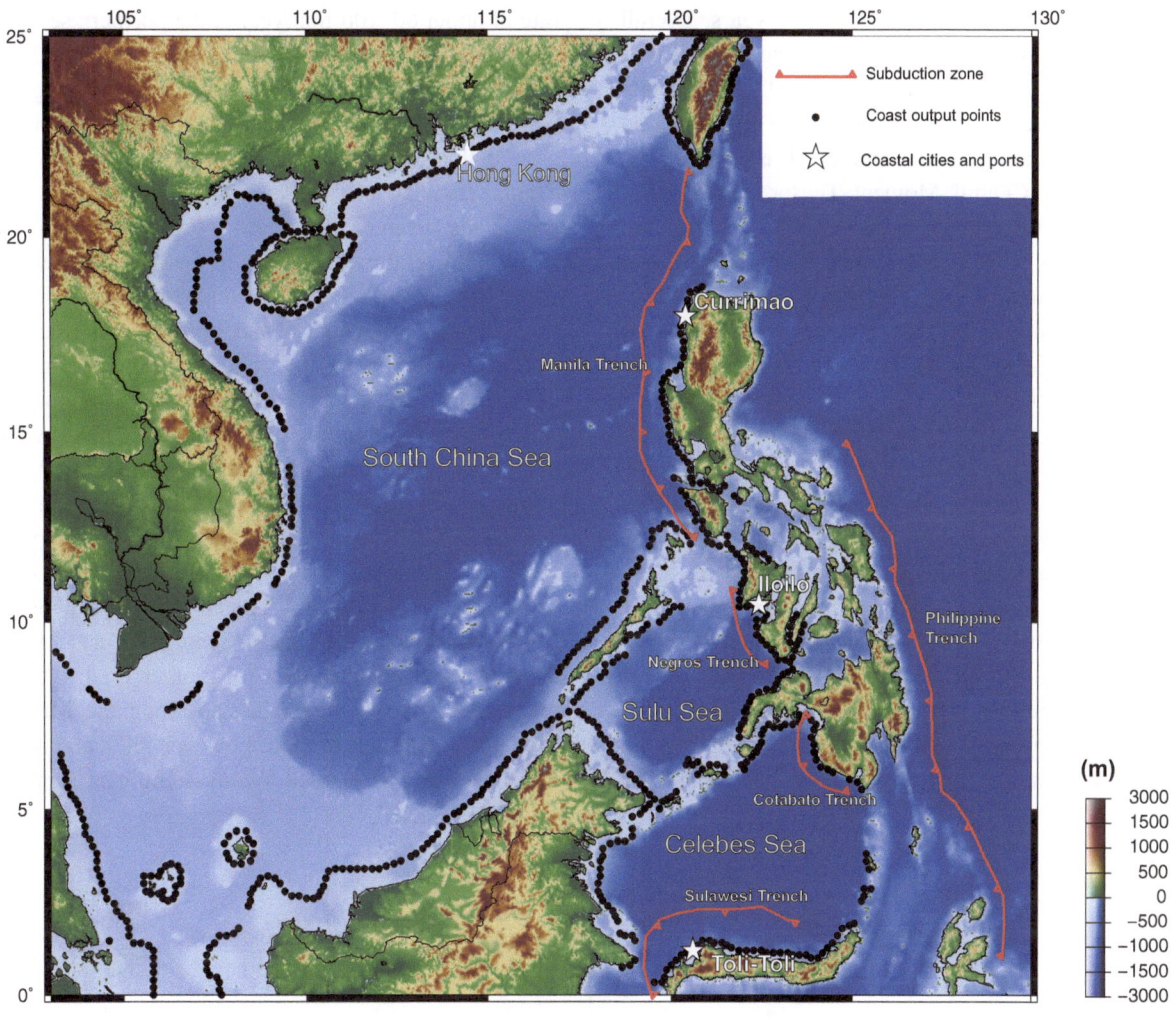

Figure 1

Bathymetry of the SCS region and major subduction zones identified in this study. *Red lines* with *triangles* indicate subduction zones; *black dots* are offshore output points of numerical model; four coastal stations are labeled as *white star* for subsequent tsunami hazard analysis, including Hong Kong, Currimao, Iloilo and Toli-Toli

of them is bordered by some active subduction systems. Previous study shows that tsunamis generated in one basin do not leak into another (Okal et al. 2011). Thus, in this study, only regional and local tsunami sources are taken into consideration. 5 potential tsunami zones are identified in Fig. 1 based on the seismicity and global plate boundary model (Bird 2003), which are Manila Trench, Negros Trench, Cotabato Trench, Sulawesi Trench and Philippine Trench. The width for each subduction zone is defined as the transverse distance perpendicular to the azimuth of the plate boundary where at least 95 % of the earthquake events fall inside. In this

section, a brief review on historical earthquake and tsunami catalogs in the SCS region is presented, with the main purpose to give a preliminary insight on plate tectonics, seismicity and tsunami hazard levels that are required for definition of magnitude distribution of each source and validation of PTHA results in the following sections.

2.1. Seismicity and Tectonic Settings

Historical earthquakes from 1900 to 2015 within 0–22°N and 103–130°E with magnitude 5.5 or above are collected from National Earthquake Information

Center (NEIC). During the past 115 years, overall 1891 events are documented, suggesting high level of seismicity in the study area (Fig. 2). Magnitudes that are not reported as moment magnitude are converted using empirical regression relation (Das et al. 2011). Furthermore, Central Moment Tensor (CMT) solutions from 1976 to 2015 for shallow earthquakes (depth ≤70 km) with magnitude larger than 5.5 are retrieved from Globe CMT Project (GCMT) to give more information on focal mechanism, and then used to determine the Tapered Gutenberg–Richter (TGR) relationship in each zone (Fig. 3).

As shown in Fig. 2, the most significant tectonic feature is the Manila Trench, a 1100-km long seismic zone between South Taiwan and Mindoro Island, along which Sunda (SU) block is subducting beneath the Luzon archipelago. According to Nguyen et al. (2012), the Manila Trench is the only candidate source that has potential to generate a basin-wide tsunami in the SCS. The oblique convergence between Sunda block and Philippine Sea (PS) Plate is absorbed along the Manila Trench (Rangin et al. 1999). Hsu et al. (2012) estimated larger subduction

rate of about 80–100 mm/year at the northern section of Manila Trench, and it decreases southward to 50–60 mm/year around 14°N. According to Fig. 3, while the majority of focal mechanism solutions around this area are classified as thrust type, some events are characterized as strike-slip and normal type. This behavior is caused by transformation of stress field as the subduction is passive (Zhu et al. 2005). The Manila Trench produced numerous small to moderate earthquakes, and only 17 earthquakes with magnitude larger than 6.5 were recorded in the last century (Fig. 2b). The two strongest earthquakes in this area occurred in 1934 and 1972 with magnitudes of 7.5. The maximum-likelihood estimate of magnitude–frequency relationship for the Manila Trench using historic catalog only yields an upper limit of 7.3. Some earlier researches presume that subduction at the Manila Trench has ceased, or at least slows down significantly (Seno and Kurita 1978; Rowlett and Kellecher 1976). Kreemer et al. (2000) revealed high level of strain rate along the Manila Trench by GPS observation; however, the compressing state was not released by any significant seismic

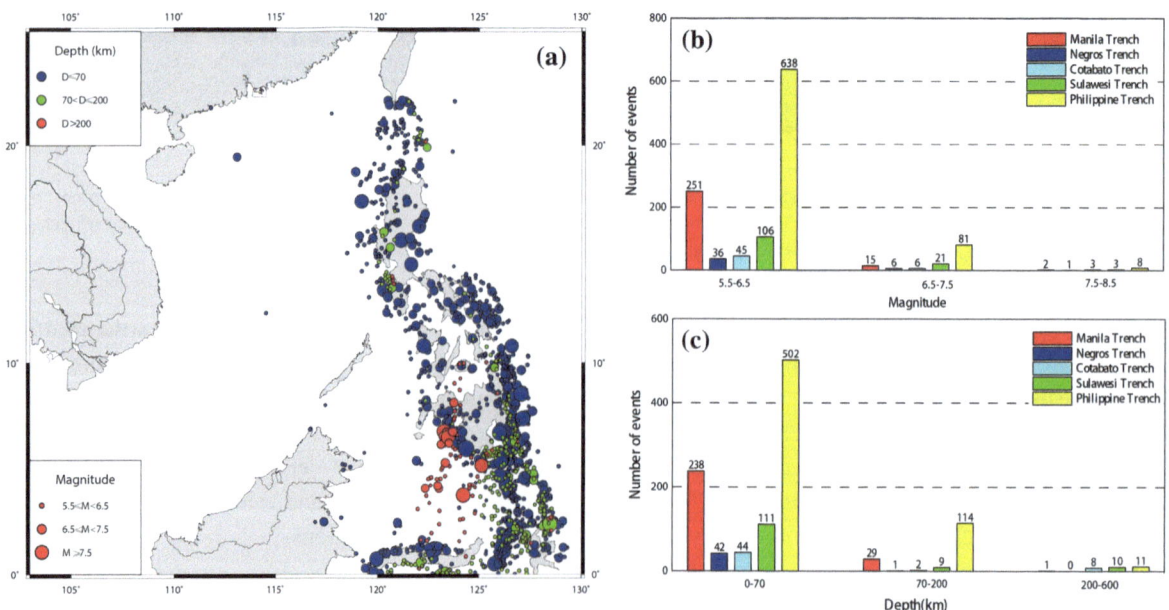

Figure 2
Historical earthquakes in the study area. **a** Spatial distribution of historical earthquakes from 1900 to 2015, magnitudes and focal depth is indicated in terms of *symbol size* and *filling colors*, respectively; **b**, **c** Histogram of earthquake magnitude and focal depth in each seismic source

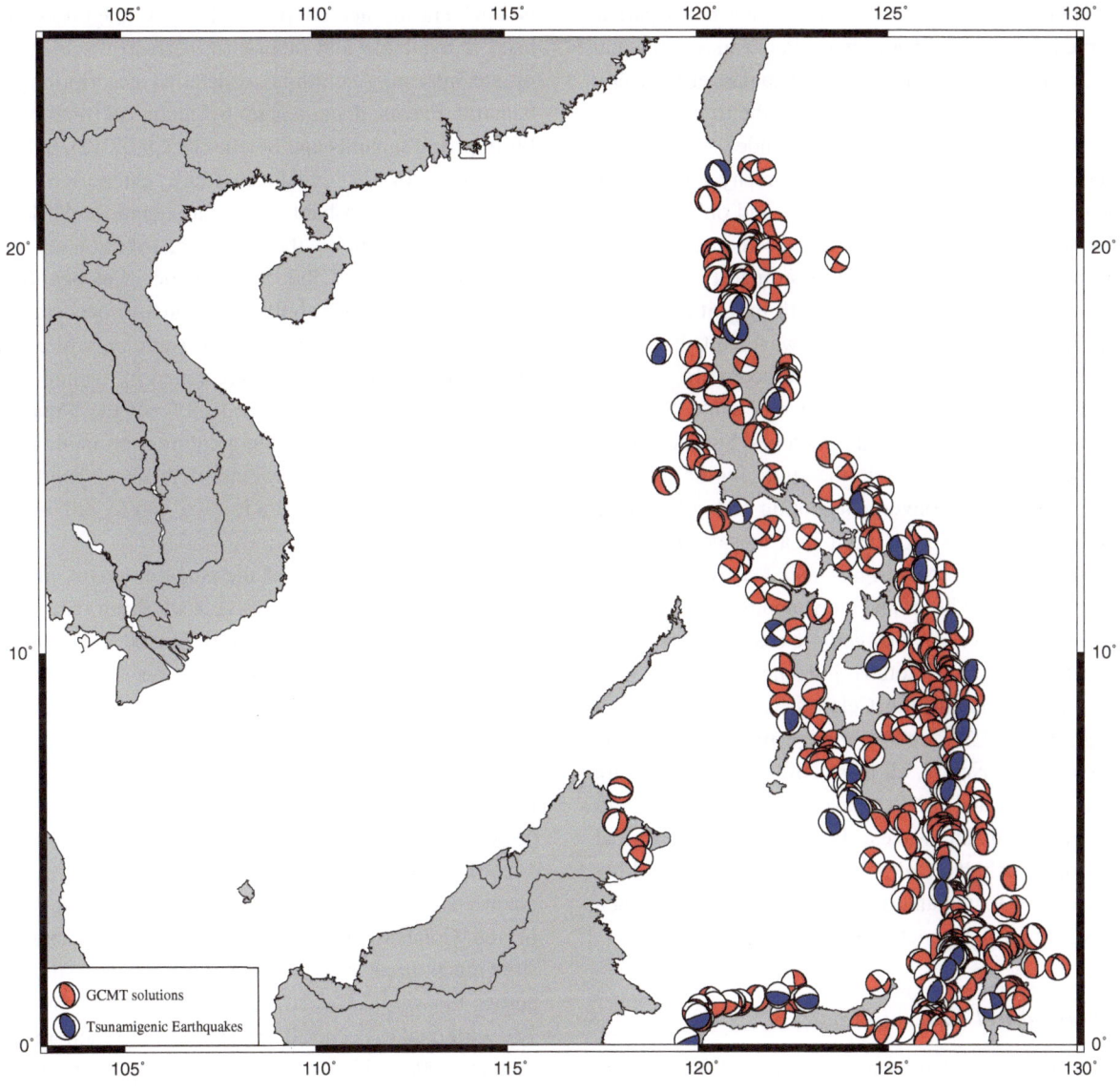

Figure 3
Historical CMT solution from 1976 to 2015 in the study area (*red balls*). overlapped with CMT solution of tsunamigenic earthquakes since 1900 (*blue balls*). Tsunami events are compiled based on NGDC and TLN tsunami database, while the CMT solutions for those events are retrieved from Selva and Marzocchi (2011) and GCMT catalog

events. Wang and Bilek (2014) attributed the paucity of giant earthquake to the creeping state of the interface along the Manila Trench.

According to Bird (2003), the subduction terminated at the middle of Manila Trench around 16.5°N and transit to the Philippine Trench on the other side of Luzon Archipelago. Unlike Manila Trench, Philippine Trench exhibits high level of seismicity and

accommodates plenty of moderate to large inter-plate thrust type earthquakes in the history (Galgana et al. 2007). Tsunamis that produced by Philippine Trench could penetrate into the SCS and Celebes Sea through Bashi Strait and open water of Karakelong Island.

Negros Trench within the Sulu Sea is considered as the southward extension of the Manila Trench along the west side of Philippine Islands; however, it is topographically discontinuous as the north part is starved by the collision with Palawan continental block on northern Panay Island (Rangin et al. 1999; Bird 2003). CMT solutions around the trench exhibit focal mechanism of thrust type blended with normal type and strike-slip type. The seismicity level of the Negros Trench is relatively lower, and only 7 events with magnitude larger than 6.5 are recorded during the past century (Fig. 2b). The strongest earthquake occurred on 24 January 1948 with magnitude reaching 7.8. The generated tsunami waves hit the Panay Island but no tsunami height was recorded.

The triple junction of the PS with SU and Australia (AU) plate in the Celebes Sea is one of the most complex areas in the study region (Rangin et al. 1999). The convergent motion and collision of multiple crustal blocks result in a mosaic of crustal fragments of diverse origins and very complicated fault systems within and surrounding the Celebes Sea. A high level of seismicity is associated with the subduction at the Cotabato Trench and Sulawesi Trench surrounding the Celebes Sea. 6 events with magnitude >7.5 were recorded with the largest one reaching 8.3. Nearly, all these strong earthquakes are thrust type and, thus, have higher potential to generate tsunami waves. Based on the historical catalog, Løvholt et al. (2012) defined the credible worst cases for Cotabato Trench and Sulawesi Trench as 8.2 and 7.9, respectively.

2.2. Historical Tsunami Events

Historical tsunami catalog in the SCS region is compiled based on the databases of NGDC and Tsunami Laboratory Novosibirsk (TLN). The merged catalog is further supplemented and validated by Philippine Tsunamis and Seiches (1589–2012) compiled by Philippine Institute of Volcanology and Seismology (Bautista et al. 2012). For NGDC database, earthquake-induced tsunami events with validity value ≥ 2 are identified and retrieved within 0–22°N and 103–130°E. All the events are categorized as questionable, probable, or definite. For TLN database, the same querying criteria as those for NGDC database are performed. The two databases are then merged to a tsunami catalog of 98 events. For events that are reported by both databases,

NGDC entries are adopted. Moment magnitude is used if available and otherwise, surface wave magnitude instead. According to detailed description of tsunami events documented by technical report of Philippine Tsunamis and Seiches (1589–2012), validity values of 24 events that are categorized as questionable or probable are re-assigned as 4 (definite). Figure 4 shows the spatial distribution of the tsunami catalog in the SCS region. Celebes Sea exhibits higher probability of tsunami occurrence corresponding to a high level of seismicity, while the SCS and Sulu Sea are less threatened by tsunami in the history. Overall 57, 15 and 28 % of the tsunami events are considered as being definite, probable and questionable. Tsunami events that correspond to higher earthquake magnitude (i.e., >6.5) are more credible and vice versa.

In more detail, most of the tsunami events along the Manila Trench are reported at its south end near the Metro Manila and Mindoro Island. The most significant tsunami event occurred on 14 February 1934 with magnitude of 7.5, and no tsunami height was reported. It seems that Sulu Sea is barely affected by tsunami hazard. The tsunami on 24 January 1948 is the most destructive event in last century. The most devastating tsunami in the Celebes Sea is the 1976 M_w 8.1 Moro Gulf earthquake and tsunami. The event occurred at a shallow portion of the subducting slab of the Cotabato Trench with its focal mechanism showing a thrust type. It produced a deadly tsunami across the whole Celebes Sea. Furthermore, an M_w 7.9 earthquake occurred in Sulawesi Trench on 1 January 1996, which generated a basin-wide tsunami either, with runup height of 3.4 m reported near the source.

According to NGDC database, tsunami observations associated with each tsunami event mainly come from three sources: eyewitness observations, tide gauge measurements and runup estimates by post-tsunami survey. Prior to mid-twentieth century, tsunami observations are reported by eyewitness and visual accounts. Tidal gauges have been extensively installed along the coast for real-time measurement of sea level variations until late 20th-century. Especially in the Philippine and Eastern Indonesia, the tsunami wave height records are very rare due to inadequate coverage of tidal gauges. The scatter plots of

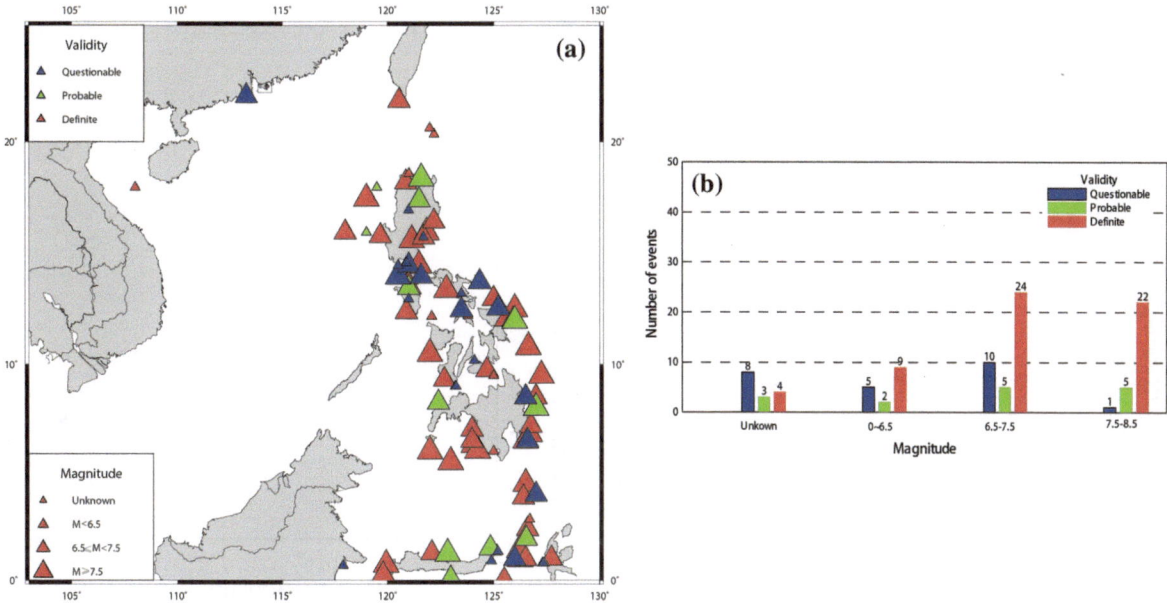

Figure 4
Historical tsunami catalog in the study area. **a** Spatial distribution of historical tsunami events from 2150 B.C. to 2015, the corresponding earthquake magnitudes and validity index values are indicated by *symbol size* and *filling color*; **b** histogram plot of validity index values in terms of magnitude of earthquakes that account for the tsunami events

tsunamigenic earthquake magnitudes with tsunami observations of three sources are presented in Fig. 5a–c. For tsunami events that have multiple records of wave height, only the largest values are chosen to represent the corresponding events. It is obvious that tidal gauge measurements yield more accurate, yet much smaller wave amplitude records. Examination of sea level records finds that the majority of wave amplitude records are registered at far field where tsunami energy attenuated considerably. Eyewitness observations are commonly collected by newspaper and interview during the post-event survey and, thus, subject to a large degree of error and uncertainty. In Fig. 5a–b, the maximum wave heights reported by eyewitness and tidal gauge measurement show a positive relationship with magnitude. It should be noted that 1859 earthquake with magnitude 7.3 was accompanied with an extreme wave height of 10 m. It is supposed as being erroneous. Another outlier can be found in Fig. 5b. This event corresponds to a series of weak earthquakes occurred southwest of Hainan Island, South China on 5 January 1992 with magnitude ranging

between 3.1 and 3.7. We assumed it as a landslide-induced tsunami due to failure of Northern SCS shelf slope.

3. Methodology of PTHA

The aim of probabilistic analysis of tsunami hazards is to quantitatively present information on the likelihood of tsunami potential in specified site or area. Typically, tsunami hazards are interpreted as probabilistic hazard curves at a particular site or hazard exceedance maps for a given probability, which are useful for insurance and land-use planning applications (Geist and Parsons 2006). For studying area that has a long period, and abundant written records of tsunami history, empirical analysis of tsunami runup and wave amplitude data could be feasible to establish tsunami probability. However, as mentioned in the Sect. 2.2, a lack of written- and instrumental-recorded tsunami observations in the SCS region makes us to resort to computationally based PTHA. Computationally based PTHA is

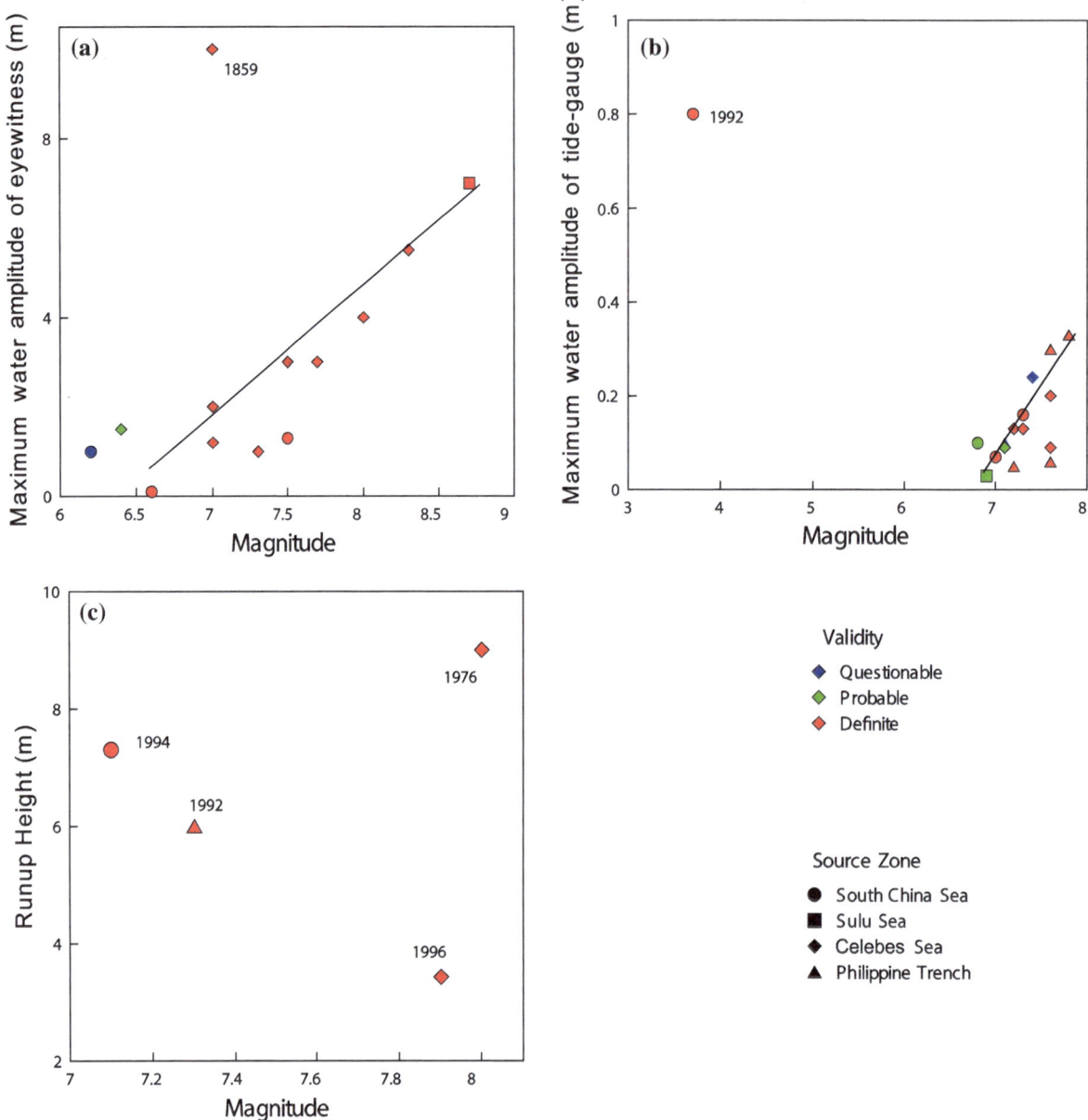

Figure 5
Relationship between tsunamigenic earthquake magnitude and tsunami observations reported by three sources: **a** eyewitness or visual accounts; **b** measurement of tidal gauges; **c** runup height by post-tsunami survey. The *solid lines* demonstrate the linear tendency of the scatter points

derived from Probabilistic Seismic Hazard Analysis (PSHA) with some modifications mainly associated with hydrodynamic modeling of tsunami waves. The general framework of PTHA can be summarized as four steps: (1) identify potential seismic sources and

establish magnitude–frequency distributions (i.e., TGR relationship) based on available earthquake catalogs; (2) generate synthetic earthquake catalog stochastically spanning long periods of time by Monte Carlo simulation with source parameters (i.e.,

TGR parameters including corner magnitude and seismicity slope); (3) establish tsunami catalogs for specified site within the study area; and (4) probabilistic calculation and interpretation.

3.1. Tapered GR Distributions for Seismic Sources

Tapered GR distribution (Kagan 1997) is employed in this study to describe the relationship between seismic moment and annual probability for each seismic source in the SCS region. The TGR relation (Kagan 2002a) is written as:

$$F(M) = \left(\frac{M_t}{M}\right)^\beta \exp\left(\frac{M_t - M}{M_c}\right) \quad (1)$$

in which M is the seismic moment, M_t is the threshold seismic moment representing catalog completeness, M_c is the seismic moment corresponding to upper-magnitude corner (m_c) that fixes the tail of the distribution, and $\beta = 2/3b$ is the seismicity slope of the distribution. Kagan (2003) compared several modern worldwide earthquake catalogs and found that GCMT is the best dataset in terms of completeness and accuracy. In this study, GCMT catalog from 1976 to 2015 is used to derive TGR relation for each source. The threshold magnitude m_t (corresponding to threshold seismic moment M_t) for catalog completeness is set to 5.5. Only shallow earthquakes with focal depth ≤ 70 km are included.

Maximum likelihood method is used to estimate β and M_c (Kagan 1997; Bird and Kagan 2004). Theoretically, the parameters β and M_c can be well estimated by this approach, if each subcatalog comprises a homogeneous distribution of small and moderate events, as well as sufficient large earthquakes that can well define upper limit of earthquake size. Figure 6 shows the contour plots of the log-likelihood function l with regard to β (horizontal axis) and m_c (vertical axis) for Manila Trench, Negros Trench, Cotabato Trench, Sulawesi Trench, and Philippine Trench. The zero contour line corresponds to the 95 % confidence area. A pair of β and m_c values can be extracted from contour plot where l reaches its highest value. It is notable that the log-likelihood contours do not close in the positive m_c direction for all the subduction zones. Only the lower bounds of

corner magnitude could be estimated from the contour. Especially for Manila Trench, the high-value area extends to the upper bound of the vertical axis due to a lack of large earthquakes in the subcatalog. Despite a global estimate of β value in the circum-Pacific subduction zones has a universal value of 0.66, the β values for seismic sources in the SCS region exhibit large variations. The contour plot for Manila Trench exhibits a best β value of 0.83, and corresponding 95 % confidence interval does not have an upper bound either. Considering the Manila Trench is the most important source in the SCS basin, we give it a universal β value of 0.66 for conservative purpose.

The best-fitting curves of TGR distributions for each seismic zone are shown in Fig. 7, and the derived source parameters are summarized in Table 1 Columns 2 and 3. It is obvious that due to a paucity of earthquake >8 in each subcatalog, the tails of each curve are not well constrained. The estimated m_c values for Manila Trench and Negros Trench are merely 7.3 and 7.1, and for Cotabato Trench and Sulawesi Trench, the m_c values can reach 8.1 and 7.9, which are comparable with the historic largest events for each trench.

3.2. Magnitude Upper Limits for Seismic Sources

One of the key issues involved in the PTHA is determining seismic moment distributions for seismic sources based on the available catalogs. As shown in Sect. 3.1, due to the limited length of earthquake catalog and paucity of large events, establishing TGR relation and associated source parameters are subject to a large degree of uncertainty. Logic-tree approach by giving different weights to candidate distributions (i.e., Characteristic and GR distribution) and magnitude upper limit were adopted by some authors (Annaka et al. 2007; Lane et al. 2013; Horspool et al. 2014) to account for this uncertainty. However, the weights for each candidate are somewhat arbitrarily determined due to our poor knowledge into earth interior. Considering a lack of consensus on magnitude upper bound on each seismic source as stated in Sect. 1, as well as its controlling effect on PTHA result, in this section we intend to estimate magnitude limits for each seismic source comprehensively by multiple approaches and principles.

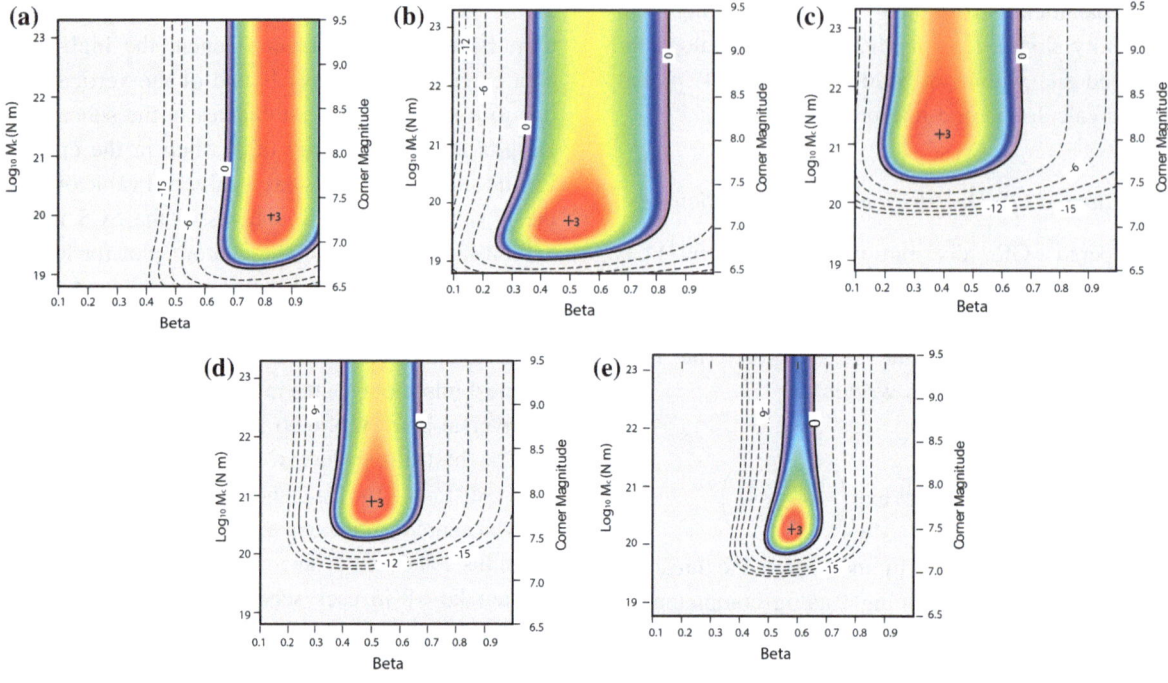

Figure 6

Contour plots of log-likelihood function for TGR distributions **a** Manila Trench; **b** Negros Trench; **c** Cotabato Trench; **d** Sulawesi Trench; and **e** Philippine Trench. The *zero contour line* corresponds to the 95 % confidence area

1. The largest earthquakes in each subcatalog since 1900. This principle is widely applied to deterministic tsunami hazard assessment to evaluate 'credible worst case'. Generally, it may underestimate the upper limit since the length of earthquake catalog is very limited.

2. Magnitude inverted from fault-length scaling law (i.e., Wells and Coppersmith 1994) based on the assumption of full rupture of entire fault zone or fault segment. For Manila Trench and Philippine Trench that have a much longer dimension, this principle tends to yield a much larger estimates of magnitude upper limit.

3. Magnitude derived from seismic and tectonic moment rate conservation principle. Based on the assumption that a certain part of tectonic deformation is released by the cumulative effects of earthquakes, M_c in the TGR relationship can be estimated as (Kagan 2002b; Kagan and Jackson 2013; Rong et al. 2014),

$$M_c \approx \left[\frac{\chi\mu WLv(1 - \beta)}{a_t M_t^\beta \Gamma(2 - \beta)}\right]^{1/(1-\beta)}, \qquad (2)$$

in which χ is the seismic coupling coefficient, μ is the rigidity set to 49 GPa, W is the down-dip width of the seismogenic zone, L is the length along the trench axis, v is the plate slip rate per annum, a_t is occurrence rate of earthquakes above the threshold moment M_t, and operator Γ denotes Gamma function. The parameters χ, W, and μ are usually not well determined and are subject to large uncertainties. The seismic coupling coefficient χ represents the fraction of plate tectonic motion within the seismogenic zone that is released by earthquakes. Scholz and Campos (2012) suggested that the coupling coefficient is usually between 0.5 and 1.0 for subduction zones. However, recent GPS observations reveal the creeping-convergence nature of Manila Trench, which yields a lower coupling value of 0.01 with standard

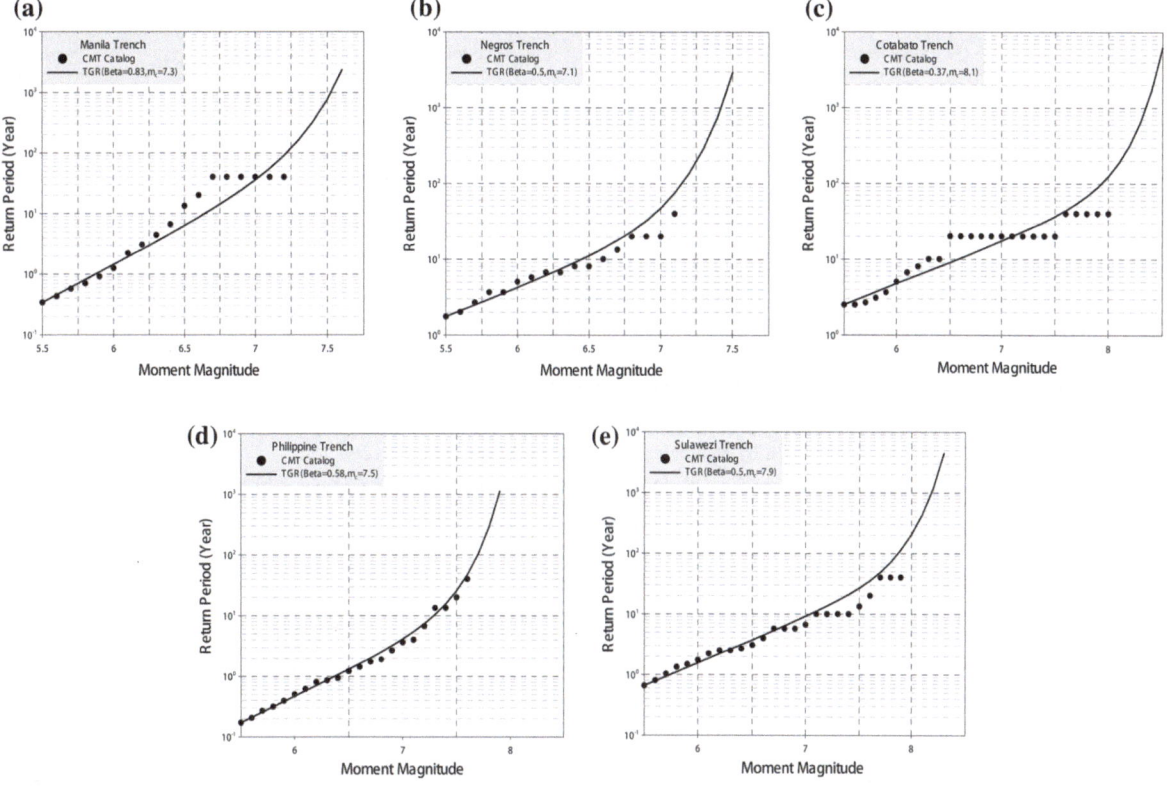

Figure 7

TGR distributions overlapped with best-fitting *curves* for five subduction zones in the SCS region: **a** Manila Trench, **b** Negros Trench, **c** Cotabato Trench, **d** Philippine Trench, **e** Sulawesi Trench. Corner magnitude m_c and β value yielded by Maximum Likelihood method are presented here and summarized in Table 2

deviation of 0.18 (Galgana et al. 2007; Hsu et al. 2012; Wang and Bilek 2014). Seismic coupling coefficient for Philippine Trench is estimated to be 0.27 with large standard deviation of 0.46. To be conservative, χ value is set to 0.2 for Manila Trench, 0.6 for Philippine Trench and 1.0 for three others. The down-dip width W is defined as the locked part of a subduction zone along the subducting slab, and typical values fall between 60 and 170 km. W values for each fault zone are determined based on the previous study by Heuret et al. (2011) and Hayes et al. (2012). All the parameters related to the TGR distribution, source geometry, and interface coupling are summarized in Table 1.

It should be noted that here we treat maximum magnitude and corner magnitude as equivalent. The maximum magnitudes derived from historic strongest earthquake and fault-length scaling law are accepted as corner magnitude in the TGR distribution, and then

serve as inputs for Monte Carlo Simulation to generate synthetic earthquake catalogs. According to the typical TGR distribution, the annual probability of an earthquake decays very quickly when the magnitude is larger than corner magnitude where the distribution start to taper. Therefore, it is appropriate to take corner magnitude as plausible upper limit for each seismic source.

3.3. Upper and Lower Bounds for Corner Magnitude m_c

As summarized in Table 1, we list 4 candidates for corner magnitude that are derived by multiple principles and approaches, including TGR regression of historic catalog ($m_{c,t}$), fault-length scaling ($m_{c,l}$), tectonic and seismic moment balance ($m_{c,b}$), and repetition of historical largest event ($m_{c,h}$). These m_c candidates show a large degree of variations,

Table 1

Parameters related to the TGR relationship, source geometry, and plate kinematics

Subduction zones	N (>5.5)	β	$M_{c,t}^a$	L (km)	$M_{c,l}^b$	W^c (km)	χ^d	V^e (cm/year^{-1})	μ (GPa)	$M_{c,b}^f$	$M_{c,h}^g$
Manila Trench	118	0.83	7.3	1061	9.1	98	0.2	90	49	**8.7**	**7.9**
Negros Trench	23	0.5	7.1	314	**8.3**	52	1.0	19.2	49	8.1	**7.8**
Cotabato Trench	16	0.37	8.1	250	**8.1**	52	1.0	18.5	49	7.7	**8.3**
Sulawesi Trench	61	0.5	7.9	590	**8.8**	62	1.0	30	49	8.2	**7.9**
Philippines Trench	235	0.58	7.5	1364	9.2	80	0.6	103	49	**8.6**	**8.0**

The corner magnitudes for each fault zone are estimated comprehensively by different approaches and principles, and summarized here (table cells highlighted with background color of gray). The upper and lower bounds of corner magnitude used for Monte Carlo simulation for each source are marked as bold

[a] $m_{c,t}$ is the corner magnitude derived by the maximum likelihood estimation of TGR relation

[b] $m_{c,l}$ denotes the magnitude inverted from empirical fault-length scaling law

[c] The values of downdip width W are compiled by Heuret et al. (2011) and Hayes et al. (2012)

[d] The values of seismic coupling coefficient χ are specified based on the result of Scholz and Campos (2012) and Galgana et al. (2007)

[e] The values of plate slip rate are retrieved from the plate-tectonic model PB2002 (Bird, 2003) and studies by Galgana et al. (2007) and Socquet et al. (2006)

[f] $m_{c,b}$ denotes corner magnitude derived by seismic and tectonic moment rate conservation principle

[g] $m_{c,h}$ denotes the largest earthquake in the NEIC catalog since 1900

Table 2

Comparison of return periods on different magnitudes in the Manila Trench with results of Ruangrassamee and Saelem (2009)

	7.0	7.5	8.0	8.5	9.0
Return periods using higher m_c-value of 8.7 (years)	9	29	98	462	1446
Return periods yielded by Ruangrassamee and Saelem (years)	6	19	63	205	667

especially for Manila Trench, Sulawesi Trench and Philippine Trench. This is expected due to distinct physical theories they hold. In this section, these m_c values are briefly reviewed, and two of them are chosen to serve as upper and lower bounds for m_c in each source zone. In the following Sect. 3.4, two sets of m_c values, along with other source parameters, will be introduced into Monte Carlo simulation to generate two synthetic earthquake catalogs for subsequent tsunami modeling.

A number of studies on seismic and tsunami hazard along the Manila Trench are conducted during the past decade. Deterministic tsunami hazard assessment based on worst scenarios along the Manila Trench assumes that the rupture breaks the whole plate interface from 13°N to 21°N and, thus, produces a worst case of 9.3 (Liu et al. 2007; Megawati et al. 2009). Based on the GPS observations, Hsu et al. (2012) proposed that fault segments extending from

the West Luzon Trough to the east of Scarborough Seamount chain are possibly aseismic. Nguyen et al. (2012) estimated that maximum earthquake magnitude for Manila Trench is among 8.3–8.7. In our study, magnitude limit $m_{c,l}$ derived by the full rupture of 1061-km long fault zone is 9.1. Manila Trench is naturally divided into two segments by Scarborough Seamount at the offshore region of Central Luzon (16.5°N). As shown in Fig. 3, the northern segment, namely North Luzon Trough, exhibits high level of seismicity, whereas at the southern segment (West Luzon Trough) the seismic activities remain at a relatively lower level. To be conservative, we take $m_{c,b} \sim 8.7$ and $m_{c,h} \sim 7.9$ as upper and lower bounds of the corner magnitude, respectively. Ruangrassamee and Saelem (2009) studied seismic occurrence rate for Manila Trench based on seismological data from 1963 through 2005. The comparison on return periods for different magnitudes between two studies

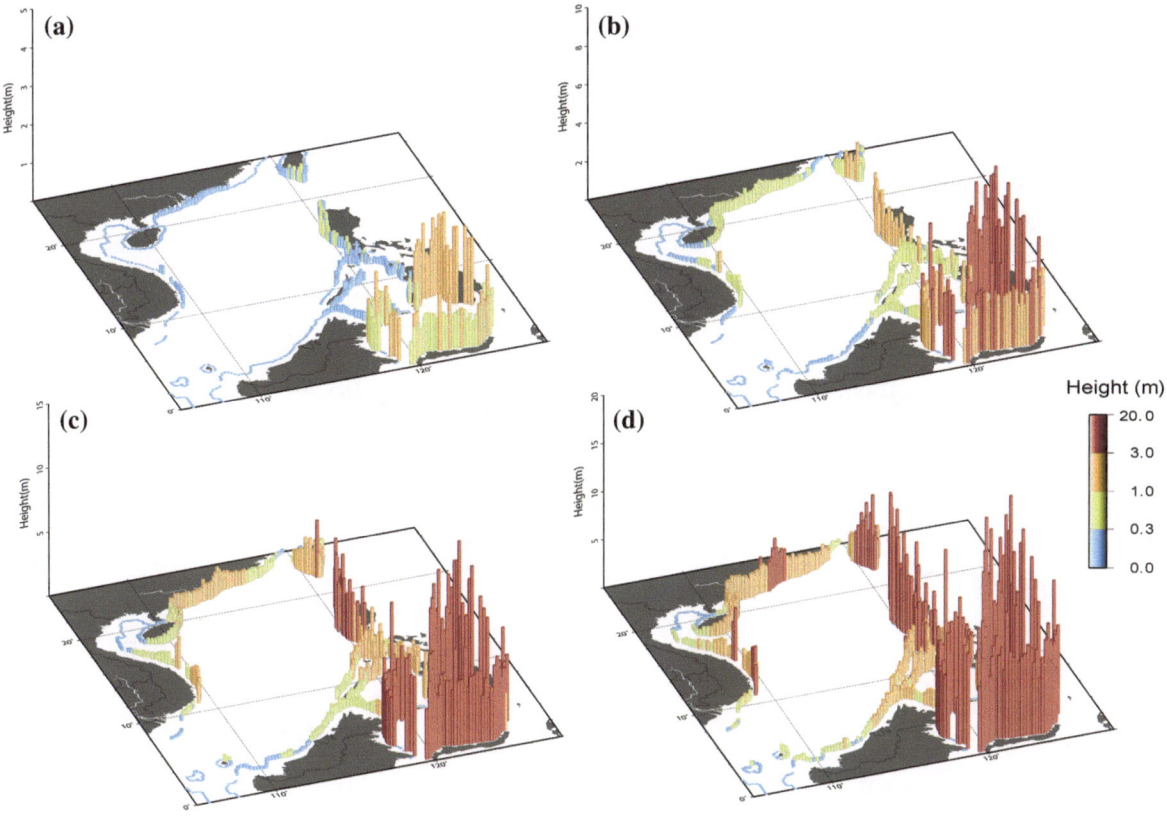

Figure 8
Coastal maximum wave amplitude for specified return periods in the SCS region using *upper bounds* of corner magnitude for each source zone: **a** 50 years; **b** 200 years; **c** 500 years; and **d** 2000 years

is shown in Table 2. The return periods for earthquake with magnitude of 7.0 and 7.5 are similar. However, the return periods of our study using higher m_c value of 8.7 are approximately 2 times of their results for magnitudes of 8.5 and 9.0. The reason is that the magnitude–frequency relation we adopt in this study is Tapered GR relationship, rather than simple GR relationship. The occurrence rate for large earthquakes begins to decrease exponentially when magnitude approaches corner magnitude m_c.

For Negros Trench, Nguyen et al. (2012) estimated that maximum magnitude for Negros Trench is among 8.0–8.4, and a full rupture of 314-km long fault segment can host a worst case of $m_{c,l} \sim 8.3$. We rule out the options of $m_{c,t} \sim 7.1$ and $m_{c,b} \sim 8.1$, and adopt $m_{c,l}$ and $m_{c,h}$ as upper and lower bounds of the corner magnitude, respectively. For Cotabato and Sulawesi Trench within the Celebes Sea, Løvholt et al. (2012) defined the credible worst cases for

Cotabato Trench and North Sulawesi Trench as 8.2 and 7.9, mostly based on the historic largest events, while Nguyen et al. (2012) suggested a possible range of 8.1–8.5. We use a pair of $m_{c,h} \sim 8.3$ and $m_{c,l} \sim 8.1$ for Cotabato Trench, and a pair of $m_{c,l} \sim 8.8$ and $m_{c,h} \sim 7.9$ for Sulawesi Trench.

3.4. Monte Carlo Simulation and Tsunami Modeling

Monte Carlo simulation provides an approach to generate long-period earthquake catalog for seismic and tsunami hazard assessment. An open-source PSHA toolbox, EQHAZ, is adopted to perform Monte Carlo simulations using prescribed TGR parameters for all source zones in the SCS region (Assatourians and Atkinson 2013). One of the merits of EQHAZ package to handle uncertainty is that it uses stochastic approach to make a weighted random draw of input parameters that will apply in generating

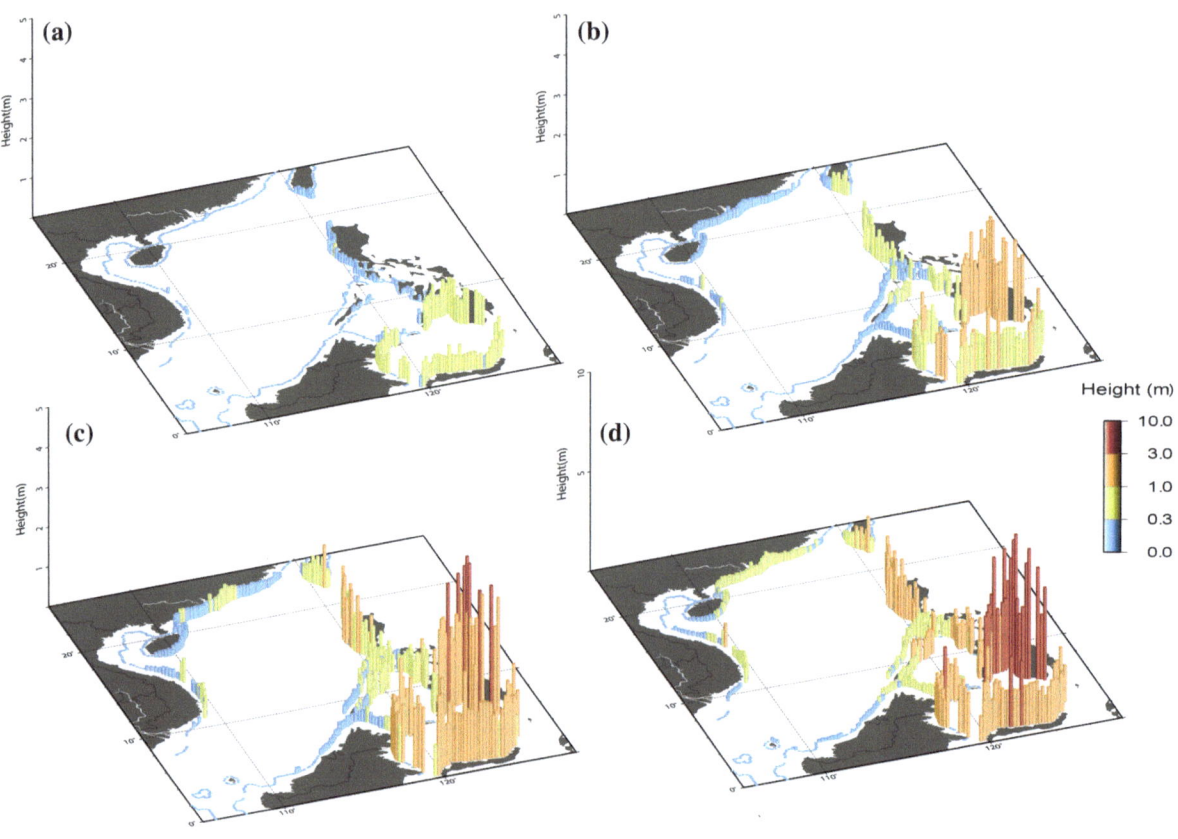

Figure 9
Coastal maximum wave amplitude for specified return periods in the SCS region using *lower bounds* of corner magnitude for each source zone: **a** 50 years; **b** 200 years; **c** 500 years; and **d** 2000 years

each event, including its location, focal depth, magnitude/frequency distribution. Thus, the generated catalog contains a long sequence that implicitly contains both the epistemic and aleatory uncertainties in these parameters. Two 50,000-year-long earthquake catalogs are generated based on two sets of corner magnitudes for each seismic source, with the purpose to figure out to what extent the tsunami hazard is dependent on the magnitude limits for seismic sources.

COMCOT v1.7 (Cornell Multi-grid Coupled Tsunami model) is used to simulate tsunami generation and propagation. The numerical scheme employed by COMCOT is the explicit leap-frog difference method. Okada's (1985) elastic finite fault theory is used to calculate vertical seafloor displacement. The source codes of COMCOT are optimized by OpenMP parallelization to improve

computation efficiency. Parameters for source geometry, such as dip and strike angles, are well estimated using historical CMT solutions, while slip angle is fixed to 90° for conservation. The scaling relationship between moment magnitude and length of rupture proposed by Wells and Coppersmith (1994) is used here. Numerical study reveals that using this scaling law can yield slightly larger maximum wave amplitude for large earthquake, which makes our PTHA results more conservative. The average amount of slip is then determined from the rupture dimensions and shear modulus. The model bathymetry covering the whole SCS region is based on NGDC ETOPO1 global model that is resampled to grid interval of 2 arc min. Overall 767 offshore output points are carefully picked along the 100–200 m isobaths for shelves that have narrow width, and along the 50 m isobath for broad northern and southern SCS shelves

Table 3

Historical earthquakes used in the empirical analysis

Date	Time	Latitude	Longitude	Depth	Magnitude	Plane1			Plane2		
						Strike (°)	Dip (°)	Rake (°)	Strike(°)	Dip (°)	Rake (°)
1918/08/15	12:18	5.7	123.5	36	8.2	320	39	62	174	56	111
1924/04/14	16:20	6.5	126.5	47	8.1	24	63	103	177	29	66
1928/12/19	11:37	7.0	124.0	39	7.3	301	44	28	190	71	130
1929/06/13	09:24	8.5	127.0	42	7.0	23	60	104	177	33	67
1934/02/14	03:59	17.5	119.0	31	7.4	346	44	50	215	58	122
1948/01/24	17:46	10.5	122.0	31	8.0	133	78	−16	226	74	−168
1949/12/29	03:03	18.0	121.0	34	7.1	19	67	−49	133	46	−147
1952/03/19	10:57	9.5	127.3	38	7.6	174	28	72	14	64	99
1968/08/01	20:19	16.3	122.1	31	7.6	351	67	53	233	42	145
1972/12/02	00:19	6.5	126.6	33	8.0	177	30	66	25	63	103
1975/10/31	08:28	12.6	126.0	50	7.5	359	72	99	150	20	63
1976/08/16	16:11	6.3	124.02	33	8.1	311	36	51	176	63	115
1982/01/11	06:10:04.0	13.7	124.32	26	7.1	351	80	98	132	13	51
1983/08/17	12:17:55.9	18.2	120.9	30	6.6	30	71	−60	149	35	−146
1988/06/24	02:06:28.1	18.6	121.01	69	5.7	16	31	85	202	59	93
1990/02/08	07:15:32.3	9.7	124.71	31	6.8	237	36	87	61	54	92
1992/05/17	10:15:31.2	7.2	126.86	33	7.3	176	25	67	20	67	100
1994/11/14	19:15:30.7	13.5	121.09	33	7.1	339	70	−178	249	88	−20
1995/04/21	00:34:47.3	12.1	125.93	23	7.2	153	22	61	3	71	101
1996/01/01	08:05:11.9	0.7	119.98	33	7.9	27	36	6	54	252	85
2002/03/05	21:16:09.1	6.0	124.25	31	7.5	314	25	70	156	67	99
2006/12/26	12:26:21.1	21.8	120.55	10	7.1	165	30	−76	329	61	−98
2008/11/16	17:02:32.7	1.3	122.09	30	7.4	92	20	84	278	70	92
2012/08/31	12:47:33.4	10.8	126.64	28	7.7	347	43	67	197	51	110

Basic parameters including date/time, epicenter, and magnitude are obtained from NGDC and TLN database; the preferred CMT solutions for events prior to 1976 are collected from Selva and Marzocchi (2011), and the rest are from GCMT catalog)

(Fig. 1). Because of huge size of the stochastic earthquake catalogs, linear part of COMCOT is used to model tsunami propagation in the open seas, and the maximum wave amplitudes at offshore output points are normalized to coastal values at a nominal depth of 20 m by Green Law. We have generated two earthquake catalogs by Monte Carlo simulation, and each has 20633 and 20306 events with magnitude ≥ 7.5 in Philippine Trench and ≥ 7.0 for other 4 source zones. For both catalogs, all scenarios are simulated to generate 767 datasets of tsunami catalogs for all coastal points.

4. The Dependency of Tsunami Hazards on Magnitude Limits

4.1. Tsunami Hazard Maps

Tsunami hazard is commonly interpreted as the maximum wave amplitude that will be exceeded at a given return period for each coastal output point. To clearly illustrate the controlling effect of magnitude limits of each fault zone on tsunami hazard in the SCS and adjoining basins, two groups of tsunami hazard maps for return period of 50, 200, 500 and 2000 years are presented in Figs. 8 and 9.

If the upper bounds of corner magnitude are used for PTHA, tsunami hazard map for 50-year return period clearly exhibits the highest level of tsunami hazard surrounding the Celebes Sea. The averaging wave amplitude at Moro Gulf exceeds 1.5 m. The maximum wave amplitudes show a much lower level along the coasts of Luzon Island and Sulu Sea. The wave amplitude drops below 0.2 m along the coasts of South China, Vietnam, and north Borneo. As shown in Fig. 9a, for PTHA result using lower bound of corner magnitude, the maximum amplitude with 50-year return period decreases to 0.3–1.0 m surrounding the Celebes Sea, and 0.1–0.3 m along the

Luzon Island. For far-field area around the SCS, tsunami hazard is negligible.

At 500-year return period (Figs. 8c, 9c), tsunami hazard in the whole SCS region has a similar distribution with that at 50-year return period. By adopting upper bounds of corner magnitude as PTHA input, tsunami hazard along the Celebes Sea ranges from 4.0 to 12.0 m with the extreme value occurring at Moro Gulf. In the SCS basin, the Manila Trench could produce the highest hazard level of 3.0–6.0 m along the coast of Luzon Island, and moderate tsunami hazard of approximately 1.0–3.0 m along the coasts of the South China and central Vietnam. The tsunami hazard along the north coast of Borneo is negligible. On the contrary, when lower bounds of corner magnitude are used, tsunami hazard plunges down to 1.0–3.7 m along the Celebes Sea. The averaging hazard level is only one-fifth of that using upper bounds. Besides, tsunami hazard along the coast of the SCS basin is basically negligible except for west coast of Luzon Island, which is much lower than that using upper bounds either. For 2000-year return period, the even larger difference between Figs. 8d and 9d could be observed. Typically, tsunami hazard in the Celebes Sea and West Luzon coast could be 2–6 times of those of using lower-bound m_c values.

Considering complex tectonic setting and a lack of large earthquake events in the SCS region, tsunami

Figure 11
Tsunami hazard *curves* for four coastal regions by adopting upper and lower m_c-values, including SCS coasts except Luzon (i.e., South China, Central Vietnam and Borneo) (**a, b**), West Luzon (**c, d**), Sulu Sea (**e, f**), and Celebes Sea (**g, h**). Figures in the *left column* are for upper m_c values cases and *right* for lower m_c values cases. 4 representing stations within the above coastal regions, including Hong Kong, Currimao, Iloilo, and Toli–Toli, are highlighted as *black lines*, while other coastal output points showed as *gray lines*

hazards in the SCS region are subject to very large variations when adopting different magnitude limits for each fault zone. Despite the magnitude limits for major seismic sources in this region are reviewed and the preferred values are put forward in Sects. 3.2 and 3.3, how to designate these parameters more appropriately remains an open issue for further study. This conclusion highlights the imperative need of a comprehensive, multiple-disciplinary seismic and tsunami hazard study involving geodesy, tectonic kinematics, seismology, and tsunami generation in the SCS and adjoining basins.

4.2. *Tentative Validation of Tsunami Hazard by Historic Tsunamis*

As stated in Sect. 2, the written and instrumental records of historical tsunami events are very rare and incomplete in this area, and the source origin and description of some tsunami events are subject to

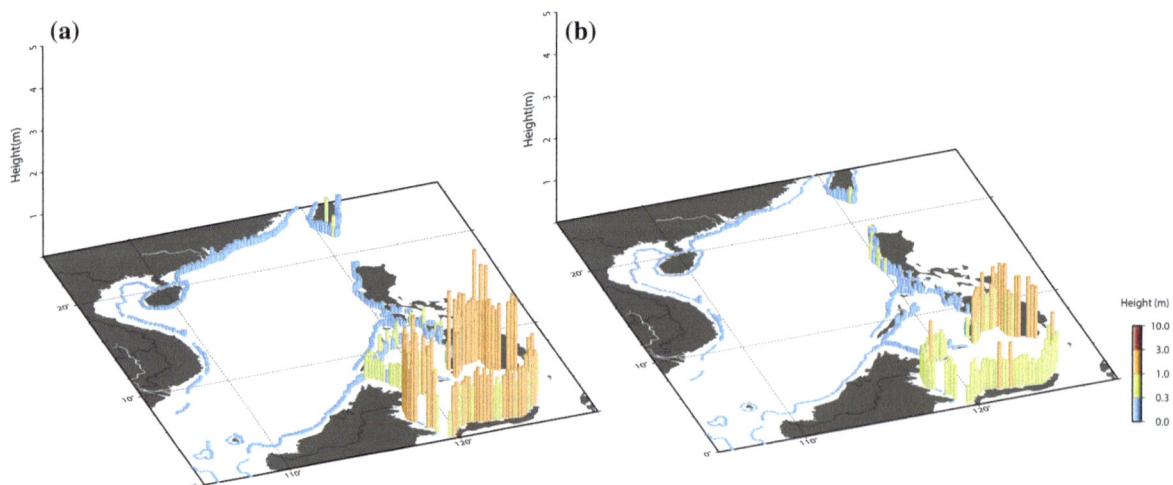

Figure 10
Comparison of **a** empirically derived tsunami hazard at 50-year return period based on 24 historic tsunami events and **b** PTHA result at 50-year return period

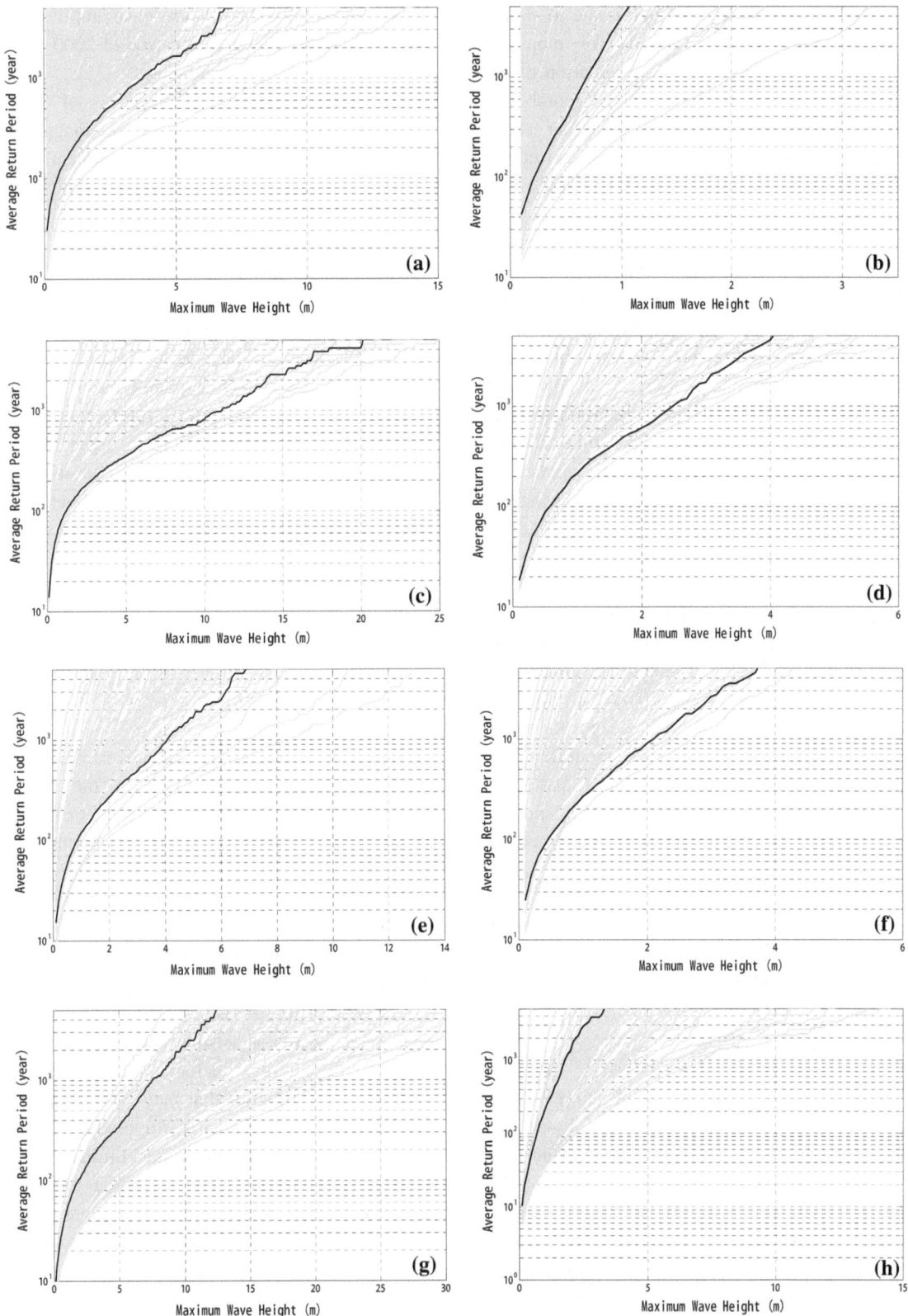

large uncertainty. Thus, it is inappropriate to simply rely on historical tsunami catalog for empirical analysis of tsunami probability, or validation of our PTHA results shown in Sect. 4.1. By combining historical records and numerical modeling skills, procedures for validation of present PTHA results are proposed: (1) reproducing 24 historic tsunami events since 1900 (listed in Table 3) that have the validity level of 'definite' by focal mechanism solutions queried from Selva and Marzocchi (2004) (1900–1975) and GCMT project (1976 to present); (2) generate tsunami catalogs for 767 coastal output points, and each catalog has 24 wave amplitude values spanning 115-year periods; (3) a power-law relation $P(h) = Ch^{-a}$ is assumed between tsunami amplitude h and annul occurrence probability P (Burroughs and Tebbens 2001), and then applied to catalogs for each point; (4) interpolate the hazard curves, and yield exceedance wave amplitudes at 50-year return period for each point.

As shown in Fig. 10, tsunami hazard at 50-year return period based on empirical analysis of 24 historical events is presented and compared with our PTHA results. It should be noted that maximum wave amplitudes shown in Fig. 10b are arithmetical average of values shown in Figs. 8a and 9a, which imply a weight value of 0.5 for each possibility. The similar distribution of tsunami hazards along the coasts of the SCS and adjoining basins lends us some confidence on the probabilistic results of the present study. An exception occurs at West Luzon coast due to a lack of credible tsunami observations along the west coast of Luzon Island; tsunami hazard based on the empirical analysis is slightly lower than that of PTHA.

4.3. Tsunami Hazard Curves

Figure 11 shows tsunami hazard curves for 4 coastal regions by adopting upper and lower m_c-values, including SCS coasts except Luzon (i.e., South China, Central Vietnam and Borneo) (a-b), West Luzon (c-d), Sulu Sea (e-f), and Celebes Sea (g-h). 4 representing stations, including Hong Kong, Currimao, Iloilo, and Toli–Toli, are highlighted as *black lines*, while other coastal output points showed as gray lines. As the length of synthetic catalog is

50,000 years, only exceedance wave amplitudes with y-axis readings below return period of 2000 years are regarded as statistical significance.

Comparison among two columns of sub-plots clearly reveals critical role of corner magnitude on PTHA through distinct levels of tsunami hazard. Tsunami hazards derived from upper limit of m_c is generally 2–6 times of those using lower limit of m_c. For coasts of the SCS basin, tsunami threat is mainly contributed by Manila Trench. The wide range between upper and lower bounds of m_c for Manila Trench leads to distinct tsunami hazard levels not only for near-field Currimao, but also Hong Kong at far field. Tsunami hazards of Hong Kong at recurrence interval of 1000 years are about 3.5 m and 0.5 m corresponding to upper and lower bounds of m_c for Manila Trench. The arithmetical average of these two values is approximately 2.0 m, which coincides with the finding of Liu et al. (2007) that the probability for a 2.0 m wave hitting Hong Kong or Macau is around 10 % for this century.

5. Closing Remarks

Based on a brief review of historical seismicity and tsunamis, tsunami hazard in the SCS and adjoining basins is analyzed by numerically based PTHA. By combining empirical method with numerical study of historical tsunami events, the present PTHA results are tentatively validated. As the main concern of this study is to clarify to what extent the magnitude limits of fault zones dominate the tsunami hazard levels, this paper puts stress on the choice of upper and lower limits of corner magnitude for major seismic sources. For the SCS region with complex tectonic features, compared with other uncertainties that may be introduced into the PTHA, it is particularly important to consider the large variations of magnitude limits. Therefore, to further clarify the tsunami hazards in the SCS and adjacent basins, the multiple-disciplinary seismic and tsunami hazard studies involving geodesy, tectonic kinematics, seismology, and tsunami generation in the SCS and adjoining basins are necessary.

Acknowledgments

We appreciate the sharing of PB2002 Global Plate Model and computing codes of Maximum Likelihood method by Peter Bird and Yan Y. Kagan. COMCOT source codes developed by Philips Liu and Xiaoming Wang can be downloaded from ceeserver.cee.cornell.edu/pll-group/comcot_down.htm. EQHAZ package for Monte Carlo simulation is downloaded from http://www.seismotoolbox.ca/EQHAZ.html. We especially thank Dr. Bautista and Dr. Narag from the Philippine Institute of Volcanology and Seismology (PHIVOLCS) to provide technical report of 'Philippine Tsunamis and Seiches (1589–2012)', which is very useful for this study. Two anonymous reviewers also provide very useful and insightful comments on the manuscript. This study is supported by Public science and technology research funds projects of ocean (No. 201405026) and GASI-GEOGE-05.

REFERENCES

Annaka, T., Satake, K., Sakakiyama, T., Yanagisawa, K., & Shuto, N. (2007). Logic-tree approach for probabilistic tsunami hazard analysis and its application to the Japanese coasts. *Pure and Applied Geophysics, 164,* 577–592.

Assatourians, K., & Atkinson, G. M. (2013). EqHaz: an open-source probabilistic seismic-hazard code based on the monte carlo simulation approach. *Seismological Research Letters, 84,* 516–524.

Bautista, M.L.P., Bautista, B.C., Salcedo. J.C. & Narag. I.C. (2012). Philippine tsunamis and seiches (1589–2012) (pp. 1–111).

Bird, P. (2003). An updated digital model of plate boundaries. *Geochemistry Geophysics Geosystems, 4*(3), 1027. doi:10.1029/2001GC000252.

Bird, P., & Kagan, Y. Y. (2004). Plate-tectonic analysis of shallow seismicity: apparent boundary width, beta, corner magnitude, coupled lithosphere thickness, and coupling in seven tectonic settings. *Bulletin of the Seismological Society of America, 94,* 2380–2399.

Burbidge, D., Cummins, P. R., Mleczko, R., & Thio, H. K. (2008). A Probabilistic tsunami hazard assessment for western Australia. *Pure and Applied Geophysics, 165,* 2059–2088.

Burroughs, S. M., & Tebbens, S. F. (2001). Upper-truncated power laws in natural systems. *Pure and Applied Geophysics, 158,* 741–757.

Ca, V. T., & Xuyen, N. D. (2008). Tsunami risk along Vietnamese coast. *Journal of Water Resources and Environmental Engineering, 23,* 24–33.

Cornell, C. A. (1968). Engineering seismic risk analysis. *Bulletin of the Seismological Society of America, 58,* 1583–1606.

Dao, M. H., Tkalich, P., Chan, E. S., & Megawati, K. (2009). Tsunami propagation scenarios in the South China Sea. *Journal of Asian Earth Sciences, 36,* 67–73.

Das, R., Wason, H. R., & Sharma, M. L. (2011). Global regression relations for conversion of surface wave and body wave magnitudes to moment magnitude. *Natural Hazards, 59,* 801–810.

Fukutani, Y., Suppasri, A., & Imamura, F. (2014). Stochastic analysis and uncertainty assessment of tsunami wave height using a random source parameter model that targets a Tohoku-type earthquake fault. *Stochastic Environmental Research and Risk Assessment, 29,* 1763–1779.

Galgana, G., Hamburger, M., McCaffrey, R., Corpuz, E., & Chen, Q. (2007). Analysis of crustal deformation in Luzon, Philippines using geodetic observations and earthquake focal mechanisms. *Tectonophysics, 432,* 63–87.

Geist, E. L., & Parsons, T. (2006). Probabilistic analysis of tsunami hazards. *Natural Hazards, 37,* 277–314.

Goda, K., Li, S., Mori, N., & Yasuda, T. (2015). Probabilistic tsunami damage assessment considering stochastic source models: application to the 2011 Tohoku earthquake. *Coastal Engineering Journal, 57,* 1550015.

Goda, K., & Song, J., (2015). Uncertainty modeling and visualization for tsunami hazard and risk mapping: A casestudy for the 2011 Tohoku earthquake. *Stochastic Environmental Research and Risk Assessment.* doi:10.1007/s00477-015-1146-x.

González, F. I., Geist, E. L., Jaffe, B., Kânoğlu, U., Mofjeld, H., Synolakis, C. E., et al. (2009). Probabilistic tsunami hazard assessment at seaside, Oregon, for near- and far-field seismic sources. *Journal of Geophysical Research: Oceans, 114,* C11023.

Hayes, G. P., Wald, D. J., & Johnson, R. L. (2012). Slab1.0: a three-dimensional model of global subduction zone geometries. *Journal of Geophysical Research: Solid Earth, 117,* B01302.

Heidarzadeh, M., & Kijko, A. (2010). A probabilistic tsunami hazard assessment for the Makran subduction zone at the northwestern Indian Ocean. *Natural Hazards, 58,* 577–593.

Heuret, A., Lallemand, S., Funiciello, F., Piromallo, C., & Faccenna, C. (2011). Physical characteristics of subduction interface type seismogenic zones revisited. *Geochemistry, Geophysics, Geosystems, 12,* Q01004.

Horspool, N., Pranatyo, I., Griffin, J., Latief, H., Natawidjaja, D. H., Kongko, W., et al. (2014). A probabilistic tsunami hazard assessment for Indonesia. *Natural Hazards and Earth System Sciences, 14,* 3105–3122.

Hsu, Y. J., Yu, S. B., Song, T. R. A., & Bacolcolk, T. (2012). Plate coupling along the Manila subduction zone between Taiwan and northern Luzon. *Journal of Asian Earth Sciences, 51,* 98–108.

Huang, Z., Wu, T. R., Tan, S. K., Megawati, K., Shaw, F., Liu, X., & Pan, T. C. (2009). Tsunami hazard from the subduction Megathrust of the South China Sea: part II. Hydrodynamic modeling and possible impact on Singapore. *Journal of Asian Earth Sciences, 36,* 93–97.

Kagan, Y. Y. (1997). Seismic moment-frequency relation for shallow earthquakes: regional comparison. *Journal of Geophysical Research: Solid Earth, 102,* 2835–2852.

Kagan, Y. Y. (2002a). Seismic moment distribution revisited: I. Statistical results. *Geophysical Journal International, 148,* 520–541.

Kagan, Y. Y. (2002b). Seismic moment distribution revisited: II. Moment conservation principle. *Geophysical Journal International, 149,* 731–754.

Kagan, Y. Y. (2003). Accuracy of modern global earthquake catalogs. *Physics of the Earth and Planetary Interiors, 135,* 173–209.

Kagan, Y. Y., & Jackson, D. D. (2013). Tohoku earthquake: a surprise? *Bulletin of the Seismological Society of America, 103*, 1181–1194.

Kreemer, C., & Holt, W. E. (2000). Active deformation in eastern Indonesia and the Philippines from GPS and seismicity data. *Journal of Geophysical Research: Solid Earth, 105*, 663–680.

Lane, E. M., Gillibrand, P. A., Wang, X., & Power, W. (2013). A probabilistic tsunami hazard study of the Auckland Region, Part I: propagation modelling and tsunami hazard assessment at the shoreline. *Pure and Applied Geophysics, 170*, 1621–1634.

Liu, Y., Santos, A., Wang, S. M., Shi, Y., Liu, H., & Yuen, D. A. (2007). Tsunami hazards along Chinese coast from potential earthquakes in South China Sea. *Physics of the Earth and Planetary Interiors, 163*, 233–244.

Lorito, S., Selva, J., Basili, R., Romano, F., Tiberti, M. M., & Piatanesi, A. (2015). Probabilistic hazard for seismically induced tsunamis: accuracy and feasibility of inundation maps. *Geophysical Journal International, 200*, 574–588.

Løvholt, F., Kühn, D., Bungum, H., Harbitz, C. B., & Glimsdal, S. (2012). Historical tsunamis and present tsunami hazard in eastern Indonesia and the southern Philippines. *Journal of Geophysical Research: Solid Earth, 117*, B09310.

Megawati, K., Shaw, F., Sieh, K., Huang, Z., Wu, T. R., Lin, Y., et al. (2009). Tsunami hazard from the subduction megathrust of the South China Sea: Part I. Source characterization and the resulting tsunami. *Journal of Asian Earth Sciences, 36*, 13–20.

Nguyen, P. H., Bui, Q. C., & Nguyen, X. D. (2012). Investigation of earthquake tsunami sources, capable of affecting Vietnamese coast. *Natural Hazards, 64*, 311–327.

Okada, Y. (1985). Surface deformation due to shear and tensile faults in a half-space. *Bulletin of the Seismological Society of America, 75*, 1135–1154.

Okal, E. A., Synolakis, C. E., & Kalligeris, N. (2011). Tsunami Simulations for Regional Sources in the South China and Adjoining Seas. *Pure and Applied Geophysics 168*, 1153–1173.

Parsons, T., & Geist, E. L. (2008). Tsunami probability in the Caribbean Region. *Pure and Applied Geophysics, 165*, 2089–2116.

Power, W., Downes, G., & Stirling, M. (2007). Estimation of Tsunami Hazard in New Zealand due to South American Earthquakes. *Pure and Applied Geophysics, 164*, 547–564.

Rangin, C., Pichon, X. L., Mazzotti, S., Pubellier, M., Chamot-Rooke, N., Aurelio, M., et al. (1999). Plate convergence measured by GPS across the Sundaland Philippine Sea plate deformed boundary: the Philippines and eastern Indonesia. *Geophysical Journal International, 139*, 296–316.

Rong, Y., Jackson, D. D., Magistrale, H., & Goldfinger, C. (2014). Magnitude limits of subduction zone earthquakes. *Bulletin of the Seismological Society of America,*. doi:10.1785/0120130287.

Rowlett, H., & Kellecher, J. (1976). Evolving seismic and tectonic patterns along the western margin of the Philippine Sea plate. *Journal Geophysical Research, 81*, 3518–3524.

Ruangrassamee, A., & Saelem, N. (2009). Effect of Tsunamis generated in the Manila Trench on the Gulf of Thailand. *Journal of Asian Earth Sciences, 36*, 56–66.

Scholz, C. H., & Campos, J. (2012). The seismic coupling of subduction zones revisited. *Journal Geophysical Research, 117*, B05310.

Selva, J., & Marzocchi, W. (2004). Focal parameters, depth estimation, and plane selection of the worldwide shallow seismicity with $M_s \geq 7.0$ for the period 1900–1976. Geochem. Geophys. Geosys. 5, Q05005.

Seno, T., & Kurita, K. (1978). Focal mechanisms and tectonics in Taiwan-Philippine region. *Journal of Physics of the Earth, 26*(Supplement), 249–263.

Socquet, A., Simons, W., Vigny, C., McCaffrey, R., Subarya, C., Sarsito, D., et al. (2006). Microblock rotations and fault coupling in SE Asia triple junction (Sulawesi, Indonesia) from GPS and earthquake slip vector data. *Journal Geophysical Research, 111*, B08409.

Sørensen, M. B., Spada, M., Babeykou, A., Wiemer, S., & Grüntha, G. (2012). Probabilistic tsunami hazard in the Mediterranean Sea. *Journal Geophysical Research, 117*, B01305.

Suppasri, A., Imamura, F., & Koshimura, S. (2012). Tsunami hazard and casualty estimation in a coastal area that neighbors the Indian Ocean and South China Sea. *Journal of Earthquake and Tsunami, 6*(2), 1250010.

Thio, H. K., Somerville, P., & Ichinose, G. (2012). Probabilistic analysis of strong motion and tsunami hazards in South East Asia. *Journal of Earthquake and Tsunami, 2*, 119–137.

Wang, K., & Bilek, S. L. (2014). Invited review paper: fault creep caused by subduction of rough seafloor relief. *Tectonophysics, 610*, 1–24.

Wells, D. L., & Coppersmith, K. J. (1994). New empirical relationships among magnitude, rupture length, rupture width, rupture area, and surface displacement. *Bulletin of the Seismological Society of America, 84*, 974–1002.

Wu, T. R., & Huang, H. C. (2009). Modeling tsunami hazards from Manila trench to Taiwan. *Journal of Asian Earth Sciences, 36*, 21–28.

Zhu, J., Qiu, X., & Zhan, W. (2005). Focal mechanism solutions and its tectionic significance in the trench of eastern South China Sea. *Acta Seismologica Sinica, 5*, 280–289.

(Received February 18, 2016, revised July 20, 2016, accepted August 2, 2016, Published online August 9, 2016)

Pure Appl. Geophys. 174 (2017), 2371–2380
© 2016 Springer International Publishing
DOI 10.1007/s00024-016-1359-z

Can Apparent Stress be Used to Time-Dependent Seismic Hazard Assessment or Earthquake Forecast? An Ongoing Approach in China

ZHONGLIANG WU,[1] CHANGSHENG JIANG,[1] and SHENGFENG ZHANG[1,2]

Abstract—The approach in China since the last 1.5 decade for using apparent stress in time-dependent seismic hazard assessment or earthquake forecast is summarized. Retrospective case studies observe that apparent stress exhibits short-term increase, with time scale of several months, before moderate to strong earthquakes in a large area surrounding the 'target earthquake'. Apparent stress is also used to estimate the tendency of aftershock activity. The concept relating apparent stress indirectly to stress level is used to understand the properties of some 'precursory' anomalies. Meanwhile, different opinions were reported. Problems in the calculation also existed for some cases. Moreover, retrospective studies have the limitation in their significance as compared to forward forecast test. Nevertheless, this approach, seemingly uniquely carried out in a large scale in mainland China, provides the earthquake catalogs for the predictive analysis of seismicity with an additional degree of freedom, deserving a systematic review and reflection.

Key words: Apparent stress, time-dependent seismic hazard, earthquake forecast, earthquake sequence, mainland China.

1. Introduction

Apparent stress, defined by the ratio of radiated energy E_S and seismic moment M_0 multiplying μ the rigidity of the source medium,

$$\sigma_a = \mu(E_S/M_0)$$

has been proposed since the 1960s (Wyss and Brune 1968; Choy and Boatwright 1995; McGarr 1999). In recent years, studies on and application of this parameter have been carried out in different fields (Choy and Boatwright 1995; Zobin 1996; McGarr 1999; Pulido and Irikura 2000; Ide and Beroza 2001; McGarr and Fletcher 2002; Choy and McGarr 2002;

Choy and Kirby 2004; Baltay et al. 2011) To much extent, this topic of study is directly related to the development of digital broadband seismological networks. Such development has made it possible to have independent estimate of radiated energy and seismic moment—before this development both radiated energy and seismic moment can only be calculated from magnitude/s (Choy and Boatwright 1995).

In the measurement and study of apparent stress, one of the approaches has been to use this parameter for time-dependent seismic hazard or earthquake forecast. Such an approach has been taken in mainland China for about 1.5 decade. Up to now, most of the works related to this approach have been published in Chinese journals (although with English translation/abstracts).[1] As a result, this approach is to much extent in lack of international visibility and international exchange. Different understandings of the concept of 'earthquake forecast/prediction' (Wu et al. 2013; Wu 2014) further block such exchange. To overcome this language/cultural barrier, with the motive similar to Wu (1997) who introduced the annual earthquake forecast in China to the international seismological community, in this paper, we try to summarize and review the related works in mainland China since the 21st century, to see what is going on, what the main results are, and what the problems are.

[1] Institute of Geophysics, China Earthquake Administration, 100081 Beijing, China. E-mail: wuzl@cea-igp.ac.cn
[2] Earthquake Administration of Shandong Province, 250014 Jinan, China.

---

[1] Some of the Chinese journals used to restart their page numbers issue by issue. To facilitate the indexing, in the reference list, for Chinese journals, the issue number is included in the parentheses next to the volume number. Furthermore, to differentiate the papers with the first authors sharing the same surname (even if with different first names), alpha–beta following the year of publication is used.

2. Pre-shock Increase of Apparent Stress?

Increase of apparent stress of small earthquakes before moderate to strong earthquakes, at the time scale of months and spatial scale which varies from near the epicenter to several hundred kilometers, were reported by several retrospective case studies. By such retrospective case study, Chen and Li (2007) obtained that apparent stress of small earthquakes near the epicenter ($\sim$50 km) of the July 4, 2006, Wen'an, Hebei, earthquake ($M_S = 5.1$) exhibited an increase before the earthquake. The increase occurred in 2004 which restored at the end of 2005. Analyzing the small earthquakes before and after the June 6, 2000, Jingtai, Gansu, $M_S5.9$ earthquake, Wu and Gu (2004) showed that apparent stress had increased since November 1999. This increasing trend had been kept till the $M_S5.9$ earthquake occurred, and then decreased. Spatial range is as large as 600 km. Qin and Qian (2006) reported that apparent stress of small events increased before the July 21, 2003, Dayao, Yunnan, $M_S6.2$ earthquake. The increase started since January 2003. On October 16, 2003, in the same place there occurred another $M_S6.1$ earthquake, with apparent stress of small events increased in September. Analyzing the February 24, 2003, Bachu-Jiashi, Xinjiang, $M_S6.8$ earthquake and the February 15, 2005, Wushi, Xinjiang, $M_S6.3$ earthquake, Sun and Shi (2007) reported that apparent stress increased about several months before the two earthquakes. The spatial range subject to analysis is about 500 km long and 100 km wide. Using the data of the regional seismic network of Anhui Province, Li et al. (2012a) studied several cases of moderate earthquakes near the southern Tancheng-Lujiang fault along the eastern coast of China, and reported the pre-shock increase trend of apparent stress of small earthquakes. The one-year scale pre-shock increase of apparent stress was also reported for the April 20, 2013, Lushan, Sichuan, $M_S7.0$ earthquake (Gong et al. 2013).

Reliability varies associated with these case studies, especially in the early time. The variation of apparent stress before the January 5, 1998, Jingyang, Shaanxi, $M_L5.2$ earthquake, as reported by Wang et al. (2004, 2005b), was hard to be considered as significant. Some works (e.g., Chen et al. 2007) compared the variation of apparent stress with remote (up to 200 km distance) earthquakes. Negative cases were also reported. Although the methods of measuring apparent stress vary, the negative results are as important as the positive ones in judging whether there were changes of apparent stress before the 'target' earthquakes. For the Laoshan, Qingdao, Shandong, $M_L4.1$ earthquake sequence, it was reported that the pre-shock variation of apparent stress cannot be seen (Zheng et al. 2006). This might be due to the magnitude of the 'target' earthquake which is apparently smaller. Yang and Zhang (2009) investigated the spatial distribution of apparent stress in the Beijing Capital Circle region where there have not occurred strong earthquakes since the 21st century, and found that the relation between moderate earthquake/s and apparent stress is vague. From the retrospective case studies of the 2008 Wenchuan earthquake independently by Yi et al. (2011) who mapped the apparent stress of small earthquakes along the Longmenshan-Minshan tectonic zone, Cheng et al. (2011) who mapped the apparent stress of small earthquakes in Sichuan Province, and Li et al. (2012c) who mapped the apparent stress of small earthquakes in the Sichuan-Yunnan region, it is hard to be persuaded that there was any significant pre-earthquake change of apparent stress (or even the spatial distribution of the higher apparent stress) which could lead to the alarm of the mid-to-north Longmenshan fault zone at any time scale.

3. Apparent Stress Associated with an Earthquake Sequence?

Judgment of the type of an earthquake sequence, that is, whether it is a foreshock swarm, is one of the important topics in earthquake forecast. Apparent stress was also used to such judgment, although questionable to some extent (Zhong et al. 2005; Li et al. 2006). A more persuasive result of Liu et al. (2006) showed that for four strong earthquakes in Yunnan Province (the January 15, 2000, Yao'an $M_S5.9$ and $M_S6.5$, the April 12, 2001, Shidian $M_S5.9$, the October 27, 2001, Yongsheng $M_S6.0$, and the July 21 and October 16, 2003, Dayao $M_S6.2$ and $M_S6.1$, earthquakes), if there are at least one earthquake in the sequence with apparent stress exceeding a threshold value (in their case, 1 MPa), then the sequence is more likely a foreshock swarm; Otherwise the swarm is not followed by strong earthquakes. By checking more data, Liu et al. (2007) further concluded that in Yunnan Province, 0.9 MPa is the threshold value indicating the likelihood of strong earthquakes. In their work, they used the R-value to test the statistical significance of the forecast. The R-value (Xu 1989), which has been widely used in mainland China for the test of earthquake forecast schemes (Shi et al. 2001), can be considered as a simplified version of the well-known ROC test (Swets 1973; Molchan 1997).

Examining Bath's law, Chen et al. (2003) used some strong to great earthquakes to show that if the apparent stress of the mainshock is small, then the magnitude difference between the mainshock and the largest aftershock will be large. They used eight cases (the February 3, 1996, Lijiang $M_S7.0$; the March 9, 1996, Atushi $M_S6.9$; the November 19, 1996, Kalakunlun $M_S7.1$; the November 8, 1997, Mani $M_S7.5$; the August 27, 1998, Jiashi $M_S6.6$; the November 14, 2001, Kokoxili/Kunlunshan Mountain Pass $M_S8.1$; the March 24, 2003, Bachu-Jiashi $M_S6.8$; and the April 7, 2003, Delingha $M_S6.6$, earthquakes) to show that the apparent stress of these earthquakes are less than 2 MPa, and the magnitude difference between the mainshock and the largest aftershock are larger than 0.9. The November 14, 2001, Kokoxili $M_S8.1$ earthquake seems an exception among these eight earthquakes, with its higher apparent stress 5.3 MPa, but still with small aftershocks followed. Focal mechanisms of these earthquakes, and the might-be super-shear rupture related to strike-slip faulting penetrating ground surface, and its relation with aftershocks, were not considered in their approach, which may be one of the reasons for such an exception. As a matter of fact, the investigation of Zhong et al. (2004a) showed that the relation between the apparent stress of the mainshock and the type of earthquake sequence is complicated, depending on several factors such as focal mechanism.

4. Problems of the Case Studies and Ongoing Improvements

Although the application of apparent stress to time-dependent seismic hazard seems promising as shown by the case studies, partly due to that the approach has been just started, several problems apparently exist, which needs an ongoing improvement. First, majority of the works are retrospective case studies. As an improvement dealing with such limitation, Chen et al. (2015) reported that before the July 22, Minxian, Gansu, $M_S6.6$ earthquake there was an increasing trend of apparent stress of the earthquakes larger than $M_L4.0$ in that region, but the occurrence of this earthquake did not change the increasing trend of apparent stress, i.e., after the $M_S6.6$ earthquake the apparent stress kept increasing. Thus, they indicated that the seismic hazard in this region still exists, although without clear indication of the time scale, which possesses to some extent the characteristics of forward forecast, to be tested by real earthquakes. Based on the variation of apparent stress, Li et al. (2015b) indicated that seismic hazard in Shandong Province seems increasing. Similarly, Gong et al. (2015) used apparent stress to claim the increasing seismic hazard near Zigong, Sichuan Province.

Using apparent stress as an indicator, Ruan et al. (2010) pointed out that the Mabian-Daguan area near the border of Sichuan and Yunnan Province exhibited increasing seismic hazard. But considering the empirical time duration of the pre-shock variation of apparent stress, whether this predictive conclusion could be related to the August 3, 2014, Ludian $M_S6.5$ earthquake is questionable, and this conclusion, if could be considered as a forecast, seems to be a false alarm. However, in the retrospective case studies there are few reports on false alarms; therefore, this work is noticeable in recording a case of false alarm.

In the predictive mapping of apparent stress, one of the difficulties is that the correlation between the variation of apparent stress and the seismic hazardous region seems complicated. The case of the July 4, 2006 Wen'an, Hebei, $M_S5.1$ earthquake (Chen et al. 2011) showed that although this earthquake is associated with pre-shock increase of apparent stress within ~50 km near the epicenter (Chen and Li 2007), inspecting a wider range, the epicenter is actually not associated with the largest increase of apparent stress. Additionally, up to now, resolution of the mapping of apparent stress is still limited. Only a few reports (e.g., Ruan et al. 2011; Li and Yao 2015) dealt with focal depths in the measurement and using of apparent stress.

The calculation of radiated energy and seismic moment, especially radiated energy, further contributes to the uncertainty of, and debates on the related results. In some works, radiated energy was calculated through the energy-magnitude scaling relation (e.g., Wu and Gu 2004; Zhong et al. 2004a; Wang et al. 2004, 2005a; Zhang et al. 2007; Gao 2009). The scattered feature of this empirical relation may affect the estimate of radiated energy, and in turn affect the estimate of apparent stress. In some works, both radiated energy and seismic moment are deduced from the energy-magnitude relation and the seismic moment-magnitude relation (e.g., Cao et al. 2005; Huang et al. 2005; Tang et al. 2007; Zhang and Yang 2007; Liu et al. 2009). In this case, the apparent stress becomes another expression of magnitude, which deviates from the original sense of the measurement of apparent stress. These cases are therefore not included in the summary of the earthquake cases in the above sections.

5. Scaling of Apparent Stress with Magnitude, and its Implication for using the Variation of Apparent Stress in Earthquake Forecast

Scaling of apparent stress with magnitude or seismic moment also affects the results, which has been paid attention to by several works (e.g., Cheng et al. 2006). According to the study of Li and Chen (2007) for the sequence of the 1999 Xiuyan, Liaoning, $M_S5.4$ earthquake, of Wang et al. (2013) for the small earthquakes in Shandong Province of eastern China, of Zhang et al. (2015) for the water injection induced earthquakes in Sichuan Province of southwestern China, and of Xie et al. (2013) for the small earthquakes in the reservoir region, among others, apparent stress scales clearly with seismic moment or magnitude. Importantly, without the consideration of such a scaling and the associated correction, some ambitious approaches such as using apparent stress to estimate the probability of major earthquakes within the next 3 years (Yi et al. 2007), may be seriously affected. The case of the September 4, 2013, Xianyou $M_L5.0$ earthquake in Fujian Province (Cai 2015)

indicated that the increasing trend of apparent stress before the earthquake can be attributed to the increasing of the magnitude of small earthquakes. Li et al. (2014) reported that before the May 28, 2012, Tangshan $M_S4.8$ earthquake, apparent stress of local small events increased, and after the earthquake apparent stress decreased, but this variation seems related to the variation of the magnitude of small earthquakes. In the work of Li et al. (2015a), it is hard to discriminate whether the conclusion that the apparent stress of the mainshock is larger than that of the aftershocks is simply due to that the magnitude of the mainshock is apparently larger. In the early work studying the 2007 Ning'er, Yunnan, $M_S6.4$ earthquake sequence, Qian et al. (2007) reported that the average apparent stress of aftershocks within the first 4 days after the mainshock was 0.95 MPa, which decreased to 0.27 MPa (the background level of apparent stress in Yunnan Province) after the fourth day. Even if in an aftershock sequence of a $M_L4.1$ earthquake (Chen et al. 2005), it was found that the apparent stress of the aftershocks (being about 0.05 MPa on average) are different from that of the mainshock (being 5 MPa). Seen in present time, however, what is the role of the scaling relation in such observation apparently needs a reevaluation. Yue et al. (2015) also noticed this problem, and discussed qualitatively the effect of magnitude on the variation of apparent stress.

Coping with this problem, Qiao et al. (2006) and Sun et al. (2015) used seismic events with similar magnitudes to compare the apparent stress. Wang et al. (2005a, b) pointed out that inspecting the variation of apparent stress, the relation between earthquake magnitude and apparent stress has to be taken into account. As a solution, they used only the small events with magnitude less than $M_L4.5$. Considering the scaling between apparent stress and magnitude, Liang et al. (2014) used the data of small earthquakes with different magnitude ranges. In the study of Pan et al. (2015)[2] discussing the difference of apparent stress of small earthquakes within and surrounding the reservoir, and before and after the

---

[2] Should be 'Yunnan Province', printed as 'Yunna' in the journal.

impounding, earthquakes with similar seismic moment were used for the comparison.

As a more sophisticated solution to this problem, Li et al. (2012b) first fit the scaling relation between apparent stress and magnitude, and then calculate the difference between the measured apparent stress and the 'theoretical' apparent stress deduced from the magnitude-apparent stress relation. They found that before the June 3, 2007, Ning'er, Yunnan, $M_S6.4$ earthquake, this value showed significant increase since September 2006. This increase terminated at the end of 2006. Spatial range is about 150 km centered at the epicenter of the 'target' earthquake. The result of Zhang et al. (2009) showed that such 'differential apparent stress' had an apparent increase in the Zipingpu Reservoir from 2005, exhibiting clear correlation with the water level. Wang and Zheng (2014) took similar approach to the analysis of apparent stress in eastern Shandong Province, and used the result in a testing mode to the annual forecast of seismic hazard. Yi et al. (2013) obtained that the October 19, 2012, Tianquan-Lushan $M_L4.6$ earthquake, although hard to be considered as a foreshock of the April 20, 2013, Lushan, Sichuan, $M_S7.0$ earthquake in the perspective of seismicity, has significantly higher apparent stress as compared to the empirical apparent stress-magnitude scaling relation. Wang and Zheng (2015) took similar approach for the Yutian, Xinjiang, $M_S7.3$ earthquake on February 12, 2014, and observed the pre-shock increase of 'differential apparent stress'.

Remarkably, the scaling relation between apparent stress and magnitude or seismic moment is dependent on the focal mechanisms of the earthquakes (Zhong et al. 2004a, b), which can also affect other physical properties such as the performance of precursory anomalies (Wu and Wang 2004). Such dependence was even used to obtain the information of dynamic friction (Wu 2001a, b). Also interesting is that reservoir-induced earthquakes and tectonic earthquakes have similar scaling relations between apparent stress and magnitude, but tectonic earthquakes have higher level of apparent stress (Zhou et al. 2012). Yang and Zhang (2010) investigated the seismicity in the Zipingpu Reservoir region (which became well-known after the 2008 Wenchuan earthquake due to the debate on its role in triggering

the Wenchuan earthquake) and found that the scaling relation between seismic moment and apparent stress changed before and after the impounding of the reservoir; before the impounding the correlation between apparent stress and seismic moment was weak, but after the impounding a scaling law emerged that apparent stress increases with seismic moment.

6. Radiated Energy as the Key Issue for Understanding and Measuring Apparent Stress

Physical picture of the relation between the variation of apparent stress and seismic hazard is not very clear and is at best conceptual or even intuitive, being discussed by most of the works only in the introduction section. In the simplest case, radiated energy during an earthquake can be expressed by (Savage and Wood 1971)

$$E_S = \left[\frac{(\sigma_0 + \sigma_1)}{2} - \sigma_f\right]M_0/\mu$$

in which σ_0, σ_1, and σ_f are the initial stress, the final stress, and the stress associated with dynamic friction, respectively. Apparent stress, having the dimension of stress and expressed by the ratio of radiated energy E_S and seismic moment M_0 multiplying the rigidity of material μ, is straightforwardly the difference between the average stress and the frictional dynamic stress (Pulido and Irikura 2000). By this interpretation, higher apparent stress is associated with either the higher level of average stress (which may lead to the failure of the material) or the lower level of frictional dynamic stress (which can also 'facilitate' the seismic slip). The picture of higher average stress is generally associated with the traditional understanding of the mechanism of earthquake and faulting. Meanwhile, seen in the perspective of recent studies, the lowered friction, as can be indexed by a series of keywords, such as 'frictional melting', 'silica gel lubrication', 'flash heating', 'powder lubrication', and 'thermal pressurization', might be equally if not more important.

It might be interesting that apparent stress, as an observational quantity of an earthquake, has another advantage. In seismology, one of the theoretical debates is whether seismic moment, defined by the product of D the slip, A the area of the seismic fault, and μ the rigidity of the source medium

$$M_0 = \mu DA$$

is ambiguous at an interface of materials crossing which μ has different values (e.g., Vavrycuk 2013). It is even suggested (Ben-Zion 2001) that the use of seismic moment should be replaced by the directly observable potency (or geometric moment) P defined by DA. Apparent stress, at least in its form, seems not affected by such an ambiguity. As a matter of fact, Mori et al. (2003) expressed apparent stress as

$$\sigma_a = (E_S/A)/D$$

that is, 'the seismic energy radiated per unit fault area per unit fault slip'.

Indeed, radiated energy plays the key role in obtaining the apparent stress. Depending on the description of the propagation of seismic waves from the source to the receiver, radiated energy can be calculated through source spectra (e.g., Ide and Beroza 2001), source time function (e.g., Vassiliou and Kanamori 1982), moment rate (e.g., Vassiliou and Kanamori 1982; Pulido and Irikura 2000), or even rupture process model of a finite source (e.g., McGarr and Fletcher 2001, 2002). Comparing with seismic moment, the 'zero-frequency description' of the seismic source process, in the measurement of radiated energy, the integration of the source spectra over a wide frequency band, there are more details to be considered in modeling the seismic wave propagation and attenuation. Seen in this perspective, it seems that at the present time, most if not all the problems in the study of apparent stress in China, probably in other places as well, are associated with the measurement of radiate energy.

7. Discussion with(out) Conclusion/s

Chinese scientists had started the measurement of apparent stress as early as in the 1980s (e.g., Ye et al. 1980; Yan et al. 1994). Relating apparent stress to seismic hazard, Huang and Yi (2000) discussed the relation between 'apparent strain' and seismic hazard at a 5-year time scale. Widespread use of apparent stress in seismic hazard analysis (or in another word, earthquake forecast) in mainland China was to some extent stimulated by the paper of Wu (2001a, b) and Wu et al. (2002) which were cited by majority of the early stage papers published—to much extent this is the motive for the authors of this article to conduct the review which has dual goals as to review the works going on, and to reflect whether such a stimulation has been a blessing or cursing of the earthquake forecast study in mainland China. Sheng and Wan (2008) summarized some of the results and discussed their physical significance, while in this article more recent works were included. No matter successful or not, this endeavor carried out by several groups of scientists based on the updated observational facilities over the past 15 years deserves to be recorded either as a review (if successful and prospective) or as a piece of history of seismology (if not as successful as expected). Noticeably, in some countries speaking different languages or with different systems of scientific development, there had been some works not well-known by international scientific community. China is only one of the examples of such countries, and apparent stress applied to earthquake forecast is only one of the examples of such works. Collecting and reviewing such works is by no means superficially for 'copyright claiming'. In fact, it is a record of the development of science, reflecting the diversity of scientific approaches, and the wide participation in the development of researches, especially facing to some difficult scientific problems such as earthquake forecast.

As indicated by the title, this paper covers only the works in mainland China which directly related apparent stress to time-dependent seismic hazard or earthquake forecast. Other works on apparent stress (for example, Chen 2005; Meng et al. 2013, 2014) are not considered. From the present works in progress, it seems that the following three aspects deserve further investigation:

1. Pre-shock increase of apparent stress of small events, with time scale of several months, and varying spatial scale surrounding the epicenter of the 'target' earthquake;

2. Relation between the apparent stress of an earthquake and the property of the earthquake, as per whether it is a foreshock;
3. Relation between the apparent stress of the mainshock and the magnitude of the largest aftershock.

Somehow interesting is that there were quite a few works studying reservoir-induced seismicity, providing the physics of faulting with some observational clues, while conceptually, apparent stress was also proposed to be used to monitoring the hazard of mining-induced seismicity (Tang et al. 2011). On the other hand, at present time, controversies still exist mainly on the following three aspects:

1. Reliability of the estimation of radiated energy, and in turn reliability of the estimation of apparent stress;
2. Scaling of apparent stress with the size of earthquakes, and the 'real' variation of apparent stress associated with seismic hazard;
3. Statistical significance of the apparent variation of apparent stress associated with the 'target' earthquakes.

All these three aspects indicate the direction of future studies. Related to these pros and cons, accumulating and digesting more observational data, as well as the standardization and systematization of the works, plays the central role.

From the present status quo, it is apparently too early to be optimistic on the application of apparent stress to seismic hazard assessment or earthquake forecast. However, the scientific merit of this approach, no matter whether successful or not, lies in providing time-dependent seismic hazard analysis or earthquake forecast with new earthquake catalogs of more degrees of freedom, and in turn a new look at earthquake sequences. As an ongoing approach, it is almost inevitable to have problems of various kinds from methodology to measurement, which are to be overcome with the progress of the study. As a matter of fact, in recent years, there have been the works on the recalculation and reevaluation of the apparent stress as well as its performance as an indicator of the time-dependent seismic hazard, for the earthquakes or earthquake sequences investigated before (e.g.,

Wang et al. 2014). We are not confident, therefore, whether it is fair to conduct a critical review of some of the works. Nevertheless, it is time to summarize the related results for the past 1.5 decade, and communicate with international seismological community, as reflected by this article which tries to provide an overview of such an approach in a way similar to the IUGG national report (e.g., Rundle and Klein 1995), with the hope of calling for interests in, and probably cooperation on, this ongoing approach.

Acknowledgments

Thanks to Guest Editor Dr. Yongxian (Angie) Zhang (CENC, Beijing) for invitation and suggestion to write a review article, to Prof. Dr. Peter Bormann (GFZ, Potsdam), Prof. Yun-tai Chen (CEA, Beijing), Dr. G. L. Choy (NEIC, USGS, Golden), and Prof. Yaolin Shi (UCAS, Beijing) for valuable advices related to apparent stress during the past two decades, to Drs. Yan'e Li, Yingchun Li, Tengfei Ma, and Xuerong Yao for assistance in collecting the data, and to the anonymous reviewers for their encouragement and valuable suggestions for the revision.

REFERENCES

Baltay, A., Ide, S., Prieto, G., & Beroza, G. (2011). Variability in earthquake stress drop and apparent stress. *Geophysical Research Letters, 38,* L06303. doi:10.1029/2011GL046698.

Ben-Zion, Y. (2001). On quantification of the earthquake source. *Seismological Research Letters, 72,* 151–152.

Cai, X.-H. (2015). Study on apparent stress of small earthquakes before and after the Fujian Xianyou $M_L5.0$ earthquake. *Journal of Geodesy and Geodynamics, 35*(6), 984–986, 1006. **(in Chinese with English abstract)**.

Cao, F.-J., Yin, T., Ma, L., & Wu, D. (2005). Time-space variational characteristics of the apparent stress before and after Xiuyan $M_S5.4$ earthquake. *Seismological Research of Northeast China, 21*(1), 14–19. **(in Chinese with English abstract)**.

Chen, X.-Z. (2005). Estimation of the stress levels in the focal region before and after the 2001 $M = 8.1$ Western Kunlun Mountain Pass earthquake. *Acta Seismologica Sinica, 27*(6), 605–609. **(in Chinese with English abstract; English version: Acta Seismologica Sinica, English edition, 18(6): 651–655)**.

Chen, X.-Z., & Li, Y.-E. (2007). Temporal changes in apparent stresses of small earthquakes around the epicentral area prior to the July 4, 2006 Wenan, Hebei earthquake (M_S = 5.1). *Earthquake Research in China, 23*(4), 327–336. **(in Chinese with English abstract)**.

Chen, X.-Z., Wang, X.-P., Wang, L.-Y., & Zhang, T.-Z. (2003). Possibility of earlier judgement on seismic tendency after strong earthquakes upon their apparent stresses. *Recent Developments in World Seismology, (7)*, 1–4. **(in Chinese with English abstract)**.

Chen, X.-Z., Xu, X.-T., & Zhai, W.-J. (2005). Variation of stress during the rupture process of the 1995 $M_L = 4.1$ Shacheng, Hebei, China, earthquake sequence. *Acta Seismologica Sinica, 27*(3), 276–281. **(in Chinese with English abstract; English version, Acta Seismologica Sinica, English edition, 18(3): 297–302)**.

Chen, Y.-W., Zhang, J., Qing, M., Wang, X.-Z., & Zhang, B. (2007). Study on the time-varying characteristics of kinematic parameters of small earthquake sequence in 'Huoshan seismic window'. *Earthquake, 27*(1), 26–32. **(in Chinese with English abstract)**.

Chen, X.-Z., Li, Y.-E., & Guo, X.-Y. (2011). Temporal-spatial variations of focal dynamic parameters in the Capital Circle and surrounding areas before the 2006 Wen'an $M_S5.1$ earthquake in Hebei Province. *Earthquake, 31*(4), 15–25. **(in Chinese with English abstract)**.

Chen, L.-J., Li, Y.-E., Yang, L.-M., Chen, J.-F., & Chen, X.-Z. (2015). Temporal-spatial variations of apparent stress before and after the 22 July 2013 Minxian $M_S6.6$ earthquake. *Earthquake, 35*(4), 30–42. **(in Chinese with English abstract)**.

Cheng, W.-Z., Chen, X.-Z., & Qiao, H.-Z. (2006). Research on the radiated energy and apparent strain of the earthquakes in Sichuan province. *Progress in Geophysics, 21*(3), 692–699. **(in Chinese with English abstract)**.

Cheng, W.-Z., Zhang, Y.-J., Ruan, X., Qiao, H.-Z., & Zhang, Z. W. (2011). Research on regional stress field dynamics before the Wenchuan $M_S8.0$ earthquake. *Earthquake Research in China, 27*(3), 215–225. **(in Chinese with English abstract)**.

Choy, G. L., & Boatwright, J. L. (1995). Global patterns of radiated seismic energy and apparent stress. *Journal Geophysical Research, 100*, 18205–18228.

Choy, G. L., & Kirby, S. (2004). Apparent stress, fault maturity and seismic hazard for normal-fault earthquakes at subduction zones. *Geophysical Journal International, 159*, 991–1012.

Choy, G. L., & McGarr, A. (2002). Strike-slip earthquakes in the oceanic lithosphere: observations of exceptionally high apparent stress. *Geophysical Journal International, 150*, 506–523.

Gao, L.-X. (2009). Analysis on the characteristics of apparent stress in midwest Inner Mongolia area. *Seismological Research of Northeast China, 25*(4), 1–7. **(in Chinese with English abstract)**.

Gong, Y., Long, F., Yi, G.-X., Zhao, M., Qiao, H.-Z., & Wang, Y.-X. (2013). Study of apparent stress on the southern segment of Longmenshan fault zone around the Lushan earthquake. *Technology for Earthquake Disaster Prevention, 8*(4), 389. **(in Chinese with English abstract)**.

Gong, Y., Long, F., Qiao, H.-Z., Zhao, M., & Wang, Y.-X. (2015). Spatial and temporal distribution of seismic apparent stress in the southeast of Sichuan Province. *Seismological and Geomagnetic Observation and Research, 36*(2), 6–12. **(in Chinese with English abstract)**.

Huang, F.-M., & Yi, Z.-G. (2000). Relation between the evolution of seismic apparent strain field and the region of strong earthquake occurrence. *Acta Seismologica Sinica, 22*(6), 577–587. **(in Chinese with English abstract; English version: Acta Seismologica Sinica, 13(6): 616–627)**.

Huang, X.-L., Liu, D.-W., Shen, X.-Q., & Qing, M. (2005). The study on the apparent stress field in East China. *Seismological and Geomagnetic Observation and Research, 26*(6), 29–32. **(in Chinese with English abstract)**.

Ide, S., & Beroza, G. C. (2001). Does apparent stress vary with earthquake size? *Geophysical Reseach Letters, 28*, 3349–3352.

Li, Y.-E., & Chen, X. Z. (2007). Determination and scaling relationship analysis of source parameters of the Nov. 29, 1999 Xiuyan 5.4 earthquake sequence. *Earthquake, 27*(4), 59–67. **(in Chinese with English abstract)**.

Li, G.-Y., & Yao, J.-J. (2015). Apparent stress features of two strong earthquake sequences in Qinghai Da Qaidam. *Seismological and Geomagnetic Observation and Research, 36*(3), 53–57. **(in Chinese with English abstract)**.

Li, F., Li, Y.-T., & Liu, Y.-F. (2006). Research on the apparent stress method applying to judge the characteristics of earthquake swarm. *Earthquake, 26*(4), 45–51. **(in Chinese with English abstract)**.

Li, F., Ge, N., Wang, X.-Z., Ling, X.-S., & Zhang, B. (2012a). Temporal and spatial characteristics of small to moderate earthquake apparent stresses in the south part of Tanlu fault belt and adjacent areas. *Earthquake, 32*(4), 53–61. **(in Chinese with English abstract)**.

Li, Y.-E., Chen, X.-Z., & Fu, H. (2012b). Temporal and spatial variation of focal dynamic parameters in Southwest Yunnan before the 2007 $M_S6.4$ Ning'er earthquake. *Earthquake, 32*(1), 28–39. **(in Chinese with English abstract)**.

Li, Y.-E., Chen, X.-Z., & Wang, H.-X. (2012c). Temporal and spatial variation of apparent stress in Sichuan area before the $M_S8.0$ Wenchuan earthquake. *Earthquake, 32*(4), 113–122. **(in Chinese with English abstract)**.

Li, H., Liu, S.-Q., Liu, W.-B., Wang, Y.-X., Shang, X.-Q., Tan, Y.-P., et al. (2014). Variations in apparent stress prior to and after the Tangshan $M_S4.8$ earthquake. *China Earthquake Engineering Journal, 36*(4), 1019–1025. **(in Chinese with English abstract)**.

Li, H., Liu, W.-B., Wang, Y.-X., Liu, S.-Q., Gong, Y.-J., Shang, Q.-Q., et al. (2015a). Apparent stress changes of the 2010 Luanxian earthquake sequence. *Earthquake, 35*(1), 47–54. **(in Chinese with English abstract)**.

Li, Y.-E., Chen, L.-J., Wang, S.-W., Guo, X.-Y., & Chen, X.-Z. (2015b). Temporal and spatial variations of apparent stress in Shandong Province, China. *Earthquake, 35*(2), 80–90. **(in Chinese with English abstract)**.

Liang, X.-J., Li, L., Song, M.-Q., Meng, X.-Q., & Liu, L.-F. (2014). Space-time characteristics analysis of the apparent stress in Shanxi. *Earthquake Research in China, 30*(1), 102–110. **(in Chinese with English abstract)**.

Liu, H.-G., Liu, J., Ding, Y.-L., Sun, Y.-J., & Yu, X. (2006). Precursory specialties of apparent stresses in Yunnan earthquake series. *Acta Seismologica Sinica, 28*(5), 462–471. **(in Chinese with English abstract; English version: Acta Seismologica Sinica, English edition, 19(5): 497–506)**.

Liu, H.-G., Wang, P.-L., Yang, C.-X., Xu, G., Sun, Y.-J., Chen, Z.-L., et al. (2007). Application of apparent stress in earthquake prediction. *Acta Seismologica Sinica, 29*(4), 437–445. **(in Chinese with English abstract)**.

Liu, F., Sun, H., & Zhang, L.-R. (2009). Study on temporal and spatial distribution of apparent stress in midwestern Inner Mongolia. *North China Earthquake Sciences, 27*(1), 7–12. **(in Chinese with English abstract)**.

McGarr, A. (1999). On relating apparent stress to the stress causing earthquake fault slip. *Journal Geophysical Research, 104*, 3003–3011.

McGarr, A., & Fletcher, J. B. (2001). A method for mapping apparent stress and energy radiation applied to the 1994 Northridge earthquake fault zone—revisited. *Geophysical Reseach Letters, 28*, 3529–3532.

McGarr, A., & Fletcher, J. B. (2002). Mapping apparent stress and energy radiation over fault zones of major earthquakes. *Bulletin Seismological Society of America, 92*, 1633–1646.

Meng, L.-Y., Zhou, L.-Q., & Liu, J. (2013). Source parameters of the 2013 Lushan M_S7.0 earthquake and the characteristics of the near fault strong ground motion. *Acta Seismologica Sinica, 35*(5), 632–641. (**in Chinese with English abstract**).

Meng, L.-Y., Zhou, L.-Q., & Liu, J. (2014). Source parameters and estimation of intensity distribution for the 2013 M_W7.7 Pakistan earthquake. *Earthquake, 34*(4), 12–19. (**in Chinese with English abstract**).

Molchan, G. M. (1997). Earthquake prediction as a decision-making problem. *Pure and Applied Geophysics, 149*, 233–247.

Mori, J., Abercrombie, R. E., & Kanamori, H. (2003). Stress drops and radiated energies of aftershocks of the 1994 Northridge, California, earthquake. *Journal Geophysical Research, 108*, 2545. doi:10.1029/2001JB000474.

Pan, Y., Zhang, X.-D., & Fu, H. (2015). Seismic activity and source parameters in the Nuozhadu Reservoir region, Yunna Province. *Earthquake, 35*(3), 31–43. (**in Chinese with English abstract**).

Pulido, N., & Irikura, K. (2000). Estimation of dynamic rupture parameters from the radiated seismic energy and apparent stress. *Geophysical Reseach Letters, 27*, 3945–3948.

Qian, X.-D., Li, Q., & Qin, J.-Z. (2007). Apparent stress of the 2007 Ning'er, Yunnan, M_S6.4 earthquake sequence. *Journal of Seismological Research, 30*(4), 311–317. (**in Chinese with English abstract**).

Qiao, H.-Z., Cheng, W.-Z., & Chen, X.-Z. (2006). Study on the apparent stress of the earthquakes on Anninghe-Zemuhe fault. *Journal of Seismological Research, 29*(2), 125–130. (**in Chinese with English abstract**).

Qin, J.-Z., & Qian, X.-D. (2006). On temporal and spatial distribution of seismic apparent stresses in Yunnan area. *Acta Seismologica Sinica, 28*(3), 221–229. (**in Chinese with English abstract; English version: Acta Seismologica Sinica, English edition, 19(3): 233–242**).

Ruan, X., Cheng, W.-Z., Qiao, H.-Z., Zhang, Z.-W., & Fu, Y. (2010). Research of source parameter and stress state of Mabian-Daguan tectonic zone. *Journal of Seismological Research, 33*(4), 294–300. (**in Chinese with English abstract**).

Ruan, X., Cheng, W.-Z., Qiao, H.-Z., Zhang, Y.-J., & Zhang, Z.-W. (2011). Research on source parameters and stress state in Anninghe-Zemuhe earthquake belt. *Northwestern Seismological Journal, 33*(1), 46–50. (**in Chinese with English abstract**).

Rundle, J. B., & Klein, W. (1995). New ideas about the physics of earthquakes. *Reviews Geophysics, 33*(suppl., US National Report to IUGG 1991–1994), 283–286.

Savage, J. C., & Wood, M. D. (1971). The relation between apparent stress and stress drop. *Bulletin Seismological Society of America, 61*, 1381–1388.

Sheng, S.-Z., & Wan, Y.-G. (2008). Apparent stress and the discussion of its physical implication. *Seismological and Geomagnetic Observation and Research, 29*(1), 36–43. (**in Chinese with English abstract**).

Shi, Y., Liu, J., & Zhang, G. (2001). An evaluation of Chinese annual earthquake predictions, 1990–1998. *Journal of Applied Probability, 38A*, 222–231.

Sun, Y.-P., & Shi, Y.-J. (2007). Apparent stress variation before and after two strong earthquakes in Xinjiang. *Seismological and Geomagnetic Observation and Research, 28*(4), 20–28. (**in Chinese with English abstract**).

Sun, Y.-J., Liu, H.-G., Jiang, H.-L., Zhan, X.-Y., Ju, H.-H., & Yang, Y. (2015). Research on the characteristics of earthquake apparent stress in Jiangsu and its adjacent area. *Journal of Seismological Research, 38*(1), 35–41. (**in Chinese with English abstract**).

Swets, J. A. (1973). The relative operating characteristic in psychology. *Science, 182*, 990–1000.

Tang, L.-R., Yang, R., Yang, Y.-Q., Zeng, X.-F., & Zhou, H.-Y. (2007). A study on the variations in apparent stress of Jiujiang-Ruichang seismic sequence. *South China Journal of Seismology, 27*(4), 19–27. (**in Chinese with English abstract**).

Tang, L.-Z., Wang, L.-H., Zhang, J., & Li, X.-B. (2011). Seismic apparent stress and deformation in a deep mine under large-scale mining and areal hazardous seismic prediction. *Chinese Journal of Rock Mechanics and Engineering, 30*(6), 1168–1178. (**in Chinese with English abstract**).

Vassiliou, M. S., & Kanamori, H. (1982). The energy release in earthquakes. *Bulletin Seismological Society of America, 72*, 371–387.

Vavrycuk, V. (2013). Is the seismic moment tensor ambiguous at a material interface? *Geophysical Journal International, 194*, 395–400. doi:10.1093/gji/ggt084.

Wang, P., & Zheng, J.-C. (2014). Temporal and spatial variation of apparent stress in eastern Shandong Province. *Earthquake, 34*(4), 70–77. (**in Chinese with English abstract**).

Wang, P., & Zheng, J.-C. (2015). Research of source parameter of Yutian M_S7.3 earthquake in Xinjiang on February 12, 2014. *Earthquake Research in China, 31*(2), 262–270. (**in Chinese with English abstract**).

Wang, W.-D., Zhang, Y.-Z., Shao, G.-Z., Liu, G.-H., Zou, Z.-W., & Di, X.-L. (2004). Study on apparent stress in the middle and southern parts of Shaanxi Province. *Northwestern Seismological Journal, 26*(4), 343–346. (**in Chinese with English abstract**).

Wang, Q., Chen, X.-Z., & Wang, L.-Y. (2005a). Change of apparent stress of Xiuyan earthquake sequence and its prediction significance. *Earthquake, 25*(2), 91–97. (**in Chinese with English abstract**).

Wang, W.-D., Peng, J.-B., Zhang, Y.-Z., Liu, G.-H., Shao, G.-Z., & Li, S.-R. (2005b). Features of apparent stress changing before and after the Jingyang, Shaanxi, M_L5.2 earthquake. *Journal of Seismological Research, 28*(3), 274–276. (**in Chinese with English abstract**).

Wang, P., Zheng, J.-C., & Zhao, J.-H. (2013). Status of seismic source spectral parameters in the Jiaodong Peninsula. *China Earthquake Engineering Journal, 35*(2), 360–365. (**in Chinese with English abstract**).

Wang, S.-W., Li, Y.-E., Guo, X.-Y., & Chen, L.-J. (2014). Recalculation of apparent stresses of the Nov. 29, 1999 Xiuyan M_S 5.4 earthquake sequence. *Earthquake, 34*(3), 50–61. (**in Chinese with English abstract**).

Wu, F. T. (1997). The Annual Earthquake Prediction Conference in China (National Consultative Meeting on Seismic Tendency). *Pure and Applied Geophysics, 149*, 249–264.

Wu, Z.-L. (2001a). Apparent stress obtained from broadband radiated energy catalogue and seismic moment catalogue and its seismological significance. *Earthquake Research in China, 17*(1), 8–15. (**in Chinese with English abstract**).

Wu, Z. L. (2001b). Scaling of apparent stress from broadband radiated energy catalogue and seismic moment catalogue and its focal mechanism dependence. *Earth, Planets and Space, 53*, 943–948.

Wu, Z. L. (2014). Chapter 16: Duties of earthquake forecast: cases and lessons in China. In M. Wyss (Ed.), *Earthquake Hazard, Risk, and Disasters* (pp. 431–448). Amsterdam: Elsevier.

Wu, J., & Gu, J.-P. (2004). Discussion on apparent stress of small earthquakes before and after Jingtai earthquake with M_S5.9 in Gansu Province. *Earthquake, 24*(1), 170–175. **(in Chinese with English abstract)**.

Wu, Z.-L., & Wang, L.-Y. (2004). Statistical property of candidate earthquake precursors and its apparent focal mechanism dependence. *Acta Seismologica Sinica, 26*(supp), 58–63. **(in Chinese with English abstract; English version: Acta Seismological Sinica, 17(supp.): 61–66)**.

Wu, Z.-L., Huang, J., & Lin, B.-C. (2002). Distribution of apparent stress in western China. *Acta Seismologica Sinica, 24*(3), 293–301. **(in Chinese with English abstract; English version: Acta Seismologica Sinica, English edition, 15(3): 309–317)**.

Wu, Z. L., Ma, T. F., Jiang, H., & Jiang, C. S. (2013). Multi-scale seismic hazard and risk in the China mainland with implication for the preparedness, mitigation, and management of earthquake disasters: an overview. Review Article. *International Journal of Disaster Risk Reduction, 4*, 21–33.

Wyss, M., & Brune, J. N. (1968). Seismic moment, stress, and source dimensions for earthquakes in the California-Nevada region. *Journal Geophysical Research, 73*, 4681–4694.

Xie, R.-H., Ruan, X., Zhang, Z.-W., & Du, Y. (2013). Research of source parameter of moderate-small earthquakes before the impoundment of the Pubugou reservoir. *Seismological and Geomagnetic Observation and Research, 34*(1/2), 1–9.

Xu, S. X. (1989). The evaluation of earthquake prediction ability. In Department of Science, Technology and Monitoring, State Seismological Bureau (Ed.), *The Practical Research Papers on Earthquake Prediction Methods* (Seismicity Section) (pp. 586–589). Beijing: Seismological Press. **(in Chinese)**.

Yan, Z.-G., Xue, J.-R., & Geng, A.-L. (1994). A study on source parameters of earthquakes in the Danjiangkou and its nearby regions. *South China Journal of Seismology, 14*(3), 24–32. **(in Chinese with English abstract)**.

Yang, Z.-G., & Zhang, X.-D. (2009). Computation of apparent stress and its spatial distribution in the capital region of China. *Earthquake, 29*(4), 32–43. **(in Chinese with English abstract)**.

Yang, Z.-G., & Zhang, X.-D. (2010). The transition of apparent stress scaling law before and after impounding in Zipingpu reservoir region. *Chinese Journal of Geophysics, 53*(12), 2861–2868. **(in Chinese with English abstract)**.

Ye, H., Zhang, W.-Y., Yu, Z.-S., & Xia, Q. (1980). On the source tectonics of 1979 Liyang earthquake of magnitude 6. *Seismology and Geology, 2*(4), 27–38. **(in Chinese with English abstract)**.

Yi, Z.-G., Song, M., Yang, X.-H., & Zhang, B. (2007). Application of apparent stress to macroseism tendency prediction in western China's mainland. *Earthquake, 27*(4), 110–120. **(in Chinese with English abstract)**.

Yi, G.-X., Wen, X.-Z., Xin, H., Qiao, H.-Z., Long, F., & Wang, S.-W. (2011). Distributions of seismicity parameters and seismic apparent stresses on the Longmenshan-Minshan tectonic zone before the 2008 M_S8.0 Wenchuan earthquake. *Chinese Journal of Geophysics, 54*(6), 1490–1500. **(in Chinese with English abstract)**.

Yi, G.-X., Wen, X.-Z., Xin, H., Qiao, H.-Z., Wang, S.-W., & Gong, Y. (2013). Stress state and major-earthquake risk on the southern segment of the Longmen Shan fault zone. *Chinese Journal of Geophysics, 56*(4), 1112–1120.

Yue, X.-Y., Wu, A.-X., Feng, G., Wu, M.-J., & Li, H. (2015). Temporal and spatial variations of apparent stress before the 2012 Tangshan M_S4.8 earthquake in the eastern Capital Circle. *Earthquake, 35*(2), 91–100. **(in Chinese with English abstract)**.

Zhang, B., & Yang, X.-H. (2007). On distribution of apparent seismic stress in Chinese mainland. *Journal of Geodesy and Geophysics, 27*(5), 26–30. **(in Chinese with English abstract)**.

Zhang, B., Yang, X.-H., & Yi, Z.-G. (2007). Time-space evolution of the seismic apparent strain field before Songpan-Pingwu earthquakes. *Northwestern Seismological Journal, 29*(3), 235–239. **(in Chinese with English abstract)**.

Zhang, Z.-W., Cheng, W.-Z., Zhang, Y.-J., Xie, R.-H., & Fu, Y. (2009). Research on seismicity and source parameters of small earthquakes in the Zipingpu Dam before Wenchuan M_S8.0 earthquake. *Earthquake Research in China, 25*(4), 367–376. **(in Chinese with English abstract)**.

Zhang, Z.-W., Qiao, H.-Z., Wu, P., Gong, Y., & Zhao, X.-Y. (2015). Study on correlation coefficient of spectral amplitude and apparent stress of water injection induced earthquake. *Journal of Seismological Research, 38*(1), 42–50. **(in Chinese with English abstract)**.

Zheng, J.-C., Zhang, Y.-X., Pan, Y.-S., Wu, Y.-J., & Wan, L.-C. (2006). Analysis on variation of ambient shear stress and apparent stress at Laoshan, Qingdao area. *Earthquake, 26*(3), 123–130. **(in Chinese with English abstract)**.

Zhong, Y.-Y., Zhu, X.-Y., & Zhang, Z. F. (2004a). Study on relations between seismic moment and magnitude for various types of earthquake sequence. *Northwestern Seismological Journal, 26*(1), 57–61. **(in Chinese with English abstract)**.

Zhong, Y.-Y., Zhang, F., Zhang, Z.-F., & Yang, G.-Y. (2004b). Possibility of earlier judgement on seismic tendency after strong earthquake based on apparent stress and stress drop. *Journal of Disaster Prevention and Mitigation Engineering, 24*(1), 8–14. **(in Chinese with English abstract)**.

Zhong, Y.-Y., Zhang, Z.-F., & Zhang, F. (2005). Research on characteristic of different types of earthquake sequence. *Journal of Disaster Prevention and Mitigation Engineering, 25*(2), 205–209. **(in Chinese with English abstract)**.

Zhou, X., Yang, F.-P., Zhong, Y.-Y., & Gong, J. (2012). A study on stress parameter characteristics of reservoir earthquakes and tectonic earthquakes in Zhejiang. *South China Journal of Seismology, 32*(suppl), 52–63. **(in Chinese with English abstract)**.

Zobin, V. M. (1996). Apparent stress of earthquakes within the shallow subduction zone near Kamchatka peninsula. *Bulletin Seismological Society of America, 86*, 811–820.

(Received January 20, 2016, revised April 30, 2016, accepted July 20, 2016, Published online August 2, 2016)

Pure Appl. Geophys. 174 (2017), 2381–2399
© 2016 Springer International Publishing
DOI 10.1007/s00024-016-1344-6

An Ensemble Approach for Improved Short-to-Intermediate-Term Seismic Potential Evaluation

HUAIZHONG YU,[1] QINGYONG ZHU,[2] FAREN ZHOU,[1] LEI TIAN,[1] and YONGXIAN ZHANG[1]

Abstract—Pattern informatics (PI), load/unload response ratio (LURR), state vector (SV), and accelerating moment release (AMR) are four previously unrelated subjects, which are sensitive, in varying ways, to the earthquake's source. Previous studies have indicated that the spatial extent of the stress perturbation caused by an earthquake scales with the moment of the event, allowing us to combine these methods for seismic hazard evaluation. The long-range earthquake forecasting method PI is applied to search for the seismic hotspots and identify the areas where large earthquake could be expected. And the LURR and SV methods are adopted to assess short-to-intermediate-term seismic potential in each of the critical regions derived from the PI hotspots, while the AMR method is used to provide us with asymptotic estimates of time and magnitude of the potential earthquakes. This new approach, by combining the LURR, SV and AMR methods with the choice of identified area of PI hotspots, is devised to augment current techniques for seismic hazard estimation. Using the approach, we tested the strong earthquakes occurred in Yunnan–Sichuan region, China between January 1, 2013 and December 31, 2014. We found that most of the large earthquakes, especially the earthquakes with magnitude greater than 6.0 occurred in the seismic hazard regions predicted. Similar results have been obtained in the prediction of annual earthquake tendency in Chinese mainland in 2014 and 2015. The studies evidenced that the ensemble approach could be a useful tool to detect short-to-intermediate-term precursory information of future large earthquakes.

Key words: Multi-method, combination, Yunnan–Sichuan region, Chinese mainland, seismic hazard evaluation.

1. Introduction

A series of recent studies suggest that earthquakes might depend sensitively on the heterogeneity of seismogenic media setting and stress transfer in the crust (Stein 1999; Curran et al. 1997; Garcimartin et al. 1997). So, a possible strategy to identify clues for prediction of a large earthquake is to explore the evolution pattern of seismicity associated with background stress field to evaluate the long-term earthquake hazard in any tectonically active region (Tiampo et al. 2002). Based on this idea, Rundle et al. (2000, 2003) proposed the pattern informatics (PI) method. They believed that 25 earthquakes occurred either on area of hotspots or within the margin of error of ±11 km when they applied the method to examine the 27 $M > 5$ earthquakes in southern California between 2000 and 2010. Similar results were obtained by Holliday et al. (2005) when they studied the $M > 7$ earthquakes in the world during 2000–2010. In spite of the well-determined spatial pattern of seismic hotspots that resulted from the previous earthquakes, this method, however, offers little information about the short-to-intermediate-term variation of regional seismic activity, making it difficult to assess the short-to-intermediate-term earthquake potential.

One approach to solving the problem is to combine forecasting models with different time horizons (Rhoades and Gerstenberger 2009). Bowman and King (2001) have described a physical approach that links static stress modeling to accelerating moment release before a large event. They demonstrated that the forecasting effectiveness of current techniques could be enhanced by substituting the circular critical region with the areas of increased Coulomb stress. Keilis-Borok et al. (2004) and Levin et al. (2006) also used this technique to the evolution of seismicity before large earthquakes. Nevertheless, the requirement of the forehand information of the fault or fault segments that will rupture in a future event may

---

[1] Department of Earthquake Prediction, China Earthquake Networks Center, No. 5, Sanlihe Nanhengjie, Xicheng District, Beijing 100045, China. E-mail: yuhz750216@sina.com

[2] School of Engineering, Sun Yat-sen University, Guangzhou 510275, China. E-mail: mcszqy@mail.sysu.edu.cn

hinder the use of this combination technique as the method of earthquake prediction. In fact, combinations of expected earthquake rates, mostly using Bayesian approach, were developed by Marzocchi et al. (2012), Zechar et al. (2010) and others. The physical basis of these pattern recognition methods for combining various precursory phenomena was previously developed by Gelfand et al. (1976), Keilis-Borok (1982), Sobolev et al. (1991) and Kossobokov and Carlson (1995). Recently, Shebalin et al. (2012, 2014) have suggested a combination method based on differential probability gains; the method gives a tool to combine rate-based and alarm-based forecasting models.

Following these works, we experiment an ensemble approach to assess earthquake potential by applying the short-to-intermediate-term earthquake prediction methods: the load/unload response ratio (LURR), the state vector (SV), and the accelerating moment release (AMR), to the areas where there are noticeable PI hotspots (Yu et al. 2013a). The underlying physics of this approach is from Shebalin et al. (2012, 2014), while the method for determining the time and magnitude of an earthquake is similar to the methodology established by Bowman and King (2001). To show the validity of the approach, the strong earthquakes ($M > 5.5$) that occurred in Yunnan–Sichuan region, China during 2013–2014 were tested in this paper. Moreover, the predictions of annual earthquake tendency in Chinese mainland in 2014 and 2015 are also displayed as the examples.

2. Models and Their Combination

Pattern informatics (PI), load/unload response ratio (LURR), state vector (SV), and accelerating moment release (AMR) are sensitive in different ways to the earthquake's source (see Appendix). The PI technique is founded on the premise that the change in the seismicity rate is a proxy for the change in the underlying stress. The use of this method implicitly assumes that earthquake fault systems are in an unstable equilibrium state and can be treated linearly about their equilibrium points. For a given located fault patch, if the part of its surrounding regions, which manifests its stress state relatively

more effectively, is loaded with anomalously high stress, it would tend to be driven toward failure. In this sense, the PI method can be applied to predict the occurrence of future large earthquakes (Tiampo et al. 2002). In the previous studies of Rundle et al. (2002), the PI hotspots seem to fit location of the earthquakes well; however, we do not know how robust it is, because the high rate of failures-to-forecast and lack of short-to-intermediate-term earthquake potential evaluation may sometimes compromise the reliability of the forecast.

The LURR and SV methods are used to detect the occurrence of large earthquakes by identifying the stable state of source media, while the AMR method is based on the hypothesis that earthquake can be regarded as a critical point. For mainland China, more than 80 % of the events with magnitude greater than 6.0 since 1980 were preceded with obvious LURR precursory anomalies (Yin et al. 2000); more than 2/3 of the events exhibit clear pre-shock AMR property (Jiang and Wu 2006); about 70 % of the events were preceded with SV anomalies (Yu et al. 2006). Even so, the debate on whether these methods produce significant precursory phenomena was still reflected by many publications (Mignan 2011; Yin et al. 2000; Smith and Sammis 2004; Yu et al. 2006). Taking AMR for example, the method has been widely applied to almost all the tectonically active regions, such as the California (Jaume and Sykes 1999), Greece (Chouliaras 2009), China (Jiang and Wu 2011), Italy (Di Giovambattista and Tyupkin 2001), Mexico (Sammis et al. 2004), Taiwan (Chen 2003), New Zealand (Robinson 2000) and Japan (Papazachos et al. 2010), but the negative results and criticisms have been gradually increasing since 2004 (Mignan 2011). Since the time period, area, and magnitude range analyzed before the main shocks were optimized to produce the strongest AMR signal, spurious cases of AMR could arise from data fitting (Hardebeck et al. 2008). Meanwhile, optimizing the search criteria may identify apparent AMR in a region, even if no significant event exists (Hardebeck et al. 2008). Hence, improvements on the method are expected to provide robust signal for identification of AMR phenomenon. With regard to this issue, Sornette and Sammis (1995) imposed the log periodicity fluctuation on AMR; Bowman et al. (1998)

introduced the parameter C, and applied it to optimize the spatial and temporal scales of data input. Nevertheless, the effectiveness for seismic hazard evaluation is still controversial (Hardebeck et al. 2008), especially in the practical prediction of earthquakes (Jiang and Wu 2005): it is hard to pre-determine the epicentral location and extent of the fault rupture prior to an earthquake. In this regard, some recent AMR studies have used spatial regions based on hypothetical stress loading patterns, rather than circles, to select the data (Bowman and King 2001; Jiang and Wu 2011). Since the methods require sufficient knowledge of the tectonic history and loading mechanism in the region, they are difficult to be applied to practical real-time prediction of earthquakes presently.

For these reasons, we attempt to establish a physically reasonable framework to apply the PI method, combining with the LURR, SV, and AMR methods, for seismic hazard evaluation. This is similar to what Bowman and King (2001) did in their seismicity and Coulomb stress accumulation analysis before large earthquakes. The major difference lies in that the regions where there are noticeable increases of Coulomb stress are replaced by the regions derived from the PI hotspots. Also, the LURR, SV and AMR methods are applied to assess if the stress change is likely to bring the regional crust closer to failure with the idea of overlapping phenomenology (Yin et al. 2000). Using this strategy, we can screen out the false alarms made by the PI method, without obviously reducing the hit rate, and, therefore, effectively augment the reliability of prediction. More importantly, the predictive outputs of these methods are valid for different timescales, months, years, or decades, making it possible to combine them for better prediction of the time-dependent seismic hazard. This can be supported by studies of Yin et al. (2002) and Zhang et al. (2006) which demonstrated that the spatial extent of the stress perturbation caused by an earthquake scales with the moment of the event. Because the anomalous stress field in the surrounding regions of the large event is detected by the PI hotspots, earthquakes that occurred outside of the regions (where the stress change may be less effective to induce a large earthquake) are excluded from seismic hazard evaluation, the predictive power of

current techniques (the LURR, the SV, and the AMR) is, therefore, enhanced.

3. Methods

In practice, using the combination approach, the first step is to apply the PI method to search for the seismic hotspots. The forecasting time window adopted in this study is 5 years. Other relevant parameters, such as the side length of Δx, time step of Δt, initial time of t_0, and seismic magnitude cutoff, are determined according to regional seismicity activity (Rundle et al. 2003).

Next, we delineate areas where large earthquake should be expected by covering the hotspots with the circular regions of a certain radius (r). The radius is decided by seismic magnitude of the ensuing main shock, in which the statistical slope of critical regions and size–magnitude is about 0.36 (Ben-Zion and Lyakhovsky 2002; Yin et al. 2002). The hotspots where the distance between any two of them is less than $2r$ are attributed to a potential critical region, and the midpoint coordinate of the hotspots is set to be the center of the region. Here, the distance between two critical regions should meet:

$$d \geq 2rc, \tag{1}$$

where r is the radius of the critical region, and c is correlation coefficient (usually $c = 0.8$), denoting the coincidence degree of two critical regions.

Then, we evaluate the short-to-intermediate-term seismic potential in each of the critical regions using the LURR and SV methods. If LURR and two or more SV scalars changed anomalously within a time frame from months to years before the forecasting time periods, the region is retained, otherwise, removed. An anomalous signal means an LURR value above the threshold of 1.0, while the anomaly in the SV time series is decided by subtracting the average value of the whole time series. If the difference is greater than 0, it is an anomalous one.

Finally, the AMR analysis is adopted to estimate time and magnitude of the potential earthquakes. We fixed the z value to fit A, B and t_c in Eq. (15) (Appendix). According to the definition, the asymptote time, t_c, is the occurrence time of the ensuing main shock. A is the cumulative Benioff strain at time t_c, which can be used to assess the magnitude of the earthquake by the following expression:

$$M_s = \left\{ \lg\left[A - \varepsilon(t_p)\right]^2 - 4.8 \right\} \Big/ 1.5, \qquad (2)$$

where t_p is the time of the prediction made. The relationship between energy and magnitude is derived from Gutenberg and Richter (1956) and Kanamori (1977). The exponent z is a focus of this stage. Sornette (1992) found that $z = 1/2$ is associated with a critical transition. Rundle et al. (2000) used scaling arguments to show that power-law time-to-failure buildup of cumulative Benioff strain may represent the scaling regime of a spinodal phase transition, with an exponent $z = 1/4$. Ben-Zion and Lyakhovsky (2002) have listed the z value of seismological observations from various authors. They concluded that the exponents fall in the range 0.1–0.55 and $z = 1/3$ for the damage rheology model of Lyakhovsky et al. (1997). Similar results were obtained using the fiber bundle model of Turcotte et al. (2003). According to these studies, the average value of $z = 1/3$ is adopted in this study.

4. Application to Seismic Data

As an example, we apply the approach to examine the large earthquakes with magnitude greater than 5.5 that occurred in the Yunnan–Sichuan region, China during 2013–2014. The region is located in the southeast corner of the Qinghai–Tibet plateau (Fig. 1) which is derived from the collision between the Indian and Eurasian plates. The continuous squeezing after the collision dominates the Cenozoic and present-day tectonic deformation in Chinese mainland, causing about 3–4 strong earthquakes of $M6$ every year on average, and two $M7$ earthquakes every 3 years. Due to the sustained collision and compression between the Indian and Eurasian plates, the tectonic activity in the Yunnan–Sichuan region is very violent. 32 large earthquakes with magnitude greater than 7.0 have been recorded in this region, including the 2008 $M8$ Wenchuan earthquake.

In our study, the earthquake catalog is retrieved from the China Earthquake Networks Center (http://www.ceic.ac.cn/), in which the $M > 3.0$ earthquakes that occurred in the Chinese mainland since 1970 have been recorded. The earthquakes were recorded in the Beijing time and on the local magnitude scale. Note that, for some special areas, such as the capital circle region, smaller earthquakes with magnitude of 1.0 can also be observed.

Figure 1 shows the PI hotspots distribution in the Yunnan–Sichuan region, with $\Delta x = 0.15°$ and $\Delta t = 10$ days. The magnitude cutoff (M_c) is 3.5. Detailed parameters for the PI calculation are: $t_0 = 01\text{-}01\text{-}1970$, $t_1 = 01\text{-}01\text{-}2008$, $t_2 = 01\text{-}01\text{-}2013$, $t_3 = 01\text{-}01\text{-}2018$, i.e., the forecast and reference periods are 2013–2018 and 2008–2013, respectively. To clearly show the spatial change of the hotspots, value of ΔP in each of the bins (Eq. 8 in the Appendix) is calculated using the logarithm of $\log_{10}(\Delta P / \Delta P_{max})$, where ΔP_{max} is the maximum value of the bins. Also shown in the figure are the circular critical regions derived from the PI hotspots. The radius of the circle is 110 km, corresponding to the areal size of an $M5.5$ earthquake.

Figure 2 shows the evaluation of LURR anomalies (LURR >1.0) from March 1, 2013 to December 31, 2014 within the critical regions displayed in Fig. 1, with a temporal window of 1 year at a sliding step of 2 months. The Benioff strain of the earthquakes with magnitude less than 4.0 within the critical regions was computed as the loading/unloading response. The inner frictional coefficient we adopted to calculate ΔCFS is 0.4 (Yin et al. 2000). The scanning radius of the circle is 110 km, and the slippage is 25 km.

Figure 3 displays the evolution of four SV scalars time series in each of the circular critical regions shown in Fig. 1, with a time horizon of 10 years. In the calculation, each circular critical region was regarded as a SV target area, which was subdivided into the 50×50 (km^2) subregions, and the small earthquakes of $M < 4.0$ are used to compute SV time series, with a temporal window of 1 year at a sliding step of 2 months.

Figure 4 displays the final seismic hazard estimates produced by the combination approach. Corresponding precursory accelerating seismicity and

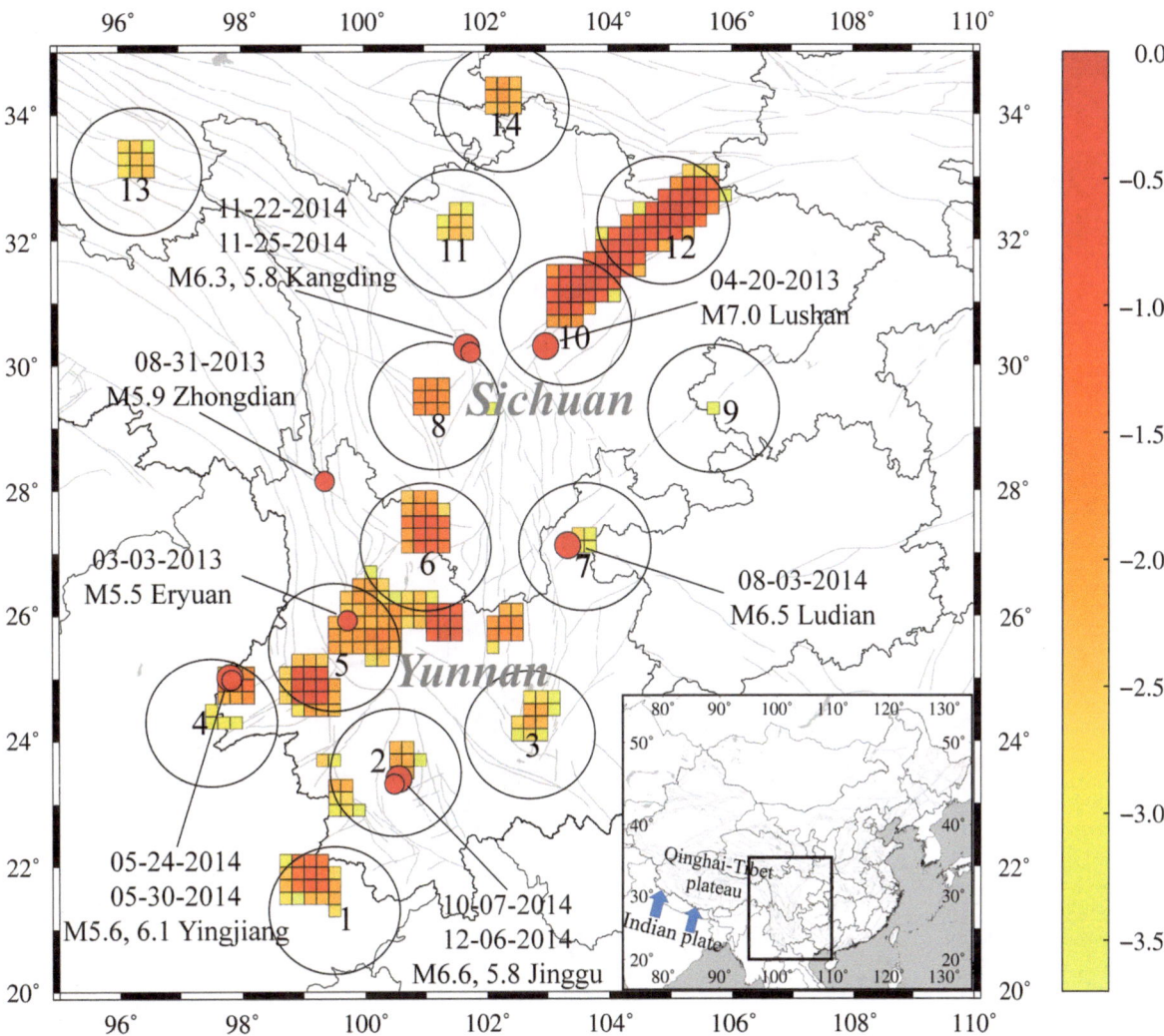

Figure 1

Seismic potential evaluated by the PI method in Yunnan–Sichuan region, China from January 1, 2013 to January 1, 2018. Values of the PI hotspots are given by *different colors*, and the *color code* is explained according to a *color bar right side of the figure*. Critical regions are delineated by the *circular black contours*. *Red dots* mark locations of the $M > 5.5$ events between January 1, 2013 and December 31, 2014. *Black and gray lines* denote the national and provincial borders and the main active faults. *Rectangular area* in the indexing map indicates the location of the study region

the asymptote of power-law time-to-failure functions are shown in Fig. 5. The magnitude cutoff is 3.5. The earthquakes with magnitude greater than 5.5 that occurred during January 1, 2013–December 31, 2014 are also plotted in the figure. Before the occurrence of the earthquakes, anomalously high LURR and SV values have often been observed (Figs. 2, 3). The asymptote time and magnitude made by the AMR method correspond to that of the actual events (Fig. 5).

5. Comparison and Discussion

By forming a mixture of the PI, LURR, SV, and AMR, we aim to create a more informative forecasting approach. Comparing the predictions shown in Figs. 1 and 4, it is clear that both the PI and the combination approaches have detected most of the large events that occurred during the forecasting period, 2013–2014 (just only the M5.9 Zhongdian earthquake was missed). More detailed comparison

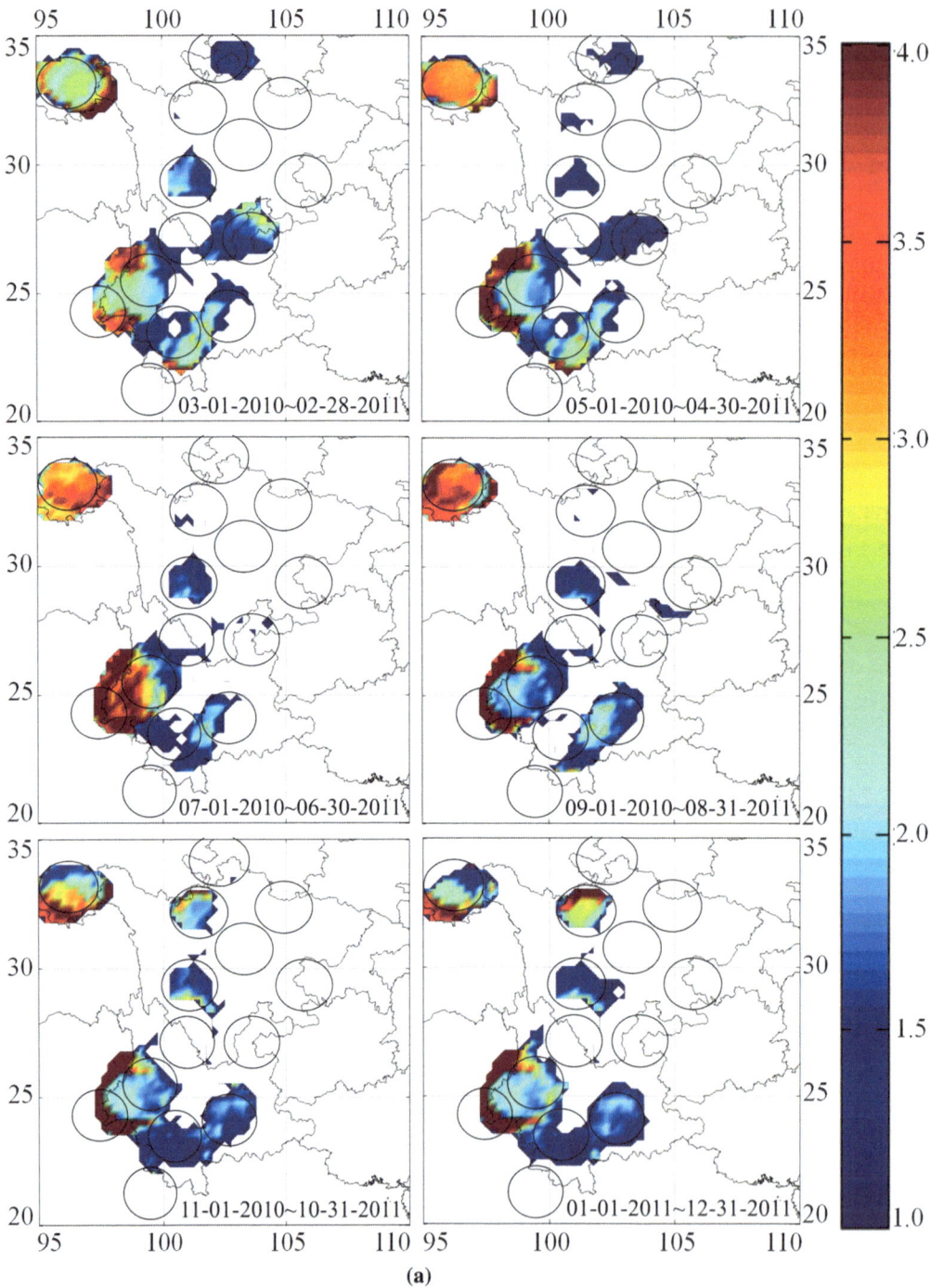

(a)

Figure 2

a, b Distribution of LURR anomalies within the critical areas shown in Fig. 1 during the period of 2010–2012. The LURR in each map is calculated using the time window (a year) shown at the *lower right corner* of the map with the sliding step of 2 months. Value of the anomalies is indicated by the *color bar* at *right side* of the maps

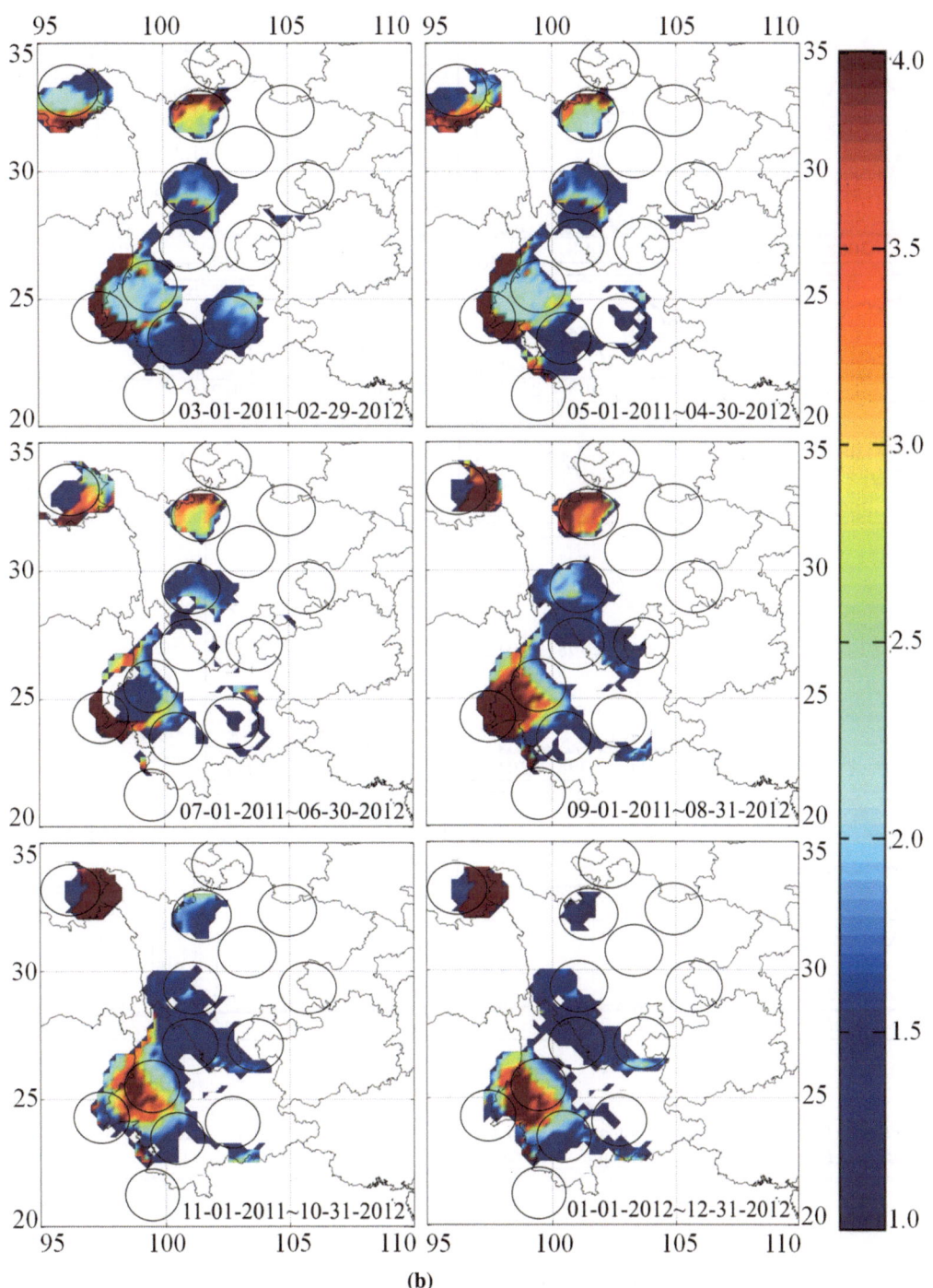

Figure 2
continued

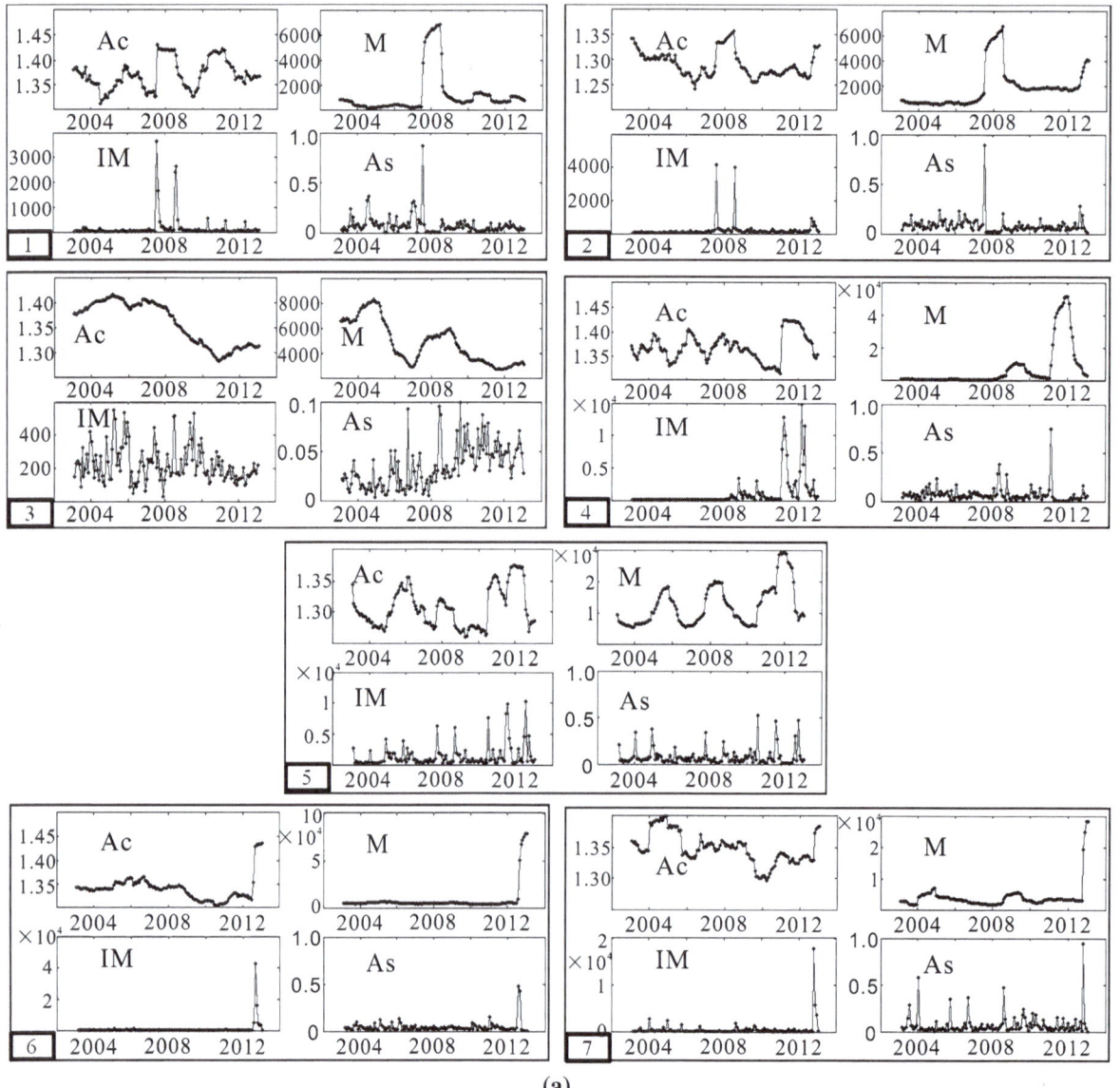

(a)

Figure 3
a, b Time series of the four SV scalars for each of the critical regions displayed in Fig. 1. The maps are sorted by the numbers at the center of each critical region. *Four dotted lines* in each map plot the traces of four SV scalars between January 1, 2003 and December 31, 2012

between the results obtained using the two approaches, however, does show some noticeable differences. They are:

1. The combination method seems to be more effective to detect future large earthquakes of particular magnitude range and location than the PI method. For some large earthquakes such as the

2013 M7.0 Lushan, and 2014 M6.3 Kangding earthquakes, significant precursors are found if circular areas are used, and are not found if just PI hotspots are used for seismic hazard evaluation. The result is consistent with the regional stress accumulation pattern during the establishment of the criticality (Bowman and King 2001): the

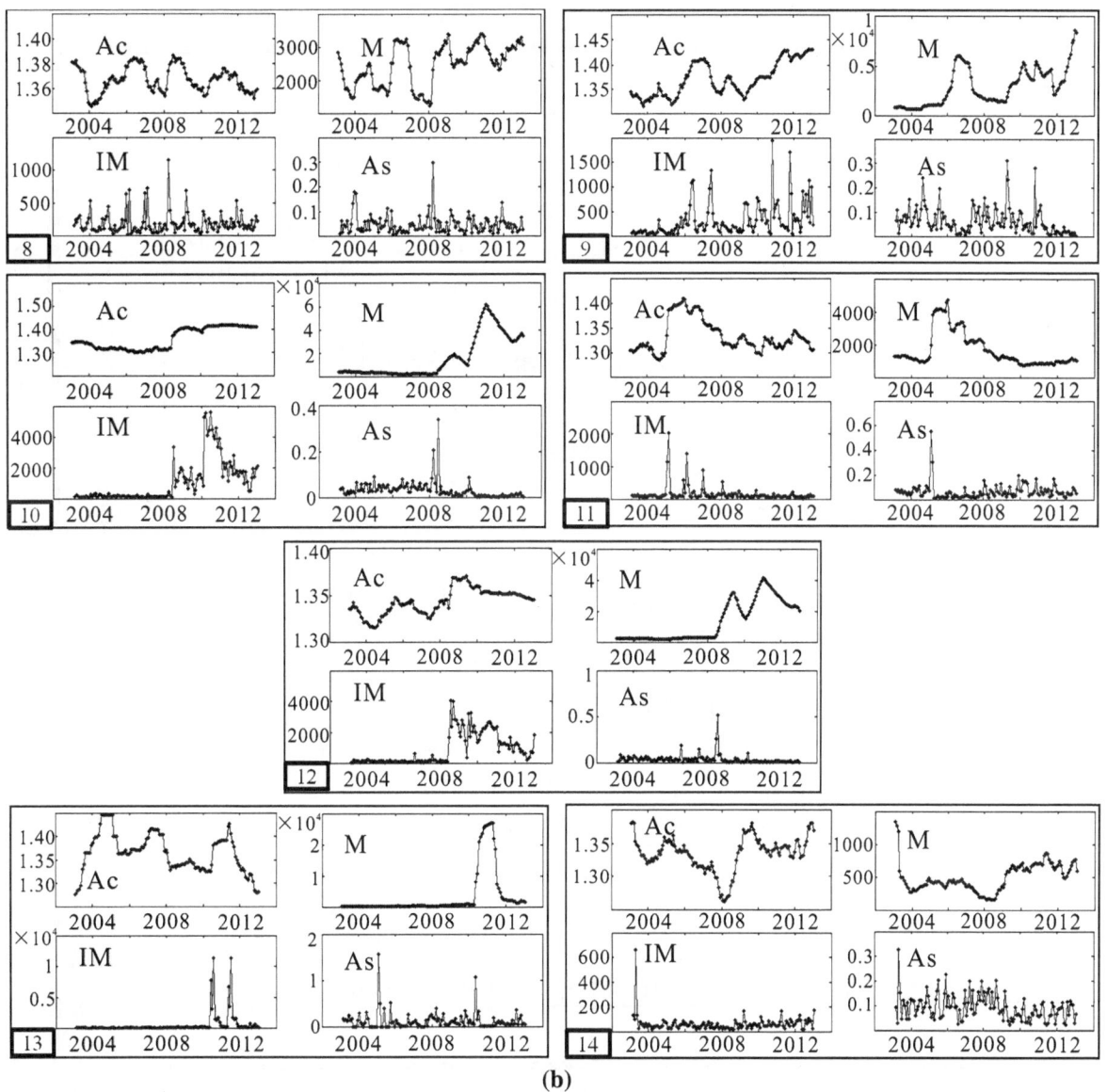

(b)

Figure 3
continued

tectonic stress is not only accumulated at the fault surface but also in a region surrounding the fault. It is reasonable to replace the PI hotspots with the circular regions of 110 km radius, and the lower rate of failures-to-forecast would be achieved.

2. Although most of the events occurred in the circular hazard regions produced by the two approaches, the one produced by the combination approach looks more prominent than the other one produced by the PI method. The numbers of the circular regions are 14 vs. 8 for the PI and combination approaches (Figs. 1, 4). Such contrast makes the prediction stand out more clearly in Fig. 4, and would represent less chance of false alarm for the large events.

To show the comparisons in a more quantitative fashion, we use a method of H score to evaluate

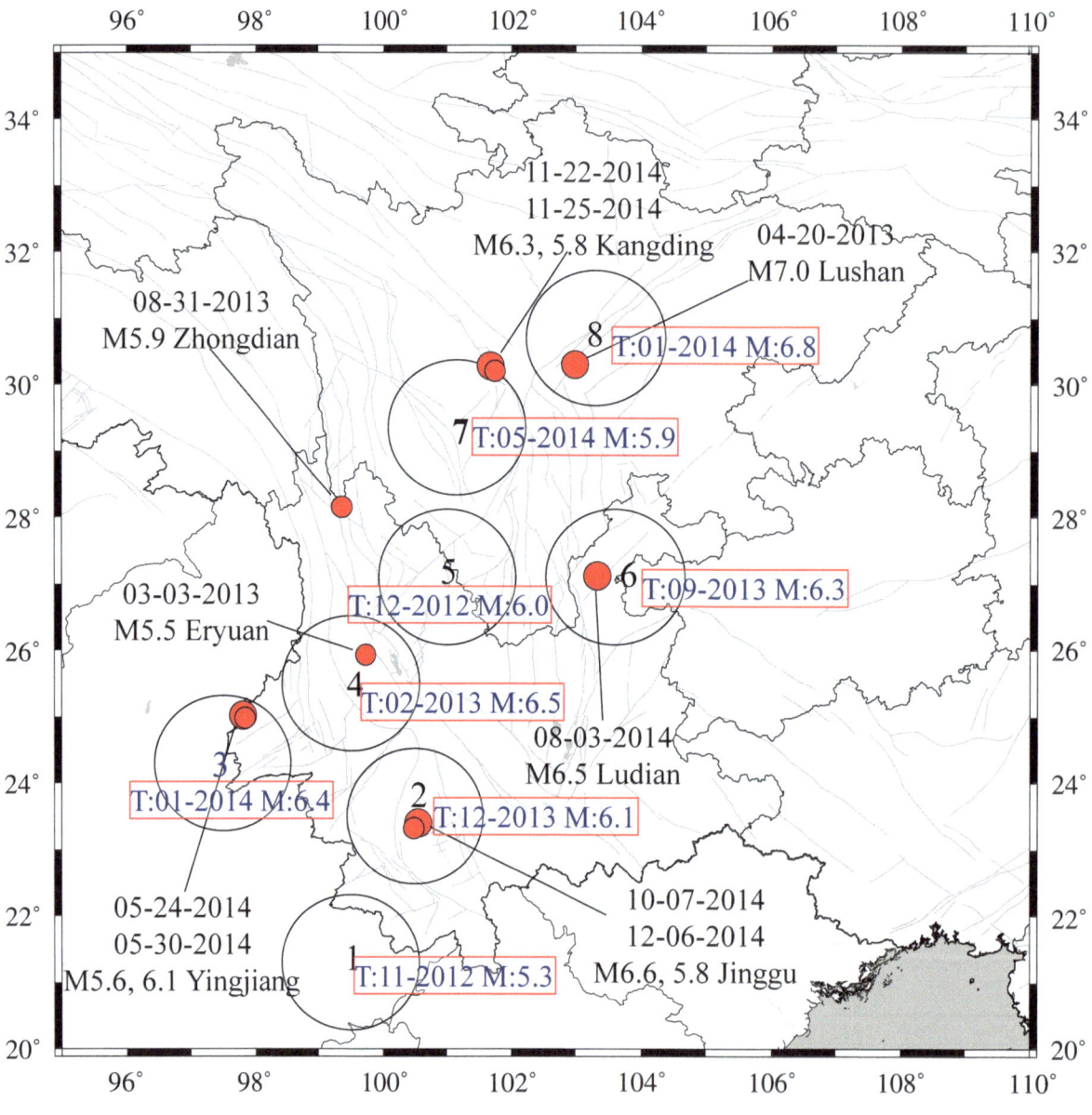

Figure 4
Seismic potential evaluated by the combination approach in Yunnan–Sichuan region, China from January 1, 2013 to December 31, 2014. Time and magnitude of the potential earthquake in each of the regions is displayed in a *red box at the center of the region*

statistically the performance of the forecasts. This method is similar to the methodology of *R* score given by Ma et al. (2004). Table 1 lists the hits, the false alarms, and the *H* scores of the seismic hazard regions shown in Fig. 4 as the detection threshold is altered. The value of H_C, indicating the predictive power of the approach, is defined as:

$$H_C = \frac{E_H}{E_T} - \frac{S_T}{S_A}, \tag{3}$$

where E_H and E_T denote, respectively, numbers of the detected earthquakes and the total major events, and S_T and S_A represent areas of the circles and the total area of Yunnan–Sichuan provinces, respectively. In

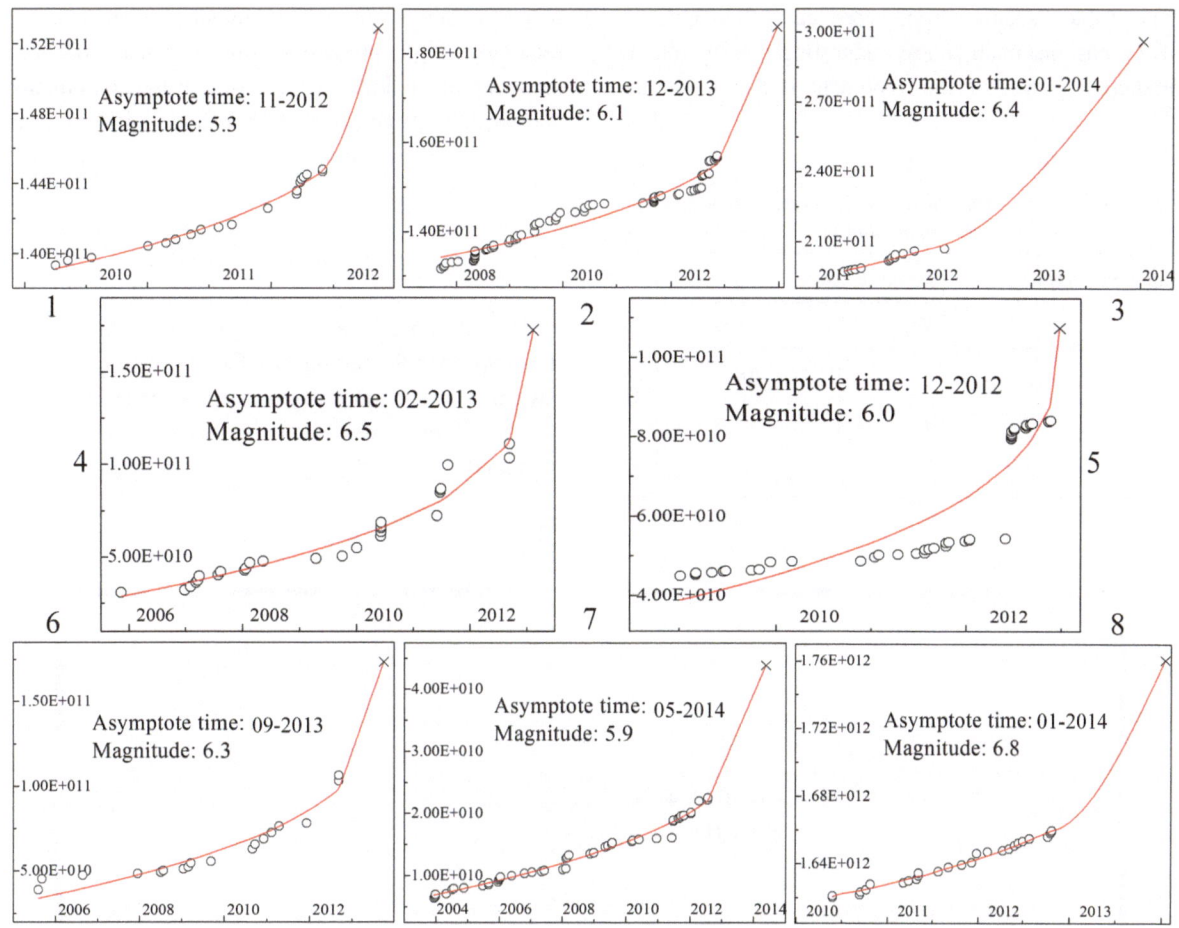

Figure 5

Cumulative Benioff strain of the earthquakes in each of the seismic hazard regions shown in Fig. 4. The maps are sorted by the numbers at the center of each critical region. *Solid curve* in each map is the fit of the data to the power-law time-to-failure equation, with the time of the predicted event denoted by a *cross symbol*

fact, this value is equal to the $1 - v - \tau$, one of possible 'loss functions' connected to the Molchan error diagram (Molchan 1991), in which v is the rate of misses and τ is the rate of false alarms, representing, respectively, the $1 - E_H/E_S$ and S_T/S_A in Eq. (3). In Table 2, we show the statistical prediction performances of the predictions displayed in Fig. 1, in which the value of H_{PI} is calculated as the H_C in Eq. (3).

The examples we show above indicate that for detection of the earthquakes with magnitude greater than 5.5, the H scores of the PI (H_{PI}) and the combination (H_C) approaches are about 0.5 vs. 0.6, while for the $M > 6.0$ earthquakes, the values are about 0.5 vs. 0.7. It is clear that the combination approach manifests the

Table 1

Statistics of the prediction performances of the seismic hazard regions displayed in Fig. 3 as a function of the detection threshold

Threshold (magnitude)	Observed events			H_C
	Yes	No	Total	
5.5	9	1	10	0.65
5.6	8	1	9	0.64
5.8	7	1	8	0.62
5.9	5	1	6	0.58
6.0	5	0	5	0.74

more preferable ability in prediction of future large earthquakes, especially for the earthquakes with magnitude greater than 6.0. On the other hand, the value of

*H*c in Table 1 seems to highly correlate with magnitude of the ensuing main shock, suggesting the significant predictive power of the approach: it has moderate

predictive power when the magnitude is greater than 5.5, relatively high predictive power when $M > 6.0$. Knowing this unique characteristic of the combination approach, we might use that for testing various kinds of earthquakes in the future.

Table 2

Statistics of the prediction performances of the seismic hazard regions displayed in Fig. 1

Threshold (magnitude)	Observed events			H_{PI}
	Yes	No	Total	
5.5	6	4	10	0.58
5.6	5	4	9	0.53
5.8	4	4	8	0.48
5.9	3	3	6	0.48
6.0	3	2	5	0.58

6. Prediction Attempts

In addition to the tests on the large earthquakes in some specific regions in Chinese mainland, we have performed more rigorous tests of the approach on the annual earthquake tendency in Chinese mainland.

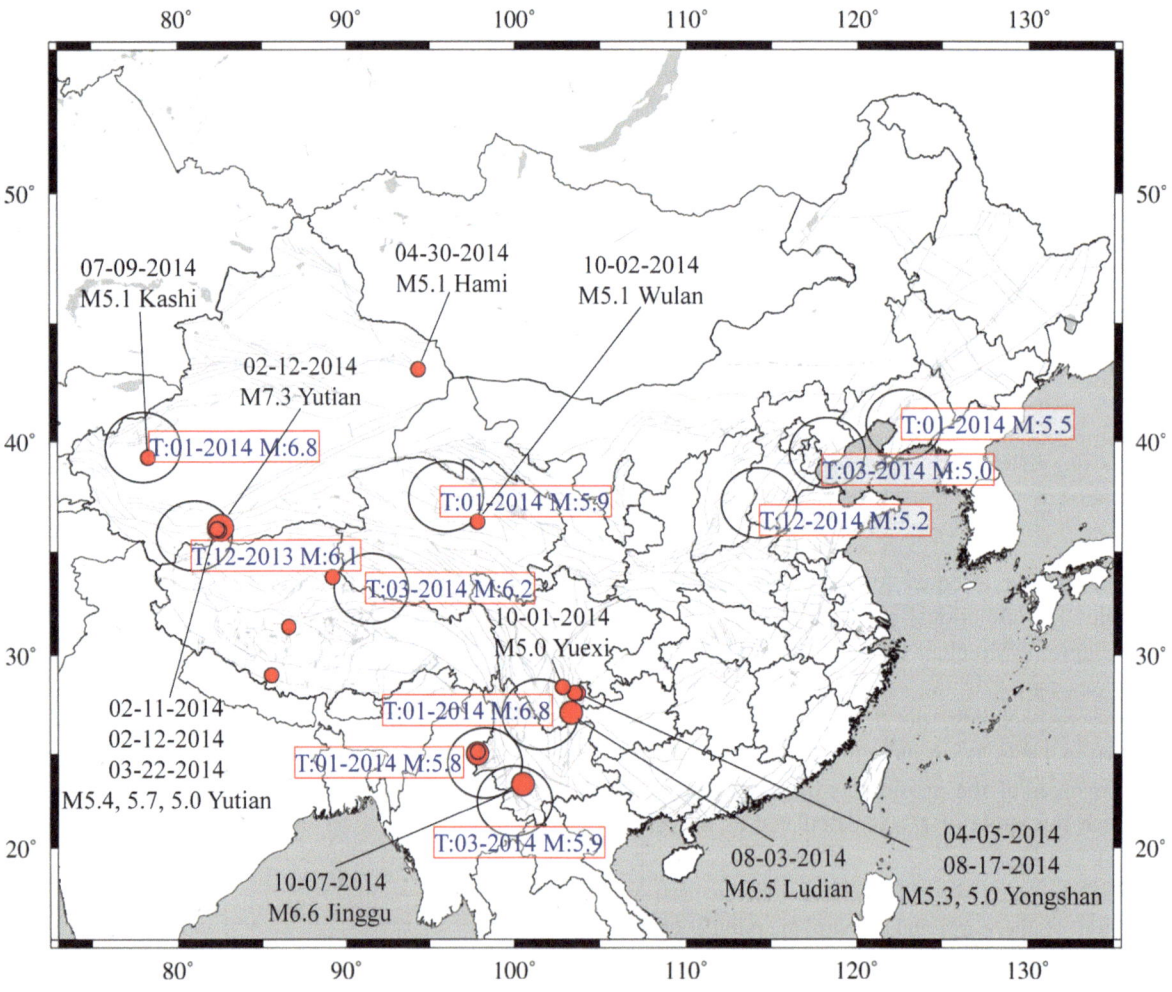

Figure 6

Seismic potential evaluated by the combination approach in Chinese mainland in 2014. *Red dots* are the $M > 5$ earthquakes during the period of forecasting, January 1, 2014–October 31, 2014

Table 3

Statistics of the prediction performances of the seismic hazard regions displayed in Fig. 4

Threshold (magnitude)	Observed events			H_C
	Yes	No	Total	
5.0	15	3	18	0.70
5.1	12	2	14	0.73
5.3	8	1	9	0.76
5.4	7	1	8	0.74
5.5	6	1	7	0.73
5.6	6	0	6	0.87
5.7	5	0	5	0.87
6.0	4	0	4	0.87

Figure 6 shows the predicted earthquake tendency in Chinese mainland in 2014. All the descriptions are the same as for Fig. 4 except some specific parameters used: PI: $\Delta x = 0.5°$, $\Delta t = 10$, $M_c = 3.0$, $t_0 = 01\text{-}01\text{-}1970$, $t_1 = 01\text{-}01\text{-}2009$, $t_2 = 01\text{-}01\text{-}2014$, $t_3 = 01\text{-}01\text{-}2019$; LURR: temporal window $= 1$ year, sliding step $= 1$ month, scanning radius $= 200$ km, slippage $= 25$ km; AMR: $M_c = 3.0$. Radius of the circular hazard areas is 200 km, corresponding to the areal size of an $M > 6$ earthquake. Also shown in the figure are the $M > 5.0$ earthquakes that occurred during the forecasting period. The prediction was made on October 1, 2013, and relevant report was submitted to Annual Consultation, a special kind of short-to-intermediate-term earthquake forecast regularly undertaken in China Earthquake Networks Center, on October 18 (Yu et al. 2014).

Results show that most of the major events, especially the four $M > 6.0$ earthquakes (the 12 February $M7.3$ Yutian, 30 May $M6.1$ Yingjiang, 3 August $M6.5$ Ludian, and 7 October $M6.6$ Jinggu earthquakes), occurred in the seismic hazard regions predicted, while just only a few $M5$ earthquakes are missed: the 31 March $M5.5$ Nima, 3 August $M5.0$ Jilong, Xizang and 30 April $M5.1$ Hami, Xinjiang earthquakes. In fact, the Hami earthquake is near the national border of China, while the Nima and Jilong earthquakes are located in the weak areas of seismic monitoring. The catalog's quality of the earthquakes is questionable. The statistical analysis of the prediction performances of the seismic hazard regions as the detection threshold is altered is listed in Table 3.

The results, once again, show the remarkable prediction power of the combination approach.

Figure 7 is the prediction of earthquake tendency in Chinese mainland in 2015. Result was submitted to Annual Consultation of China Earthquake Networks Center on October 14, 2014 (Yu et al. 2015). Actually, all the two $M > 6.0$ earthquakes in the forecasting period (the July 3, 2015 $M6.5$ Pishan and November 22, 2014 $M6.3$ Kangding earthquakes) were predicted right by using the combination approach.

Finally, it should be pointed out that, besides the 2014 $M7.3$ Yutian earthquake (displayed in Fig. 6), using this idea, we have predicted other two large earthquakes with magnitude greater than 7.0 in Chinese mainland since 2010. Predictions of the 2010 $M7.1$ Yushu and 2013 $M7.0$ Lushan earthquakes are shown in Fig. 8, which were submitted to Annual Consultation of China Earthquake Networks Center on November 11, 2009 and October 15, 2012, respectively (Yu et al. 2010, 2013b). The results had revealed the critical extent of source media before the quakes, and further demonstrated the validity of the combination approach.

7. Conclusion

Through our application practices, we find that by combining the PI method with the LURR, SV and AMR methods, the prediction power of current techniques can be augmented. While we illustrate the approach using seismicity in Chinese mainland, the technique is general and might be applied to any tectonically active region. Combination of the models which is valid for different space–time scales facilitates the analysis of the complex temporal and spatial evolution of seismic system. Statistically, the combination approach outperforms its component models, with the lower rate of false alarm and even the higher hit rate. Although more data testing is needed to further verify the approach, it is ready to be employed in the real world for quantitative prediction of future large earthquakes. If sufficient knowledge of the regional seismicity pattern is known, the approach presented in this paper allows us to achieve gradually approximation of the crucial parameters of future large earthquakes, such as the time, location and magnitude.

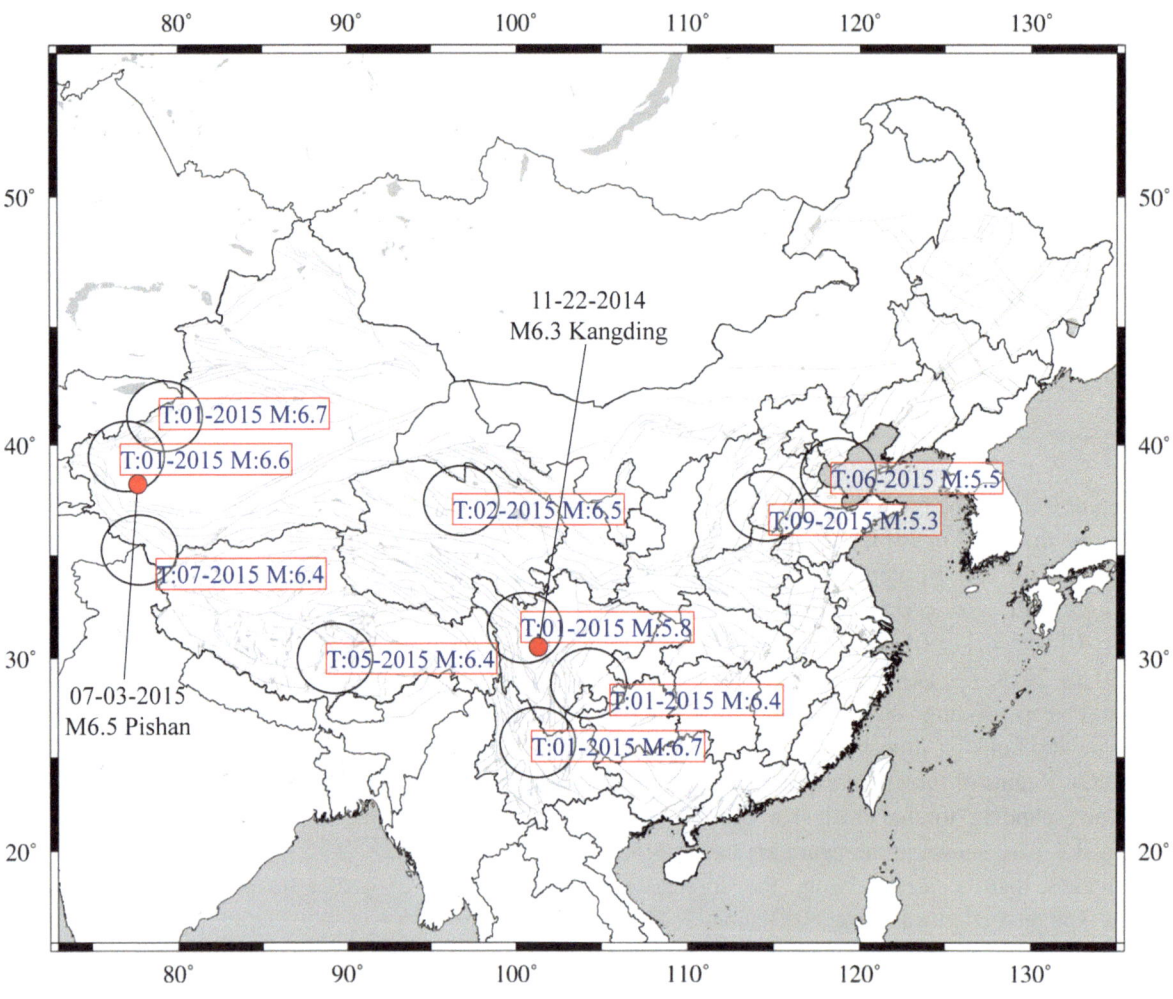

Figure 7

Seismic potential evaluated by the combination approach in Chinese mainland in 2015. All the descriptions are the same as for Fig. 6 except some specific parameters used: seismic data: 01-01-1970 to 10-01-2014; earthquake exhibitions, $M > 6.0$. Periods for PI evaluation are: $t_1 = $ 01-01-2010, $t_2 = $ 01-01-2015, $t_3 = $ 01-01-2020

Acknowledgments

The research was supported by the Spark Program of Earthquake Science of China (Grant No. XH12058) and the Grant support from the Chinese NSFC (No. 91230114).

Appendix

Pattern Informatics

The pattern informatics (PI) method, which is an example of a phase dynamical measure, was proposed to detect the characteristic precursory patterns before large earthquakes (Rundle et al. 2000, 2003). Detailed procedures of this method can be outlined as follows:

1. To divide the study region into square bins with side length of Δx.
2. To define average rate of occurrence of earthquakes in the i-th bin over the period t_b to t

$$I_i(t_b, t) = \frac{1}{t - t_b} \sum_{t'=t_b}^{t} N_i(t'), \qquad (4)$$

where t_b varies between t_0 and t_1 at a time step of Δt, and t_0 is the initial time. The time interval t_b–t_1 is the

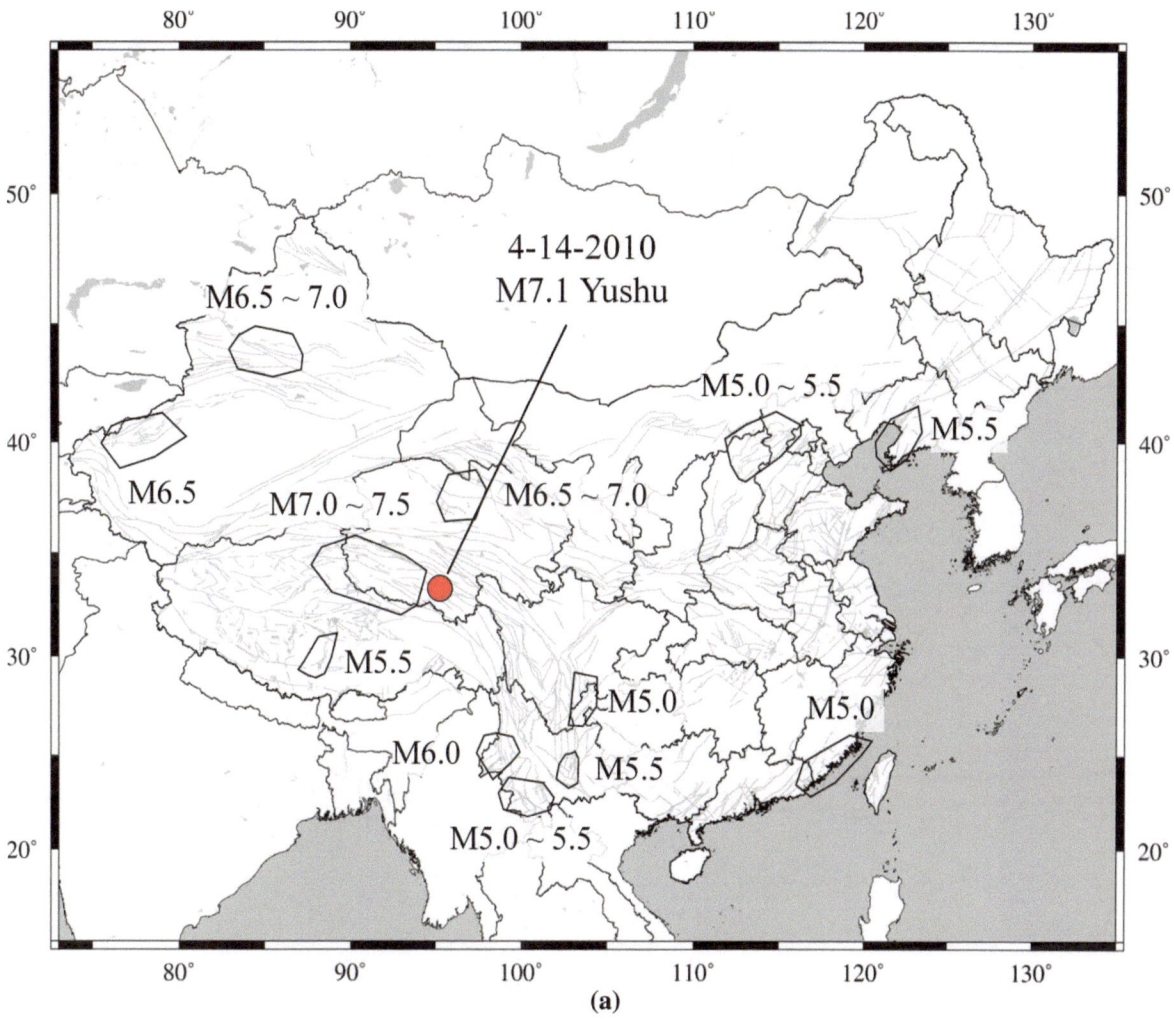

Figure 8
Seismic potential evaluated by the combination approach in Chinese mainland in 2010 (**a**) and 2013 (**b**). *Red dot* in each map represents, respectively, the 2010 *M*7.1 Yushu and 2013 *M*7.0 Lushan earthquakes. The maps are retrieved from Yu et al. (2010, 2013b)

reference period. $N_i(t)$ is the number of earthquakes with magnitude greater than M_c in the i-th bin. M_c is the magnitude cutoff.

3. To normalize the activity rate function,

$$\widehat{I}_i(t_b, t) = \frac{I_i(t_b, t) - \langle I_i(t_b, t) \rangle}{\sigma(t_b, t)}, \qquad (5)$$

where $\langle I_i(t_b, t) \rangle$ and $\sigma(t_b, t)$ are the average activity rate function and its spatial standard deviation over all the bins at time t.

4. To assess the change of normalized activity rate function for the time period $t_1 – t_2$,

$$\Delta I_i(t_b, t_1, t_2) = \widehat{I}_i(t_b, t_2) - \widehat{I}_i(t_b, t_1) \qquad (6)$$

5. To calculate the probability of change of activity in the i-th bin,

$$P_i(t_0, t_1, t_2) = \overline{\Delta I_i(t_0, t_1, t_2)}^2, \qquad (7)$$

where $\overline{\Delta I_i(t_0, t_1, t_2)} = \frac{1}{t_1 - t_0} \sum_{t_b=t_0}^{t_1} \Delta I_i(t_b, t_1, t_2)$. In phase dynamical systems, probabilities are related to the square of the associated vector phase function (Rundle et al. 2000).

6. To evaluate the difference between the $P_i(t_0, t_1, t_2)$ and its spatial mean $\langle P_i(t_0, t_1, t_2) \rangle$, representing

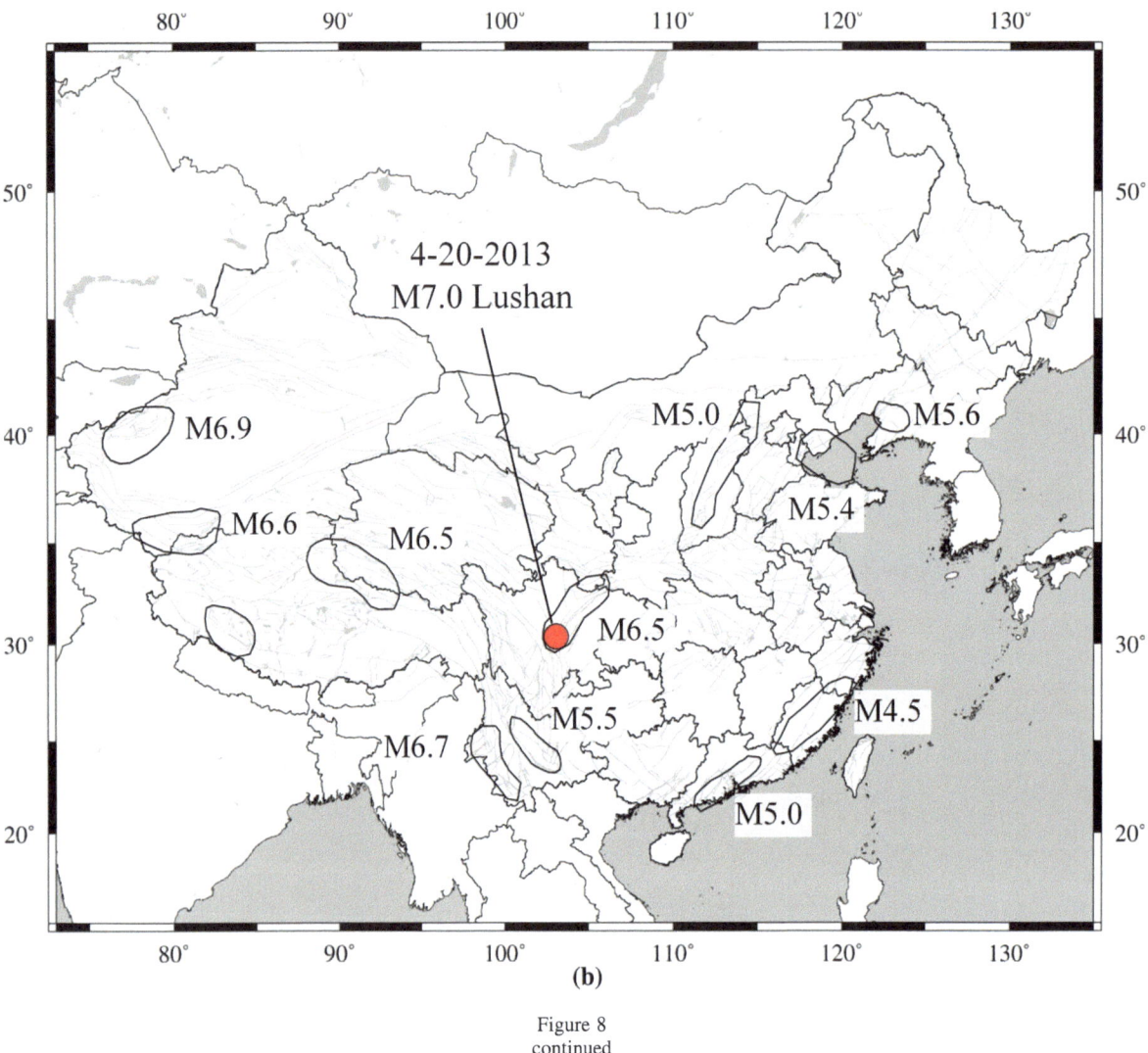

Figure 8
continued

the probability of change in activity relative to the background,

$$\Delta P_i(t_0, t_1, t_2) = P_i(t_0, t_1, t_2) - \langle P_i(t_0, t_1, t_2) \rangle. \quad (8)$$

The hotspots are defined to be the bins (or the regions) where $\Delta P_i(t_0, t_1, t_2)$ is positive.

Load/Unload Response Ratio

Over the past decade, an earthquake prediction method named the load/unload response ratio (LURR) has been developed by Yin and others (Yin et al. 2000; Zhang et al. 2006). In earthquake prediction practice using the method, the seismic

energy release within certain temporal and spatial windows is usually used as data input. Loading and unloading periods are determined by calculating the earth tide-induced Coulomb failure stress change along a tectonically favored rupture direction on a specified fault plane (Hainzl et al. 2010; Harris 1998). The Coulomb failure stress is defined as:

$$CFS = \tau_n + f\sigma_n, \quad (9)$$

where f, τ_n and σ_n stand for inner frictional coefficient, shear stress and normal stress (positive in tension), respectively, and n is the normal of the fault plane on which the CFS reaches its maximum. When the change of Coulomb failure stress (ΔCFS) >0, it is in a loading state; and when ΔCFS <0, it is in an unloading state. The LURR is, thus, expressed as a ratio between energy released during loading and that released during unloading periods:

$$Y_m = \frac{\left(\sum_{i=1}^{N+} E_i^m\right)_+}{\left(\sum_{i=1}^{N-} E_i^m\right)_-}, \qquad (10)$$

where E_i is seismic energy released by the i-th event, and $N+$ or $N-$ represents the numbers of events that occurred during the loading and unloading stages, respectively. When $m = 1/2$, E^m denotes the Benioff strain. Note that the focal mechanisms of the small earthquakes are assumed in agreement with that of the main shock to contribute positively to ΔCFS for the main shock. This assumption is supported by studies of Hauksson (1994), Hauksson et al. (2002), and Hardebeck and Hauksson (2001), which demonstrated that the focal mechanisms of regional small earthquakes prior to the Landers and Hector Mine earthquakes were quite consistent with that of the ensuing main shocks. To avoid volatile fluctuations due to poor statistics, the loading and unloading periods are usually summed over many load–unload cycles within the time window. Circular region is usually adopted as the spatial window, and the optimal critical region scale for LURR evaluation is determined by computing the LURR anomaly within differently sized regions centered at epicenter of the upcoming large event to reach the maximum LURR precursory anomaly (Yin et al. 2000). The forecasting time window, from months to years, is determined by the magnitude of the ensuing earthquake: the larger the earthquake, the longer the time.

State Vector

The idea of SV is adopted from statistical physics into seismology to characterize the spatial and temporal evolution of seismic activities (Yu et al. 2006). This method is defined by dividing the target region into n uniform square subregions, and the sum of seismic magnitudes in each subregion within certain temporal window is computed as the component of an n-dimensional vector. If a series of vectors at different times are known, the temporal and spatial evolution of seismicity may be obtained. Previous studies show that anomalously high SV peaks have

often been observed months to years before large earthquakes (Yu et al. 2006). Generally, four scalars are defined to measure evolution of the vectors.

1. Modulus of the vectors,

$$M = |\vec{V}_k|, \qquad (11)$$

where $\vec{V}_k$ is the state vector at time t_k ($k = 1, 2...n$), whose temporal window is T at a sliding step of Δt.

2. Angle between two consecutive vectors,

$$A_s = \arccos\left(\frac{\vec{V}_{k+1}\vec{V}_k}{|\vec{V}_{k+1}||\vec{V}_k|}\right). \qquad (12)$$

3. Modulus of the increment vector:

$$IM = |\vec{V}_{k+1} - \vec{V}_k|. \qquad (13)$$

4. Angle between vector $\vec{V}_k$ and equalized vector $\vec{V}_e$,

$$A_c = \arccos\left(\frac{\vec{V}_e\vec{V}_k}{|\vec{V}_e||\vec{V}_k|}\right), \qquad (14)$$

where the equalized vector $\vec{V}_e$ consists of equal components.

Accelerating Moment Release

Prior to the occurrence of large or great earthquakes, the accelerating moment release (AMR) is usually observed (Jaume and Sykes 1999). Bufe and Varnes (1993) suggested that a simple power-law time-to-failure equation derived from damage mechanics could be used to model the observed seismicity. This hypothesis is an outgrowth of efforts to characterize large earthquakes as a critical phenomenon (Rundle 1989). The function has the following form:

$$\varepsilon_p(t) = A + B(t_c - t)^z, \qquad (15)$$

where t_c is the time of the large event, B is negative and z is the exponent. A is the value of $\varepsilon(t)$ when $t = t_c$ (i.e., the final Benioff strain up to and including the largest event). The cumulative Benioff strain at time t is defined as:

$$\varepsilon(t) = \sum_{i=1}^{N(t)} E_i(t)^{\frac{1}{2}}, \qquad (16)$$

where E_i is the energy of the i-th event and $N(t)$ is the number of events at time t.

REFERENCES

Ben-Zion, Y., & Lyakhovsky, V. (2002). Accelerated seismic release and related aspects of seismicity patterns on earthquake faults. *Pure and Applied Geophysics, 159*(10), 2385–2412.

Bowman, D. D., & King, G. C. P. (2001). Accelerating seismicity and stress accumulation before large earthquake. *Geophysical Research Letters, 28*(21), 4039–4042.

Bowman, D. D., Ouillon, G., Sammis, C. G., Sornette, A., & Sornette, D. (1998). An observational test of the critical earthquake concept. *Journal of Geophysical Research, 103*(10), 24359–24372.

Bufe, C. G., & Varnes, D. J. (1993). Predictive modelling of the seismic cycle of the greater San Francisco bay region. *Journal of Geophysical Research, 98*, 9871–9883.

Chen, C. C. (2003). Accelerating seismicity of moderate-size earthquakes before the 1999 Chi-Chi, Taiwan, earthquake: Testing time-prediction of the self-organizing spinodal model of earthquakes. *Geophysical Journal International, 155*, F1–F5.

Chouliaras, G. (2009). Seismicity anomalies prior to 8 June 2008, Mw = 6.4 earthquake in Western Greece. *Natural Hazards and Earth Systems Sciences, 9*, 327–335.

Curran, D. R., Seaman, L., & Shockey, D. A. (1997). Dynamic failure of solids. *Physical Review, 147*, 253–388.

Di Giovambattista, R., & Tyupkin, Y. S. (2001). An analysis of the process of acceleration of seismic energy emission in laboratory experiments on destruction of rocks and before strong earthquakes on Kamchatka and in Italy. *Tectonophysics, 338*, 339–351.

Garcimartin, A., Guarino, A., Bellon, L., & Ciliberto, S. (1997). Statistical properties of failure precursors. *Physical Review Letters, 79*, 3202–3205.

Gelfand, I. M., Guberman, S. A., Keilis-Borok, V. I., Knopoff, L., Press, F., Ranzman, E. Y., et al. (1976). Pattern recognition applied to earthquake epicenters in California. *Physics of the Earth and Planetary Interiors, 11*, 227–283.

Gutenberg, B., & Richter, C. F. (1956). Earthquake magnitude, intensity, energy and acceleration. *Bulletin of the Seismological Society of America, 46*, 105–145.

Hainzl, S., Zoller, G., & Wang, R. (2010). Impact of the receiver fault distribution on aftershock activity. *Journal of Geophysical Research, 115*, B05315. doi:10.1029/2008JB006224.

Hardebeck, J. L., Felzer, K. R., & Michael, A. J. (2008). Improved tests reveal that the accelerating moment release hypothesis is statistically insignificant. *Journal of Geophysical Research, 113*, B08310. doi:10.1029/2007JB005410.

Hardebeck, J. L., & Hauksson, E. (2001). Crustal stress field in southern California and its implications for fault mechanics. *Journal of Geophysical Research, 106*, 21859–21882.

Harris, R. A. (1998). Introduction to special section: Stress triggers, stress shadows, and implication for seismic hazard. *Journal of Geophysical Research, 103*, 24347–24358.

Hauksson, E. (1994). State of stress from focal mechanisms before and after the 1992 Landers earthquake sequence. *Bulletin of the Seismological Society of America, 84*(3), 917–934.

Hauksson, E. L., Jones, M., & Hutton, K. (2002). The 1999 Mw7.1 Hector Mine, California, earthquake sequence: Complex conjugate strike-slip faulting. *Bulletin of the Seismological Society of America, 92*, 1154–1170.

Holliday, J. R., Rundle, J. B., Tiampo, K. F., Klein, W., & Donnellan, A. (2005). Systematic procedural and sensitivity analysis of the Pattern Informatics method for forecasting large ($M > 5$) earthquake events in Southern California. *Pure and Applied Geophysics, 163*(11–12), 2433–2454.

Jaume, S. C., & Sykes, L. R. (1999). Evolution toward a critical point: A review of accelerating seismic moment/energy release prior to large great earthquakes. *Pure and Applied Geophysics, 155*(2–4), 279–305.

Jiang, C. S., & Wu, Z. L. (2005). Accelerating strain release before strong earthquakes: More complex in the real world. *Journal of the Graduate School of the Chinese Academy of Sciences, 22*(3), 286–291.

Jiang, C. S., & Wu, Z. L. (2006). Benioff strain release before earthquakes in China: Accelerating or not? *Pure and Applied Geophysics, 163*(2006), 1965–1976.

Jiang, C. S., & Wu, Z. L. (2011). Intermediate-term medium-range accelerating moment release (AMR) prior to the 2010 Yushu M 7.1 earthquake. *Chinese Journal of Geophysics, 54*(6), 1501–1510. **(in Chinese).**

Kanamori, H. (1977). The energy release in great earthquakes. *Journal of Geophysical Research, 82*, 2981–2987.

Keilis-Borok, V. I. (1982). A worldwide test of three long-term premonitory seismicity patterns: A review. *Tectonophysics, 85*, 47–60.

Keilis-Borok, V. I., Shebalin, P., Gabrielov, A., & Turcotte, D. (2004). Reverse tracing of short-term earthquake precursors. *Physics of the Earth and Planetary Interiors, 145*(1–4), 75–85.

Kossobokov, V. G., & Carlson, J. M. (1995). Active zone size vs. activity: A study of different seismicity patterns in the context of the prediction algorithm M8. *Journal of Geophysical Research, 100*, 0431–0441.

Levin, S. Z., Sammis, C. G., & Bowman, D. D. (2006). An observational test of the stress accumulation model based on seismicity preceding the 1992 Landers, CA earthquake. *Tectonophysics, 413*, 39–52.

Lyakhovsky, V., Ben-Zion, Y., & Agnon, A. (1997). Distributed damage, faulting, and friction. *Journal of Geophysical Research, 102*(B12), 27635–27649.

Ma, H. S., Liu, J., Wu, H., & Li, J. F. (2004). Scientific evaluation of annual earthquake prediction efficiency based on *R*-value. *Earthquake, 24*(2), 31–37. **(In Chinese)**.

Marzocchi, W., Zechar, J. D., & Jordan, T. H. (2012). Bayesian forecast evaluation and ensemble earthquake forecasting. *Bulletin of the Seismological Society of America, 102*(6), 2574–2584.

Mignan, A. (2011). Retrospective on the accelerating seismic release (ASR) hypothesis: Controversy and new horizons. *Tectonophysics, 505*, 1–16.

Molchan, G. M. (1991). Structure of optimal strategies in earthquake prediction. *Tectonophysics, 193*, 267–276.

Papazachos, B. C., Karakaisis, G. F., Scordilis, E. M., Papazachos, C. B., & Panagiotopoulos, D. G. (2010). Present patterns of decelerating–accelerating seismic strain in South Japan. *Journal of Seismology, 14*(2), 273–288.

Rhoades, D. A., & Gerstenberger, M. C. (2009). Mixture models for improved short-term earthquake forecasting. *Bulletin of the Seismological Society of America, 154*(2A), 636–646.

Robinson, R. (2000). A test of the precursory accelerating moment release model on some recent New Zealand earthquakes. *Geophysical Journal International, 140*, 568–576.

Rundle, J. B. (1989). A physical model for earthquakes III. *Journal Geophysical Research, 94,* 2839–2855.

Rundle, J. B., Klein, W., Tiampo, K. F., & Gross, S. J. (2000). Linear pattern dynamics in nonlinear threshold systems. *Physical Review E, 61,* 2418–2432.

Rundle, J. B., Tiampo, K. F., Klein, W., & Martins, J. S. S. (2002). Self-organization in leaky threshold systems: The influence of near-mean field dynamics and its implications for earthquakes, neurobiology, and forecasting. *Proceedings of the National Academy of Sciences of the United States of America, 99,* 2514–2521.

Rundle, J. B., Turcotte, D. L., Shcherbakov, R., Klein, W., & Sammis, C. (2003). Statistical physics approach to understanding the multiscale dynamics of earthquake fault systems. *Reviews of Geophysics, 41*(4), 1019. doi:10.1029/2003RG000135.

Sammis, C. G., Bowman, D. D., & King, G. (2004). Anomalous seismicity and AMR preceding the 2001 and 2002 Calexico Mexico earthquakes. *Pure and Applied Geophysics, 161,* 2369–2378.

Shebalin, P., Narteau, C., & Holschneider, M. (2012). From alarm-based to rate-based earthquake forecast models. *Bulletin of the Seismological Society of America, 102*(1), 64–72.

Shebalin, P., Narteau, C., Zechar, J. D., & Holschneider, M. (2014). Combining earthquake forecasts using differential probability gains. *Earth Planets and Space, 66,* 37.

Smith, S., & Sammis, C. (2004). Revisiting the tidal activation of seismicity with a damage mechanics and friction point of view. *Pure and Applied Geophysics, 161,* 2393–2404.

Sobolev, G. A., Chelidze, T. L., & Zavyalov, A. D. (1991). Map of expected earthquakes based on a combination of parameters. *Tectonophysics, 193,* 255–266.

Sornette, D. (1992). Mean-field solution of a block-spring model of earthquake. *Journal of Physics I France, 2,* 2089–2096.

Sornette, D., & Sammis, C. G. (1995). Critical exponent from renormalization group theory of earthquakes: Implication for earthquake prediction. *Journal de Physique I, 5,* 607–619.

Stein, R. S. (1999). The role of stress transfer in earthquake occurrence. *Nature, 402,* 605–609.

Tiampo, K. F., Rundle, J. B., McGinnis, S. A., & Klein, W. (2002). Pattern dynamics and forecast methods in seismically active regions. *Pure and Applied Geophysics, 159*(10), 2429–2467.

Turcotte, D. L., Newman, W. I., & Shcherbakov, R. (2003). Micro- and macro-scopic models of rock fracture. *Geophysical Journal International, 152,* 718–728.

Yin, X. C., Mora, P., Peng, K. Y., Wang, Y. C., & Weatherly, D. (2002). Load–unload response ratio and accelerating moment/energy release, critical region scaling and earthquake prediction. *Pure and Applied Geophysics, 159,* 2511–2524.

Yin, X. C., Wang, Y. C., Peng, K. Y., Bai, Y. L., Wang, H. T., & Yin, X. F. (2000). Development of a new approach to earthquake prediction-load/unload response ratio (LURR) theory. *Pure and Applied Geophysics, 157*(11–12), 2365–2383.

Yu, H. Z., Cheng, J., Zhang, X. T., Li, G., Liu, J., & Zhang, Y. X. (2010). *Multi-method combined analysis of earthquake trend in Chinese mainland in 2010. Study of earthquake trend in China (2010)* (pp. 497–505). China: China Earthquake Press. **(in Chinese)**.

Yu, H. Z., Cheng, J., Zhang, X. T., Zhang, L. P., Liu, J., & Zhang, Y. X. (2013a). Multi-methods combined analysis of future earthquake potential. *Pure and Applied Geophysics, 170*(1–2), 173–183.

Yu, H. Z., Yin, X. C., Zhu, Q. Y., & Yan, Y. D. (2006). State vector: A new approach to prediction of the failure of brittle heterogeneous media and large earthquakes. *Pure and Applied Geophysics, 163*(11–12), 2561–2574.

Yu, H. Z., Zhang, X. T., & Cheng, J. (2013b). *PI-LURR-SV-AMR combined analysis of 2013 China earthquake potential. Study of earthquake trend in China (2013)* (pp. 208–210). China: China Earthquake Press. **(in Chinese)**.

Yu, H. Z., Zhang, X. T., Cheng, J., & Zhang, Y. X. (2014). *Multi-method combined analysis of earthquake potential in Chinese mainland in 2014. Study of earthquake trend in China (2014)* (pp. 226–235). China: China Earthquake Press. **(in Chinese)**.

Yu, H. Z., Zhang, X. T., Cheng, J., & Zhang, Y. X. (2015). *Multi-method combined analysis of seismicity activities in Chinese mainland in 2015, Study of earthquake trend in China (2015)* (pp. 205–210). China: China Earthquake Press. **(in Chinese)**.

Zechar, J. D., Gerstenberger, M. C., & Rhoades, D. A. (2010). Likelihood-based tests for evaluating space-rate-magnitude earthquake forecasts. *Bulletin of the Seismological Society of America, 100*(3), 1184–1195.

Zhang, H. H., Yin, X. C., Liang, N. G., Yu, H. Z., Li, S. Y., Wang, Y. C., et al. (2006). Acoustic emission experiments of rock failure under load simulating the hypocenter condition. *Pure and Applied Geophysics, 163*(11–12), 2389–2406.

(Received December 14, 2015, revised June 20, 2016, accepted June 24, 2016, Published online July 11, 2016)

Pure Appl. Geophys. 174 (2017), 2401–2410
© 2016 Springer International Publishing
DOI 10.1007/s00024-016-1318-8

Pure and Applied Geophysics

Reducing False Alarms of Annual Forecast in the Central China North–South Seismic Belt by Reverse Tracing of Precursors (RTP) Using the Pattern Informatics (PI) 'Hotspots'

Shengfeng Zhang,[1,2] Zhongliang Wu,[1] and Changsheng Jiang[1]

Abstract—The annual consultation on the likelihood of earthquakes in the next year, the 'Annual Consultation Meeting', has been one of the most important forward forecast experiments organized by the China Earthquake Administration (CEA) since the 1970s, in which annual alarm regions are identified by an expert panel considering multi-disciplinary 'anomalies'. In such annual forecasts, one of the problems in need of further technical solution is its false alarms. To tackle this problem, the concept of 'reverse tracing of precursors (RTP)' is used to the annual consultation, as a temporal continuation and spatial extension of the work of Zhao et al. (Pure Appl Geophys 167:783–800, 2010). The central China north–south seismic belt (in connection to the CSEP testing region) is selected as the testing region of such an approach. Applying the concept of RTP, for an annual alarm region delineated by the Annual Consultation Meeting, the distribution of 'hotspots' of the pattern informatics (PI), which targets the 5-year-scale seismic hazard, is considered. The 'hit', or successful forecast, of the annual seismic hazard is shown to be related to the sufficient coverage of the 'hotspots' within the annual alarm region. The ratio of the areas of the 'hotspots' over the whole area of the annual alarm region is thus used to identify the false alarms which have few 'hotspots'. The results of the years 2004–2012 show that using a threshold of 17 % can reduce 34 % (13 among 38) of the false alarms without losing the successful hit (being 6 in that period).

Key words: Annual consultation, false alarm, reverse tracing of precursors (RTP), pattern informatics (PI) hotspots, the central China north–south seismic belt.

1. Introduction

Among the time-dependent seismic hazard assessment or earthquake forecast with variable

Electronic supplementary material The online version of this article (doi:10.1007/s00024-016-1318-8) contains supplementary material, which is available to authorized users.

[1] Institute of Geophysics, China Earthquake Administration, Beijing 100081, China. E-mail: wuzl@cea-igp.ac.cn
[2] Earthquake Administration of Shandong Province, Jinan 250014, China.

definitions of their spatio-temporal ranges, annual earthquake forecast plays a unique and interesting role by balancing the capability of time-dependent seismic hazard assessment and the annual plan for earthquake preparedness. In China, such an approach is implemented by the Annual Consultation Meeting on the Likelihood of Earthquakes organized by the China Earthquake Administration (CEA). By combining tectonic, seismic, and other geophysical information, an expert panel of the Annual Consultation draws conclusions about the seismic tendency in the next year and identifies the areas with higher seismic risk (Wu et al. 2007, 2013; Wu 2014). Figure 1 shows an example of the output of the Annual Consultation[1], from which successful forecast (hit), false alarm, and failure (miss-to-predict) can be seen clearly. The Annual Consultation Meeting, although having several problems regarding to its science, technology, and organization, has its scientific merit that it is a real forward forecast test, and, most remarkably, it has been persistently conducted for nearly four decades. It is natural that, in recent years, this activity has caused increasing attention and has been discussed in different aspects (Wu 1997; Zheng et al. 2000; Shi et al. 2000, 2001, 2004; Zhang et al. 2002; Ma et al. 2004; Wang 2005; Wu et al. 2007; Zhao et al. 2010; Zhuang and Jiang 2012).

One of the problems of the Annual Consultation is its false-alarm rate. From Fig. 1 it can be seen that among the 10 annual alarm areas in 2004 the ratio of false alarm is up to 50 %. To solve this problem, one of the techniques available is the reverse tracing of precursors (RTP) proposed by Keilis-Borok et al. (2004), which traces long-term anomalies within the

[1] Data from the Department for Earthquake Monitoring and Prediction, China Earthquake Administration, 2014.

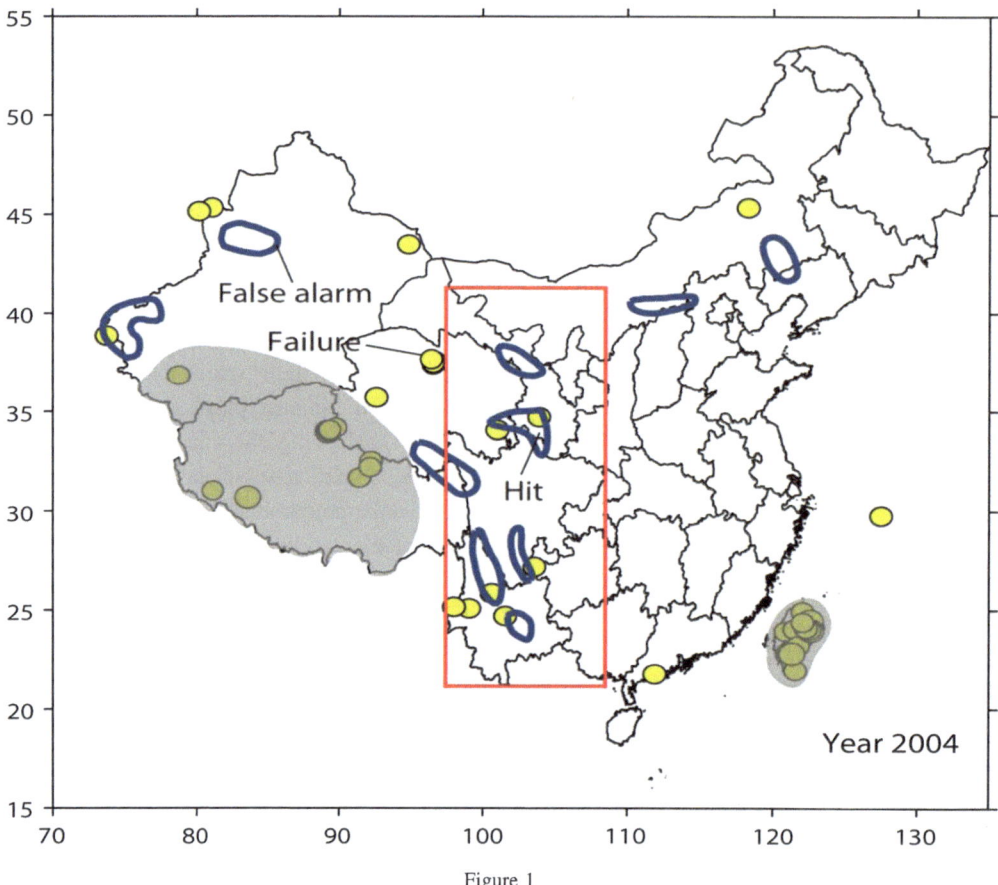

Figure 1
Example of the output of the annual consultation: the result for the year 2004. *Blue closed solid lines* delimitate the regions with increased probability of earthquakes. The *gray shaded area* in Tibet with low monitoring capability and Taiwan with a different system are not included in the annual consultation. *Yellow dots* show the epicenters of earthquakes over magnitude 5 occurred in 2004. Case examples of successful forecast (marked as 'Hit' in the figure), false alarm, and miss-to-hit (marked as 'Failure') are shown in the figure. *Red box* depicts the central China north–south seismic belt which is the study region of this paper

regions identified by short-term precursors (Shebalin *et al.* 2004, 2006). The RTP method assumes that a 'real' short-term anomaly region should be preceded by intermediate-term precursors formed in its vicinity. According to Keilis-Borok *et al.* (2004), an earthquake is generated by two interacting processes in a fault network including an accumulation of energy that the earthquake will release, and a rise of instability which triggers such abrupt release—both can be detected by the 'premonitory' patterns of seismicity. Zechar and Zhuang (2010) discussed the statistical aspects of such an approach.

There are several options for choosing the intermediate-term anomalies in the RTP consideration.

Our approach is the pattern informatics (PI) algorithm proposed by Rundle *et al.* (2003) which have been applied to the analysis of the pattern of seismicity in several places (Chen *et al.* 2005; Nanjo *et al.* 2006a, b; Holliday *et al.* 2005, 2006a, b, 2007; Wu and Chen 2007; Wu *et al.* 2008a, b; Jiang and Wu 2008, 2010, 2011a, b; Cho and Tiampo 2012; Zhang *et al.* 2010, 2013, 2014, 2016; Jiang *et al.* 2013; Kawamura *et al.* 2013; Xia *et al.* 2015). The RTP approach in this paper, as a continuation (in time) and extension (in space) of that of Zhao *et al.* (2010), is based on the analysis of PI anomalies with 5-year time scale in an 'alarm region' of the increase of probability of earthquakes for one-year time scale.

2. Region for Test and Data Used

Partly due to the disastrous great earthquakes, such as the 1920 Haiyuan $M8\frac{1}{2}$ earthquake with fatality up to about 200,000 and the recent 2008 Wenchuan $M_S8.0$ earthquake with fatality and missing up to 87,000, one of the focuses of special attention in China for earthquake preparedness is the central China north–south seismic belt. This region has been studied in the perspective of seismology, geodesy, geology, and geodynamics (Yi *et al.* 2002, 2006; Xu *et al.* 2005; Zhang *et al.* 2014), and has been proposed to be one of the testing regions for the CSEP (Collaboratory for the Study of Earthquake Predictability) project (Mignan *et al.* 2013; Zhang *et al.* 2016). This belt, located in between the Tibetan Plateau, the Ordos block, the Sichuan basin, and the South China block, is featured by complicated fault systems and heterogeneous seismicity.

Somehow arbitrarily, in the present study, the region under consideration is delimited by the ranges of latitude and longitude 21.0°–41.5°N and 97.5°–107.5°E, as shown in Fig. 1, with the reference of the CSEP testing region proposed by Mignan *et al.* (2013). The earthquake catalogue used in this study is the Monthly Earthquake Catalogue from January 1, 1970, to December 31, 2013, provided by the China Earthquake Networks Center (CENC). A homogeneous monitoring capability of completeness magnitude 3.0 has been exhibited since 1970 in this region (Su *et al.* 2003). Zhang *et al.* (2016) discussed in detail the magnitude issues related to this region, following which we use only 'magnitude' for the size of the events in the catalogue. Also following convention that the cutoff magnitude of the earthquake catalogue under study should be no less than the completeness magnitude and is generally taken as about 2 magnitude units less than that of the 'target' earthquakes, in our analysis, we take the cutoff magnitude as 3.0. In the Annual Consultation, the 'target magnitude' is generally $M_S5.0+$, satisfying the conditions for the PI calculation. Additionally, one of the advantages of the PI algorithm is that it is basically a number-counting algorithm, therefore the uncertainty and ambiguity of magnitudes has little effect on its performance.

Up to now the systematic data of the annual consultation (classified for 3 years, with part of them published by the Seismological Press in Beijing) are only reflected in the open file reports (in Chinese) of the China Earthquake Administration (CEA). The data of the annual consultation used in this study are from the China Earthquake Networks Center (CENC), with the permission of the Department for Earthquake Monitoring and Prediction of the CEA.

3. Method of the RTP to Reduce False Alarms

3.1. The RTP Approach

The original version of the RTP approach uses 'earthquake chains' to identify the long-range short-term anomalous region. The 'earthquake chains', which depict an area of a long-range short-term activation of seismicity, have false alarms because not all chains precede large earthquakes. To eliminate false alarms, intermediate-term precursors identified by pattern recognition were used (Keilis-Borok *et al.* 2004). This idea has shown to be able to apply to the case of the Annual Consultation (Zhao *et al.* 2010). In such a 'modified' approach, the alarm regions identified by the Annual Consultation is considered as the anomalous region with shorter time scale, i.e., the 'candidate' (for the annual alarms). Eliminating the false alarm is implemented by considering whether or not intermediate-term anomalies have occurred within each candidate alarm region within several years, say, 5 years, preceding its appearance. The similarity and difference between the original RTP proposed by Keilis-Borok *et al.* (2004) and our approach are listed in Table 1.

Physically, the RTP consideration has a deep root from the concept of fractional Brownian motion (see, e.g., Turcotte 1992). The variation of stress level, characterized by any physically measurable quantity such as the fluctuation detected in an earthquake catalogue, has the property that the longer the time scale, the larger the amplitude of the variation. Short-term anomalies reflect the variation at a shorter time-scale, or in another word, the 'local' scale, while intermediate-term anomalies reflect the variation at longer time scale, or in another word, the 'global'

Table 1

Comparison of anomalies used in the RTP approach

	Short-term ('candidate')		Intermediate-term anomaly	
	Time scale	Anomaly	Time scale	Anomaly
KEILIS-BOROK et al. (2004)	6 months	Earthquake chain	Several years	TIP
ZHAO et al. (2010), and this study	1 year	Annual consultation	5 years	PI

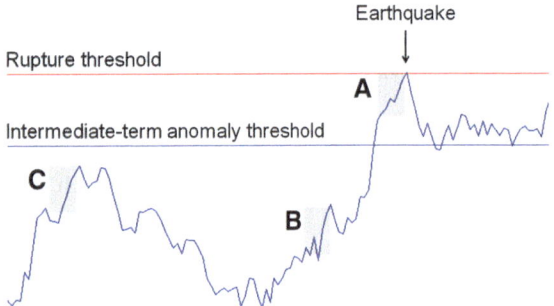

Figure 2

Stress level described by fractional Brownian motion (fBm) (produced by Matlab© via command FBM = wfbm(H, L, 'plot'), with Hurst exponent $H = 0.6$), as a sketch map demonstrating the physical picture of the RTP approach. The *horizontal line in red* shows the rupture threshold (as an extreme simplification since in the real world it is not necessary to be a straight line). The *horizontal line in blue* (also as an extreme simplification) shows the threshold above which intermediate-term anomaly appears. Case A (increasing of stress level with a given slop) is the short-term anomaly which is associated with an earthquake. Cases B and C (with the same type as case A) are the short-term anomalies associated with false alarms. RTP is to discriminate case A from cases B and C by considering whether there are intermediate-term anomalies within the region which possesses short-term anomaly

scale. If the 'match' of such a stress level with the 'rupture threshold' determined by the strength of materials can be considered as the criteria for an earthquake to occur, as shown in Fig. 2, then the reverse tracing of precursors (RTP) is to discriminate the cases that short-time anomalies occur with, or without, the background intermediate-term anomalies, that is, case A, or cases B and C, in Fig. 2. In our study, 'short-term' is 1 year, and 'intermediate-term' is 5 years.

3.2. The Pattern Informatics (PI) Analysis

Also from the consideration of a threshold system, the PI algorithm seems one of the most

qualified candidates to characterize the 'intermediate-term' anomaly, with the input data being regional earthquake catalogues; and the output being the relative increase of the probability of earthquakes, or the distribution of 'hotspots' described by the top 30 % of the normalized probability increase.

To detect the fluctuation, the PI algorithm uses counts of earthquakes within spatial–temporal cells to describe the averaged seismic activity and its variation, through the following steps: (1) the whole region under study is binned into boxes or 'pixels' with size $D \times D$ centered at a point x_i. (2) Each point x_i is associated with a time series $N_i(t)$ where $N_i(t)$ is the time-dependent average rate of earthquakes with magnitude greater than the cutoff magnitude M_c in box i and its Moore neighborhood. $N_i(t)$ is calculated for box i within a period starting from time t_b to time t. (3) The 'seismic activity intensity' functions of box i is defined as the average rate of occurrence of earthquakes:

$$I_i(t_b, t) = \frac{1}{t - t_b} \sum_{t'=t_b}^{t} N_i(t').$$ (1)

(4) The probability of a future strong earthquake in box i is defined as the square of the average intensity fluctuation:

$$P_i(t_0, t_1, t_2) = \overline{\Delta I_i(t_0, t_1, t_2)}^2,$$ (2)

in which t_1 is the starting time of the 'anomaly identification window', t_2 the ending time of the 'anomaly identification window' and the starting time of the 'forecast window', and t_3 the ending time of the 'forecast window'. The 'sliding window' for PI calculation is selected to start from t_0. (5) Subtracting the mean probability over all boxes and denoting this change as the probability-increase of future earthquakes via

$$\Delta P_i(t_0, t_1, t_2) = P_i(t_0, t_1, t_2) - <P_i(t_0, t_1, t_2)> .$$

$$(3)$$

The 'hotspots' are defined to be the boxes where the probability function is larger than the background level.

In our approach we use the PI code of Rundle's group to analyze the fluctuation of seismicity, with parameter settings as follows: the spatial grid is taken as 0.2° and depth of earthquakes is not considered; 'anomaly training window' is 5 years; the starting point of the catalogue subject to analysis is taken as 15 years before the starting time of the 'forecast window'; logarithm amplitude log $(\Delta P/\Delta P_{max})$ is used to represent the relative probability increase for the 'target' earthquakes; only the top 30 % values are considered as the 'hotspots' and shown in the PI mapping. Figure 3 gives an example of the application of the PI forecast, which shows the 'hotspot' distribution indicating the seismic hazard for the period from January 1, 2006, to December 31, 2010 (the 'forecast window'). As a reference, the relative intensity (RI) 'hotspot' map is also presented, with the same parameter settings as the PI map. In the RI algorithm, it is assumed that strong earthquakes will occur in the regions where earthquakes occurred before.

3.3. The ROC Test of the Forecasts

As a test of the performance of the algorithm, receiver operating characteristic (ROC) test (SWETS 1973; MOLCHAN 1997) is conducted against real earthquake activity, by changing the 'alarm threshold' of the 'forecast region' and counting the 'hit rate' and 'false alarm rate' relative to real earthquake activity. Here 'hit rate' means the number of 'forecasted' events divided by the total number of 'target' events. 'Hit' or 'forecasted' is defined as the case that the 'target event' occurs within any 'alarmed cell' or one of its nearest neighbors. In the ROC test, the PI forecast is compared with a random guess, the null hypothesis. The larger area under the ROC curve when compared to the area of the triangle

for random forecast, the better the performance of the forecast will be.

Figure 4 shows the result of the ROC test based on the false alarm rate and the hit rate. In this region, its seems that PI and RI algorithm have comparable performance, apparently due to the fact that seismic activation plays a more important role. From the figure it may be seen that both PI and RI method outperform random guess and have the capability of indicating the increase of the probability of strong earthquakes within the next 5 years.

4. RTP Approach Applied to Annual Forecast: 2004–2012

4.1. RTP with PI for Annual Forecast

Figure 5 shows the 'Hotspot' distribution using the PI method by analyzing the seismicity from January 1, 1997, to December 31, 2006, as the background window, and from January 1, 2007, to December 31, 2011, as the anomaly identification window, targeting at the seismic hazard within the period from January 1, 2012, to December 31, 2016. In the same figure, the result of the Annual Consultation for the year 2012 is shown by blue closed lines delimitating the annual alarm regions, together with the blue circles showing the 'target earthquakes' of the annual consultation (i.e., earthquakes with magnitude larger than $M_S5.0$ occurred in 2012). In the figure, the overlapping area between the PI 'hotspots' and the alarm region from the Annual Consultation is indicated by the numbers marked to the annual alarm areas, together with the results of the annual forecast, that is, 'hit', or false alarm. From the figure it may be seen that 'hits' are apparently associated with widespreading (and coherent) distribution of hotspots. That is, the areas with wide-spreading hotspots are still possible to be false alarms, but only the widespreading hotspots are associated with 'hits'. In another word, those alarm areas with very low coverage of hotspots tend to be false alarms, which is the case that RTP works.

In the calculation, the areas of an annual alarm region and the PI 'hotspot' coverage within it are given by summing up the number of discrete pixels,

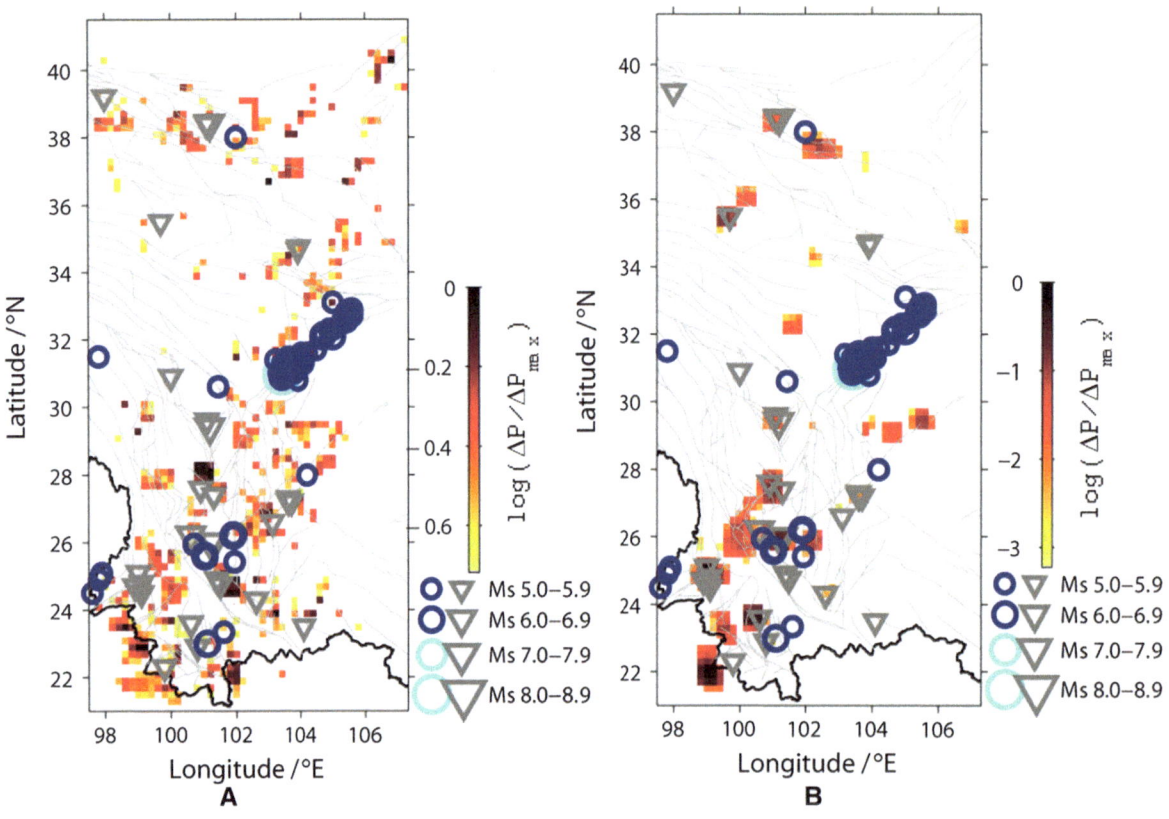

Figure 3
A show-case example of the application of PI/RI forecast, with 5-year forecast window from January 1, 2006, to December 31, 2010. *Color-coded* 'hotspots' indicate the relative probability increase for the 'target' earthquakes (above $M_S5.0$). *Blue circles* are the 'target' earthquakes in the 'forecast window', and *gray reverse triangles* are the earthquakes within the same magnitude range as the 'target' earthquakes that occurred in the 'anomaly training window'. **a** PI method. **b** RI method

using a regular grid, on the map which presents the PI hotspots. For the annual alarm region, all the pixels covered by its boundary are considered as the 'boundary' and not included in the calculation of the area. For the annual alarm regions which are 'cut' by the border of the study region, only the part within the study region (which has the PI hotspot result) is taken in the calculation. The inhomogeneous distribution of hotspots may affect the elimination of false alarms associated with these 'incomplete' regions, such as the first alarm region from the north in Fig. 5.

4.2. Retrospective Determination of the Overlapping Threshold

Figure 6 plots the overlap ratio of the 'hotspots' in the annual alarm regions of each year. In the figure,

false alarms and hits are presented by circles and solid dots, respectively. The red dashed line shows the reference level of the threshold proportion for eliminating the false alarms. In this approach, we aim at correctly removing the false-alarm regions without sacrificing the hits due to the removal. The result from 2004 to 2012 shows that 0.17 can be selected as such a threshold value. From Figs. 5 and 6 as well as the electronic Appendix, it can be seen that 13 among the 38, that is, about 34 %, false alarms, can be eliminated by this RTP approach. Up to now, this result is only empirical, with limited statistical significance due to the limitation of samples, and without a clear picture of physics. However, inspecting in Fig. 6 (the right column) the distribution of the overlapping ratio, and keeping in mind not to over-interpret the result, it may be seen that the

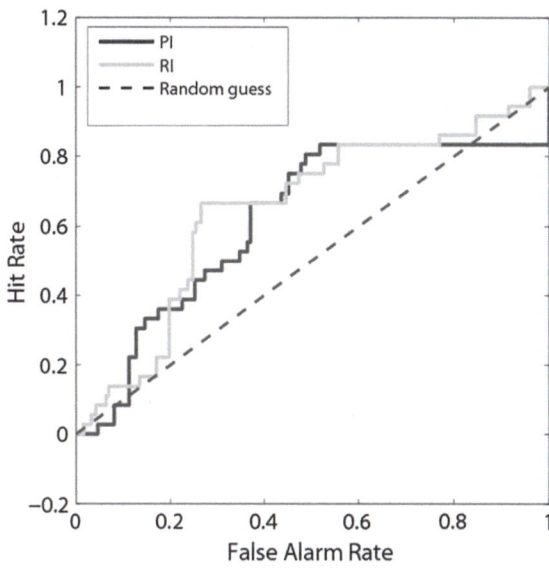

Figure 4

ROC test associated with Fig. 3a, b. *Black solid line* is for the PI forecast, and *gray solid line* for the RI forecast

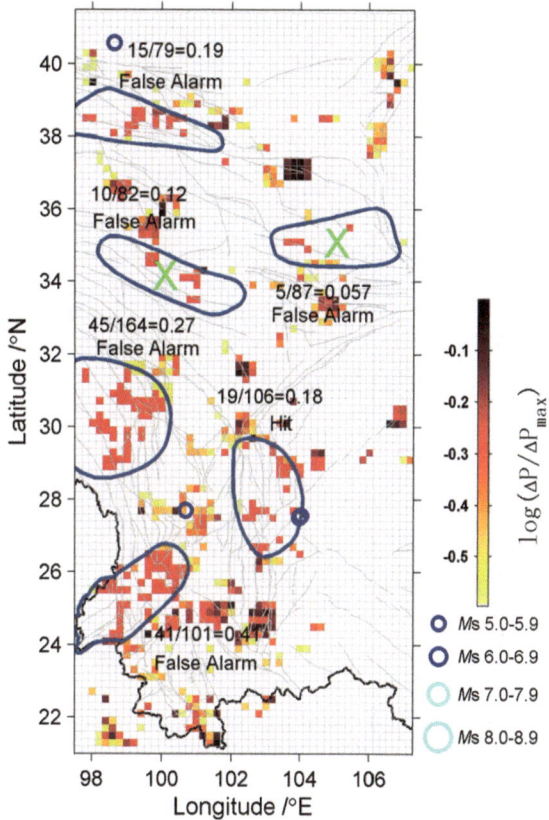

Figure 5

'Hotspot' distribution using the PI method by analyzing the earthquake catalogue from January 1, 1997, to December 31, 2006, as the background window, and from January 1, 2007, to December 31, 2011, as the anomaly identification window, together with the result of the annual consultation for the year 2012 (delimited by *blue solid lines*). *Blue* and *cyan circles* show the 'target earthquakes' occurred in the period from January 1 to December 31, 2012. *Text* shows the proportion of the overlapped area between the PI 'hotspots' and the annual alarm region. Hits and false alarms are marked in the figure. The false alarm removed by RTP is marked by *green crosses*

distribution appears to be 'bi-mode' with two peaks, and the empirically selected threshold is just in between the two peaks of the distribution.

Annual consultation provides a kind of alarm-based forecast, as shown by Figs. 1 and 5. For such 'black-and-white' forecasts, one of the indexes evaluating the performance is the *R*-value, the hit rate minus the false alarm rate (XU 1989), which has been used in China for the test of annual earthquake forecast (SHI *et al.* 2001). The *R*-value can be considered as a simplified version of the ROC test, or the difference between the vertical and horizontal axes on the ROC diagram. Seen in this perspective, the above threshold is equivalent to the selection that with the hit rate unchanged, the *R*-value increases with the decrease of the false alarm rate.

5. Conclusions and Discussion

We use the concept of 'reverse tracing of precursors (RTP)' to tackle the problem of false alarms of the Annual Consultation. We consider the central China north–south seismic belt which has been selected as the CSEP testing region in China. For an annual alarm region delineated by the Annual Consultation Meeting, the distribution of PI 'hotspots' targeting at the 5-year-scale seismic hazard is considered. It was shown that successful forecast is generally related to the sufficient coverage of the PI 'hotspots' within the alarm region. The ratio of the areas of the PI 'hotspots' over the whole area of the annual alarm region, therefore, is used to identify the false alarms which have few 'hotspots'. The result of the years 2004–2012 obtains the threshold ratio of the overlap area 17 %, for the central China north–south seismic belt. This leads to the following suggestive procedure for eliminating the false alarms in the

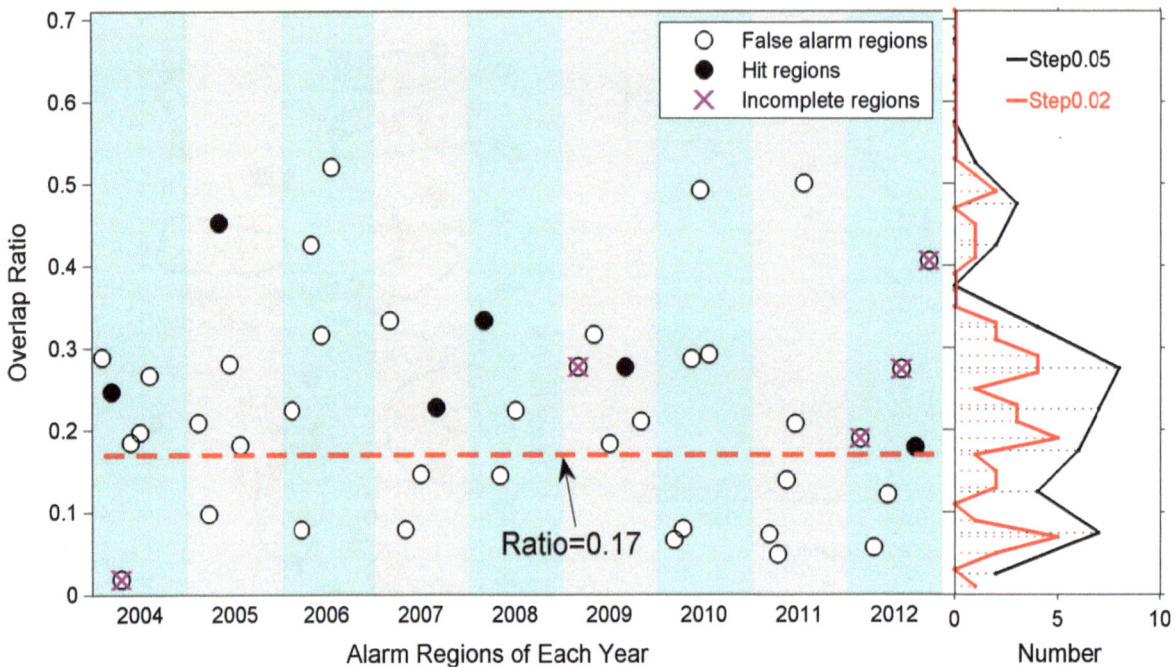

Figure 6
The overlap ratio of the 'hotspots' in the annual alarm regions of each year. Within each year, the alarm regions are ordered from the north to the south. *Circles* and *solid dots* show the false alarms and the 'hits', respectively. *Crosses* are marked to the 'incomplete alarm regions' such as the first alarm region from the north in Fig. 5. The *red dashed line* shows the reference level of the threshold proportion for eliminating the false alarms. In this approach 'conservative' decision is made to correctly remove the false-alarm regions, without sacrificing the hits due to the removal. The *right column* presents the distribution of the overlap ratio, using different steps for making the histograms

annual consultation using the PI algorithm as the intermediate-term anomaly indicator: (1) plotting on the same map the alarm regions produced by the Annual Consultation and the PI 'hotspots'; (2) calculating the proportion of overlapped area between each alarm region and the PI 'hotspots'; and (3) eliminating the regions which have a proportion of 'overlap area' lower than the threshold—for the present case, 17 %. For the period 2004–2012, using such an approach, 34 % of the false alarms can be eliminated, without losing the 6 successful hits, as shown in Fig. 6.

To better test this approach, it is necessary to put it to the forward forecast. To this end, it might be useful to compare the work of ZHAO *et al.* (2010) and this study to inspect the stability of the result as well as the threshold. In ZHAO *et al.* (2010), a smaller region (20.8°–33.0°N, 97.2°–105.0°E, corresponding to the southern part of the north–south seismic belt) and an earlier time (1990–2003) was the target for

investigation. With the threshold 7 %, on average 1–2 false alarm regions per year were removed. To some extent, this work, as a continuation in time and extension in space, of the work of ZHAO *et al.* (2010), is to investigate the question proposed in that paper that due to the region dependence of the characteristics of seismic activity, whether the PI-based RTP approach for annual forecast holds for other regions. In this work, a larger region (21.0°–41.5°N, 97.5°–107.5°E), mainly extended to the north (corresponding to the CSEP testing region) was selected, and a later time range, 2004–2012, was considered. With 17 % being the threshold, on average 1–2 false alarm regions per year were removed without losing the 'hits'. It seems that, firstly, the PI-based RTP approach has stability to some extent, at least for the central China north–south seismic belt; secondly, the larger region, the higher the threshold value of the 'overlapping area', which might be understood by the picture of fractional Brownian motion (fBm). This also hints that to remove the false alarms correctly without losing the 'hits', it is better to consider smaller regions with relatively homogeneous seismicity inside. Last but not the least, although in China the result of the Annual Consultation is classified for 3 years due to the scientifically experimental nature of the Annual Consultation and the social sensitivity of short-term earthquake forecasts, for the CSEP-testing region, an inside forward test can be considered from now on.

Acknowledgments

Thanks to Prof. J.B. Rundle and Prof. C.C. Chen, for guidance and helps in the PI calculation, to Dr. J.C. Zhuang for helpful discussion in revising the manuscript, and to the China Earthquake Networks Center (CENC) for providing the earthquake catalogue and the data of the Annual Consultation. Thanks are also to Prof. X.C. Yin and Prof. Y.X. (Angie) Zhang for invitation to participate in the 2015 ACES Workshop in Chengdu, and to the CSEP-China project for financial support.

REFERENCES

CHEN, C. C., RUNDLE, J. B., HOLLIDAY, J. R., NANJO, K. Z., TURCOTTE, D. L., LI, S. C. and TIAMPO, K. F., 2005, *The 1999 Chi-Chi, Taiwan, earthquake as a typical example of seismic activation and quiescence*. Geophys. Res. Lett., *32*: L22315, doi:10.1029/2005GL023991.

CHO, N. F. and TIAMPO, K. F., 2012, *Effects of location errors in pattern informatics*. Pure Appl. Geophys., *170*: 185–196, doi:10.1007/s00024-011-0448-2.

HOLLIDAY, J. R., NANJO, K. Z., TIAMPO, K. F., RUNDLE, J. B. and TURCOTTE, D. L., 2005, *Earthquake forecasting and its verification*. Nonlin. Processes Geophys., *12*: 965–977, doi:arXiv:cond-mat/0508476.

HOLLIDAY, J. R., RUNDLE, J. B., TIAMPO, K. F. and TURCOTTE, D. L., 2006a, *Using earthquake intensities to forecast earthquake occurrence times*. Nonlin. Processes Geophys., *13*: 585–593.

HOLLIDAY, J. R., RUNDLEA, J. B., TIAMPO, K. F., KLEIND, W. and DONNELLANE, A., 2006b, *Modification of the pattern informatics method for forecasting large earthquake events using complex eigenfactors*. Tectonophysics, *413*: 87–91.

HOLLIDAY, J. R., CHEN, C.C., TIAMPO, K. F., RUNDLE, J. B., TURCOTTE, D. L. and DONNELLAN, A., 2007, *A RELM earthquake forecast based on pattern informatics*. Seism. Res. Lett., *78*: 87–93.

JIANG, C. S. and WU, Z. L., 2008, *Retrospective forecasting test of a statistical physics model for earthquakes in Sichuan-Yunnan region*. Sci. China Ser. D, *51*: 1401–1410.

JIANG, C. S. and WU, Z. L., 2010, *PI forecast for the Sichuan-Yunnan region: retrospective test after the May 12, 2008, Wenchuan earthquake*. Pure Appl. Geophys., *167*: 751–761, doi:10.1007/s00024-010-0070-8.

JIANG, C. S. and WU, Z. L., 2011a, *PI forecast with or without declustering: an experiment for the Sichuan-Yunnan region*. Nat. Hazard Earth Sys., *11*: 697–706, doi:10.5194/nhess-11-697-2011.

JIANG, C. S. and WU, Z. L., 2011b, *Intermediate-term medium-range precursory accelerating seismicity prior to the 12 May 2008, Wenchuan earthquake*. Pure Appl. Geophys., *170*: 209–219, doi:10.1007/s00024-011-0413-0.

JIANG, H., WU, Z. L., MA, T. F. and JIANG, C. S., 2013, *Retrospective test of the PI forecast: case study of the April 20, 2013, Lushan, Sichuan, China, MS7.0 earthquake*. Physics, *42*: 334–340, doi:10.7693/wl2013504 (in Chinese with English abstract).

KAWAMURA, M., WU, Y. H., KUDO, T. and CHEN, C. C., 2013, *Precursory migration of anomalous seismic activity revealed by the pattern informatics method: A case study of the 2011 Tohoku earthquake, Japan*. Bull. Seismol. Soc. Amer., *103*: 1171–1180.

KEILIS-BOROK, V., SHEBALIN, P., GABRIELOV, A. and TURCOTTE, D., 2004, *Reverse tracing of short-term earthquake precursors*. Phys. Earth Planet. Inter., *145*: 75–85.

MA, H. S., LIU, J., WU, H. and LI, J. F., 2004, *Scientific evaluation of annual earthquake prediction efficiency based on R-value*. Earthquake, *24*: 31–37 (in Chinese with English abstract).

MIGNAN, A., JIANG, C. S., ZECHAR, D. J., WIEMER, S., WU, Z. L. and HUANG, Z. B., 2013, *Completeness of the mainland China earthquake catalog and implications for the setup of the China earthquake forecast testing center*. Bull. Seismol. Soc. Amer., 103: 845–859, doi:10.1785/0120120052.

MOLCHAN, G. M., 1997, *Earthquake prediction as a decision-making problem*. Pure Appl. Geophys., *149*: 233–247.

NANJO, K. Z., RUNDLE, J. B., HOLLIDAY, J. R. and TURCOTTE, D. L., 2006a, *Pattern informatics and its application for optimal forecasting of large earthquakes in Japan*. Pure Appl. Geophys, **163**: 2417–2432, doi:10.1007/s00024-006-0130-2.

NANJO, K. Z., HOLLIDAY, J. R., CHEN, C. C., RUNDLE, J. B. and TURCOTTE, D. L., 2006b, *Application of a modified pattern informatics method to forecasting the locations of future large earthquakes in the central Japan*. Tectonophysics, *424*: 351–366.

RUNDLE, J. B., TURCOTTE, D. L., SHCHERBAKOV, R., KLEIN, W. and SAMMIS, C., 2003, *Statistical physics approach to understanding the multiscale dynamics of earthquake fault systems*. Rev. Geophys., *41*: 1019, doi:10.1029/2003RG000135.

SHEBALIN, P., KEILIS-BOROK, V., ZALIAPIN, I., UYEDA, S., NAGAO, T. and TSYBIN, N., 2004, *Advance short-term prediction of the large Tokachi-oki earthquake, September 25, 2003, M = 8.1 A case history*. Earth Planets Space, *56*: 715–724.

SHEBALIN, P., KEILIS-BOROK, V., GABRIELOV, A., ZALIAPIN, I. and TURCOTTE, D., 2006, *Short-term earthquake prediction by reverse analysis of lithosphere dynamics*. Tectonophysics, *413*: 63–75.

SHI, Y. J., WU, Z. L. and BAI, L., 2004, *Evaluation of annual prediction of seismicity tendency using Pearson test*, Journal of the Graduate School of the Chinese Academy of Sciences, *21*: 248–253 (in Chinese with English abstract).

SHI, Y. L., LIU, J. and ZHANG, G. M., 2000, *The evaluation of Chinese annual earthquake prediction in the 90 s*. Journal of Graduate School Academia Sinica, *17*: 63–69 (in Chinese with English abstract).

SHI, Y. L., LIU, J. and ZHANG, G. M., 2001, *An evaluation of Chinese annual earthquake predictions*, 1990–1998. J. Appl. Probab., *38A*: 222–231.

SU, Y. J., LI, Y. L., LI, Z. H., YI, G. X. and LIU, L. F., 2003, *Analysis of minimum complete magnitude of earthquake catalog in Sichuan-Yunnan region*. Journal of Seismological Research, *26*: 10–16 (in Chinese with English abstract).

SWETS, J. A., 1973, *The relative operating characteristic in psychology*. Science, *182*: 990–1000.

TURCOTTE, D. L., 1992, *Fractals and Chaos in Geology and Geophysics*. Cambridge: Cambridge Univ. Press.

WANG, H. T., 2005, *Some preliminary think on prediction of annual seismic risk region*, Recent Developments in World Seismology, *317*: 103–105 (in Chinese with English abstract).

WU, F. T., 1997, *The Annual Earthquake Prediction Conference in China (National Consultative Meeting on Seismic Tendency)*. Pure Appl. Geophys., *149*: 249–264.

WU, Y.-M. and CHEN, C. C., 2007, *Seismic reversal pattern for the 1999 Chi-Chi, Taiwan, M_W 7.6 earthquake*. Tectonophysics, *429*: 125–132, doi:10.1029/2008GL035215.

WU, Y. H., CHEN, C. C. and RUNDLE, J. B., 2008a, *Detecting precursory earthquake migration patterns using the pattern informatics method*. Geophys. Res. Lett., *35*: L19304, doi:10.1029/2008GL035215.

WU, Y. H., CHEN, C. C. and RUNDLE, J. B., 2008b, *Precursory seismic activation of the Pingtung (Taiwan) offshore doublet earthquakes on 26 December 2006: A pattern informatics analysis*. Terre. Atmos. Ocean Sci., *19*: 743–749.

WU, Z. L., 2014, Chapter16: Duties of earthquake forecast: cases and lessons in China. In: Wyss, M. (eds.), *Earthquake Hazard, Risk, and Disasters*, Amsterdam: Elsevier, 431–448.

WU, Z. L., LIU, J., ZHU, C. Z., JIANG, C. S. and HUANG, F. Q., 2007, *Annual consultation on the likelihood of earthquakes in continental China: Its scientific and practical merits*. Earthquake Research in China, *21*: 365–371.

WU, Z. L., MA, T. F., JIANG, H. and JIANG, C. S., 2013, *Multi-scale seismic hazard and risk in the China mainland with implication for the preparedness, mitigation, and management of earthquake disasters: An overview*. International Journal of Disaster Risk Reduction, *4*: 21–33.

XIA, C. Y., ZHANG, Y. X., ZHANG, X. T. and WU, Y. J., 2015, *Predictability test for pattern information method by two $M_S7.3$ Yutian, Xinjiang, earthquakes*. Acta Seismologica Sinica, *37*: 312–322. doi:10.11939/jass.2015.02.011 (in Chinese with English abstract).

XU, S. X., 1989, The evaluation of earthquake prediction ability. In: Department of Science, Technology and Monitoring, State Seismological Bureau (eds.), *The Practical Research Papers on Earthquake Prediction Methods (Seismicity Section)*. Beijing: Seismological Press, 586–589 (in Chinese).

XU, X. W., ZHANG, P. Z., WEN, X. Z., QIN, Z. L., CHEN, G. H. and ZHU, A. L., 2005, *Features of active tectonics and recurrence behaviors of strong earthquake in the western Sichuan Province and its adjacent regions*. Seismology and Geology, *27*: 446–461 (in Chinese with English abstract).

YI, G. X., WEN, X. Z. and XU, X. W., 2002, *Study on recurrence behaviors of strong earthquakes for several entireties of active fault zones in Sichuan-Yunnan region*. Earthquake Research in China, *18*: 267–276 (in Chinese with English abstract).

YI, G. X., WEN, X. Z., WANG, S. W., LONG, F. and FAN, J., 2006, *Study on fault sliding behaviors and strong-earthquake risk of the Longmenshan-Minshan fault zones from current seismicity parameters*. Earthquake Research in China, *22*: 117–125 (in Chinese with English abstract).

ZECHAR, J. D. and ZHUANG, J. C., 2010, *Risk and return: evaluating Reverse Tracing of Precursors earthquake predictions*. Geophysical Journal International, *182*: 1319–1326.

ZHANG, G. M., LIU, J. and SHI, Y. L., 2002, *An scientific evaluation of annual earthquake predication ability*. Acta Seismologica Sinica, *15*: 550–558.

ZHANG, S. F., WU, Z. L. and JIANG, C. S., 2016, *The central China north-south seismic belt: seismicity, ergodicity, and five-year PI forecast in testing*. Pure Appl. Geophys., *173*: 245–254, doi:10.1007/s00024-015-1123-9.

ZHANG, X. T., ZHANG, Y. X., XIA, C. Y., WU, Y. J. and YU, H. Z., 2014, *Anomalous seismic activities in the Sichuan-Yunnan region and its adjacent areas before the Lushan $M_S7.0$ earthquake by the pattern informatics method*. Acta Seismologica Sinica, *36*: 780–789, doi:10.3969/j.issn.0253-3782.2014.05.003.

ZHANG, Y. X., ZHANG, X. T., YIN, X. C. and WU, Y. J., 2010, *Study on the forecast effects of PI method to the north and southwest China*. Concurrency and Computation: Practice & Experience, *22*: 1559–1568.

ZHANG, Y. X., ZHANG, X. T., WU, Y. J. and YIN, X. C., 2013, *Retrospective study on the predictability pattern informatics to the Wenchuan M8.0 and Yutian M7.3 Earthquakes*. Pure Appl. Geophys., *170*: 197–208.

ZHAO, Y. Z., WU, Z. L., JIANG, C. S. and ZHU, C. Z., 2010, *Reverse tracing of precursors applied to the annual earthquake forecast: Retrospective test of the Annual Consultation in the Sichuan-Yunnan region of southwest China*. Pure Appl. Geophys., *167*: 783–800, doi:10.1007/s00024-010-0077-1.

ZHENG, Z. B., LIU, J., LI, G. F., QIAN, J. D. and WANG, X. Q., 2000, *Statistical simulation analysis of the correlation between the annual estimated key regions with a certain seismic risk and the earthquakes in China*. Acta Seismologica Sinica, *13*: 575–584.

ZHUANG, J. C. and JIANG, C. S., 2012, *Scoring annual earthquake predictions in China*. Tectonophysics, *524–525*: 155–164.

(Received January 20, 2016, revised May 16, 2016, accepted May 17, 2016, Published online May 30, 2016)

Pure Appl. Geophys. 174 (2017), 2411–2426
© 2017 Springer International Publishing
DOI 10.1007/s00024-017-1551-9

Test of the Predictability of the PI Method for Recent Large Earthquakes in and near Tibetan Plateau

YONGXIAN ZHANG,[1] ⓘ CAIYUN XIA,[2] CHENG SONG,[3] XIAOTAO ZHANG,[1] YONGJIA WU,[1] and YAN XUE[1]

Abstract—Five large earthquakes of $M \geq 7.0$ (based on the magnitude scale of the China Earthquake Networks Center) occurred in and near the Tibetan Plateau during 2008–2014, including the Wenchuan $M8.0$ earthquake on May 12, 2008 (BJT). In this paper, the Tibetan Plateau was chosen to be the study region, and calculating parameters of pattern informatics (PI) method with grid of $1° \times 1°$ and forecasting time interval of 8 years were employed for the retrospective study according to the previous studies for $M7$ earthquake forecasting. The sliding step of forecasting interval was 1 year, and the hotspot diagrams of each forecasting interval since 2008 were obtained year by year. The relationships among the hotspots and the $M \geq 7.0$ earthquakes that occurred during the forecast intervals were studied. The predictability of PI method was tested by verification of receiver-operating characteristic curve (ROC) and R score. The results show that the successive obvious hotspots occurred during the sliding forecasting intervals before four of the five earthquakes, while hotspots only occurred in one forecasted interval without successive evolution process before one of the five earthquakes, which indicates that four of the five large earthquakes could be forecasted well by PI method. Test results of the predictability of PI method by ROC and R score show that positive prospect of PI method could be expected for long-term earthquake forecast.

Key words: PI method, Earthquake predictability, ROC test, R score test, Earthquake-forecasting efficacy, Tibetan Plateau.

1. Introduction

From 2008, five $M \geq 7.0$ earthquakes attacked the West Continental China successively. They are Yutian $M7.3$ earthquake (35.6°N, 81.6°E) (based on the catalogue from China Seismic Network, same in the

following) on March 21, 2008 (BJT, same in the following); Wenchuan $M8.0$ earthquake (31.0°N, 103.4°E) on May 12, 2008; Yushu $M7.1$ earthquake (33.2°N, 96.6°E) on April 14, 2010; Lushan $M7.0$ earthquake (30.3°N, 103.0°E) on April 20, 2013; and Yutian $M7.3$ earthquake (31.0°N, 103.4°E) on Feb 12, 2014. All these five large earthquakes occurred in and near the Tibetan Plateau (Fig. 1). This provides an opportunity to test the predictability of the pattern informatics (PI) method (Rundle et al. 2000a, b, 2002, 2003; Tiampo et al. 2002a, b; Holliday et al. 2005, 2006a) to all these five large earthquakes in this region.

PI method has proven to be an efficacious approach to earthquake forecasting in medium-term time scale (several years to 10 years) in different tectonic regions, such as California region (Rundle et al. 2002; Tiampo et al. 2002b, Holliday et al. 2005), Japan region (Nanjo et al. 2006a, b; Kawamura et al. 2013, 2014), Taiwan region (Chen et al. 2005, 2006; Wu et al. 2008a, b; Chang et al. 2013), China mainland (Jiang and Wu 2008; Zhang et al. 2009, 2013; Zhang et al. 2014; Xia et al. 2014), and the worldwide region (Holliday et al. 2005). The results of ROC test show that the PI method outperforms not only the random guess method but also the simple number-counting approach based on the clustering hypothesis of earthquakes (the RI forecast) (Rundle et al. 2000a, b, 2002, 2003; Tiampo et al. 2002a, b; Holliday et al. 2005, 2006a). The forecasting efficacy of PI method has also been tested through the investigation of the parameters and conditions, including time spans, for optimizing seismicity-based forecasts by ensuring that the mean activity rate remains constant (Tiampo et al. 2010; Migan and Tiampo 2010; Jiang and Wu 2011; Tiampo and Shcherbakov 2013; Tiampo et al. 2013;

---
[1] China Earthquake Networks Center, Beijing 100045, China. E-mail: yxzhseis@sina.com
[2] Liaoning Earthquake Administration, Shenyang 110034, China.
[3] Institute of Earthquake Science, China Earthquake Administration, Beijing 100036, China.

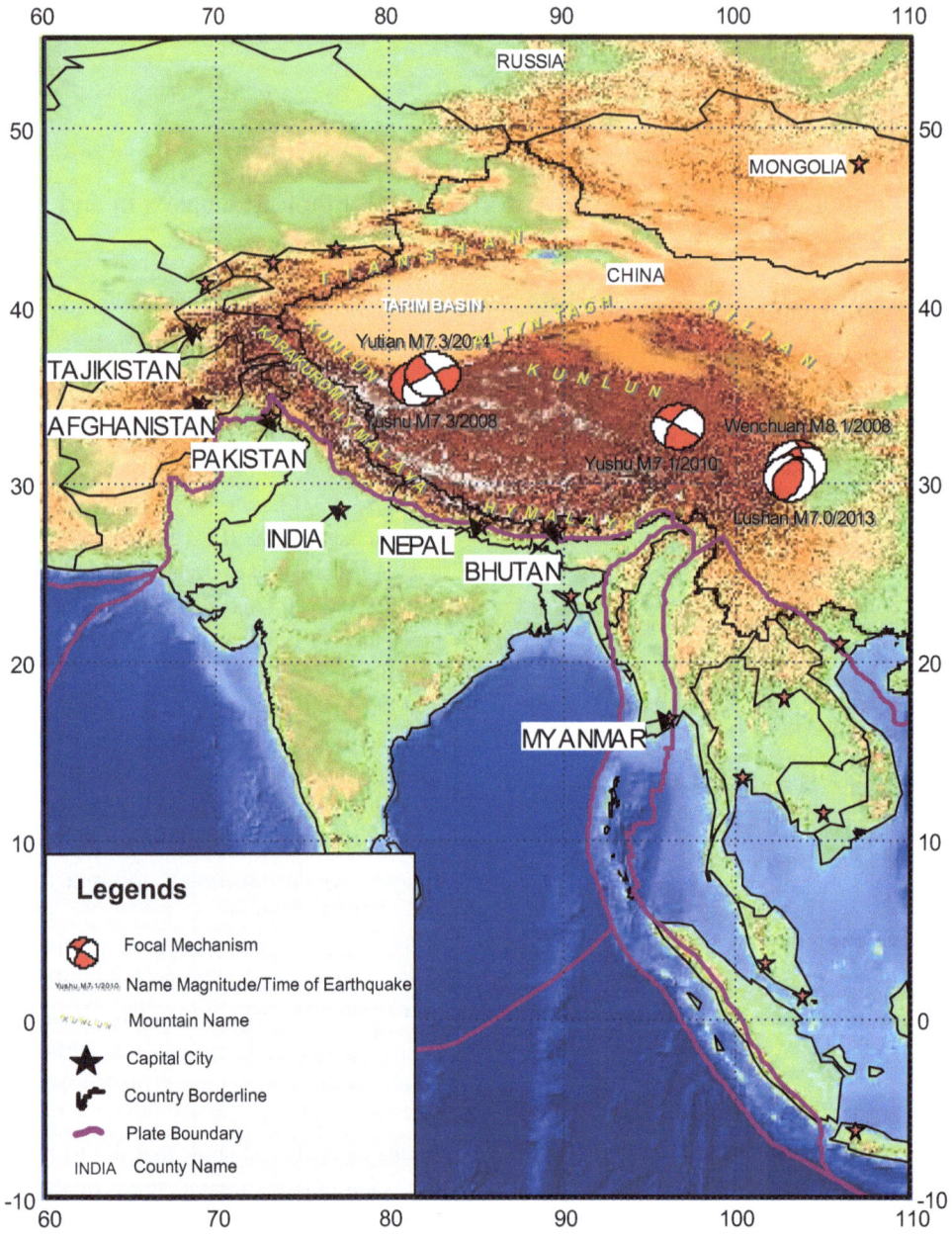

Figure 1
Geographic map of the Tibetan Plateau and its recent large earthquakes during 2008–2014

Zhang et al. 2013). Tiampo et al. (2010) found that the ergodicity defined by magnitude and time period provides more reliable forecasts of future events in both natural and synthetic catalogues by PI method. Tiampo and Shcherbakov (2012) applied the TM metric and threshold optimization for forecasting parameter estimation to the PI method, and the combined application of these techniques is found successful in forecasting those large events that occurred in Haiti, Chile, and California in 2010 on both global and regional scales. However, when Mignan and Tiampo (2010) tested the PI index by synthetic catalogues where a realistic spatiotemporal clustering has been added on top of the theoretical precursory seismicity, they found that the PI index is found successful in identifying the precursory quiescent signal, but fails in identifying precursory accelerating seismicity directly, because of being more sensitive to aftershock sequences of background events than to the activation-like behavior of the acceleration. Jiang and Wu (2011) also found that the PI forecasts seem to be affected by the aftershock sequence included in the "anomaly identifying interval," and the PI forecast approach using "background events" seems to display a better performance when they de-clustered the catalogue of the Sichuan-Yunnan region of southwest China by the epidemic-type aftershock sequences (ETAS) model and investigated the effects of de-clustering on the PI forecasts. Cho and Tiampo (2013) studied the effects of locational errors in PI by generating a series of perturbed catalogues by adding different levels of noise to epicenter locations based on the Southern Californian dataset. Their results showed that the maximum performance of the PI technique with respect to both skill scores did not decrease systematically for any of the noise levels used, indicating that the PI performance is not sensitive to the locational errors for catalogues with large numbers of events as a result of their dependence on seismic clustering for its forecasting skill. Zhang et al. (2013) studied the forecasting effects of the calculating parameters on the PI method by varying the calculating parameters of the grid size and the reference time scale, and the results showed that the forecasting efficacy could be improved by choosing approximate parameters after a systematic retrospective study on Wenchuan $M8.0$ and Yutian $M7.3$ earthquakes that occurred in 2008.

In fact, five large earthquakes of $M \geq 7.0$ occurred in and near the Tibetan Plateau during 2008–2014, including Wenchuan $M8.0$ and Yutian $M7.3$ earthquakes in 2008; Yushu $M7.1$ earthquake in 2010; Lushan $M7.0$ earthquake in 2013; and Yutian $M7.3$ earthquake in 2014. This supplies the samples to test if the optimal calculating parameters from Zhang et al. (2013) are effective to the subsequent earthquakes. In this paper, the Tibetan Plateau was chosen to be the study region, and calculating parameters of PI method with grid of $1° \times 1°$ and forecasting time interval of 8 years were employed

for the retrospective study for $M7$ earthquake forecasting. The sliding step of forecasting interval was 1 year, and the snapshots of hotspot map of each forecasting interval since 2008 were obtained on yearly basis. Based on the results, the relationships among the hotspots and the $M \geq 7.0$ earthquakes that occurred during the forecast intervals were studied, and the predictability of PI method was tested by verification of receiver-operating characteristic curve (ROC) and R score.

2. The PI Method

PI method was invented by Rundle et al. (2000a, b, 2002, 2003) and developed by Tiampo et al. (2002a, b, c) and Holliday et al. (2005, 2006b). It is an earthquake-forecasting method for quantifying the temporal variations in seismicity based on the statistical mechanics of complex systems. The result is a map of areas with high probability of earthquake potential (hotspots) where earthquakes are likely to occur during a specified period in the future. Rundle et al. (2002) published a forecast map of hotspots of $M \geq 5$ earthquakes for California during the period of 2000–2010 (http://quakesim.jpl.nasa.gov/scorecard. html). The testing results show that 17 out of the 19 earthquakes that have occurred between the beginning date of this forecast (January 2000) and September 2006 were coincident with the forecast anomalies within the ± 11 km margin of error that is the coarse-graining box size (Tiampo et al. 2008). Nanjo et al. (2006a, b) modified the PI method for use with the Japanese catalogues, and their retrospective study showed that the $M = 6.8$ Niigata earthquake that occurred on October 23, 2004 could be successfully forecasted. Chen et al. (2005) modified the PI method for use with Taiwan catalogues and found the Chi–Chi $Ms7.6$ earthquake located in the hotspot area. In this paper, we applied the algorithm of PI method described by Holliday et al. (2005) and realized by Zhang et al. (2009).

Following Holliday et al. (2005), the algorithm of PI method is performed as follows:

(1) First, we divide the region of interest into N_B grids with linear dimension Δx. Grids are

identified by a subscript i and are centered at x_i. For each grid, the number of earthquakes per unit time at time t larger than the lower cutoff magnitude M_c constructs the time series $N_i(t)$, The time series in grid i is defined between a base time t_b and the present time t.

(2) All earthquakes in the region of interest with magnitudes greater than M_c are included, and M_c is specified in order to ensure completeness of the data through time, from an initial time t_0 to a final time t_2.

(3). A reference time interval from t_b to t_1 (t_b lies between t_0 and t_1), a change time interval from t_1 to t_2 ($t_2 > t_1$), and a forecast time interval from t_2 to t_3 are defined. The objective is to quantify anomalous seismic activity in the change interval from t_1 to t_2 relative to the reference interval t_b to t_1, and the forecast time intervals from t_2 to t_3 are defined. The change and forecast time intervals are taken to have the same length.

(4) $I_i(t_b, t) = \frac{1}{t - t_b} \sum_{t'=t_b}^{t} N_i(t')$ is the seismic intensity in grid i, between the two times $t_b < t$, which means the average number of earthquakes with magnitudes greater than M_c occur in the grid per unit time during the specified time interval from t_b to t.

(5) The statistically normalized seismic intensity of grid i during the time interval from t_b to t is then defined as $\hat{I}_i(t_b, t) = \frac{I_i(t_b,t) - <I_i(t_b,t)>}{\sigma(t_b,t)}$, where $<I_i(t_b, t)>$ is the mean intensity averaged over all the grids, and $\sigma(t_b, t)$ is the standard deviation of intensity over all the grids.

(6) The measure of anomalous seismicity in grid i is the difference between the two normalized seismic intensities: $\Delta I_i(t_b, t_1, t_2) = \hat{I}_i(t_b, t_2) - \hat{I}_i(t_b, t_1)$.

(7) To reduce the relative importance of random fluctuations (noise) in seismic activity, the average change in intensity $(\overline{\Delta I_i(t_0, t_1, t_2)})$ over all possible pairs of normalized intensity maps is defined to have the same change interval: $\overline{\Delta I_i(t_0, t_1, t_2)} = \frac{1}{t_1 - t_0} \sum_{t_b=t_0}^{t_1} \Delta I_i(t_b, t_1, t_2)$, where the sum is performed over increments of the time series, which here are considered as days.

(8) The probability of a future earthquake in grid i, $P_i(t_0, t_1, t_2)$, is defined as the square of the average intensity change: $P_i(t_0, t_1, t_2) = \overline{\Delta I_i(t_0, t_1, t_2)}^2$.

(9) The change in the probability in grid i, relative to the background mean probability over all grids, is $\Delta P_i(t_0, t_1, t_2) = P_i(t_0, t_1, t_2) - <P_i(t_0, t_1, t_2)>$, where $<P_i(t_0, t_1, t_2)>$ is the background probability for a large earthquake.

Hotspots are initially defined to be the regions where $\Delta P_i(t_0, t_1, t_2)$ is positive. In these regions, $P_i(t_0, t_1, t_2)$ is larger than the average value for all grids (the background level). In order to get the normalized hotspot map and reduce the number of hotspots for reduction of false alarm rate, we calculate the value of $\log \Delta P_i(t_0, t_1, t_2) / \Delta P_{max}(t_0, t_1, t_2)$, and set a threshold of $\log \Delta P_i(t_0, t_1, t_2) / \Delta P_{max}(t_0, t_1, t_2)$ to get the final hotspot map. The number of hotspots depends on the threshold of $\log \Delta P_i(t_0, t_1, t_2) / \Delta P_{max}(t_0, t_1, t_2)$. Since the intensities are squared in defining probabilities, the hotspots may be due to either increases of seismic activity during the change time interval (activation) or due to decreases (quiescence).

There exists a common hypothesis in seismicity-based earthquake-forecasting techniques that future large earthquakes tend to occur in the locations where the activity of small events changed abnormally, such as AMR, LURR, PI, M8, b-value, RTL, RI, etc. (Tiampo and Shcherbakov 2012). For example, Bufe et al. (1993) regarded that before a major earthquake (M_f) occurs, the seismicity of smaller earthquakes with magnitude of $M_f - 2$ often show abnormal acceleration or decrease; Jaume et al. (1999) concluded that the seismicity of smaller earthquakes with magnitude from $M_f - 2$ to $M_f - 3$ often show activity of abnormal acceleration or decrease before a major earthquake (M_f). Kossobokov et al. (1999) employed smaller earthquakes, $\sim M4$, to calculate the "Time of Increased Probability," or TIP, for the forecasting of a larger event of approximately $M6.5-8$; Papazachos et al. (2005) determined that the seismicity events of smaller earthquakes with magnitude from $M_f - 1.5$ to $M_f - 2.0$ often show activity of abnormal acceleration or decrease before a major earthquake (M_f). In our study, we hypothesize that earthquakes with magnitudes larger than $M_c + 2.0$ will occur preferentially in hotspots during the forecast time interval from t_2–t_3 the same as that Hollidays et al. (2005).

3. The Tibetan Plateau and the Its Seismicity

3.1. Location of the Tibetan Plateau and Its Recent Large Earthquakes during 2008–2014

The Tibetan Plateau is chosen to be the study region. This region includes most part of western China and some parts of Bhutan, Nepal, India, Pakistan, Afghanistan, etc. (Fig. 1). The plateau is bordered to the south by the inner Himalayan range, to the north by the Kunlun Range which separates it from the Tarim Basin, and to the northeast by the Qilian Range which separates the plateau from the Hexi Corridor and Gobi Desert. To the east and southeast, the plateau gives way to the forested gorge and ridge geography of the mountainous headwaters of the Salween, Mekong, and Yangtze rivers in western Sichuan (the Hengduan Mountains) and southwest Qinghai. In the west, the curve of the rugged Karakoram range of northern Kashmir embraces it. It is bounded on the north by a broad escarpment where the altitude drops from around 5000 to 1500 m in less than 150 km. Along the escarpment is a range of mountains. In the west, the Kunlun Mountains separate the plateau from the Tarim Basin. About half way across the Tarim, the bounding range becomes the Altyn-Tagh, and the Kunluns, by convention, continue somewhat to the south. In the 'V' formed by this split lies the western part of the Qaidam Basin. To the west are short ranges called the Danghe, Yema, Shule, and Tulai Nanshans. The easternmost range is the Qilian Mountains. The line of mountains continues further east of the plateau as the Qin Mountains which separate the Ordos Region from Sichuan (Li 1987; Zhang et al. 2002).

Five large earthquakes of $M \geq 7.0$ occurred in and near the Tibetan Plateau during 2008–2014 (Fig. 1), including the serious deadly Wenchuan $M8.0$ earthquake. The focal mechanisms of these five earthquakes are listed in Table 1.

3.2. Earthquakes Above M7.0 in and Near the Tibetan Plateau Since 1900

Due to the strong collision between the Indian Plate and Eurasian Plate, a number of active tectonic blocks including Tibetan active block have formed since the Cenozoic era (Zhang et al. 1999), and hence the complex tectonics and high seismicity in this region (e.g., Xu et al. 2005). We select a square region of (21.0°–41.0°N, 74.0°–106.0°E) including the Tibetan Plateau for the study of the predictability of PI method to recent large earthquakes that occurred in this region. According to the earthquake catalogue supplied by China Earthquake Administration (http://10.5.202.22/bianmu), 83 large earthquakes of $M \geq 7.0$ occurred in the selected region during 1900–2014, including 10 tremendous earthquakes with magnitude above $M8.0$ (Fig. 2), and the largest one is Chayu, Tibet (28.4°N, 96.7°E) $M8.6$ earthquake that occurred on August 15, 1950 (BJT).

The year of 1970 is a milestone for Chinese micro-earthquake catalogue. Nationwide earthquake catalogues have been produced by the constantly improved Chinese seismic network since 1970. Hence, those large earthquakes covered by the seismic monitoring network could be employed to study the predictability of PI method. As shown in Fig. 2, there were a total of 83 large earthquakes with $M7.0$ and higher during the

Table 1

Focal mechanisms of the five large earthquakes of $M \geq 7.0$ in the Tibetan Plateau during 2008–2014

No.	Date (BJT)	Latitude/°	Longitude/°	Magnitude/°	Depth/km	Place	Nodal plane A			Nodal plane B		
							Strike/°	Dip/°	Rake/°	Strike/°	Dip/°	Rake/°
1	20080321	35.60	81.60	7.3	12	Yutian	358	41	−110	203	52	−74
2	20080512	30.95	103.40	8.0	14	Wenchuan	231	35	138	357	68	63
3	20100414	33.20	96.60	7.1	14	Yushu	210	67	178	300	88	23
4	20130420	30.30	103.00	7.0	13	Lushan	212	42	100	19	49	81
5	20140212	36.13	82.52	7.3	10	Yutian	332	85	−176	242	86	−5

Date, latitude, longitude, magnitude, and depth are from China Earthquake Networks Center. Strike, dip, and rakes of nodal A and nodal B are from Harvard University

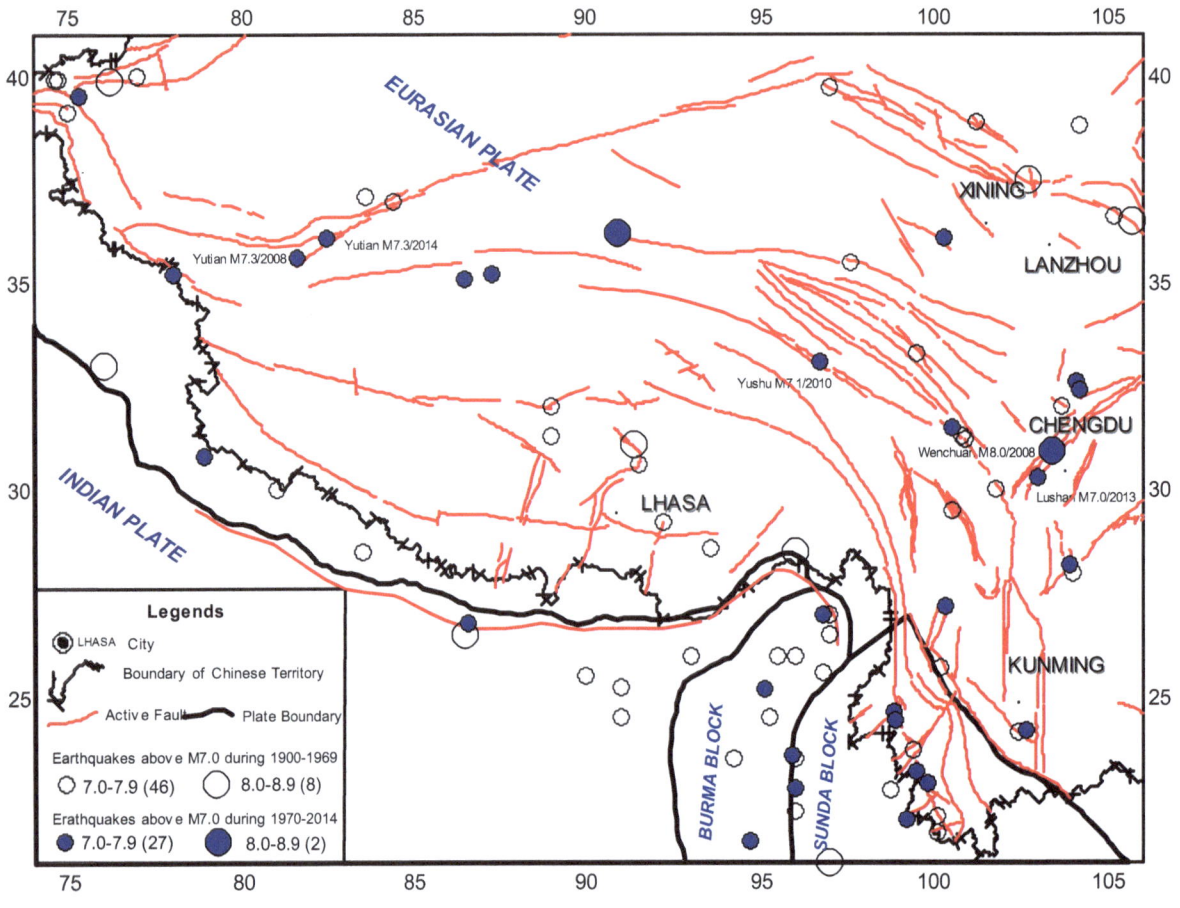

Figure 2
Large earthquakes with $M \geq 7.0$ that occurred in and near the Tibetan Plateau during 1900–2014

period of 1900–2014. 29 out of these 83 earthquakes occurred after 1970 in the selected region, and 21 out of the 29 quakes occurred in the Chinese territory boundary. The algorithm of PI requires a reference time interval from t_b to t_1 (initial time $t_0 = 1970$ in our study, and t_b lies between t_0 and t_1), and a change interval from t_1 to t_2 ($t_2 > t_1$). The forecasting interval is from t_2 to t_3, where t_3 should be the year 2014 or before in our retrospective study. Therefore, the earthquakes that occurred during the interval from t_1 to t_2 could be used for the retrospective study, and those that occurred during the interval from t_2 to t_3 could be used for the test of the predictability of PI forecasting method. For example, if the forecasting interval from t_2 to t_3 is 10 years, then those large earthquakes that occurred during 2005–2014 could be employed for the test. In general, $t_1 - t_0$ should be much longer than $t_2 - t_1$ to retain the higher robustness of the results. According to China Earthquake Networks Center (CENC), there were five $M \geq 7.0$ earthquakes that occurred in the selected region during 2005–2014 as shown in Fig. 2 and listed in Table 2.

3.3. The Monitoring Ability and Completeness of Earthquake Catalogue in and near the Tibetan Plateau

In 1970s and 1980s, earthquakes less than $M_L 4.5$ could not be recorded completely in the western Continental China when evaluated by Gutenburg-Richter law, which was mainly caused by the lower monitoring ability in the Tibetan Plateau at that time (Zhang et al. 2013). With the

Table 2

Forecasting efficacy E_f of PI for seven forecasting windows

Forecasting window	2001–2008 (Fig. 5a)	2002–2009 (Fig. 5b)	2003–2010 (Fig. 5c)	2004–2011 (Fig. 5d)	2005–2012 (Fig. 5e)	2006–2013 (Fig. 5f)	2007–2014 (Fig. 5g)
$E_f - 0.5$	0.43	0.39	0.39	0.46	0.43	0.42	0.38

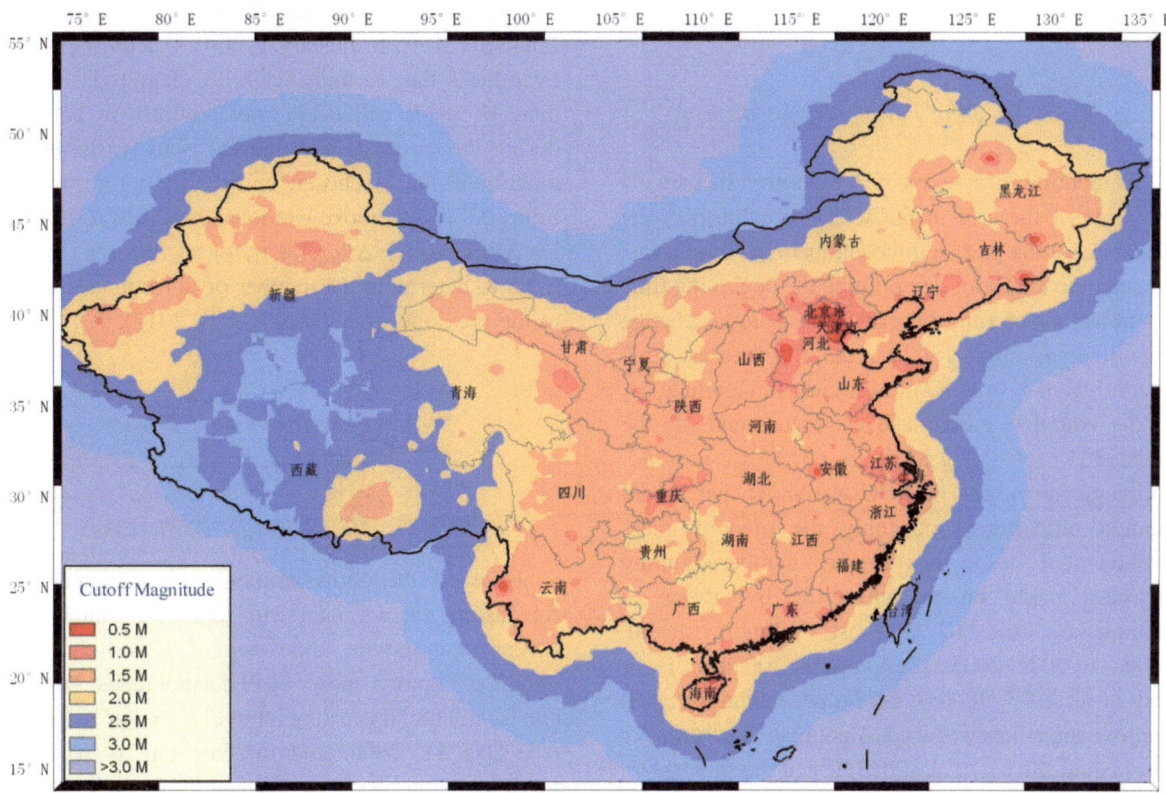

Figure 3

Map of seismic monitoring ability in and near Continental China (Provided by the Department of China Seismic Network, CENC. Contact: huangzhibin@seis.ac.cn)

gradual advancement in China Seismic Network, the lower cutoff of complete earthquake catalogue in this region reached to $M_L 4.0$ since (Zhang et al. 2013). By now, the advanced China Digital Seismic Network is able to record all earthquakes larger than $M_L 3.0$ in and near the Tibetan Plateau[1] (Fig. 3). In our study, we need to employ the earthquake catalogue of the selected region over the entire time duration from 1970 (the initial time

t_0) to 2014, so the lower cutoff magnitude M_c of the complete earthquake catalogue in and near the Tibetan Plateau should be $M_L 4.5$.

Figure 4 shows the distribution map of earthquakes larger than $M_L 4.5$ recorded by China Seismic Network during the period from 1970 to 2014. From this figure, we can see that the recorded earthquakes cover the territory belonging to China and its neighborhood. The complete earthquake catalogue with lower cutoff of $M_L 4.5$ could meet the requirement of the retrospective study to test the predictability of PI method for the recent large

---

[1] Provided by the Department of China Digital Seismic Network, CENC. Contact: huangzhibin@seis.ac.cn.

earthquakes in this region covered by the recorded earthquakes. From Fig. 4, we see that some of the earthquakes outside the Chinese territory boundary could not be covered by the complete earthquake catalogue recorded by Chinese Seismic Network; these earthquakes could not be employed as the target earthquakes for the test in our study due to lack of data or incomplete catalogue.

4. The Computing Parameters for the Test

Computing parameters of PI algorithm include (1) the scale of the grid in the selected region, say the length Δx of the grid, and the number N_B of the grids; (2) the time nodes for determining the beginning time t_b, the reference time interval from t_b to t_1, change time interval from t_1 to t_2, and the forecast time interval from t_2 to t_3; (3) lower cutoff magnitude M_c of the complete earthquake catalogue; and (4) the threshold of $\log \Delta P_i(t_0, t_1, t_2)/\Delta P_{\max}(t_0, t_1, t_2)$. Although the results of ROC test show that the PI method outperforms not only the random guess method but also the simple number-counting approach based on the clustering hypothesis of earthquakes (the RI forecast) (Rundle et al. 2000a, b, 2002, 2003; Tiampo et al. 2002a, b; Holliday et al. 2005, 2006a), the proper parameters will improve the efficacy of the hotspot map in earthquake forecasting. Zhang et al. (2013) tested the forecasting efficacy of the PI method applied to the Wenchuan M8.0 and Yutian M7.3 earthquakes in West Continental China in 2008 by fixing t_b(June 1, 1970) and t_3(June 1, 2008) under different values of t_1, t_2, and Δx, and the results of the ROC test and R score test show that the models with forecasting intervals of 8–10 years under grid size of $\Delta x = 1°$ and those with forecasting intervals of 7–10 years under grid size of $\Delta x = 2°$ are better. In order to test if the these parameters also work well for the succeeding earthquakes during 2009–2014 in western China mainland after Wenchuan M8.0 earthquake, all actually occurring in the Tibetan Plateau, we select the square region of (21.0°–41.0°N, 74.0°–106.0°E) including the Tibetan Plateau for the study. The grid size of $\Delta x = 1.°$ This means the total number of grids is 640. The forecasting interval from t_3–$t_2 = 8$ years. t_b is

fixed to January 1, 1970. In order to keep the target earthquakes during 2008–2014 in the forecasting interval t_3–t_2, t_3 changes from January 1, 2008 to December 31, 2021 with the moving step of 12 months, and t_2 changes from January 1, 2001 to December 31, 2008 with the moving step of 12 months. According to the hypothesis of t_3–$t_2 = t_2$–t_1(Holliday et al. 2005), t_1 changes from January 1, 1993 to January 1, 2001. According to our hypothesis that earthquakes with magnitudes larger than M_c +2.0 will occur preferentially in hotspots during the forecast time interval t_2 to t_3, the cutoff magnitude M_c is chosen to be 5.0 to forecast the future M7.0 and above earthquakes. The threshold of $\log \Delta P_i(t_0, t_1, t_2)/\Delta P_{\max}(t_0, t_1, t_2)$ is chosen to be -0.6 to determine the number of hotspots according to the previous work (Zhang et al. 2013).

5. Results of Retrospective Tests for the Predictability of PI Method

5.1. PI Patterns and the Target Earthquakes

Base on the above computing parameters, the forecasting hotspot maps with forecasting interval of 8 years and moving time step of 12 months were obtained. Figure 5 shows 14 PI hotspot maps. The first one is with the forecasting interval $t_2 = $ Jan 1, 2001 to $t_3 = $ Dec 31, 2008, and the last one is with the forecasting interval from $t_2 = $ Jan 1, 2014 to $t_3 = $ Dec 31, 2021. Earthquakes that occurred during each forecasting interval were also dotted in the maps.

1. Yutian M7.3 and Wenchuan M8.0 earthquakes in 2008.
   These two earthquakes were covered by eight PI forecasting intervals, as shown in Fig. 5a–h. For Yutian M7.3 earthquake, it occurred in the intersection of Altyn-Tagh and the Kunluns, northwestern boundary of the Tibetan Plateau (Fig. 1). Yutian M7.3 earthquake dropped in a hotspot or its Moore neighborhood grids (the eight grids surrounding the hotspot grid are defined as the Moore neighborhood (Moore 1962)) in each succeeding forecasting interval (Fig. 5a–h). This means that Yutian M7.3 could be forecasted by PI method. For Whenchuan M8.0 earthquake, it

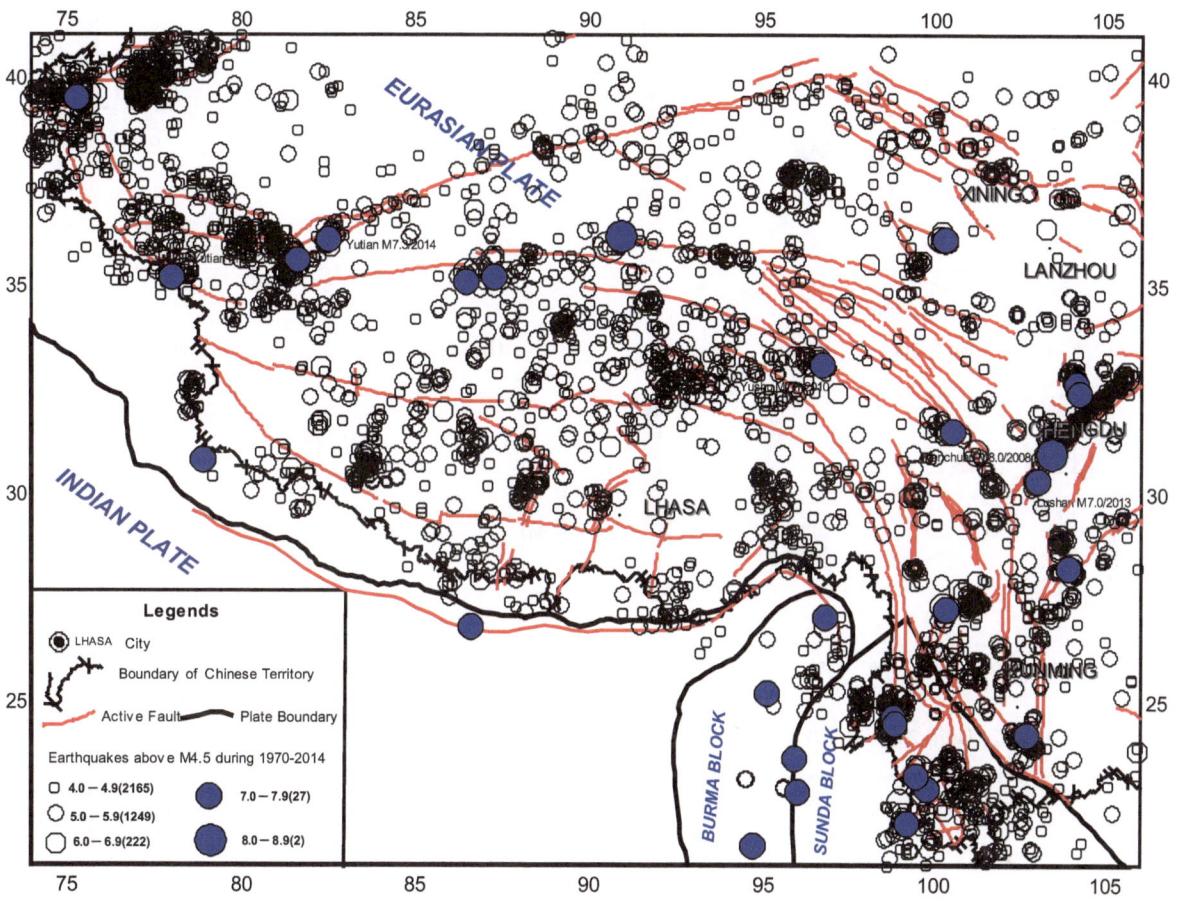

Figure 4

Earthquake distribution map in and near the Tibetan Plateau recorded by China Seismic Network. (The earthquake catalogue is from the CENC, with all recorded earthquakes larger than $M_L4.5$ during the period of 1970–2014.)

occurred in the Longmenshan Faults, at the eastern boundary of the Tibetan Plateau (Fig. 1). Hotspot only existed in the epicenter grid during the forecasting interval from January 1, 2008 to December, 2015 (Fig. 5h). That is to say that Wenchuan M8.0 could be forecasted in only one forecasting interval among the total of eight such ones.

2. Yushu M7.1 earthquake in 2010.

Yushu M7.1 earthquake occurred in the central area of the Tibetan Plateau (Fig. 1). This earthquake was covered by eight forecasting intervals, as shown in Fig. 5c–j. Hotspots existed in the epicenter grid in three forecasting intervals as shown in Fig. 5d–f, and existed in the Moore neighborhood grids in the forecasting interval as

shown in Fig. 5j. Therefore, this earthquake could also be forecasted by the PI method.

3. Lushan M7.0 earthquake in 2013.

Lushan M7.0 earthquake also occurred in the Longmenshan Faults, eastern boundary of the Tibetan Plateau (Fig. 1). It is located about 80 km southwest from the Wenchuan M8.0 earthquake. This earthquake was covered by eight forecasting intervals, as shown in Fig. 5f–m. Hotspots existed in the epicenter grid or the Moore neighborhood grids in four forecasting intervals as shown in Fig. 5h, j, k, m. Therefore, this earthquake could also be forecasted by the PI method.

4. Yutian M7.3 earthquake in 2014.

This earthquake occurred about 100 km northeast from the Yutian M7.3 earthquake in 2008 (Fig. 1).

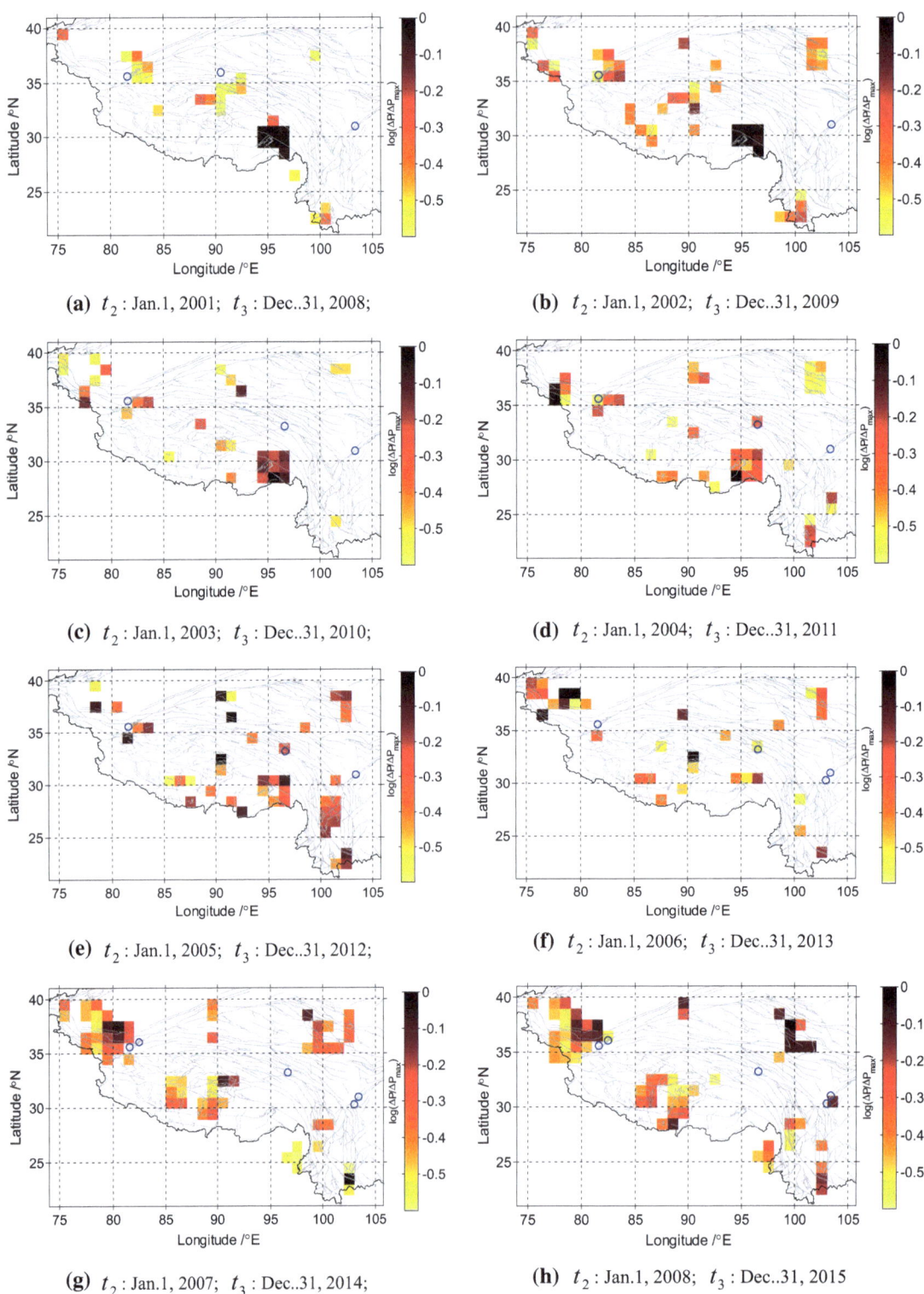

(a) t_2 : Jan.1, 2001; t_3 : Dec..31, 2008;

(b) t_2 : Jan.1, 2002; t_3 : Dec..31, 2009

(c) t_2 : Jan.1, 2003; t_3 : Dec..31, 2010;

(d) t_2 : Jan.1, 2004; t_3 : Dec..31, 2011

(e) t_2 : Jan.1, 2005; t_3 : Dec..31, 2012;

(f) t_2 : Jan.1, 2006; t_3 : Dec..31, 2013

(g) t_2 : Jan.1, 2007; t_3 : Dec..31, 2014;

(h) t_2 : Jan.1, 2008; t_3 : Dec..31, 2015

Figure 5

Hotspot maps of each forecasting interval with the threshold possibility of $\log_{10}(\Delta P_i(t_0,t_1,t_2)/\Delta P_{\max}(t_0,t_1,t_2)) = -0.6$. The *blue circles* denote the positions of target earthquakes that occurred during the forecasting intervals

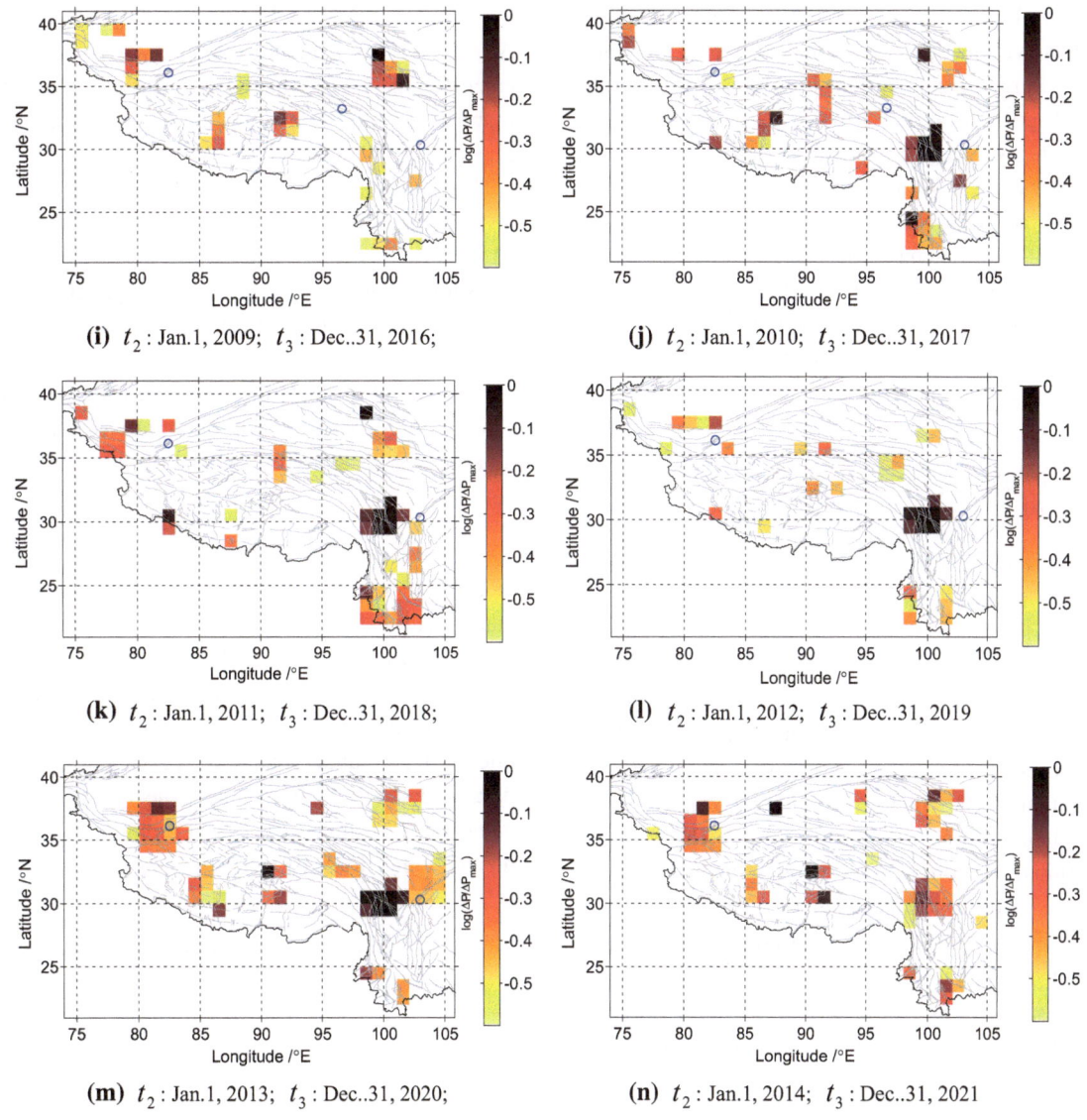

(i) t_2: Jan.1, 2009; t_3: Dec..31, 2016;

(j) t_2: Jan.1, 2010; t_3: Dec..31, 2017

(k) t_2: Jan.1, 2011; t_3: Dec..31, 2018;

(l) t_2: Jan.1, 2012; t_3: Dec..31, 2019

(m) t_2: Jan.1, 2013; t_3: Dec..31, 2020;

(n) t_2: Jan.1, 2014; t_3: Dec..31, 2021

Figure 5
continued

This earthquake was covered by eight forecasting intervals, as shown in Fig. 5g–n. Hotspots existed in the epicenter grid or the Moore neighborhood grids in seven forecasting intervals as shown in Fig. 5g, h, j–n. Therefore, this earthquake could also be well forecasted by the PI method.

From Fig. 5, we see that, following Zhang's tested computing parameters (Zhang et al. 2013), all of these five large earthquakes could be forecasted by the PI method in different degrees. The Yutian $M7.3$ earthquake could be forecasted very well with the consistent appearance of the hotspot in the epicentral grid or the Moore Neighbor grids in the succeeding time forecasting intervals. However, the Wenchuan $M8.0$ earthquake could be forecasted in only one forecasting interval.

From Fig. 5, we also see that there are many hotspots without large earthquakes; theses are false alarms. Further some earthquakes did not occur in the hotspot grid or in the Moore neighborhood grids. Therefore, we need to evaluate the efficacy of the PI method by some kind of quantitative test. Here, we employ the Receiver Operating Characteristic-ROC (Swets 1973; Molchan 1997) test as Rundle et al. (2000c, 2003), Tiampo et al. (2002c) and Holliday et al. (2005) did. We also employ the R score evaluation method proposed by Xu et al. (1989) and Shi et al. (2000). The evaluation results are as follows.

5.2. Retrospective Evaluation for the Predictability of PI Method by ROC Test

ROC test is conducted by systematically changing the 'alarm threshold' of the 'forecast region' and counting the 'hit rate' and 'false alarm rate' compared with real earthquake activity. The 'alarm threshold' here is the cutoff value of $\log_{10}(\Delta P_i(t_0, t_1, t_2)/\Delta P_{max}(t_0, t_1, t_2))$. Following Rundle et al. (2000c, 2003), Tiampo et al. (2002) and Holliday et al. (2005), we define that: during the forecast interval from t_2 to t_3, if an earthquake larger than $M7.0$ occurs in a hotspot grid or within the Moore neighborhood of the grid, this is a success. If no earthquake occurs in a non-hotspot grid, this is also a success; If no earthquake occurs in a hotspot grid or within the Moore neighborhood of the hotspot grid, this is a false alarm; If earthquake occurs in a grid, which is not hotspot grid or the Moore neighborhood of the hotspot grid, this is a failure to forecast.

According to the above definitions, we can obtain the values a (Forecast = yes, Observed = yes), b (Forecast = yes, Observed = no), c (Forecast = no, Observed = yes), and d (Forecast = no, Observed = no), respectively, for the hotspot maps. The fraction of colored grids, also called the probability of forecast of occurrence, is $r = (a + b)/N$, where the total number of grids is $N = a + b + c + d$. The hit rate is $H = a/(a + c)$ and is the fraction of large earthquakes that occur on a hotspot. The false alarm rate is $F = b/(b + d)$ and is the fraction of non-observed earthquakes that are incorrectly forecast.

For each forecasting interval, we obtain the PI hotspot maps under different thresholds of

$\log_{10}(\Delta P_i(t_0, t_1, t_2)/\Delta P_{max}(t_0, t_1, t_2))$; here $\Delta P_i(t_0, t_1, t_2)$ is from 0 to $\Delta P_{max}(t_0, t_1, t_2)$. Then we calculate the hit rate and false rate of each hotspot map under different thresholds according to the above-mentioned method.

For example, Fig. 6 shows the diagrams of ROC test for forecasting interval from Jan 1, 2004 to Dec 31, 2011. This figure shows that no matter how the threshold changes, Hit rate is always greater than False alarm rate. That means PI method outperforms the random forecast definitely.

In order to evaluate the degree of forecast efficacy of each forecasting interval quantitatively, we define a parameter E_f as follows (Zhang et al. 2013):

$$E_f = \sum_i H_i \cdot \Delta F_i \qquad (1)$$

where H_i denotes the hit rate associated with the false alarm rate F_i. The essence of E_f is the area surrounded by the curve of $H(F)$, the line of $H = 0$, and the line of $F = F_{max}$. Here F_{max} is the false alarm rate under the threshold of $\Delta P_i(t_0, t_1, t_2) = 0$. If the hit rate is bigger under the same False alarm rate, the parameter E_f is bigger; hence, the forecast efficacy could be determined by E_f. The bigger value of E_f means higher forecast efficacy.

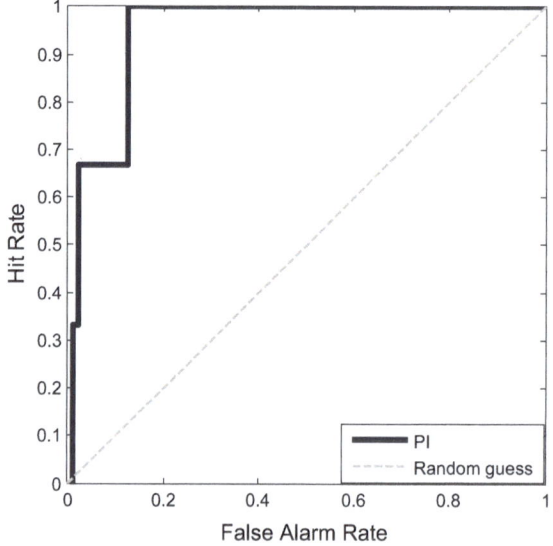

Figure 6
ROC test for forecasting interval from Jan 1, 2004 to Dec 31, 2011. Hit rates and false rates are calculated by changing the threshold possibility $\Delta P_i(t_0, t_1, t_2)$ from 0 to $\Delta P_{max}(t_0, t_1, t_2)$

Table 3

R scores of PI forecasting map for seven forecasting windows

Forecasting Window	2001–2008 (Fig. 5a)	2002–2009 (Fig. 5b)	2003–2010 (Fig. 5c)	2004–2011 (Fig. 5d)	2005–2012 (Fig. 5e)	2006–2013 (Fig. 5f)	2007–2014 (Fig. 5g)
R score	0.24	0.36	0.24	0.54	0.53	0.39	0.21

We only test the forecasting maps of Fig. 5a, b, c, d, e, f, g because the ending time of the forecasting intervals shown in Fig. 5i, j, k, l, m has yet to reach completion. Although the ending time of the forecasting interval in Fig. 5h has reached completion, we did not perform the test because the Nepal $M8.1$ earthquake (28.3°N, 84.7°E) that occurred on April 25, 2015 was very near the Chinese territory boundary, and we need to do further study to verify if this quake could be taken as a target earthquake.

According to the definition of E_f, when $E_f - 0.5$ is positive, the forecasting of PI method will outperform the random guess method. The values of the $E_f - 0.5$ corresponding to the seven forecasting intervals shown in Fig. 5a–g are listed in Table 2, respectively. All these values are positive, which means PI forecasting efficacy under the selected parameters is stable during different forecasting intervals.

5.3. Retrospective Evaluation for the Predictability of PI Method by R Score Test

Xu et al. (1989) developed an evaluation method for the efficacy of earthquake forecast method in China, the so-called R score. Shi et al. (2000) developed the algorithm of R score for testing the annual potential seismic hazard regions forecasted by China Earthquake Administration (CEA). Compared with the ROC test, "the Moore neighborhood" is not considered in the R score test, that is to say that, during t_2–t_3, if an earthquake lager than $M7.0$ occurs in a hotspot grid, this is considered a success; If no earthquake occurs in a non-hotspot grid, this is also considered a success; If no earthquake occurs in a hotspot grid, this is considered a false alarm; If earthquake occurs in a grid, which is not hotspot grid,

this is considered a failure to forecast. So R score test is more rigorous than ROC test.

According to the above definitions, values a (Forecast = yes, Observed = yes), b (Forecast = yes, Observed = no), c (Forecast = no, Observed = yes), and d (Forecast = no, Observed = no) are obtained for the hotspot map. The hit rate is $H = a/(a + c)$. The false alarm rate is $F = b/(b + d)$. R score = $H - F$.

The R scores corresponding to the seven forecasting intervals (Fig. 5a–g) are listed in Table 3, respectively. The average value of R scores in the seven forecasting intervals is about 0.36. Positive R score means the forecasting efficacy outperforms the random guess method (Shi et al. 2000). The results in Table 3 show that the PI method outperforms the random guess method.

6. Discussions

6.1. Are the Computing Parameters Reasonable?

The computing parameters in this study were selected from the previous work on Wenchuan $M8.0$ and Yutian $M7.3$ earthquakes (Zhang et al. 2013). Based on the systematic verification of different models with different forecasting intervals and different grid scales by ROC and R score tests, Zhang et al. (2013) regarded that the forecasting intervals of 8–10 years under grid size $\Delta x = 1°$ and those with forecasting intervals of 7–10 years under grid size $\Delta x = 2°$ are better. Our retrospective study verifies that these parameters also work very well for the succeeding Yushu $M7.1$, 2010; Lushan $M7.0$, 2013, and Yutian $M7.3$, 2014 earthquakes. So the computing parameters in this study are reasonable for all these five large earthquakes. However, the relationship between the grid size of PI and the critical seismogenic size (e.g., Bufe and Varnes 1993) or the rupture scale (e.g., Kanamori and Anderson 1975) is an issue that needs to be studied in the future. More earthquake cases need to be studied to achieve a solid conclusion.

6.2. What else will Affect the Hotspot Pattern?

In addition to the computing parameters we considered in this study, the selected range for study might be a factor to affect the hotspot pattern. The Tibetan Plateau is included in the western China Mainland. We compared the hotspot patterns under the same computing parameters in Tibetan Plateau region and Western China Mainland region (Zhang et al. 2013), and found that the patterns are different for Wenchuan M8.0 earthquake. This is caused by the algorithm of PI method that all grids are considered for calculation. Further study needs to be carried out to include the effects by the range of the selected region.

6.3. Need we Use the Original Earthquake Catalogue or De-clustered Catalogue for Forecasting Map?

Aftershocks occupy a significant proportion of the total number of earthquakes, especially during the period with large earthquake sequences. Jiang and Wu (2011) de-clustered the catalogue of the Sichuan-Yunnan region of southwest China by the epidemic-type aftershock sequences (ETAS) model and investigated the effects of de-clustering on the PI forecasts. Their results showed that PI forecasts seem to be affected by the aftershock sequence included in the "anomaly identifying interval," and the PI forecast using "background events" seems to have a better performance. Further study regarding the influence of the aftershock sequences on the PI forecasting map needs to be conducted in the future.

7. Conclusions and Prospects

In this paper, the Tibetan Plateau (included in the region of 21°–41°N, 74°–106°E) was taken as the study region to verify the predictability of the PI

method by the receiver-operating characteristic (ROC) test and R Score test. The results show that

1. Successive obvious hotspots occurred in the sliding forecasting intervals before four of the five earthquakes, while hotspots only occurred in one forecasted interval without successive evolution process before one of the five earthquake, which indicates four of the five large earthquakes could be forecasted well by PI method;
2. Test results of the Predictability of PI method by ROC and R score show that positive prospect of PI method could be expected for long-term earthquake forecast.

The effects of changeable selected regions, grid sizes, and de-clustering of catalogue were not considered in our current study, and they will be examined in the future. The future study is suggested to work out the proper parameters to keep the PI pattern stable and make the forecasting map the most optical.

Acknowledgements

The authors gratefully acknowledge the support from the Chinese Ministry of Science and Technology under Grants No. 2010DFB20190 and No. 2012BAK19B02-05. The authors also thank Prof. J.B. Rundle for certain valuable comments on PI methods and the anonymous reviewers for their constructive comments on the paper. The authors also offer their thanks to CENC (China Earthquake Networks Center) for the earthquake catalogue.

REFERENCES

Bufe, C. G., & Varnes, D. J. (1993). Predictive modeling of the seismic cycle of the greater San Francisco Bay region. *Journal of Geophysical Research, 98*, 9871–9883.

Chang, L., Chen, C., & Wu, Y. (2013). A study on the pattern informatics and its application to earthquake prediction in Taiwan[C]//AGU Fall. *Meeting Abstracts, 1*, 2341.

Chen, C. C., Rundle, J. B., Holliday, J. R., Nanjo, K. Z., Turcotte, D. L., Li, S. C., et al. (2005). The 1999 Chi-chi, Taiwan, earthquake as a typical example of seismic activation and quiescence. *Geophysical Research Letters, 32*, L22315.

Chen, C. C., Rundle, J. B., Li, H. C., et al. (2006). From tornadoes to earthquakes: forecast verification for binary events applied to the 1999 Chi-Chi, Taiwan, earthquake. *Terrestrial Atmospheric and Oceanic Sciences, 17*(3), 503–516.

Cho, N. F., & TIAMPO, K. F. (2013). Effects of location errors in pattern informatics[J]. *Pure Applied Geophysics, 170*(1–2), 185–196.

Holliday, J. R., Nanjo, K. Z., Tiampo, K. F., Rundle, J. B., & Turcotte, D. L. (2005). Earthquake forecasting and its verification. *Nonlinear Processes in Geophysics, 12,* 965–977.

Holliday, J. R., Rundle, J. B., Tiampo, K. F., Klein, W., & Donnellan, A. (2006a). Modification of the pattern informatics method for forecasting large earthquake events using complex eigenfactors. *Tectonophysics, 413,* 87–91.

Holliday, J. R., Rundle, J. B., Tiampo, K. F., Klein, W., & Donnellan, A. (2006b). Systematic procedural and sensitivity analysis of the pattern informatics method for forecasting large ($M \geq 5$) earthquake events in southern California. *Pure Applied Geophysics, 163,* 2433–2454.

Jaume, S. C., & Sykes, L. R. (1999). Evolving towards a critical point: a review of accelerating seismic moment/energy release prior to large and great earthquakes. *Pure and Applied Geophysics, 155,* 279–306.

Jiang, C. S., & Wu, Z. L. (2008). Retrospective forecasting test of a statistical physics model for earthquakes in Sichuan-Yunnan region. *Science in China, Series D: Earth Sciences, 51*(10), 1401–1410. doi:10.1007/s11430-008-0112-6.

Jiang, C. S. & Wu, Z. L.(2011). PI forecast with or without declustering: an experiment for the Sichuan-Yunnan region. *Natural Hazards* and *Earth System* Sciences, 11, 697–706. doi:10.5194/nhess-11-697-2011. www.nat-hazards-earth-syst-sci.net/11/697/2011/

Kanamori, H., & Anderson, D. L. (1975). Theoretical basis of some empirical relations in seismology. *Bulletin of the Seismological Society of America, 65*(5), 1073–1095.

Kawamura, M., Wu, Y. H., Kudo, T., & Chen, C. C. (2013). Precursory migration of anomalous seismic activity revealed by the pattern informatics method: a case study of the 2011 Tohoku earthquake, Japan. *Bulletin of the Seismological Society of America, 103*(2B), 1171–1180.

Kawamura, M., Wu, Y. H., Kudo, T., & Chen, C. C. (2014). A statistical feature of anomalous seismic activity prior to large shallow earthquakes in Japan revealed by the pattern informatics method. *Natural Hazards and Earth System Sciences, 14*(4), 849.

Kossobokov, V. G., Romashkova, L. L., & Keilis-Borok, V. I. (1999). Testing earthquake prediction algorithms: statistically significant advance prediction of the largest earthquakes in the Circum-Pacific, 1992–1997. *Physics of the Earth and Planetary Interiors, 111,* 187–196.

Li, B. Y. (1987). On the Extent of the Qinghai-Xizang (Tibet) Plateau. *Geographical Research (in Chinese with English abstract), 6*(3), 57–64.

Migan, A., & Tiampo, K. F. (2010). Testing the pattern informatics index on synthetic seismicity catalogues based on the non-critical PAST. *Tectonophysics, 483,* 255–268. doi:10.1016/j.tecto.2009.10.023.

Molchang, M. (1997). Earthquake prediction as a decision-making problem. *Pure and Applied Geophysics, 149,* 233–247.

Moore, E. F. (1962). Machine models of self reproduction. In *Proceedings of the fourteenth symposium on applied mathematics* (pp. 17–33). American Mathematical Society.

Nanjo, K. Z., Holliday, J. R., Chen, C. C., Rundle, J. B., & Turcotte, D. L. (2006a). Application of a modified pattern informatics method to forecasting the locations of future large earthquakes in the central Japan. *Tectonophysics, 424,* 351–366.

Nanjo, K. Z., Rundle, J. B., Holliday, J. R., et al. (2006b). Pattern informatics and its application for optimal forecasting of large earthquakes in Japan. *Pure and Applied Geophysics, 163*(11–12), 2417–2432.

Papazachos, C. B., Karakaisis, G. F., Scordilis, E. M., & Papazachos, B. C. (2005). Global observational properties of the critical earthquake model. *Bulletin of the Seismological Society of America, 95,* 1841–1855.

Rundle, J. B., Klein, W., Gross, S. J., & Tiampo, K. F. (2000a). Dynamics of seismicity patterns in systems of earthquake faults. In J. B. Rundle, D. L. Turcotte, & W. Klein (Eds.) *Geo-complexity and the Physics of Earthquakes*, vol. 120 of Geophysics *Monograph Series* (pp. 127–146). Washington, D. C: AGU.

Rundle, J. B., Klein, W., Tiampo, K. F., & Gross, S. J. (2000b). Linear pattern dynamics in nonlinear threshold systems. *Physical Review E, 61,* 2418–2432.

Rundle, J. B., Klein, W., Turcotte, D. L., et al. (2000c). Precursory seismic activation and critical-point phenomena. *Pure and Applied Geophysics, 157,* 2165–2182.

Rundle, J. B., Tiampo, K. F., Klein, W., & Martins, J. S. S. (2002) Self-organization in leaky threshold systems: The influence of near-mean field dynamics and its implications for earthquakes, neurobiology, and forecasting. *Proceedings* of the *National Academy* of Sciences of the *United States* of America, 99(Suppl. 1), 2514–2521.

Rundle, J. B., Turcotte, D. L., Shcherbakov, R., Klein, W., & Sammis, C. (2003). Statistical physics approach to understanding the multiscale dynamics of earthquake fault systems. *Reviews of Geophysics, 41,* 1019–1038.

Shi, Y. L., Liu, J., & Zhang, G. M. (2000). The evaluation of Chinese annual earthquake prediction in the 90s. *J Graduate School Academia Sin (in Chinese with English abstract), 17,* 63–69.

Swets, J. A. (1973). The relative operating characteristic in psychology. *Science, 182,* 990–1000.

Tiampo, K. F., Bowman, D. D., Colella, H. & Rundle, J. B. (2008). The stress accumulation method and the pattern informatics index: Complementary approaches to earthquake forecasting. *Pure Applied. Geophysics,* 165, 693–709, 0033–4553/08/030693–17. doi:10.1007/s00024-008-0329-5.

Tiampo, K. F., Klein, W., Li, H.-C., Migan, A., Toya, Y., Kohen-Kadosh, S. L. Z., et al. (2010). Ergodicity and earthquake catalogs: forecast testing and resulting implications. *Pure and Applied Geophysics,* 167, 763–782. doi:10.1007/s00024-010-0076-2.

Tiampo, K. F., Rundle, J. B., McGinnis, S., Gross, S. J., & Klein, W. (2002a). Eigenpatterns in southern California seismicity. *Journal of Geophysical Research,* 107, 2354.

Tiampo, K. F., Rundle, J. B., McGinnis, S., & Klein, W. (2002b). Pattern dynamics and forecast methods in seismically active regions. *Pure Applied Geophysics,* 159, 2429–2467.

Tiampo, K. F., Rundle, J. B., McGinniss, S., et al. (2002c). Mean-field threshold systems and phase dynamics: an application to earthquake fault systems. *Europhysics Letters,* 60, 481–487.

Tiampo, K. F., & Shcherbakov, R. (2012). Seismicity-based earthquake forecasting techniques: ten years of progress. *Tectonophysics,* 522–523, 89–121. doi:10.1016/j.tecto.2011.08.019.

Tiampo, K. F., & Shcherbakov, R. (2013). Optimization of seismicity-based forecasts. *Pure and Applied Geophysics, 170,* 139–154. doi:10.1007/s00024-012-0457-9.

Wu, Y. H., Chen, C. C. & Rundle, J. B. (2008a) Detecting precursory earthquake migration patterns using the pattern informatics method. *Geophysical Research Letters* 35 (19) (Art. No. L19304).

Wu, Y. H., Chen, C. C., & Rundle, J. B. (2008b). Precursory Seismic Activation of the Pingtung (Taiwan) Offshore Doublet Earthquakes on 26 December 2006: A Pattern Informatics Analysis. *Terrestrial Atmospheric And Oceanic Sciences*, 19(6): 743–749.

Xia, C. Y., Zhang, Y. X., Zhang, X. T., & Wu, Y. J. (2014). Test of the predictability of PI method by two Ms7.3 earthquakes in Yutian County, Xinjiang Uygur Autonomous Region. *Journal of Seismologica Sinica (In Chinese with English abstract)*, 37(1), 192–201.

Xu, S. X. (1989). Mark evaluation for earthquake prediction efficacy. In: Department of Science and Technology Monitoring, State Seismological Bureau (Ed.) *Collected Papers of Research on Practical Methods of Earthquake Prediction (Volume of Seismology) (in Chinese)* (pp. 586–590). Beijing: Academic Books and Periodical Press.

Xu, X. W., Zhang, P. Z., Wen, X. Z., Qin, Z. L., Chen, G. H., & Zhu, A. L. (2005). Features of active tectonics and recurrence behaviours of strong earthquakes in the western Sichuan Province and its adjacent regions[J]. *Seismology and Geology, 27*(3), 446–461. **(in Chinese with English abstract)**.

Zhang, P. Z. (1999). Late quaternary tectonic deformation and earthquake hazard in continental China[J]. *Quaternary Sciences*, 19(5), 404–413 **(in Chinese with English abstract)**.

Zhang, Y. L., Li, B. Y., & Zheng, D. (2002). A discussion on the boundary and area of the Tibetan Plateau in China. *Geographical Research (in Chinese with English abstract), 21*(1), 1–8.

Zhang, Y. X., Zhang, X. T., Wu, Y. J., & Yin, X. C. (2013). Retrospective study on the predictability of pattern informatics to the Wenchuan M8.0 and Yutian M7.3 earthquakes. *Pure and Applied Geophysics, 170*((1-2)), 197–208. doi:10.1007/s00024-011-0444-6. **(Published online 2012)**.

Zhang, X. T., Zhang, Y. X, Xia, C. Y., Wu, Y. J., & Yu H. Z. (2014). A study on the PI anomaly before the Lushan M7.0 earthquake in the Sichuan-Yunnan region and neighboring areas. *Journal of Seismologica Sinica (In Chinese with English abstract)*, 36(5), 780–789.

Zhang, Y. X., Zhang, X. T., Yin, X. C. & Wu, Y. J. (2009). *Study on the forecast effects of PI method to the North and Southwest China, Currency and Computation: Practice and Experience*. New York: Wiley. doi:10.1002/cpe.1515 (http://www.interscience.wiley.com) **(Published online)**.

(Received April 22, 2016, revised April 6, 2017, accepted April 8, 2017, Published online May 8, 2017)

Pure Appl. Geophys. 174 (2017), 2427–2442
© 2016 The Author(s)
This article is published with open access at Springerlink.com
DOI 10.1007/s00024-016-1415-8

Long-Term Seismic Quiescences and Great Earthquakes in and Around the Japan Subduction Zone Between 1975 and 2012

Kei Katsumata[1]

Abstract—An earthquake catalog created by the International Seismological Center (ISC) was analyzed, including 3898 earthquakes located in and around Japan between January 1964 and June 2012 shallower than 60 km with the body wave magnitude of 5.0 or larger. Clustered events such as earthquake swarms and aftershocks were removed from the ISC catalog by using a stochastic declustering method based on Epidemic-Type Aftershock Sequence (ETAS) model. A detailed analysis of the earthquake catalog using a simple scanning technique (ZMAP) shows that the long-term seismic quiescences lasting more than 9 years were recognized ten times along the subduction zone in and around Japan. The three seismic quiescences among them were followed by three great earthquakes: the 1994 Hokkaido-toho-oki earthquake (M_w 8.3), the 2003 Tokachi-oki earthquake (M_w 8.3), and the 2011 Tohoku earthquake (M_w 9.0). The remaining seven seismic quiescences were followed by no earthquake with the seismic moment $M_0 \geq 3.0 \times 10^{21}$ Nm (M_w 8.25), which are candidates of the false alarm. The 2006 Kurile Islands earthquake (M_w 8.3) was not preceded by the significant seismic quiescence, which is a case of the surprise occurrence. As a result, when limited to earthquakes with the seismic moment of $M_0 \geq 3.0 \times 10^{21}$ Nm, four earthquakes occurred between 1976 and 2012 in and around Japan, and three of them were preceded by the long-term seismic quiescence lasting more than 9 years.

Key words: Seismic quiescence, ZMAP, Seismicity, Earthquake prediction.

1. Introduction

No one knows whether a precursory seismicity exists several years or dozen of years before great earthquakes. According to the seismic quiescence hypothesis, the occurrence rate of small earthquakes starts to decrease several years or dozen of years before a great earthquake in and around the focal area

ruptured by a subsequent main shock (Inouye 1965; Utsu 1968; Mogi 1969; Ohtake et al. 1977). Along the Kurile, the Japan and the Ryukyu Trenches, which are one of the active subduction zones in the world, great earthquakes have occurred repeatedly. In this region, four earthquakes shallower than 60 km are listed on the Global CMT catalog (Dziewonski et al. 1981; Ekström et al. 2012) with the seismic moment larger than $M_0 = 3.0 \times 10^{21}$ Nm ($M_w = 8.25$); the 1994 Hokkaido-toho-oki earthquake ($M_0 = 3.05 \times 10^{21}$ Nm, $M_w = 8.3$), the 2003 Tokachi-oki earthquake ($M_0 = 3.05 \times 10^{21}$ Nm, $M_w = 8.3$), the 2006 Kurile Islands earthquake ($M_0 = 3.51 \times 10^{21}$ Nm, $M_w = 8.3$), and the 2011 Tohoku earthquake ($M_0 = 53.1 \times 10^{21}$ Nm, $M_w = 9.1$). Previous studies reported the seismic quiescence prior to these recent great earthquakes. Takanami et al. (1996) pointed out that the seismic quiescence started 3 years before the 1994 Hokkaido-toho-oki earthquake. Katsumata and Kasahara (1999) found that the 1994 event followed the seismic quiescence starting 5–6 years before the main shock. Takahashi and Kasahara (2004) recognized that the seismicity rate decreased in the early 1990s in and around the source volume of the 2003 Tokachi-oki earthquake. Katsumata (2011a) found that the seismic quiescence started 5 years before the 2003 main shock within the Pacific slab subducting beneath the continental plate and suggested that the seismic quiescence was caused by the local stress drop due to a precursory quasi-static pre-slip on the plate boundary. Wyss and Habermann (1979) expressed an intermediate-term earthquake prediction based on the seismic quiescence identified: the location is 45.5°–49.2°N and 153°–155°E, the length of the seismic fault is from 200 to 400 km, that is, the magnitude is larger

[1] Institute of Seismology and Volcanology, Hokkaido University, North-10 West-8, Sapporo 060-0810, Japan. E-mail: kkatsu@mail.sci.hokudai.ac.jp

than 8.0, and the occurrence time is from 1979 to 1994. The predicted location was labeled as L1 and L2 in Fig. 7 of Wyss and Habermann (1979). In this area the 2006 Kurile Islands earthquake occurred on the plate boundary. Consequently the prediction expressed by Wyss and Habermann (1979) was correct in terms of the location and the magnitude; the occurrence time was, however, not correct. Katsumata (2011b) found that a long-term seismic quiescence started 23 years before the 2011 Tohoku earthquake.

Ogata (1992) investigated the seismicity by using the Epidemic Type Aftershock-Sequences (ETAS) model and found that all great earthquakes ($M \approx 8$) in and around Japan, which occurred between the 1923 Kanto earthquake and the 1968 Tokachi-oki earthquake, were preceded by the statistically significant seismic quiescences. Moreover, he found no significant seismic quiescence between 1968 and 1990 in and around Japan. This is the only previous study that the seismic quiescences associated with great earthquakes were searched systematically in and around Japan. In the time period after 1990, there is no study on the systematic search of the seismic quiescence in and around Japan. Therefore, the purpose of this study is to investigate systematically how often the long-term seismic quiescence is observed between 1976 and 2012 along the Kurile, the Japan, and the Ryukyu Trenches and how many the seismic quiescence is followed by a subsequent great earthquake. The word of "long-term" means the time period of ~ 10 years or longer in this study.

2. Data

The study area consists of three areas along trenches: the Kurile, the Japan, and the Ryukyu Trenches (Fig. 1; Table 1). The ISC earthquake catalog (International Seismological Centre 2013) was analyzed between 1 January 1964 and 30 June 2012 for the Kurile Trench area and the Ryukyu Trench area and between 1 January 1964 and 28 February 2011 for the Japan Trench area. Since the Tohoku earthquake (M_w 9.1) occurred on 11 March 2011 in the Japan Trench area and it was followed by many aftershocks, I did not use

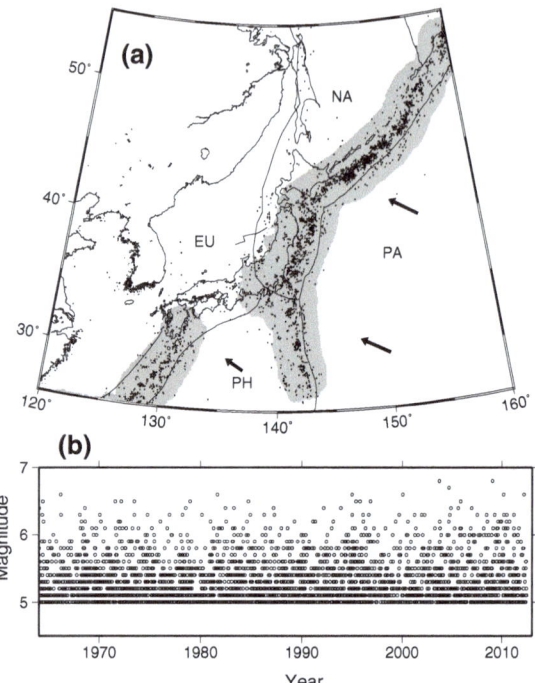

Figure 1
a Map of study area in and around the Japan subduction zone and epicentral distribution of earthquakes used for calculating Z values in this study (1 January 1964–30 June 2012, $m_b \geq 5.0$, $0 \leq$ depth (km) ≤ 60), which is the ISC earthquake catalog after a declustering process is applied. *Hatched areas* indicate the nodes with the resolution circle of $r_{max} \leq 200$ km and the Z values were calculated. *Arrows* indicate the direction of plate motion relative to the Eurasian plate (DeMets et al. 1994). EU, PA, PH, and NA indicate the Eurasian, the Pacific, the Philippine Sea, and the North American plates, respectively. **b** A magnitude–time plot for all earthquakes shown in **a**

earthquakes in the Japan Trench area after the 2011 Tohoku earthquake. When analyzing temporal change in seismicity, it is important to select an adequate kind of the magnitude of earthquake and its range. The surface wave magnitude and the moment magnitude were not used in this study. Since the body wave magnitude m_b has been determined and reported by ISC between 1964 and 2012 continuously, I used the earthquake list with the body wave magnitude m_b in this study. Wiemer and Wyss (2000) developed a method to estimate the magnitude completeness, M_c, by plotting the cumulative number of earthquakes versus magnitude. As a result of applying the method to the ISC catalog, M_c is approximately 5.0 for all of the three areas from 1964 to 1970

Table 1

Parameters for ZMAP in this study

Trench	Kurile	Japan	Ryukyu
Area	140°–160°E, 39°–55°N	135°–146°E, 26°–43°N	120°–140°E, 26°–40°N
T_{start}	1 January 1964	1 January 1964	1 January 1964
T_{end}	30 June 2012	28 February 2011	30 June 2012
m_b	≥ 5.0	≥ 5.0	≥ 5.0
Depth (km)	≤ 60.0	≤ 60.0	≤ 60.0
N_{eq}	1641	1207	1050
Grid size	$0.1° \times 0.1°$	$0.1° \times 0.1°$	$0.1° \times 0.1°$
N_{grid}	10,510	8519	8517
N_{zmap}	40	40	40
r_{max} (km)	≤ 200	≤ 200	≤ 200
Δt (year)	0.1	0.1	0.1
ΔT (year)	9.0	9.0	9.0
t_s (year)	$1975.0 \leq t_s \leq 2003.5$	$1975.0 \leq t_s \leq 2002.1$	$1975.0 \leq t_s \leq 2003.5$
Z values	4,138,970	3,244,596	1,910,112
$Z \geq +6.0$	3771	308	229

Table 2

List of parameters of the ETAS Model for the ISC catalogue

Trench	A, events/day	α, M^{-1}	c, day	p	D^2, degree^2	q
Kurile	0.340	1.764	0.0170	1.176	6.567×10^{-3}	1.919
Japan	0.316	1.618	0.0161	1.134	6.477×10^{-3}	1.979
Ryukyu	0.161	1.903	0.0140	1.168	4.702×10^{-3}	2.189

and decreases gradually. Thus, I used earthquakes with $m_b \geq 5.0$, which are located without fail between 1964 and 2012, and with a depth of 60 km and shallower.

A stochastic declustering method developed by Zhuang et al. (2004), which is to remove clustered earthquakes such as swarms and aftershocks, was applied to the ISC catalog. The method of Zhuang et al. (2004) separates seismicity into the background events and the clustered events based on the ETAS model. Since the three study areas extend over a long distance, the declustering process was applied separately for the three areas. Parameters estimated by fitting the ETAS model to the ISC catalog was listed on Table 2. As a result 1641, 1207, and 1050 earthquakes are remained as background events in the Kurile Trench, the Japan Trench, and the Ryukyu Trench areas, respectively, and they are used in the following analysis (Fig. 1).

3. Method

A simple space–time scanning technique ZMAP (Wiemer and Wyss 1994) was used to find significant rate changes in seismicity. The ZMAP parameters for the analysis are shown in Table 1. The study areas are covered by spatial grid points with an interval of $0.1° \times 0.1°$. The epicentral distance between all pairs of epicenters and the nodes was calculated and $N_{zmap} = 40$ earthquakes were selected around each node in the order that the epicentral distance is short. The spatial resolution r_{max} for the node is defined by the largest value among the $N_{zmap} = 40$ epicentral distances. The parameter N_{zmap} is the same value for all nodes in order to compare the statistical significance.

In case of the Kurile Trench and the Ryukyu Trench areas, the $N_{zmap} = 40$ earthquakes took place in a time period between T_{start} (1 January 1964) and

T_{end} (30 June 2012), thus the average rate of occurrence is 40 earthquakes/48.5 years = 0.82 earthquakes/year. The Z values are calculated as follows based on the N_{zmap} earthquakes selected for each node. The time period from T_{start} and T_{end} is divided into $N_{\Delta t}$ short-term (ST) time windows at regular intervals with a length of Δt, which is $\Delta t = 0.1$ years and thus $N_{\Delta t} = 485$ in case of the Kurile Trench and the Ryukyu Trench areas. The number of earthquakes $n_i(i = 1, \ldots, N_{\Delta t})$ was counted in each ST time window. The background seismicity rate R_{bg} is then calculated as follows:

$$R_{bg} = \frac{1}{n_{bg}} \left(\sum_{i=1}^{N_1} n_i + \sum_{i=N_2+1}^{N_{\Delta t}} n_i \right),$$

where n_{bg} is $N_1 + (N_{\Delta t} - N_2)$, $(T_{start} + N_1 \Delta t) \equiv t_s$ is a starting time of a long-term (LT) time window and $(T_{start} + N_2 \Delta t)$ is an ending time of the LT time window. The width of the LT time window $\Delta T = (N_2 - N_1)\Delta t$ is set to be $\Delta T = 9$ years in this study. The seismicity rate R_w was calculated in the LT time window as follows:

$$R_w = \frac{1}{n_w} \sum_{i=N_1+1}^{N_2} n_i,$$

where $N_2 - N_1$ is equal to $n_w = \Delta T / \Delta t$, which is $n_w = 90$ in this study. R_w was compared with R_{bg} by using the Z value in the following equation:

$$Z(x_i, y_j, t_s) = (R_{bg} - R_w)\left(\frac{S_{bg}}{n_{bg}} + \frac{S_w}{n_w}\right)^{-\frac{1}{2}},$$

where $x_i(i = 1, \ldots, n_{lon})$ and $y_j(j = 1, \ldots, n_{lat})$ are longitude and latitude of a node, respectively. t_s is the starting time of the LT time window and $T_{start} \leq t_s \leq (T_{end} - \Delta T)$. S_{bg} and S_w are the variances defined by the following equations:

$$S_{bg} = \frac{1}{n_{bg}} \left\{ \sum_{i=1}^{N_1} (n_i - R_{bg})^2 + \sum_{i=N_2+1}^{N_{\Delta t}} (n_i - R_{bg})^2 \right\},$$

$$S_w = \frac{1}{n_w} \sum_{i=N_1+1}^{N_2} (n_i - R_w)^2.$$

Taking the spatial resolution into account the nodes with $r_{max} \leq 200$ km are selected and the Z values are calculated. The number of these nodes is

10,510, 8519, and 8517 in the Kurile Trench, the Japan Trench, and the Ryukyu Trench areas, respectively (Fig. 1). Katsumata (2011a) also described the ZMAP method with a concrete example.

4. Results

Figure 2 shows the high-Z anomalies equal to $Z = +6.0$ or larger detected in and around the Japan subduction zone. The positive Z value indicates the seismic quiescence that corresponds to the decrease of seismicity rate in the LT time window with a length of 9 years when comparing with the background rate. The Z values were calculated for the node of $r_{max} \leq 200$ km, and then the total number of Z values are 9,293,678 and the number of Z values with $Z \geq +6.0$ are 4308. These 4308 Z values were divided into ten groups of anomaly taking the spatial and temporal distribution into account (Table 3). Each group is characterized by the location of seismic quiescence, the spatial extent of the seismic quiescence area, the start time and the duration time of the seismic quiescence, the value of Z, and the probability that the seismic quiescence is observed by chance if earthquakes occur in random.

The probability by chance was calculated as follows. Assumed that earthquakes occur in random as the Poisson's process, n earthquakes take place in the time period T, and h earthquakes take place in the time period S. The probability P that this seismicity is observed is calculated by using the following equation (Shimazaki 1973):

$$P = \binom{r + k - 1}{k} p^r q^k,$$

$$p = \frac{T}{T+S}, \quad q = \frac{S}{T+S}, \quad r = n+1, \quad k = h.$$

For example, in the case of Area 2, $n = 20$, $T = 17.7$ years, $h = 1$, and $S = 13.2$ years, thus $P = 0.00008$, which is the probability that the seismic quiescence is observed by chance when the earthquakes occur in random. The probability is small significantly for all of the ten anomaly groups, and thus the ten seismic quiescence identified in this

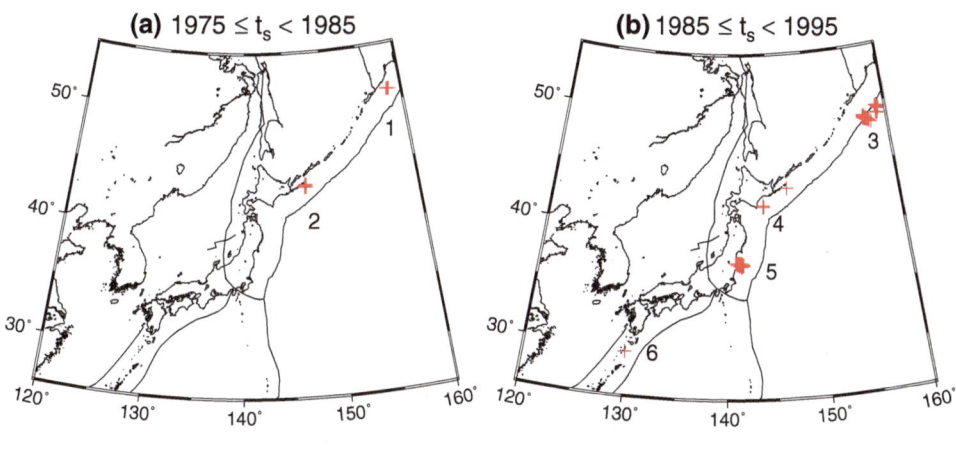

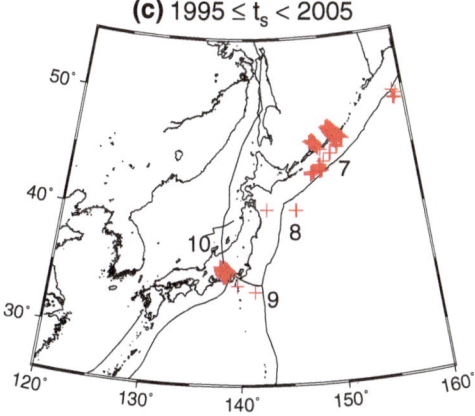

Figure 2

Long-term seismic quiescences found in this study. *Red crosses* indicate anomalous nodes with the Z value of $+6.0$ or larger. A positive Z value represents a decrease in the seismicity rate. The long-term time window with a length of $\Delta T = 9$ years was used for calculating the Z values. The anomalous nodes are shown in the time period of **a** $1975 \leq t_s < 1985$, **b** $1985 \leq t_s < 1995$, and **c** $1995 \leq t_s < 2005$. t_s is the starting time of the long-term time window. The numbers from *1* to *10* indicate the seismic quiescence area shown in Figs. 3, 4, 5, 6, 7 and 8 and Table 3

study are not explained by a random fluctuation in seismicity.

5. Discussions and Concluding Remarks

5.1. Characteristic of Seismic Quiescence

Figure 3 shows two anomalies in Area 2 and Area 4. In Area 2, the seismic quiescence started in 1981.6, lasting 13.2 years, and ended at the same time as the occurrence of the Hokkaido-toho-oki earthquake (M_w 8.3) in 1994, which is not an interplate thrust

earthquake but an intraplate earthquake within the PA plate (Kikuchi and Kanamori 1995; Katsumata et al. 1995; Tanioka et al. 1995). The seismic quiescence followed by the 1994 Hokkaido-toho-oki earthquake (M_w 8.3) has been reported by Takanami et al. (1996) and Katsumata and Kasahara (1999). The duration time of the seismic quiescence is, however, found to be 13.2 years in this study, which is different from the two previous studies. Takanami et al. (1996) pointed out that the seismic quiescence started 3 years before the 1994 main shock. Katsumata and Kasahara (1999) found that the duration time was 5–6 years. Takanami et al. (1996) used earthquakes with $M \geq 3.5$

Table 3

Seismic quiescence areas identified in this study

Area	°N	°E	r_{max}(km)	Start	dt (year)	Z	Probability
1	51.0	158.5	159	1978.9	9.9	6.3	6×10^{-5}
2	43.5	147.1	31	1981.6	13.2	6.4	8×10^{-5}
3	49.6	158.8	167	1993.9	10.8	6.4	2×10^{-5}
4	42.0	144.0	51	1994.0	9.8	6.1	1×10^{-3}
5	37.3	141.5	25	1988.0	10.6	6.6	6×10^{-5}
6	29.6	129.9	75	1985.2	9.4	6.3	6×10^{-5}
7	45.9	148.6	154	1996.1	$\geq$16.4	6.4	9×10^{-7}
	46.4	151.1	99	1996.9	12.4	6.4	8×10^{-6}
8	40.5	142.7	48	1996.2	12.5	6.1	6×10^{-4}
9	33.7	141.3	46	1998.8	9.6	6.3	8×10^{-5}
10	35.5	138.0	133	1998.3	9.4	6.4	3×10^{-4}

Area the area numbers are the same as those shown in Fig. 2, *dt* the duration time of seismic quiescence

between 1 January 1984 and 8 October 1994, which is obviously too short period to recognize the seismic quiescence during 13.2 years. Katsumata and Kasahara (1999) used earthquakes located by Hokkaido University between 1 March 1985 and 3 October 1994 and located by the Japan Meteorological Agency (JMA) between 1 January 1977 and 3 October 1994, which are also too short period to detect the long-term seismic quiescence during 13.2 years found in this study. They also used earthquakes located by ISC between 1 January 1970 and 3 October 1994, which is long enough to detect the 13.2-year-long quiescence. I found a large Z value area near the epicenter of the main shock at the time slices from 1982.5 to 1985.5 in Fig. 8 of Katsumata and Kasahara (1999), which possibly corresponds to the seismic quiescence in this study.

Area 2 is defined by the resolution circle centered at (43.5°N, 147.1°E) with a radius of 31 km, which is the spatial extent of the seismic quiescence area. Hereafter a term of "the quiescence area" will be used as an abbreviation of "the spatial extent of the seismic quiescence area". In the case of Area 2, the quiescence area includes the epicenter of the main shock. Katsumata et al. (2002) presented a long-term slow slip event (SSE) model preceding to the 1994 main shock based on the tide gauge data. Since the fault plane of the long-term SSE is assumed to be in the quiescence area, the seismic quiescence might be related to the long-term SSE.

In Area 4, the seismic quiescence started in 1994.0, lasting 9.8 years, and ended at the same time of the occurrence of the Tokachi-oki earthquake (M_w 8.3) in 2003, which is an interplate earthquake on the interface between the subducting PA plate and the overriding NA plate (Yamanaka and Kikuchi 2003; Yagi 2004; Tanioka et al. 2004; Shinohara et al. 2004). The quiescence area includes the epicenter of the main shock and overlaps with the asperity ruptured by the 2003 event. The seismic quiescence followed by the 2003 Tokachi-oki earthquake (M_w 8.3) has been reported by Takahashi and Kasahara (2004) and Katsumata (2011a). The duration time of the seismic quiescence is found to be 9.8 years in this study, which is consistent with Takahashi and Kasahara (2004). Takahashi and Kasahara (2004) analyzed earthquakes with M $\geq$ 5.0 in the JMA catalog from 1952 to 2003 and they obtained the same results as those in this study even if they used the JMA catalog and they did not apply the declustering process. Katsumata (2011a) analyzed earthquakes with $M \geq 3.3$ in a re-determined earthquake catalog between 1994 and 2003 and found that the duration time of the seismic quiescence was 5 years before the 2003 main shock. The earthquake catalog used by Katsumata (2011a) was too short to identify the long-term seismic quiescence during 9.8 years, thus it is not clear whether the results in Katsumata (2011a) are consistent with those in this study.

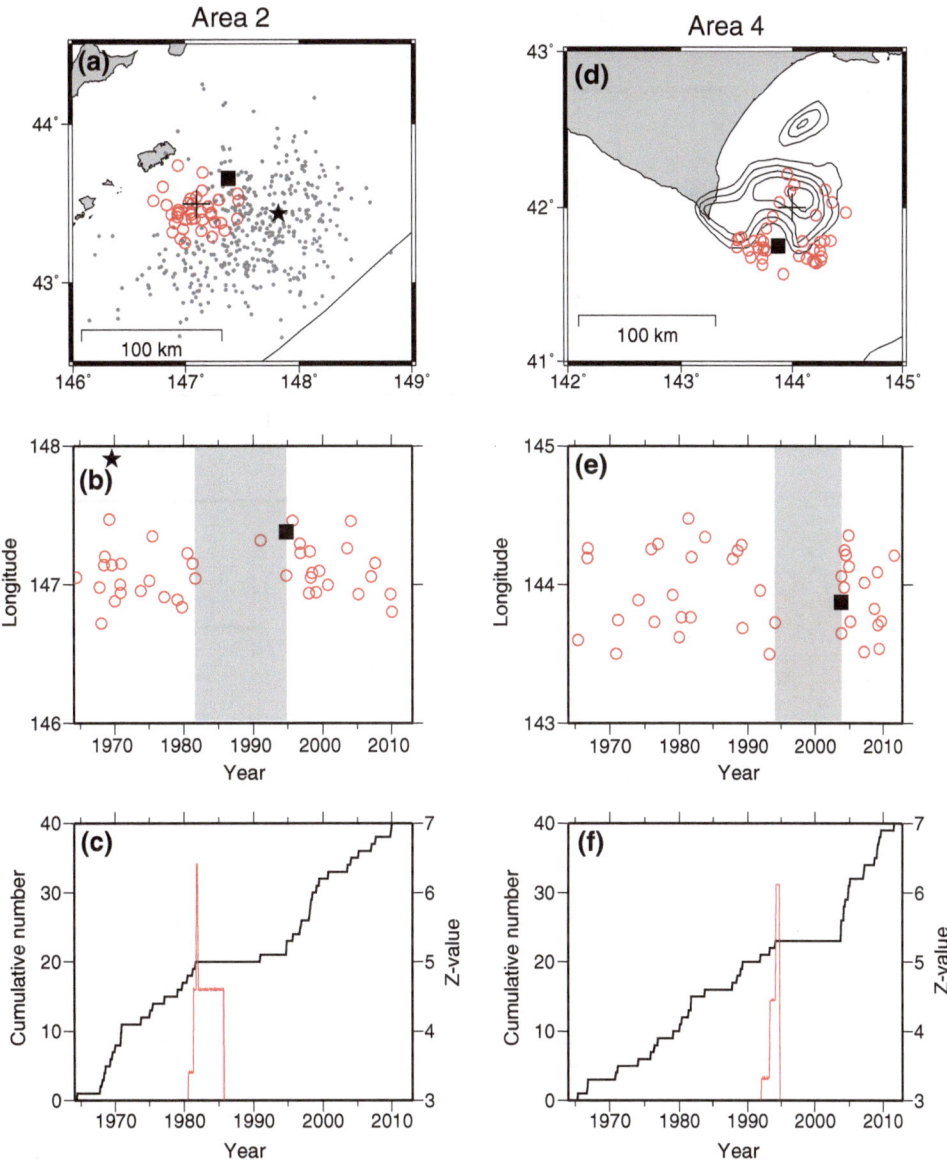

Figure 3

Seismic quiescence in Area 2 (**a–c**) and Area 4 (**d–f**). **a** *Red open circles* are $N_{zmap} = 40$ epicenters sampled around the node indicated by a *cross*. A *closed square* and a *closed star* indicate the epicenters of the 1994 and the 1969 Hokkaido-toho-oki earthquakes, respectively, determined by ISC. *Gray dots* indicate aftershocks of the 1994 event with $M \geq 4.0$ determined by JMA within 1 week after the main shock. The trench axis is shown by a *thin line*. **b** Space–time plot of the epicenters shown in **a**. A *gray zone* indicates the time period of the long-term seismic quiescence. **c** *Black lines* indicate a cumulative number *curve* of the epicenters in **a** and *red lines* indicate the Z value as a function of time. **d** *Red open circles* and a *cross* indicate the same things as those in **a**. A *closed square* is the epicenter of the 2003 Tokachi-oki earthquake determined by ISC. The co-seismic displacement of the fault is shown as contours every 1 m (Yamanaka and Kikuchi 2003). **e** Space–time plot of the epicenters in **d**. **f** Cumulative number curve and the Z value plot for the epicenters in **d**

Figure 4 shows an anomaly in Area 5. The quiescence area is located around the southwestern part of the large slip area of the 2011 Tohoku earthquake (M_w 9.0) (e.g., Yokota et al. 2011). In this case the seismic quiescence was observed in a limited area rather than in the whole area of the seismic fault ruptured by the main shock. The seismic quiescence started in 1988.0. No earthquake

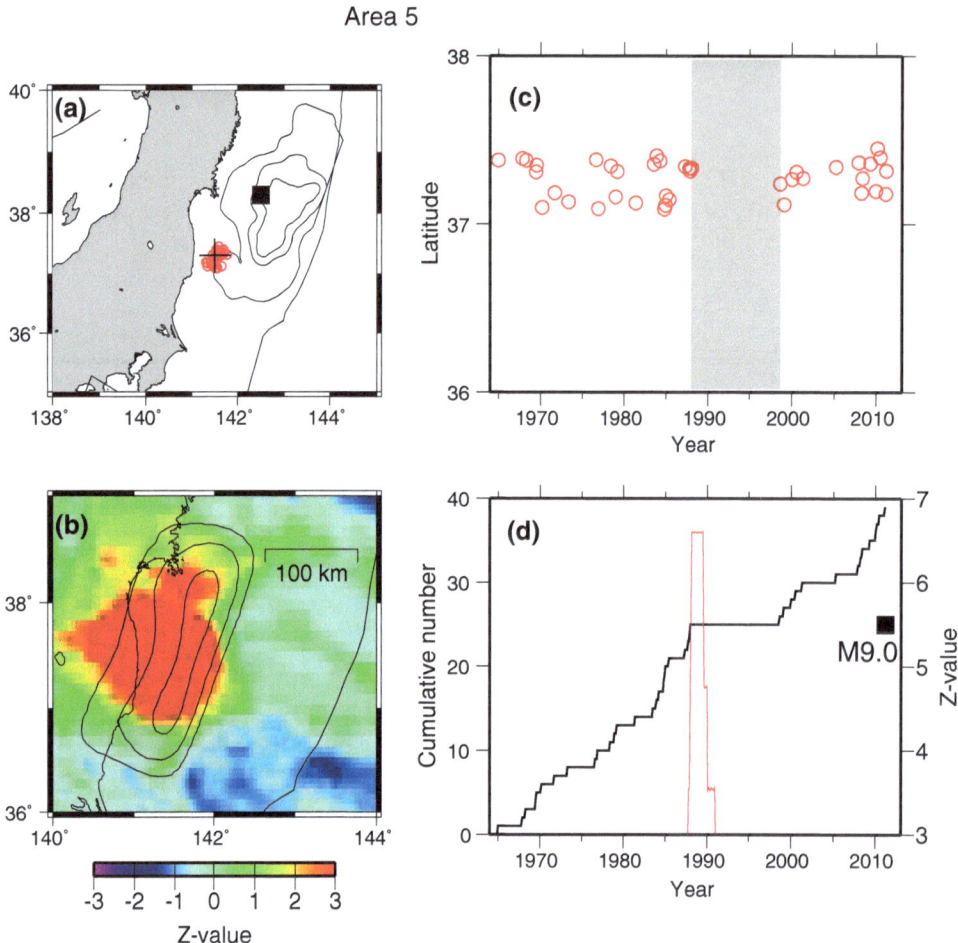

Area 5

Figure 4
Seismic quiescence in Area 5. **a** *Red open circles* are $N_{zmap} = 40$ epicenters sampled around the node indicated by a *cross*. A *closed square* indicates the epicenter of the 2011 Tohoku earthquake determined by ISC. The co-seismic displacement of the fault is shown as contours every 10 m (Yokota et al. 2011). The trench axis is shown by a *thin line*. **b** Time slice of Z value distribution and very long-term transient event. The time window starts at $t_s = 1988.8$ and ends at $t_s + \Delta T$, where $\Delta T = 9$ years. A *red color* (positive Z value) and *blue color* (negative Z value) represent a decrease and increase in the seismicity rate, respectively. The displacement of the very long-term transient event is shown as contours every 10 cm (Yokota and Koketsu 2015). **c** Space–time plot of the epicenters shown in **a**. A *gray zone* indicates the time period of the long-term seismic quiescence. **d** *Black lines* indicate a cumulative number *curve* of the epicenters in **a** and *red lines* indicate the Z value as a function of time

with $m_b \geq 5.0$ occurred from 1988.0 to 1998.6, the seismicity recovered after 1998.6 with the almost same seismicity rate as that before the seismic quiescence started, and the main shock occurred in 2011.

In the case of the 2011 Tohoku earthquake, the time when the seismic quiescence ended dose not match with the time when the main shock occurred. Therefore, we should examine more carefully whether the seismic quiescence is a long-term

precursor to the 2011 main shock or not. First, this seismic quiescence is not a man-made change caused by incomplete earthquake catalog. Katsumata (2011b) analyzed earthquakes with $M \geq 4.5$ in the JMA catalog without the declustering process and found that the seismic quiescence started in 1987.9 in the same area as Area 5 [Fig. 4d in Katsumata (2011b)], which is very consistent with the results in the present study. Second, Yokota and Koketsu (2015) revealed a very long-term transient event

Area 8

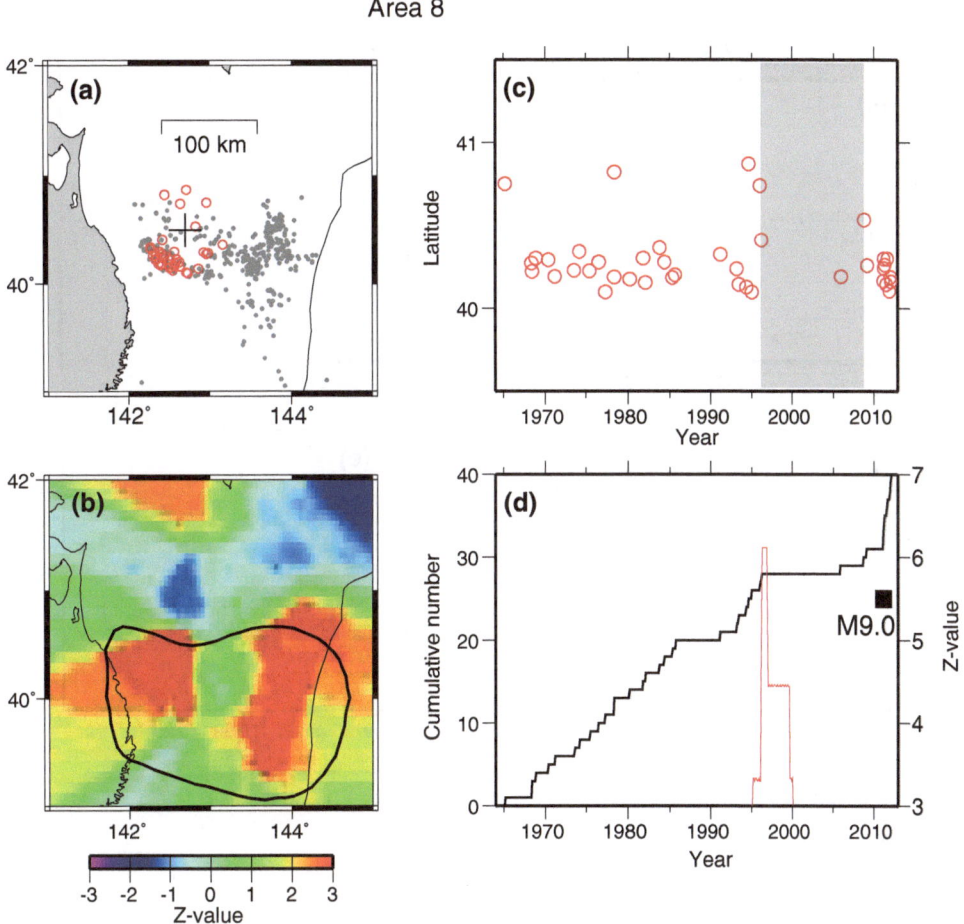

Figure 5

Seismic quiescence in Area 8. **a** *Red open circles* are $N_{zmap} = 40$ epicenters sampled around the node indicated by a *cross*. *Gray dots* indicate aftershocks of the 1994 Sanriku-oki earthquake with $M \geq 4.0$ determined by JMA within 1 month after the main shock. The trench axis is shown by a *thin line*. **b** Time slice of Z value distribution. The time window starts at $t_s = 1996.5$ and ends at $t_s + \Delta T$, where $\Delta T = 9$ years. A *red color* (positive Z value) and *blue color* (negative Z value) represent a decrease and increase in the seismicity rate, respectively. A *thick line* shows the area where the interplate coupling recovered (Ozawa et al. 2007). **c** Space–time plot of the epicenters shown in **a**. A *gray zone* indicates the time period of the long-term seismic quiescence. **d** *Black lines* indicate a cumulative number *curve* of the epicenters in **a** and *red lines* indicate the Z value as a function of time

preceding the 2011 Tohoku earthquake (M_w 9.0), they presented a fault plain model to explain the very long-term transient event. Figure 4b is the spatial distribution of Z values at $t_s = 1988.8$ and I found that the quiescence area of Area 5 is located on this fault plain. This spatial matching strongly suggests that the seismic quiescence is related to the very long-term transient event. On the other hand the temporal relationship between the seismic quiescence and the very long-term transient event is not clear. The seismic quiescence started in 1988.0 and a precursory

long-term foreshock started in 1998.6. The very long-term transient event started around 2002 (Yokota and Koketsu 2015). Mavrommatis et al. (2014) presented another model that the very long-term transient event started around 1996 and the slip was accelerated toward the main shock in 2011, whereas the fault plain is located at the place similar to that of Yokota and Koketsu (2015). Since there is no GPS data before 1996, it is not clear which time, 1996 or 2002, is plausible as the start time of the very long-term transient event.

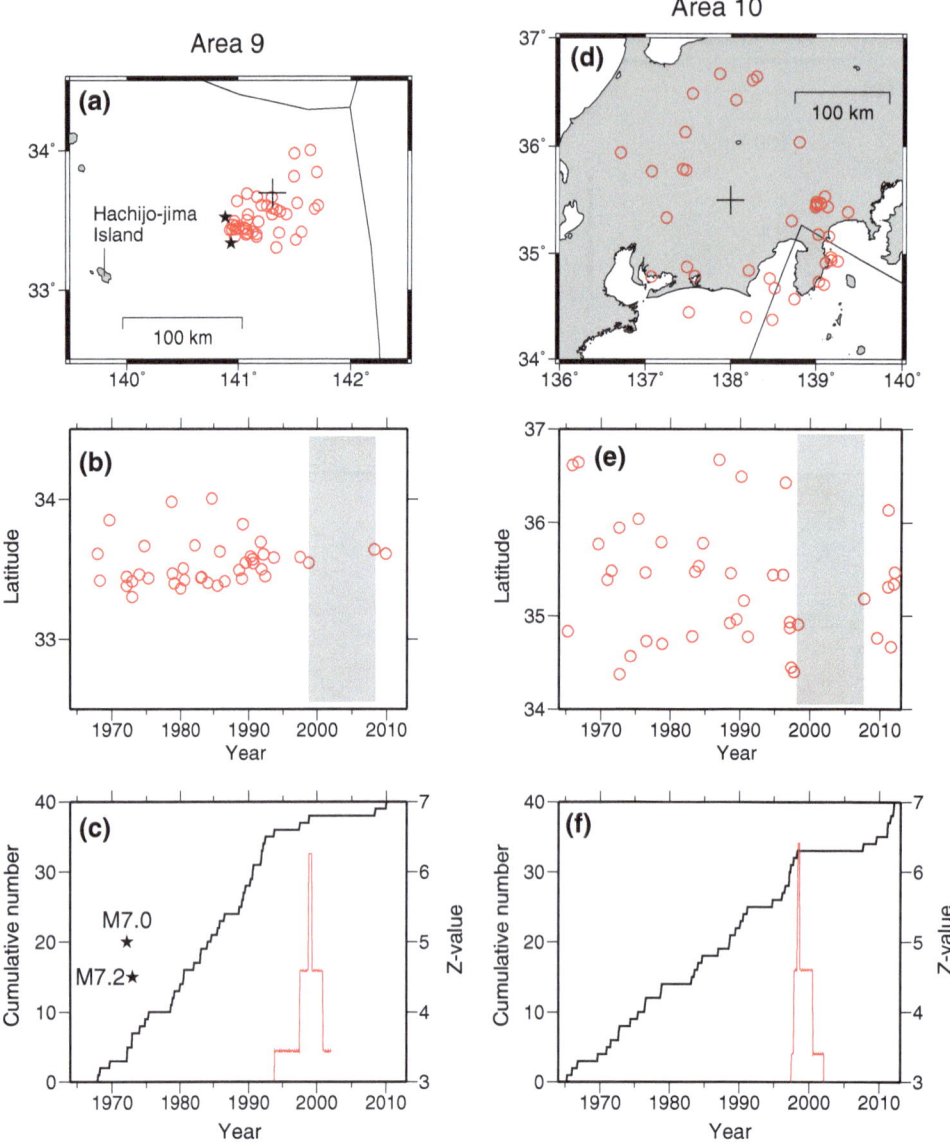

Figure 6
Seismic quiescence in Area 9 (**a–c**) and Area 10 (**d–f**). **a** *Red open circles* are $N_{zmap} = 40$ epicenters sampled around the node indicated by a *cross*. *Two closed stars* indicate the epicenters of the 1972 Hachijo-jima earthquakes with $M = 7.0$ and 7.2. The trench axis is shown by a *thin line*. **b** Space–time plot of the epicenters shown in **a**. A *gray zone* indicates the time period of the long-term seismic quiescence. **c** *Black lines* indicate a cumulative number *curve* of the epicenters in **a** and *red lines* indicate the Z value as a function of time. **d** *Red open circles* and a *cross* indicate the same things as those in **a**. **e** Space–time plot of the epicenters in **d**. **f** Cumulative number curve and the Z value plot for the epicenters in **d**

As shown in Fig. 3 the main shocks occurred in the middle of the seismic quiescence in the case of the 1994 Hokkaido-toho-oki earthquake and the 2003 Tokachi-oki earthquake, whereas some M9-class giant earthquakes were preceded by the long-term foreshock activity. Kanamori (1981) reported that the

1964 Alaska earthquake (M9.2) was preceded by the long-term foreshock during about 10 years before the main shock and the 1957 Aleutians earthquake (M9.1) was also preceded by the long-term foreshock during about 8 years before the main shock. Katsumata (2015) reported that the 2004 Sumatra

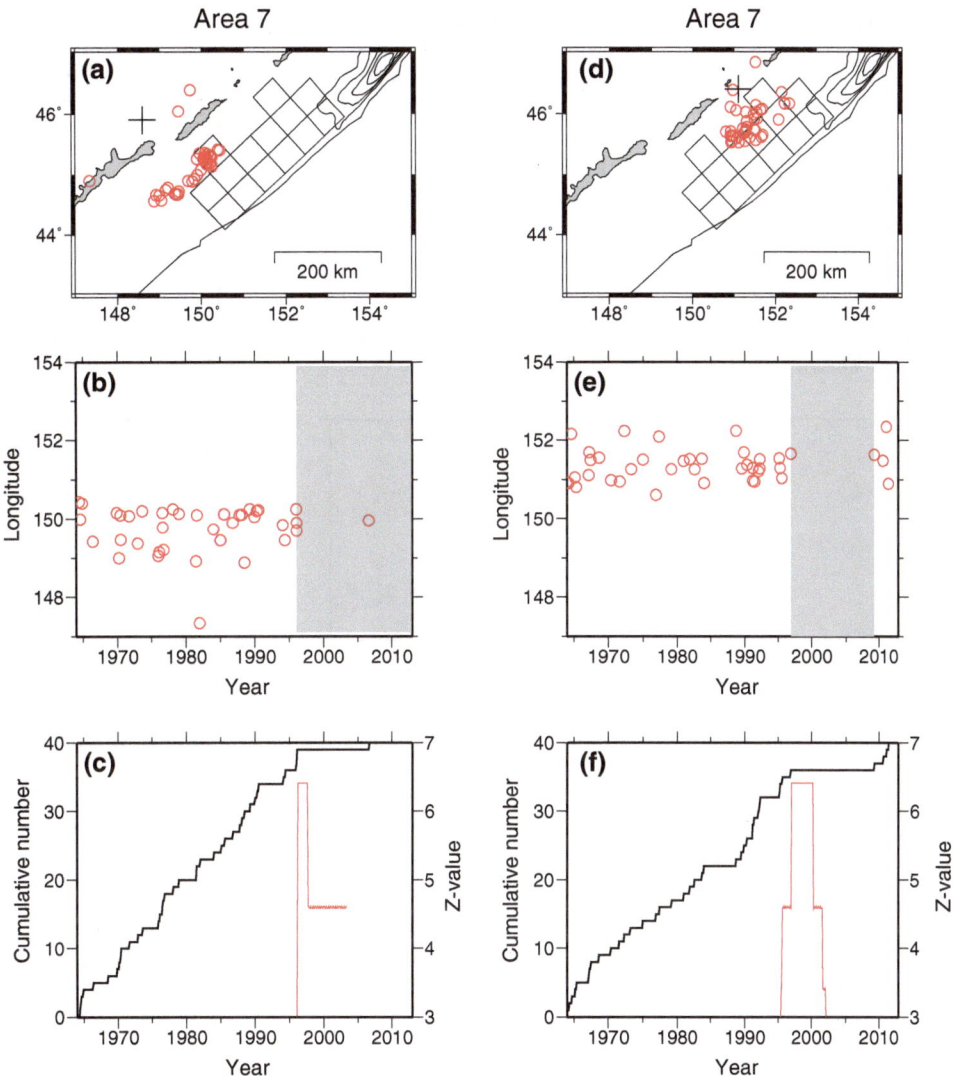

Figure 7
Seismic quiescence in Area 7. **a, d** *Red open circles* are $N_{zmap} = 40$ epicenters sampled around the node indicated by a *cross*. *Fourteen squares* are the subfaults of the 1963 Etorof earthquake with the slip of 0.5 m or larger obtained by Ioki and Tanioka (2011). The co-seismic displacement of the 2006 Kurile earthquake is shown as contours of 1, 3, 5, 7, and 9 m (Yamanaka 2006). **b, e** Space–time plot of the epicenters. A *gray zone* indicates the time period of the long-term seismic quiescence. **c, d** *Black lines* indicate a cumulative number *curve* of the epicenters and *red lines* indicate the Z value as a function of time

earthquake (M_w 9.1) was preceded by the long-term foreshock during about 5 years before the main shock. Therefore, I suggest that a sequence of the seismic quiescence during about 10 years and the subsequent foreshock activity during about 10 years was an inherent behavior of M9-class earthquakes, and the 2011 Tohoku earthquake was no exception.

Figure 5 shows an anomaly in Area 8. Area 8 includes nine nodes with $Z \geq +6.0$ and one of them

is shown in Fig. 5a and Table 3 as a typical anomalous node. The seismic quiescence began in around 1996 after the occurrence of the Sanriku-oki earthquake (M_w 7.7) on 28 December 1994, which is an interplate thrust earthquake on the upper boundary of the PA plate (Sato et al. 1996; Tanioka et al. 1996; Nakayama and Takeo 1997; Nagai and Kikuchi 2001). Figure 5b is the spatial distribution of Z values at $t_s = 1996.5$. The seismicity decreased within the

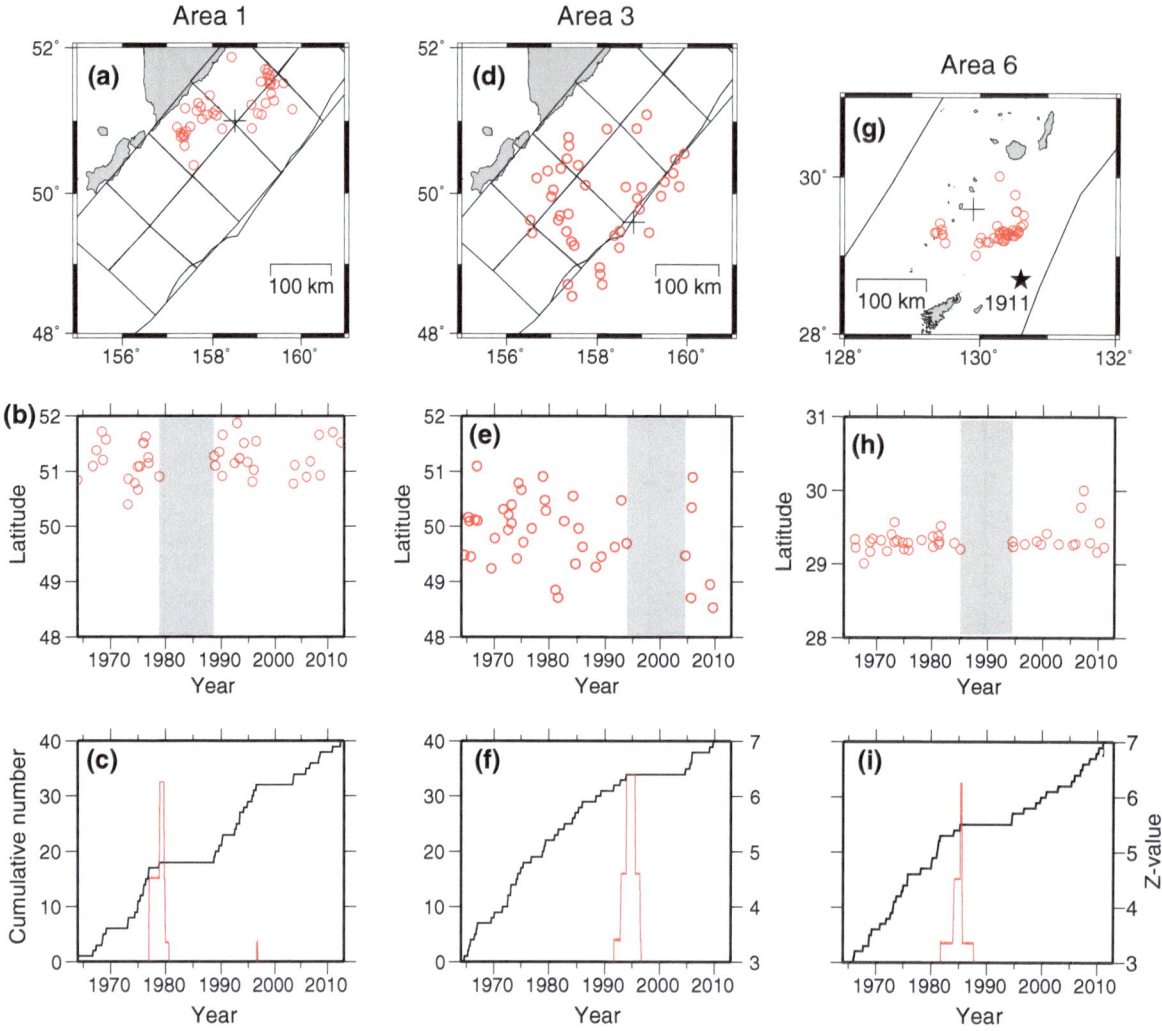

Figure 8
Seismic quiescence in Area 1 (**a–c**), Area 3 (**d–f**) and Area 6 (**g–i**). **a, d, g** *Red open circles* are $N_{zmap} = 40$ epicenters sampled around the node indicated by a *cross*. *Squares* in **a** and **d** are the subfaults of the 1952 Kamchatka earthquake (M9.0) (Johnson and Satake 1999). The trench axis is shown by a *thin line*. A *closed star* in **g** is the epicenter of the 1911 great earthquake with $M \sim 8$ (Goto 2013). **b, e, h** Space–time plot of the epicenters. A *gray zone* indicates the time period of the long-term seismic quiescence. **c, f, i** *Black lines* indicate a cumulative number *curve* of the epicenters and *red lines* indicate the Z value as a function of time

area where the interplate coupling recovered (Ozawa et al. 2007). Moreover, the timing of the recovery is well matched with that of starting of the seismic quiescence. These facts suggest that the occurrence of M5-class interplate earthquakes have been suppressed by the strong interplate coupling, which is the reason why the seismic quiescence is observed in Area 8.

Figure 6 shows anomalies in Area 9 and Area 10. In Area 9, the seismic quiescence started in 1998.8,

ended in 2008.4, no great earthquake occurred in this area, and thus this seismic quiescence is probably a false alarm. The two earthquakes (M7.0 and M7.2) occurred in 1972 off east coast of Hachijo-jima Island at a depth of around 50 km and both of the two main shocks were interplate events with a fault plane on the upper surface of the PA plate (Moriyama et al. 1989). This observation indicates that the plate coupling is strong enough to generate a large earthquake even at a depth of 50 km. The two 1972 events were located around the western boundary of the quiescence area found in this study. Katsumata (2011b) analyzed earthquakes with $M \geq 4.5$ in the JMA catalog without the declustering process and found that the seismic quiescence started in 1992 off east coast of Hachijo-jima Island, which is consistent with the result in this study. In Area 10, the seismic quiescence started in 1998.3, ended in 2007.7, no great earthquake occurred in this area, and thus this seismic quiescence is probably a false alarm. The seismicity decreased clearly within the quiescence area, especially in the south of 35.5°N. In this area a long-term SSE was observed in this period, which is referred to as the Tokai SSE (Ozawa et al. 2002; Miyazaki et al. 2006; Suito and Ozawa 2009; Ochi and Kato 2013). However the spatial resolution of Area 10 is too poor to reveal that the quiescence area corresponds to the area of the Tokai SSE. Therefore, it is not clear whether the seismic quiescence was related to the Tokai SSE.

Figure 7 shows an anomaly in Area 7. The seismic quiescence started in around 1996 or 1997. Ioki and Tanioka (2011) obtained the slip distribution of the 1963 Etorof earthquake (M8.2) by using the tsunami data. The quiescence area is well matched with the subfaults with the slip of 0.5 m or larger obtained by Ioki and Tanioka (2011). It is not clear whether this seismic quiescence ended or not.

Figure 8 shows anomalies in Areas 1, 3 and 6. In Area 1, the seismic quiescence started in 1978.9, lasting 9.9 years, ended in 1988.8, and this seismic quiescence was not followed by a great earthquake. The seismic quiescence area is located in the central part of the focal area ruptured by the 1952 Kamchatka earthquake (M9.0) (Johnson and Satake 1999). In Area 3, the seismic quiescence started in 1993.9, lasted 10.8 years, ended in 2004.7, and this seismic quiescence was not followed by a great earthquake, neither. The quiescence area is located in the southwestern part of the 1952 event. These two false alarms on the seismic fault ruptured by the 1952 event possibly suggest that the seismic quiescence occurs frequently in the interseismic period and they are not a precursor to a great earthquake. In Area 6, the seismic quiescence started in 1985.2, lasted 9.4 years, and ended in 1994.6. No large earthquake with $M \geq 7.0$ has been observed in this area from

1964 to 2012 and thus this seismic quiescence should be a false alarm. The seismicity rates are 1.3 events/year for the period between 1964.0 and 1985.2 and 0.79 events/year for the period between 1994.6 and 2012.5, that is, the seismicity rate has not recovered to the same rate as that before the seismic quiescence. Goto (2013) re-determined the hypocenter of the 1911 great earthquake with $M \sim 8$ and suggested the location of its asperity. The quiescence area in Area 6 is well matched with the northern part of the asperity ruptured by the 1911 event.

5.2. Seismic quiescences and great earthquakes

The ten seismic quiescences were identified from 1975 to 2012 in this study. In the same period four great earthquakes occurred with the seismic moment $M_0 \geq 3.0 \times 10^{21}$ Nm ($M_w \geq 8.25$). The three of ten seismic quiescences were followed by three great earthquakes: the 1994 Hokkaido-toho-oki earthquake (M_w 8.3), the 2003 Tokachi-oki earthquake (M_w 8.3), and the 2011 Tohoku earthquake (M_w 9.0). The remaining seven seismic quiescences were followed by no earthquake with the seismic moment $M_0 \geq 3.0 \times 10^{21}$ Nm ($M_w \geq 8.25$), which are candidates of the false alarm. The 2006 Kurile Islands earthquake (M_w 8.3) was an interplate thrust faulting on the upper boundary of the PA plate (Ji 2006; Yagi 2006; Yamanaka 2006). The 2006 event was not preceded by the significant seismic quiescence with $Z \geq 6.0$, which is a case of the surprise occurrence. To count the cases that no earthquake with $M_0 \geq 3.0 \times 10^{21}$ Nm ($M_w \geq 8.25$) was observed if no quiescence with $Z \geq +6.0$ is observed, the subduction zone is divided into 17 segments in the study area (Fig. 9). All segments have enough size of the area to generate M8-class earthquakes. Suppose that the earthquake catalog is as long as 40 years. If we assume that the seismic quiescence is 10 years long, the following simple three patterns occur: (1) the first 10 years (background rate)—the second 10 years (quiescence)—the third 10 years (background rate)—the last 10 years (background rate) or (2) the first 10 years (background rate)—the second 10 years (background rate)—the third 10 years (quiescence)—the last 10 years (background rate) or (3) the first 10 years (background rate)—the second

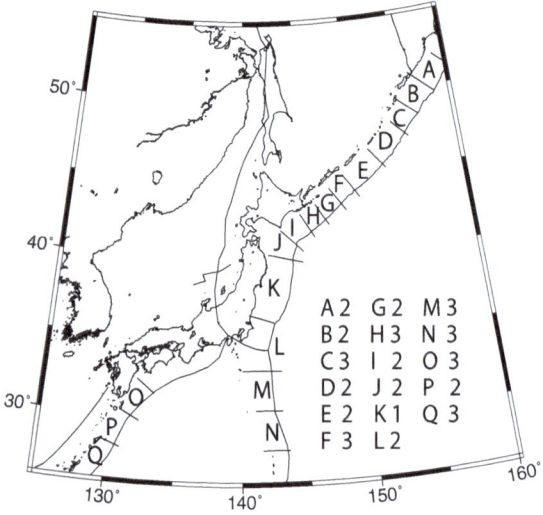

quiescence has no relation with the occurrence of great earthquake. The probability that the null hypothesis is correct was calculated as follows:

$$P = \frac{4!47!10!41!}{51!} \times \left(\frac{1}{1!3!7!40!} + \frac{1}{0!4!6!41!} \right) = 0.021,$$

Therefore the null hypothesis was rejected by 95 % of significant level, that is, the seismic quiescence is related with the occurrence of great earthquake. The probability of 0.021 is, however, not very small; thus, the relationship between the seismic quiescence and the occurrence of great earthquake is weak and further verification is needed in the future work.

Figure 9

The number of cases that no great earthquake occurs when no long-term seismic quiescence is observed. *Seventeen segments* are assumed along the subduction zone in and around Japan labeled as *A to Q*

Table 4

2 by 2 table for the seismic quiescence hypothesis

	Yes seismic quiescence	No seismic quiescence
Yes earthquake	3	1
No earthquake	7	40

10 years (background rate)—the third 10 years (background rate)—the last 10 years (quiescence). In the present study, the ISC catalog is as long as 48 years from 1964 to 2012 and this is the case of the earthquake catalog 40 years long. Therefore, there are three possible time period of the seismic quiescence in each segment and 17 segments × 3 periods = 51 periods. Actually, I identified the seismic quiescence 10 times in this study and thus the remaining 41 periods are the case of no-earthquake-with-no-seismic-quiescence. In the case of Segment K, which is the focal area of the 2011 Tohoku earthquake, I counted the seismic quiescence two times because of precursory seismicity 20 years long. Consequently the total number of the case of no-quiescence-with-no-earthquake is 40. A 2 by 2 consistency table was examined by the Fisher's exact test (Table 4). A null hypothesis is that the seismic

Acknowledgments

I thank two anonymous reviewers for valuable comments. I used a computer code for the space–time ETAS model and stochastic declustering at http://bemlar.ism.ac.jp/zhuang/software.html. GMT-SYSTEM (Wessel and Smith 1991) is used to produce figures. This study was supported by the Ministry of Education, Culture, Sports, Science and Technology (MEXT) of Japan, under its Earthquake and Volcano Hazards Observation and Research Program. This work was supported by JSPS KAKENHI Grant Number 26400445.

REFERENCES

DeMets, C., Gordon, R. G., Argus, D. F., & Stein, S. (1994). Effect of recent revisions to the geomagnetic reversal time scale on estimation of current plate motions. *Geophysical Research Letters, 21*, 2191–2194.

Dziewonski, A. M., Chou, T.-A., & Woodhouse, J. H. (1981). Determination of earthquake source parameters from waveform data for studies of global and regional seismicity. *Journal of Geophysical Research, 86*, 2825–2852.

Ekström, G., Nettles, M., & Dziewonski, A. M. (2012). The global CMT project 2004-2010: Centroid-moment tensors for 13,017 earthquakes. *Physics of the Earth and Planetary Interior, 200–201*, 1–9.

Goto, K. (2013). Re-evaluation of hypocenter of the 1911 great earthquake around Kikai-jima, Japan. *Journal of the Seismological Society of Japan (Zisin), 65*, 231–242. (**in Japanese**).

Inouye, W. (1965). On the seismicity in the epicentral region and its neighborhood before the Niigata earthquake. *Kenshin-jiho (Quarterly Journal of Seismology), 29*, 139–144. (**in Japanese**).

International Seismological Centre (2013). *On-line Bulletin.* International Seismological Center, Thatcham, United Kingdom. http://www.isc.ac.uk/iscbulletin/search/catalogue/. Accessed 7 Apr 2016.

Ioki, K., & Tanioka, Y. (2011). Slip distribution of the 1963 great Kurile earthquake estimated from tsunami waveforms. *Pure and Applied Geophysics, 168*, 1045–1052.

Ji, C. (2006). *Rupture process of the 2006 NOV 15 magnitude 8.3—KURIL Island earthquake (revised).* http://earthquake.usgs.gov/eqcenter/eqinthenews/2006/usvcam/finite_fault.php. Accessed Mar 2016.

Johnson, J. M., & Satake, K. (1999). Asperity distribution of the 1952 great Kamchatka earthquake and its relation to future potential in Kamchatka. *Pure and Applied Geophysics, 154*, 541–553.

Kanamori, H. (1981). The nature of seismicity patterns before large earthquakes. In W. Simpson & P. G. Richards (Eds.), *Earthquake prediction maurice ewing series 4 D* (pp. 1–19). Washington: American Geophysical Union.

Katsumata, K. (2011a). Precursory seismic quiescence before the $M_w = 8.3$ Tokachi-oki, Japan earthquake on 26 September 2003 revealed by a re-examined earthquake catalog. *Journal of Geophysical Research,*. doi:10.1029/2010JB007964.

Katsumata, K. (2011b). A long-term seismic quiescence started 23 years before the 2011 off the Pacific coast of Tohoku earthquake ($M = 9.0$). *Earth, Planets and Space, 63*, 709–712.

Katsumata, K. (2015). A long-term seismic quiescence before the 2004 Sumatra ($M_w 9.1$) earthquake. *Bulletin of the Seismological Society of America, 105*(1), 167–176.

Katsumata, K., Ichiyanagi, M., Miwa, M., Kasahara, M., & Miyamachi, H. (1995). Aftershock distribution of the October 4, 1994 $M_w 8.3$ Kurile islands earthquake determined by a local seismic network in Hokkaido. *Japan. Geophysical Research Letters, 22*, 1321–1324.

Katsumata, K., & Kasahara, M. (1999). Precursory seismic quiescence before the 1994 Kurile earthquake ($M_w = 8.3$) revealed by three independent seismic catalogs. *Pure and Applied Geophysics, 155*, 443–470.

Katsumata, K., Kasahara, M., Ozawa, S., & Ivashchenko, S. (2002). A five years super-slow aseismic precursor model for the 1994 M8.3 Hokkaido-Toho-Oki lithospheric earthquake based on tide gauge data. *Geophysical Research Letters,*. doi:10.1029/2002GL014982.

Kikuchi, M., & Kanamori, H. (1995). The Shikotan earthquake of October 4, 1994: lithospheric earthquake. *Geophysical Research Letters, 22*, 1025–1028.

Mavrommatis, A. P., Segall, P., & Johnson, K. M. (2014). A decadal-scale deformation transient prior to the 2011 M_w 9.0 Tohoku-oki earthquake. *Geophysical Research Letters, 41*, 4486–4494. doi:10.1002/2014GL060139.

Miyazaki, S., Segall, P., McGuire, J. J., Kato, T., & Hatanaka, Y. (2006). Spatial and temporal evolution of stress and slip rate during the 2000 Tokai slow earthquake. *Journal of Geophysical Research, 111*, B03409. doi:10.1029/2004JB003426.

Mogi, K. (1969). Some feature of recent seismic activity in and near Japan (2), Activity before and after great earthquakes. *Bulletin of Earthquake Research Institute, Tokyo University, 47*, 395–417.

Moriyama, T., Tajima, F., & Seno, T. (1989). The unusual zone of seismic coupling in the Bonin arc: the 1972 Hachijo-Oki earthquakes and related seismicity. *Pure and Applied Geophysics, 129*, 233–261.

Nagai, R., & Kikuchi, M. (2001). Comparative study on the source processes of recurrent large earthquake in Sanriku-oki region: the 1968 Tokachi-oki earthquake and the 1994 Sanriku-oki earthquake. *Journal of the Seismological Society of Japan (Zisin), 54*, 267–280. (**in Japanese**).

Nakayama, W., & Takeo, M. (1997). Slip history of the 1994 Sanriku-Haruka-Oki, Japan, earthquake deduced from strong-motion data. *Bulletin of Seismological Society of America, 87*, 918–931.

Ochi, T., & Kato, T. (2013). Depth extent of the long-term slow slip event in the Tokai district, central Japan: a new insight. *Journal of Geophysical Research, 118*, 4847–4860. doi:10.1002/jgrb.50355.

Ogata, Y. (1992). Detection of precursory relative quiescence before great earthquakes through a statistical model. *Journal of Geophysical Research, 97*, 19845–19871.

Ohtake, M., Matsumoto, T., & Latham, G. V. (1977). Seismicity gap near Oaxaca, Southern Mexico as a probable precursor to a large earthquake. *Pure and Applied Geophysics, 115*, 375–386.

Ozawa, S., Murakami, M., Kaidzu, M., Tada, T., Sagiya, T., Hatanaka, Y., et al. (2002). Detection and monitoring of ongoing aseismic slip in the Tokai region, central Japan. *Science, 298*, 1009–1012.

Ozawa, S., Suito, H., Nishimura, T., Tobita, M., & Munekane, H. (2007). Possibility of recovery of slip deficit rate between the North American plate and the Pacific plate off Sanriku, northeast Japan. *Geophysical Research Letters, 34*, L20308. doi:10.1029/2007GL030477.

Sato, T., Imanishi, K., & Kosuga, M. (1996). Three-stage rupture process of the 28 December 1994 Sanriku-Oki earthquake. *Geophysical Research Letters, 23*, 33–36.

Shimazaki, K. (1973). Statistical method of detecting unusual seismic activities. *Bulletin of Seismological Society of America, 63*, 969–982.

Shinohara, M., et al. (2004). Aftershock observation of the 2003 Tokachi-oki earthquake by using dense ocean bottom seismometer network. *Earth, Planets and Space, 56*, 295–300.

Suito, H., & Ozawa, S. (2009). Transient crustal deformation in the Tokai district—The Tokai slow slip event and postseismic deformation caused by the 2004 off southeast Kii Peninsula earthquake (in Japanese). *Journal of the Seismological Society of Japan (Zisin), 61*, 113–135.

Takahashi, H., & Kasahara, M. (2004). The 2003 Tokachi-oki earthquake off southeastern Hokkaido, Japan—seismic activity from the former 1952 Tokachi-oki earthquake, foreshock, mainshock, aftershocks, and triggered earthquakes. *Journal of Seismological Society of Japan, 57*, 115–130. (**in Japanese**).

Takanami, T., Sacks, I. S., Snoke, J. A., Motoya, Y., & Ichiyanagi, M. (1996). Seismic quiescence before the Hokkaido-Toho-Oki earthquake of October 4, 1994. *Journal of Physics of the Earth, 44*, 193–203.

Tanioka, Y., Hirata, K., Hino, R., & Kanazawa, T. (2004). Slip distribution of the 2003 Tokachi-oki earthquake estimated from tsunami waveform inversion. *Earth, Planets and Space, 56*, 373–376.

Tanioka, Y., Ruff, L., & Satake, K. (1995). The great Kurile earthquake of October 4, 1994 tore the slab. *Geophysical Research Letters, 22*, 1661–1664.

Tanioka, Y., Ruff, L., & Satake, K. (1996). The Sanriku-Oki, Japan, earthquake of December 28, 1994 (M_w7.7): Rupture of different asperity from a previous earthquake. *Geophysical Research Letters, 23*, 1465–1468.

Utsu, T. (1968). Seismic activity in Hokkaido and its vicinity. *Geophysical Bulletin of Hokkaido University, 13*, 99–103. **(in Japanese)**.

Wessel, P., & Smith, W. H. F. (1991). Free software helps map and display data. *Eos Transaction of AGU, 72*, 445–446.

Wiemer, S., & Wyss, M. (1994). Seismic quiescence before the Landers ($M = 7.5$) and Big Bare ($M = 6.5$) 1992 earthquakes. *Bulletin of Seismological Society of America, 84*, 900–916.

Wiemer, S., & Wyss, M. (2000). Minimum magnitude of completeness in earthquake catalogs: examples from Alaska, the western United States, and Japan. *Bulletin of Seismological Society of America, 90*, 859–869.

Wyss, M., & Habermann, R. E. (1979). Seismic quiescence precursory to a past and a future Kurile Island earthquake. *Pure and Applied Geophysics, 117*, 1195–1211.

Yagi, Y. (2004). Source rupture process of the 2003 Tokachi-oki earthquake determined by joint inversion of teleseismic body wave and strong ground motion data. *Earth, Planets and Space, 56*, 311–316.

Yagi, Y. (2006). http://www.geo.tsukuba.ac.jp/press_HP/yagi/EQ/Chishima/. Accessed Mar 2016.

Yamanaka, Y. (2006). http://wwweic.eri.u-tokyo.ac.jp/sanchu/Seismo_Note/2006/EIC183.html. Accessed Mar 2016.

Yamanaka, Y., & Kikuchi, M. (2003). Source process of the recurrent Tokachi-oki earthquake on September 26, 2003, inferred from teleseismic body waves. *Earth, Planets and Space, 55*(12), e21–e24.

Yokota, Y., & Koketsu, K. (2015). A very long-term transient event preceding the 2011 Tohoku earthquake. *Nature Communications, 6*, 5934. doi:10.1038/ncomms6934.

Yokota, Y., Koketsu, K., Fujii, Y., Satake, K., Sakai, S., Shinohara, M., et al. (2011). Joint inversion of strong motion, teleseismic, geodetic, and tsunami datasets for the rupture process of the 2011 Tohoku earthquake. *Geophysical Research Letters, 38*, L00G21. doi:10.1029/2011GL050098.

Zhuang, J., Ogata, Y., & Vere-Jones, D. (2004). Analyzing earthquake clustering features by using stochastic reconstruction. *Journal of Geophysical Research, 109*, B05301. doi:10.1029/2003JB002879.

(Received April 7, 2016, revised September 7, 2016, accepted October 8, 2016, Published online October 21, 2016)

Pure Appl. Geophys. 174 (2017), 2443–2456
© 2017 Springer International Publishing
DOI 10.1007/s00024-017-1569-z

Pure and Applied Geophysics

Statistical Studies of Induced and Triggered Seismicity at The Geysers, California

A. Hawkins,[1] D. L. Turcotte,[1] M. B. Yıkılmaz,[1] L. H. Kellogg,[1] and J. B. Rundle[1,2]

Abstract—This study considers the statistics of fluid-induced and remotely triggered seismicity at The Geysers geothermal field, California. Little seismicity was reported before steam extraction began in 1960. Beginning in 1980 the residual water associated with power generation was re-injected, producing induced seismicity. Beginning in 1997 large-scale injections of cold water began to enhance the generation of steam. This led to an increase in $M < 1.2$ earthquakes from approximately 5 per month to 20. Two excellent seismic networks generate two earthquake catalogs for the fluid-induced seismicity at The Geysers. Although this seismicity satisfies Gutenberg–Richter (GR) scaling to a good approximation, the scaling parameters differ. We propose a correction that eliminates this problem. We show that the seismicity at The Geysers is nearly independent of time for the period 2009–2014 and suggest that this supports our hypothesis that the seismic moment release is in near balance with the geodetic moment accumulation in the region. Our study demonstrates that aftershocks of the larger fluid-induced earthquakes also satisfy GR scaling as well as Omori's law for their time dependence. Our results support the hypothesis that the earthquakes are caused by the reduction in friction on faults due to the injected fluids. Statistics of remotely triggered earthquakes and their associated aftershocks at The Geysers are also presented. The 8/24/14 $M = 6.02$ South Napa earthquake triggered an $M \approx 4.38$ event as well as some 80 other $M > 1.25$ events. The GR and decay statistics are given. However, to separate aftershocks from remotely triggered earthquakes, an additional triggered sequence is studied. The $M = 7.2$ 4/4/10 Baja earthquake triggered some 34 $M > 1.25$ earthquakes at The Geysers in the first hour including an $M = 3.37$ event. We conclude that the remotely triggered seismicity is dominated by local aftershocks of the larger remotely triggered earthquakes.

Key words: Fluid-induced seismicity, remotely triggered seismicity, The Geysers, geothermal seismicity.

[1] Department of Earth and Planetary Sciences, University of California, Davis, CA 95616, USA. E-mail: lhkellogg@uc-davis.edu; lhkellogg@ucdavis.edu

[2] Department of Physics, University of California, Davis, CA 95616, USA.

1. Introduction

In this paper we study the statistical behavior of the seismicity at The Geysers geothermal area, California. Injection of water to enhance geothermal production has resulted in extensive induced seismicity. The Geysers geothermal field is situated near the southern edge of Clear Lake, approximately 120 km north of San Francisco, CA. It is positioned between bounding faults, the Maacama and Mercuryville faults to the southwest and the Collayami fault zone to the northeast (Fig. 1). The Geysers is a steam-dominated geothermal field. The area is heated by The Geysers Plutonic Complex, a silicic magmatic body, which was episodically emplaced from ∼1.1 to 1.8 Ma (Schmitt et al. 2003). The pluton provides structural control for The Geysers and has a northwest trend that corresponds to the current geothermal reservoir. Magmatism in this region is attributed to the formation of a slab window east of the San Andreas Fault due to the subduction of the Farallon plate (Stimac et al. 2001).

There was little localized seismicity in The Geysers region prior to the extraction of steam for geothermal power generation that began in 1960. Beginning in 1980 the water recovered from the extracted steam was re-injected into the reservoir. This injection produced a relatively low level of induced seismicity. In 1997, large-scale injection of wastewater from Lake County began to increase energy production. Additional effluent was provided from the City of Santa Rosa beginning in 2003 to further augment injection. These wastewater projects provide the approximately 20 million gallons of reclaimed water that is injected into The Geysers reservoir each day. When the wastewater is injected into the hot underlying rock, it boils producing steam and enhancing the energy production of The Geysers

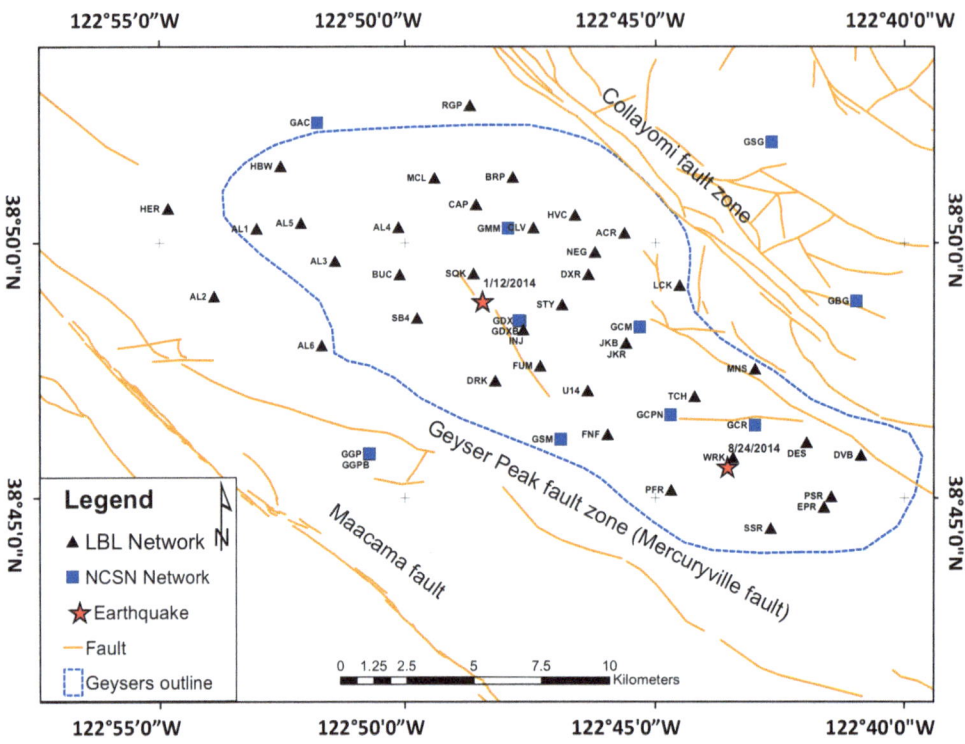

Figure 1
Map of The Geysers geothermal area (*dashed contour*), several major faults, the LBNL stations (*triangles*), the NCSN stations (*squares*), and two earthquakes we consider are shown (after Guilhem et al. 2014)

power plants. The large-scale injection of wastewater led to a large increase in seismicity. An overview of the effect of high rate fluid injection on seismicity at The Geysers has been given by Beall et al. (2010).

The western edge of The Geysers geothermal field is within 10 km of the Maacama Fault (Fig. 1), a right-lateral strike-slip fault, the northern extension of the Hayward fault system. The Maacama Fault is active and displays surface creep, with the highest measured creep rate (5.7 ± 0.1 mm/year) to the north in Willits, CA, and a lower measured creep rate (4.3 ± 0.8 mm/year) to the south in Ukiah Valley (Prentice et al. 2014). Based on geological data the estimated slip rate on the Maacama Fault is 9 ± 3 mm/year (Dawson and Weldon 2013), considerably larger than the present measured surface creep rates. Based on paleoseismic data, it is certainly possible that an $M \approx 7.0$ or larger earthquake could occur on the Maacama Fault. A major concern is the possible role of The Geysers fluid injection in the possible triggering of such an earthquake.

Although fluid injection can certainly generate induced earthquakes, there are other causes such as reservoir filling and coal mining. Suggested mechanisms for induced seismicity in geothermal fields and enhanced geothermal systems include pore pressure increase, temperature reduction, volume change, and chemical alteration (Majer et al. 2007). A particular focus of this paper is the role of fluid injection in reducing the frictional strength of faults. Geodetic (GPS) observations indicate that The Geysers region is in an area of near uniform tectonic shear strain accumulation associated with the relative plate boundary motion between the Pacific plate and the Sierra-Nevada, Central Valley plate (Prescott et al. 2001; Chéry 2008; Jolivet et al. 2009). We propose the hypothesis that this geodetic moment accumulation is in a near balance with the seismic moment release associated with The Geysers seismicity. We present seismic data to support this balance.

In this paper, we will first consider the overall statistics of the seismicity in The Geysers region. We will specifically consider the frequency–magnitude statistics of the seismicity and its temporal variability. To do this we will utilize data from Northern California Seismic Network (NCSN) catalog and the Lawrence Berkeley National Laboratory (LBL) catalog. We will note that earthquake magnitudes from the two catalogs systematically differ. We will quantify the difference and suggest why a correction should be made to the LBL data. We will then consider the aftershock statistics of several of the largest induced earthquakes that have occurred at The Geysers. We will consider the statistics of this seismicity in terms of: (1) the frequency–magnitude behavior and (2) the aftershock behavior in terms of frequency–magnitude and temporal-decay behaviors. One important question is whether the injected fluids are the direct cause of the seismicity or the indirect cause by reducing normal stresses and triggering double-couple earthquakes that release tectonic stresses. This question has been discussed in detail by Johnson (2014a, b) and Guilhem et al. (2014) who found that shear failures dominate. We also study the statistics of triggered earthquakes generated by the seismic waves of distant earthquakes, specifically, the large triggered earthquake generated at The Geysers by the $M = 6.0$ South Napa earthquake 72 km away, in August 2014. An important question is the relative roles of smaller triggered seismicity and the aftershocks generated by one or more large triggered earthquakes. To address this question we also study the statistics of the triggered earthquakes generated at The Geysers by the $M = 7.2$ El Mayor-Cucapah earthquake in April 2010. In this paper, induced earthquakes are earthquakes caused by fluid injection at The Geysers, aftershocks are aftershocks of the induced earthquakes, and triggered earthquakes are aftershocks of remote earthquakes.

2. Earthquake Catalogs

A unique feature of The Geysers seismicity is the availability of two catalogs of events. The first is the Northern California Seismic Network (NCSN) catalog (NCEDC 2014). This is the "official catalog" for earthquakes in this region and the magnitudes of smaller earthquakes are local magnitudes. The second

is the Lawrence Berkeley National Laboratory (LBL) catalog. This catalog uses data from the LBL network of some 30 short-period seismometers installed in The Geysers region. The LBL seismic network in The Geysers was initiated in 2003 by the US Department of Energy to monitor induced seismicity of enhanced geothermal systems (EGS). Magnitude determinations utilize earthquake moments determined from the low-frequency spectral content (Guilhem et al. 2014). The primary advantage of the LBL catalog data is that many more low magnitude earthquakes are included due to the high density and close proximity of seismometers to the earthquake sources. A map of The Geysers region, the locations of the geothermal area, mapped faults, and seismic stations are given in Fig. 1.

Also shown are the epicenters of the two earthquakes that will be a focus of this paper. The first is an $M = 4.53$ induced earthquake on January 12, 2014 and the second is an $M = 4.48$ triggered earthquake on August 24, 2014. A major difficulty with the LBL data is a systematic difference in the magnitudes compared with the magnitudes given in the NCSN catalog. We will consider these differences in some detail.

Our studies will consider the frequency–magnitude statistics of selected earthquakes in several ways. The cumulative number of earthquakes with magnitudes greater than M, N_C in The Geysers region are given as a function of M for different time periods. Gutenberg–Richter (GR) scaling is utilized to compare the data from the two networks (Gutenberg and Richter 1954):

$$\log_{10} N_C = a - bM, \tag{1}$$

where a is a measure of seismic intensity and the b value (slope) usually has a value near one.

Our studies of The Geysers seismicity will utilize the rectangular region given in Fig. 2. Seismic catalogs were downloaded from the Northern California Earthquake Data Center (NCEDC). To generate frequency–magnitude distributions of the earthquakes, magnitude 0+ earthquakes were downloaded for the specified region in both the LBL and NCSN catalogs for a 10-year interval from 10/30/04 to 10/31/14. Frequency–magnitude statistics were obtained for individual years, the first

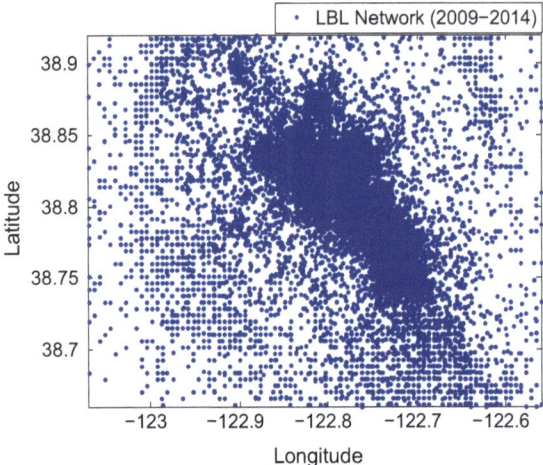

Figure 2
Location of all magnitude 0+ seismic events in the LBL catalog from 10/31/09 to 10/30/14 in our study area. *Latitude* min = 38.66, max = 38.92, *Longitude* min = −123.07, max = −122.56

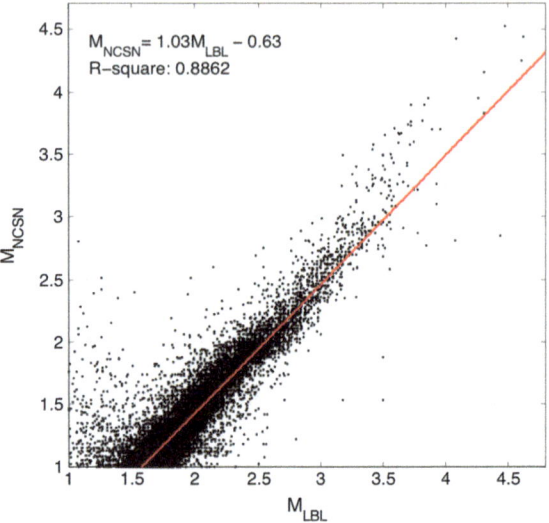

Figure 3
Dependence of the NCSN magnitude M_{NCSN} on the LBL magnitude M_{LBL} for some 18,000 earthquakes in The Geysers geothermal area during the period 10/31/09–10/30/14. The best bisquare linear fit to the data is also shown from Eq. (2). The LBL magnitudes are general 0.5 magnitude units larger that the NCSN magnitudes

5 years, second 5 years, and the 10-year interval. Boyle et al. (2011) presented frequency–magnitude data for the two catalogs of Geysers earthquakes for the period 4/22/2003–12/1/2010. For the NCSN catalog, the results of Boyle et al. (2011) are essentially identical to ours with $b = 1.14$. For the LBL catalog the results were quite different from ours with a poor Gutenberg–Richter correlation. It was for this reason we have chosen to use only the more recent data from both the LBL and NCSN catalogs in our studies.

We next carry out a systematic comparison of the magnitude data from the two networks. We consider the period 10/31/09–10/30/14. The NCSN magnitudes M_{NCSN} versus the LBL magnitudes M_{LBL} are plotted for some 18,000 events in Fig. 3. The best bisquare weight linear fit to the data is

$$M_{NCSN} = 1.03 M_{LBL} - 0.63, \qquad (2)$$

and the root mean square error to the fit is 0.1344. At $M_{NSCN} = 1$, the expected LBL magnitude is $M_{LBL} = 1.58$; at $M_{NSCN} = 3$ the expected LBL magnitude is $M_{LBL} = 3.52$. Thus, the LBL magnitudes are generally 0.5 magnitude units larger than the NCSN magnitudes. The mean deviation of 0.13 magnitude units from this fit is relatively small. A previous, more limited comparison of the magnitudes from the two networks at The Geysers was given by Edwards and Douglas (2014).

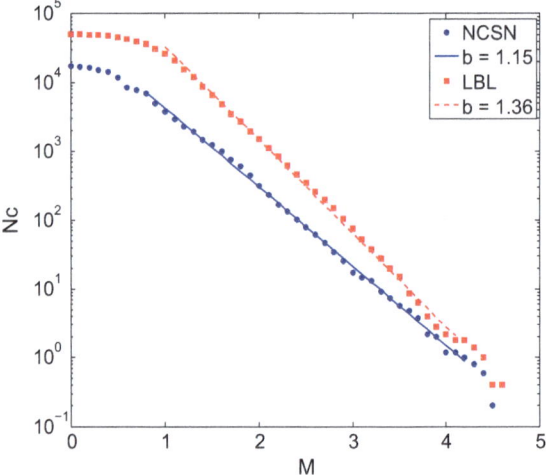

Figure 4
Cumulative frequency–magnitude distributions of earthquakes in The Geysers geothermal region for the period 10/31/09–10/30/14. The cumulative number of earthquakes N_c per year with magnitudes greater that M are given as a function of M for the NCSN data and the LBL data. The *straight line* correlations are with Gutenberg–Richter scaling Eq. (1). Note that the b-values (the *slopes*) are systematically different

We next consider the frequency–magnitude statistics for the two data sets for the 5-year period (10/31/09–10/30/14). The cumulative numbers of earthquakes per year with magnitudes greater than M (N_C) are given as a function of M in Fig. 4. We correlate the two sets of data with Gutenberg–Richter scaling Eq. (1). The deviations of the data from Gutenberg to Richter scaling at small magnitudes indicate the sensitivity limits of the two networks. The lower limits of data completeness are approximately $M_{NCSN} > 0.8$ for the NCSN network and $M_{LBL} > 1.2$ for the LBL network. In Fig. 3, we took a lower limit of $M_{NCSN} = 1.0$ to ensure completeness. During the 5-year period (2009–2014), the NCSN recorded about 7000 events per year and the LBL network about 15,000 events per year that were above these magnitude cutoffs. There are about twice as many reliable earthquakes in the LBL catalog than in the NCSN catalog. A fraction of these, about 18,000 earthquakes, are included in Fig. 3.

When a wide range of earthquake magnitudes is considered, it is standard practice to utilize the Richter magnitude scale (Utsu 2002a, b). There are many variations on how magnitudes are determined. The NCSN magnitudes use the classic local magnitude M_L for intermediate size earthquake, $M \approx 2$ to $M \approx 4$. This magnitude is based on the S-wave amplitude on a Wood–Anderson seismometer. For larger earthquakes, M larger than about 4, earthquakes moments are determined and a converted to moment magnitudes M_W using the empirical relation given by Hanks and Kanamori (1979). For small earthquakes, M smaller than about 2, the coda magnitude M_d is used. It must be emphasized the earthquake magnitude scale is empirical, whereas the earthquake moment is derived from physics.

The LBL earthquake catalog utilizes a hybrid method based on the moments of small earthquake and a conversion to moment magnitude utilizing the empirical relation given by Hanks and Kanamori (1979) for large earthquakes. They obtain the moment of small earthquakes from the low-frequency displacement spectral level at frequencies from about 1 Hz to the event corner frequency (Guilhem et al. 2014). Based on our results given in Fig. 3 we argue that a further empirical correction, as given in Eq. (2), is required.

An important question regarding the scaling shown in Fig. 4 is whether earthquakes with $M > 5$ can be expected in this region in the future. The largest recorded earthquake in the 5-year period had a magnitude $M \approx 4.5$. However, high seismicity has only been present for about 10 years and the extrapolated return period for an $M \approx 5$ earthquake, obtained from an extrapolation of the GR statistics given in Fig. 4, would be about 25 years, so it is not possible to use this data to preclude a larger earthquake.

We next utilized Gutenberg–Richter scaling to compare the 1-year intervals of seismicity to the 5-year interval in both the LBL and NCSN data sets. The 5-year interval a and b *values* were calculated and superimposed on the 1-year intervals to observe the degree to which the 5-year interval approximated the single-year intervals for the LBL catalog (Fig. 5) and NCSN Network (Fig. 6). We found that the GR scaling of the 5-year interval provided a good approximation of the GR scaling for the 1-year intervals in both the LBL and NCSN catalogs.

3. Aftershock Statistics of Major Induced Earthquakes at The Geysers

We now consider the aftershock statistics of four large induced earthquakes at The Geysers. Our objective is to compare the aftershock statistics of these induced earthquakes with the generally accepted statistics of tectonic earthquakes. In making this comparison we will use the statistical methods Shcherbakov et al. (2006) applied to the Parkfield aftershock sequence. We will give a detailed study of the most recent large induced earthquake and give tabulated statistical results for the four earthquakes. We will utilize results from both the NCSN catalog and the LBL catalog. The NCSN magnitude data have the advantage that the magnitude statistics are consistent with generally published aftershock data. As we have shown there is a systematic deviation between NCSN magnitudes and LBL magnitudes. The LBL magnitude data have the advantage of lower magnitude completeness due to the large number of local seismometers. Our studies in this section will also be used to better understand the large triggered earthquakes in the next section.

We consider in some detail the aftershock statistics for a large induced earthquake that occurred in

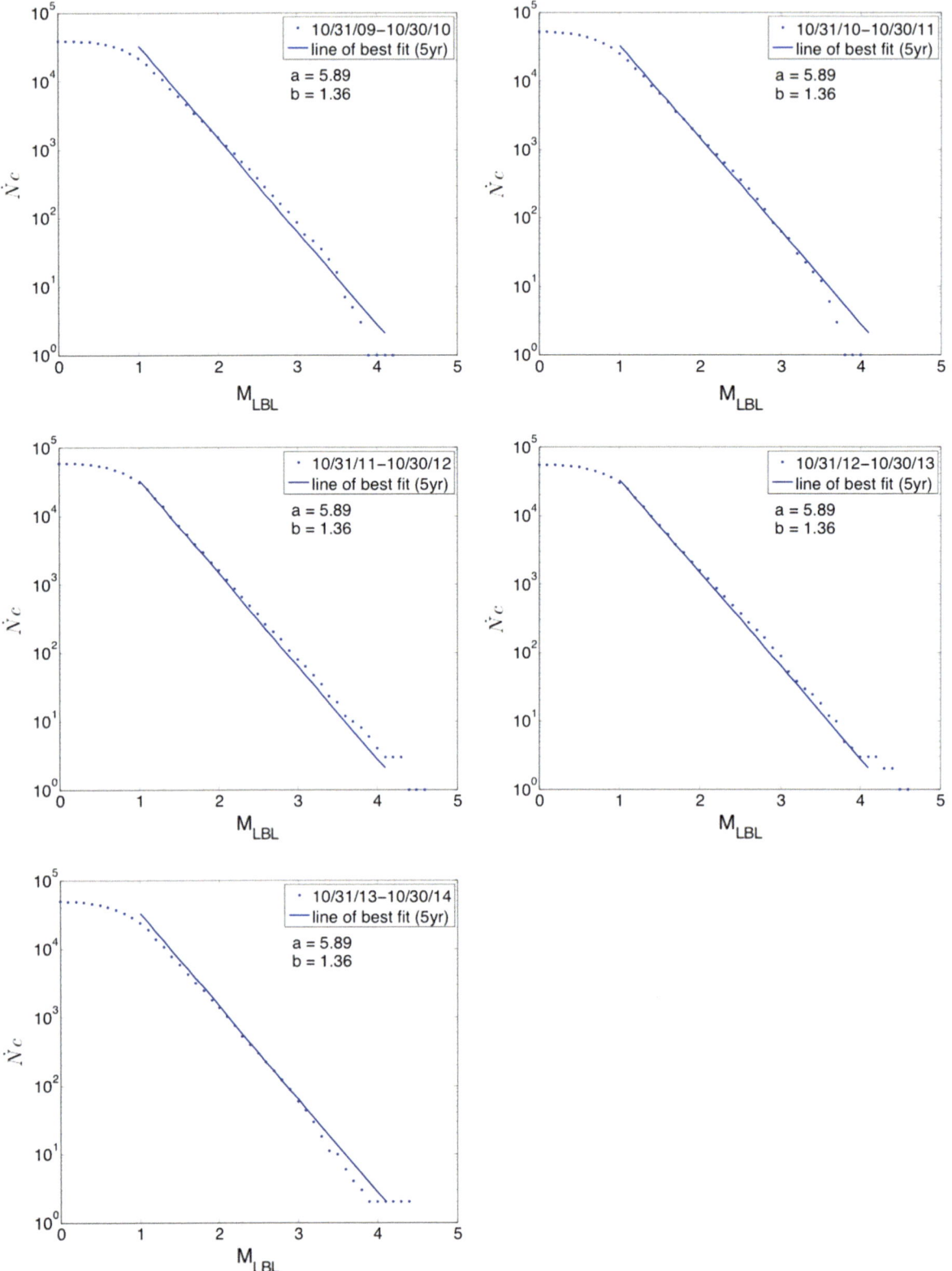

◀Figure 5
Yearly frequency–magnitude statistics from the LBL catalog (2009–2014) with the line of best fit from the 5-year frequency–magnitude data from the LBL catalog from Fig. 4

The Geysers geothermal area at 20:24:47 UTC on January 12, 2014. This earthquake had an NCSN magnitude $M_{NCSN} = 4.53$ and an LBL magnitude $M_{LBL} = 4.48$. It occurred with an epicenter at Lat. 38.81350 N, Long. 122.81360 W (location shown on Fig. 1). The frequency–magnitude statistics for the two data sets are given in Fig. 7. We again correlate the data using the GR scaling from Eq. (1). Based on the systematic divergence of the two data sets illustrated in Fig. 3, we consider aftershocks with $M_{NCSN} \geq 0.75$ and $M_{LBL} \geq 1.25$. For the LBL data, we have an excellent GR correlation with $b = 1.33$, very close to the value $b = 1.36$ obtained for all LBL earthquakes shown in Fig. 4.

Another important scaling law concerning aftershocks is Båth's law (Båth 1965). This law states that empirically it is a good approximation to assume that the difference in magnitude ΔM between a mainshock M_{ms} and its largest aftershock $M_{as}^{(max)}$ is a constant independent of the magnitude of the mainshock, that is

$$\Delta M = M_{ms} - M_{as}^{(max)} \tag{3}$$

with $\Delta M \approx 1.2$. From Fig. 7 we see that $\Delta M_{LBL} = 4.48 - 3.25 = 1.23$ and that $\Delta M_{NCSN} = 4.53 - 2.80 = 1.73$. The two values of ΔM are not in very good agreement.

A modified version of Båth's law was proposed by Shcherbakov and Turcotte (2004). The magnitude of the "largest" aftershock consistent with GR scaling M^* is obtained by setting $N(\geq M) = 1$ in Eq. (1) with the result

$$M^* = \frac{a}{b}. \tag{4}$$

The modified form of Båth's law is therefore

$$\Delta M^* = M_{ms} - M^* \tag{5}$$

that is approximately constant, again with $\Delta M \approx 1.2$. The value of ΔM^* is a direct measure of the aftershock productivity of a mainshock. From

Fig. 7 we see that $M^*_{LBL} = 3.13$ and $M^*_{NCSN} = 2.87$ so that $\Delta M^*_{LBL} = 1.35$ and $\Delta M^*_{NCSN} = 1.66$. The two values of ΔM^* are considerably closer than the two values for ΔM given above.

We next consider the decay rate of the aftershocks. A widely accepted empirical relation for the decay rate of aftershock activity is the modified form of Omori's law (Utsu 1961)

$$\frac{dN}{dt} = \frac{1}{\tau\left(1 + \frac{t}{c}\right)^p}, \tag{6}$$

where dN/dt is the rate of occurrence of aftershocks with magnitudes greater than M, t is the time that has elapsed since the mainshock, and τ and c are constant characteristic times. It is found that the power p is generally near one. Setting $p = 1$ we integrate Eq. (6) to give

$$N_C = \frac{c}{\tau}\ln\left(1 + \frac{t}{c}\right), \tag{7}$$

where N_C is the cumulative number of aftershocks at a time t after the mainshock.

The numbers of earthquakes that occurred during the 24 h after the January 12, 2014 mainshock are given in Fig. 8. Data for aftershock magnitudes $M_{LBL} \geq 1.25$ are given for the LBL data and data for aftershock magnitudes $M_{NCSN} \geq 0.75$ are given for the NCSN data. Also given are the expected numbers of background earthquakes based on the number in the 24 h prior to the event. The best fits of Omori's law from Eq. (7) are given for both data sets. Excellent correlations are obtained for both the LBL and NCSN data. Values of the scaling parameters are given in Table 1.

We have carried out similar studies of aftershock statistics for three other large induced earthquakes at The Geysers. The values of the scaling parameters for these earthquakes are also given in Table 1. In general, the aftershock statistics for the four earthquakes are quite typical for tectonic earthquakes. In our discussion we will use NCSN catalog data, since the magnitude determinations are consistent with catalog determinations of tectonic earthquakes. For b-values we have $b = 1.26, 1.16, 0.99, 1.17$ which are very typical of tectonic aftershock frequency–magnitude data. For the magnitude differences ΔM^* we have

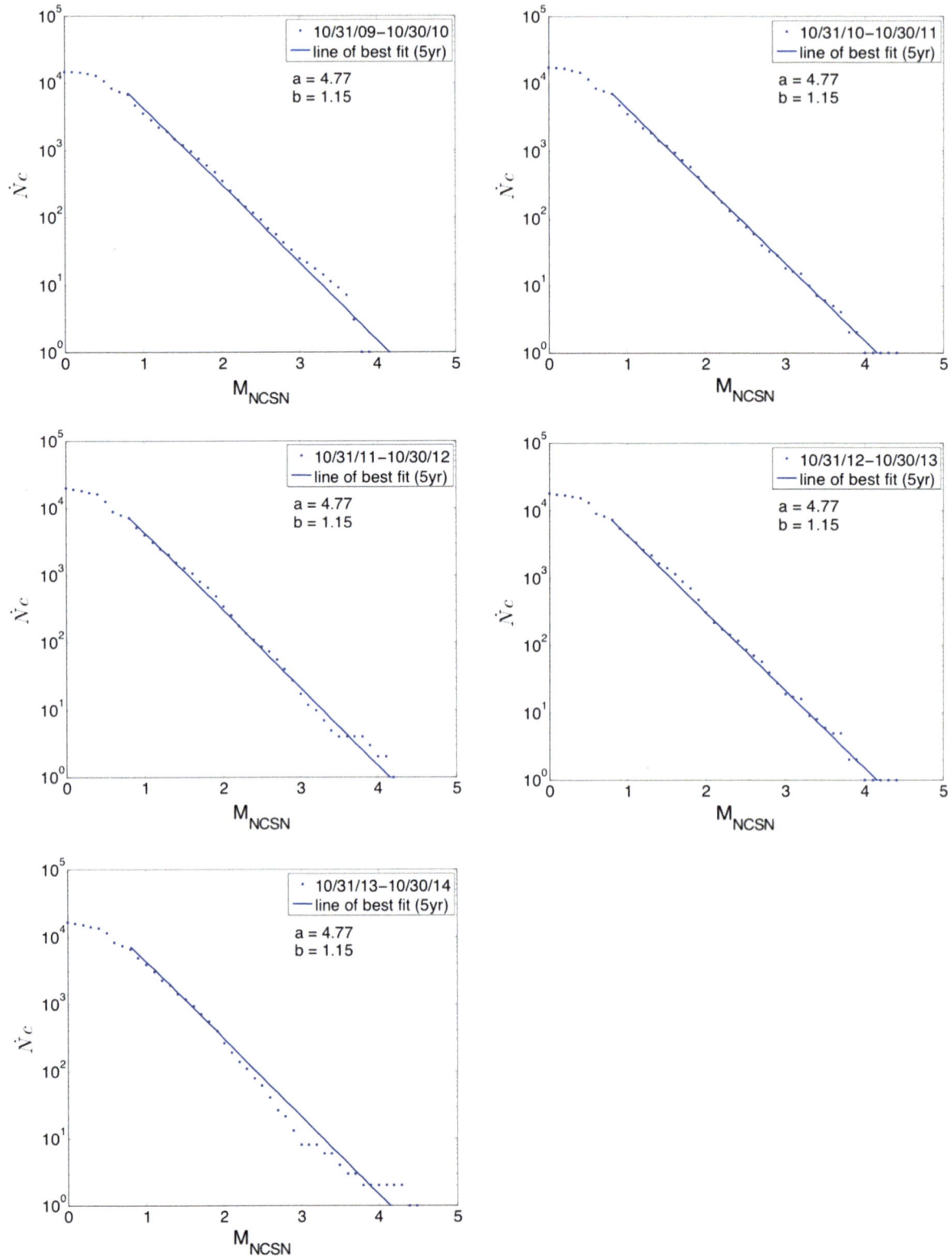

◀Figure 6
Yearly frequency–magnitude statistics from the NCSN catalog (2009–2014) with the line of best fit from the 5-year frequency–magnitude data from the NCSN catalog from Fig. 4

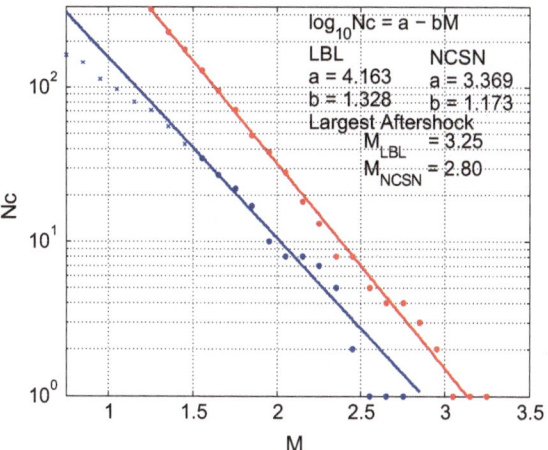

Figure 7

Cumulative frequency–magnitude distributions for the aftershocks that occurred in the 24 h after the large January 12, 2014 Geysers earthquake ($M_{LBL} = 4.48$, $M_{NCSN} = 4.53$). Data from the LBL catalog (red) and NCSN catalog (blue) are given. The straight lines are the least square fits of Eq. (1) to the data. The points with crosses were not used in the fitting

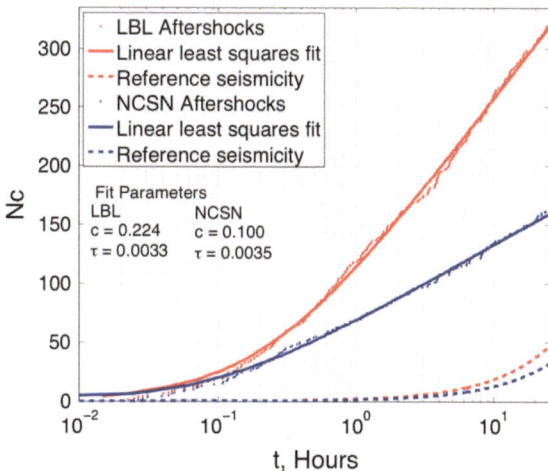

Figure 8

Cumulative number N_c of earthquakes that occurred at time t during the 24 h after the January 12, 2014 mainshock ($M_{LBL} = 4.48$, $M_{NCSN} = 4.53$). Data for $M_{LBL} \geq 1.25$ from the LBNL catalog (red) and data for $M_{NCSN} \geq 0.75$ from the NCSN catalog (blue) are given. In both cases the expected number of background earthquakes is given. Also given are the fit parameters for Omori's law from Eq. (7)

values 2.17, 2.00, 1.23, and 1.66, which again are typical of values for tectonic earthquakes (Shcherbakov and Turcotte 2004). Assuming the power law p in Omori's law, Eq. (6) has the value $p = 1$, the expected cumulative number of aftershocks at a time t after the mainshock has been given in Eq. (7). The production of aftershocks for the four induced earthquakes we consider are in quite good agreement with this relation. We strongly favor the hypothesis that both the mainshocks and the aftershocks are tectonic earthquakes that have been triggered by the reduction in effective normal stress on faults due to fluid injection. We will return to this discussion after our consideration of triggered earthquakes at The Geysers. Although the mean values of scaling parameters are quite typical of tectonic values, the scatter of the frequency–magnitude data is relatively large. We attribute this variability of seismic parameters to the variability of fluid pressures due to both spatial and temporal variations in injection rates.

4. Triggered Earthquakes and Their Aftershocks at The Geysers

Remotely triggered aftershocks at large distances from a mainshock became widely accepted after the 1992 $M_w = 7.3$ Landers earthquake, east of Los Angeles, CA (Hill et al. 1993). Many of the remotely triggered aftershocks occured in volcanic geothermal areas. Prejean et al. (2004) studied remotely triggered aftershocks generated by the 2002 $M_w = 7.9$ Denali, Alaska earthquake. They reported 19 $M > 2$ triggered earthquakes at The Geysers geothermal area, 3120 km from the mainshock. Aiken and Peng (2014) studied remotely triggered earthquakes at three geothermal regions including The Geysers. Stark and Davis (1996) studied remotely triggered aftershocks at The Geysers from seven other distant mainshocks including Landers. Freed (2005) has given a comprehensive review of remotely triggered aftershocks.

One of the largest triggered earthquakes occurred at The Geysers following the $M_w = 6.02$ South Napa earthquake which occurred on 8/24/2014 at 10:20:44 UTC (see Fig. 1). The triggered earthquake at The Geysers occurred at 10:21:11 UTC. This earthquake had an NCSN magnitude $M_{NCSN} = 4.48$ with an

Table 1

Summary of scaling parameters for the four large induced earthquakes at The Geysers that we have studied

UTC time of occurrence		Location		Magnitude		ΔM		ΔM^*		b	
MM/DD/YY	HH:MM:SS	Latitude	Longitude	NCSN	LBL	NCSN	LBL	NCSN	LBL	NCSN	LBL
03/01/11	02:19:47	38.81533 N	122.82000 W	4.43	4.09	1.88	0.97	2.17	1.03	1.26	1.19
02/13/12	04:47:13	38.79267 N	122.74316 W	4.16	4.31	1.69	1.46	2.00	1.53	1.16	1.27
03/14/13	09:09:23	38.81233 N	122.78616 W	3.95	4.26	0.49	0.36	1.23	1.13	0.99	1.19
01/12/14	20:24:47	38.81350 N	122.81360 W	4.53	4.48	1.73	1.23	1.66	1.35	1.17	1.33

Results from the NCSN and LBL catalogs are given

epicenter at Lat. 38.7602 N, Long. 122.7257 W. These values are considered more reliable than the LBL values. This remotely triggered aftershock occurred 26.78 s after the mainshock at a distance between the epicenters of 72.44 km.

We consider the earthquakes in The Geysers region in the 24 h after the large triggered earthquake. The frequency–magnitude statistics for the two data sets are given in Fig. 9. The GR scaling from Eq. (1) is also given. Reasonably good correlations are obtained. The difference in magnitude between the largest induced earthquake ($M = 4.48$) and the second largest induced earthquake ($M = 2.51$) is $\Delta M = 1.97$. The largest earthquake consistent with GR scaling is $M^* = 2.71$ so that $\Delta M^* = 1.67$. The cumulative numbers of

earthquakes that occurred in the 24 h are given in Fig. 10. Data for $M_{LBL} \geq 1.25$ are given for the LBL data and data for aftershock magnitudes $M_{NCSN} \geq 0.75$ are given for the NCSN data. Also given are the rates of occurrence of background seismicity based on the 24 h prior to the event. The best fits of Omori's law from Eq. (7) are given for both data sets. It should be noted that other smaller triggered earthquakes would also be likely to occur and these cannot be distinguished from the aftershocks after the occurrence of the large triggered earthquake. It should be noted that data analysis of the aftershocks is difficult at early times because of the interference of the Napa earthquake seismic

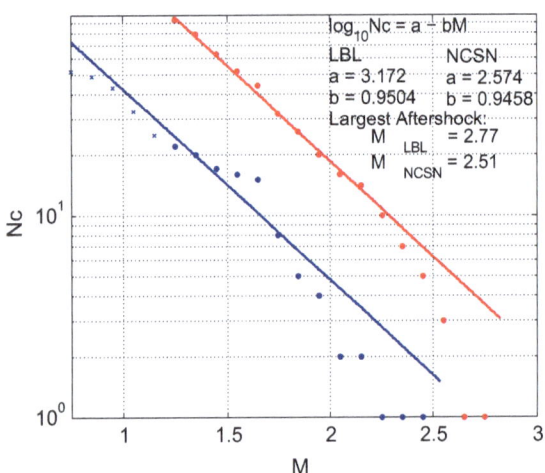

Figure 9
Cumulative frequency–magnitude distributions for the earthquakes that occurred in the 24 h after the large August 24, 2014 triggered Geysers earthquake ($M_{LBL} = 4.46$, $M_{NCSN} = 4.38$) and data from the LBL catalog (*red*) and the NCSN catalog (*blue*) are given. The *straight lines* are the least *square* fits of Eq. (1) to the data

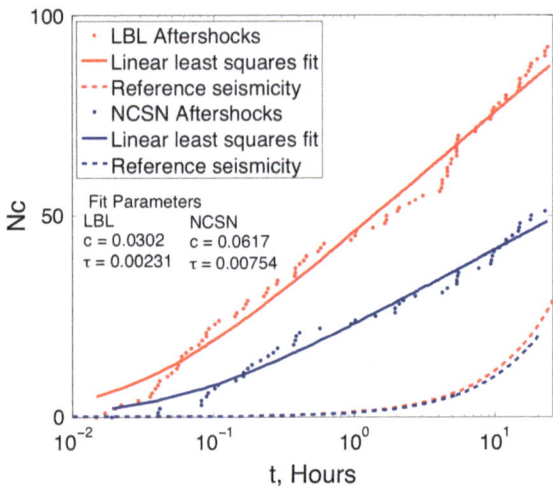

Figure 10
Cumulative number of earthquakes that had occurred at time t during the 24 h after the occurrence of the large August 24, 2014 triggered Geysers earthquake. Data for $M_{LBL} \geq 1.25$ from the LBL catalog (*red*) and data for $M_{NCSN} \geq 0.75$ from the NCSN catalog (*blue*). In both cases the expected rates of occurrence of background seismicity is given. Also given are the correlations with Omori's law from Eq. (7)

signal. Because of the large differences in magnitudes between the large induced earthquake and the subsequent earthquakes, we conclude that the smaller earthquakes are dominantly aftershocks of the large induced earthquake.

To better understand the results given above, we will consider another example of triggered seismicity at The Geysers: the triggered earthquakes following the $M = 7.2$ El Mayor-Cucapah (Baja, California) earthquake that occurred at 22:40:42 UTC April 4, 2010. The first identified triggered earthquake from NCSN data occurred at 22:49:17 UTC and had an NCSN magnitude $M_{\mathrm{NCSN}} = 3.23$ and an LBL magnitude $M_{\mathrm{LBL}} = 3.37$. This was the largest detected earthquake in the sequence. The frequency–magnitude statistics for the two data sets are given in Fig. 11. We again correlate the data using the GR scaling from Eq. (1). Based on the systematic divergence of the two data sets illustrated in Fig. 3, we consider earthquakes with $M_{\mathrm{NCSN}} \geq 0.75$ and $M_{\mathrm{LBL}} \geq 1.25$. For the LBL data we have a reasonably good GR correlation with $b = 1.00$. The NCSN data also show reasonably good GR correlation with $b = 0.75$. Again, we wish to address the question whether this seismicity is primarily induced earthquakes or aftershocks of one or more large induced earthquakes. To do this we consider the temporal

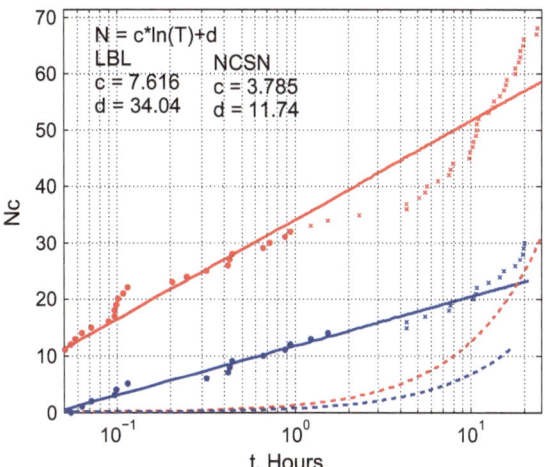

Figure 12
Cumulative number of earthquakes triggered at The Geysers by the April 4, 2010 El Mayor-Cucapah earthquake ($M = 7.2$) as a function of time in the 24 h after the arrival of the surface wave. Data for $M_{\mathrm{LBL}} \geq 1.25$ from the LBL catalog (red) and data for $M_{\mathrm{NCSN}} \geq 0.75$ from the NCSN catalog (blue). In both cases, the expected rate of occurrence of background seismicity is given. Also given are the correlations with Eq. (8)

dependence of earthquakes triggered at The Geysers by the $M = 7.2$ El Mayor-Cucapah (Baja) earthquake. The cumulative numbers N_C of earthquakes that occurred up to time t, given for the 24 h after the arrival of the P wave, are given in Fig. 12. Data for aftershock magnitudes $M_{\mathrm{LBL}} \geq 1.25$ are given for the LBL data and data for aftershock magnitudes $M_{\mathrm{NCSN}} \geq 0.75$ are given for the NCSN data. Also given are the rates of occurrence of background seismicity based on the 24 h prior to the first triggered earthquake. The largest triggered earthquake had a magnitude $M_{\mathrm{LBL}} = 3.37$ and $M_{\mathrm{NCSN}} = 3.23$; it occurred at 22:49:16 UTC, 3′49″ after the first surface wave arrivals.

It is not appropriate to correlate Omori's law defined in Eqs. (6) and (7) with the time dependence of triggered earthquakes because there is no well-defined starting point for time. For an aftershock sequence, the starting time is clearly the time of occurrence of the mainshock. For a sequence of induced earthquakes, the starting time could be the time of arrival of the P wave, time of arrival of the surface waves, or the time of occurrence of a large induced earthquake. To overcome this difficulty, Brodsky (2006) introduced the relation;

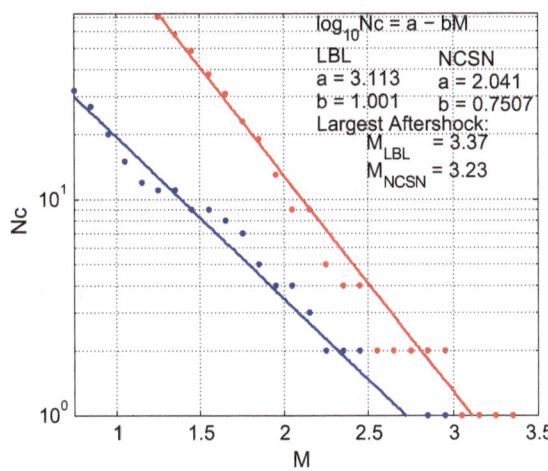

Figure 11
Cumulative frequency–magnitude distributions for the earthquakes that occurred in The Geysers region in the 24 h following the April 4, 2010 El Mayor-Cucapah earthquake ($M = 7.2$) and data from the LBL catalog (red) and the NCSN catalog (blue) are given. The straight lines are the linear least square fits of Eq. (1) to the data

$$N_c = c \ln t + d, \qquad (8)$$

where N_c is the cumulative number of earthquakes at time t. At large times with $c \ln t \gg d$, this result is identical to the large time form ($t \gg c$) of our Eq. (7). In using Eq. (8), the time delay term d is obtained by data fitting. We compare the two data sets in Fig. 12 with Eq. (8). We exclude events $t > 2$ h since they are consistent with the reference background seismicity. A reasonably good agreement is obtained for both LBL and NCSN data. Comparison of the background rates with the observed rates in Fig. 12 indicates a significant excess over the background seismicity for the first 10 h. For the NCSN data, the excess is about 15 events and for the LBL data the excess is about 40 events. The difference can be attributed to greater small magnitude sensitivity of the LBL network. The seismicity in the 24 h after the first triggered earthquake can have any of three origins: (1) background Geysers seismicity, (2) directly triggered earthquakes, and (3) aftershocks of the large ($M_{\mathrm{NCSN}} = 3.23$, $M_{\mathrm{LBL}} = 3.37$) initial triggered earthquake. The systematic time dependence shown in Figs. 10 and 12 rules out background Geysers seismicity. We interpret the relatively good agreement with an Omori law type of decay, Eq. (8), to indicate that the seismicity after about 0.1 h is primarily aftershocks, but that it is likely that some of the earlier events are directly triggered.

5. Discussion

In this study, we have analyzed the statistical properties of seismicity in The Geysers region, focusing on data from 2009 to 2014. We have considered the frequency–magnitude statistics of induced mainshocks, the frequency–magnitude time delays and Båth's law statistics of the aftershocks of the induced main shocks, and the frequency–magnitude and time delay statistics of remotely triggered Geysers seismicity. We have given a comparison of the network data from the LBL and NCSN networks. The frequency–magnitude statistics demonstrate that the data sets are complete for $M_{\mathrm{NCSN}} > 0.8$ and $M_{\mathrm{LBL}} > 1.2$. An offset of approximately 0.5 magnitude units which decreases as magnitude increases is

clearly demonstrated between the networks, with the LBL magnitudes being larger than those of the NCSN. A systematic divergence in data accounts for differences in b values, 1.15 for NCSN data versus 1.36 for LBL data.

The LBL network data have better low magnitude determinations because of the larger number of stations. Internally, the processed data are self-consistent and provide excellent frequency–magnitude correlations with GR statistics. However, the systematic deviation between the magnitudes of the earthquakes as given in the two catalogs presents a serious problem. The NCSN data processing is clearly consistent with international magnitude standards. We conclude that LBL catalog magnitudes should be corrected using the cross correlations given in Fig. 3 and Eq. (2).

An important question concerning the fluid-induced seismicity is the focal mechanisms of the seismic displacements. If the ruptures are caused only by the increased pressure of the injected water, an opening mode of failure would be expected to be oriented perpendicular to the least principal stress (σ_3). Preexisting tectonic shear stresses in the rock could activate the opening mode fracture in shear. Alternatively, the increased pressure of the injected water could reduce the effective normal stress on a preexisting fault resulting in shear failure.

Johnson (2014b) presented moment tensor solutions for 20 earthquakes at The Geysers and also summarized previous published data that demonstrate the relative importance of an opening component versus a shear component. About a half of all events studied had at least some opening component with a fraction as large as 0.3. Based on the association given above between seismicity and applied shear strain, we would argue that relief of regional accumulating tectonic shear stress due to reductions of effective stress by fluid injection and the generation of opening mode fracture reactivated in shear are the dominant causes of earthquakes at The Geysers.

We have also found that the frequency–magnitude statistics on the 1-year intervals 2009–2014 have general good agreement with the 5-year interval in both catalogs. We suggest that this seismic moment release is in near balance with the geodetic moment accumulation associated with the near uniform shear strain between the Pacific and Sierra–Nevada Central Valley plates.

We have presented studies of the aftershock statistics of four of the largest fluid-induced earthquakes. In general, we find good correlations of the frequency–magnitude statistics of the aftershock sequences with GR correlations. The LBL b-values range from 1.19 to 1.33 and the NCSN b-values range from 0.99 to 1.26. The mean value for the LBL data is $\bar{b}_{LBL} = 1.25$ and that for the NCSN data is $\bar{b}_{NCSN} = 1.15$. These values are very consistent with the values for all induced earthquakes.

Båth's law states that the difference in magnitudes between the mainshocks and their largest aftershocks are nearly constant, independent of the mainshock magnitudes (Båth 1965). The modified Båth's law proposed by Shcherbakov and Turcotte (2004) replaces the maximum observed aftershock with the inferred "largest" aftershock derived from GR scaling. The modified Båth's law states that ΔM^* is a constant. From a study of ten large California earthquakes, $\overline{\Delta M^*} = 1.11 \pm 0.9$, similar to Båth's law (Shcherbakov and Turcotte 2004). The reasoning behind Båth's law is still unknown, although it is proposed that this activity shows self-similarity in earthquake triggering. The modified Båth's law was applied to the four large fluid-induced earthquakes that occurred in The Geysers: 1 March 2011, 13 February 2012, 14 March 2013, and 12 January 2014. For these earthquakes, ΔM^*_{LBL} ranged from 1.03 to 1.52 so that $\overline{\Delta M^*_{LBL}} = 1.26$ and ΔM^*_{NCSN} ranged from 1.66 to 2.17 so that $\overline{\Delta M^*_{NCSN}} = 1.89$. Values of ΔM^*_{LBL} on average were closer to the expected values ΔM^* than those of ΔM^*_{NCSN} which on average were higher than expected.

We compared the cumulative number of aftershocks that occurred in the 24 h after the four large induced earthquakes. The values were compared with the form of Omori's law with $p = 1$ as given in Eq. (7). Generally good agreement was found. Overall, the aftershock statistics for the four large induced earthquakes are consistent with tectonic earthquakes in general. We conclude that the fluid-induced earthquakes at The Geysers are relieving tectonic stresses. They occur because the injected high-pressure water reduces the frictional resistance of slip on the preexisting faults.

We have also studied the role of remotely triggered earthquakes at The Geysers. We have emphasized the large triggered earthquake generated at The Geysers by the $M = 6.0$ South Napa earthquake in August 2014. This earthquake had a $M_{NCSN} = 4.46$. In the following 24 h there were about 50 $M_{NCSN} > 0.75$ earthquakes and about 90 $M_{LBL} > 1.25$ earthquake. The expected background rates were about 20 and 30 earthquakes, respectively. Based on the frequency–magnitude and time delay statistics of these earthquakes, we conclude that they are dominantly aftershocks of the large fluid-induced earthquake.

We reached a similar conclusion for the seismicity triggered at The Geysers by the $M = 7.2$ El Mayor-Cucapah earthquake in April 2010. The largest triggered earthquake had a magnitude $M_{NCSN} = 3.23$ and the associated seismicity was dominated by aftershocks of this earthquake. We further conclude that the seismicity directly triggered by the remote earthquake does not have a GR frequency–magnitude scaling and is dominated by aftershocks of the larger directly triggered earthquakes. A similar conclusion was reached by Brodsky (2006).

Acknowledgements

Data products and metadata for this study were accessed through the Northern California Earthquake Data Center (NCEDC), doi:10.7932/NCEDC. We thank Douglas Neuhauser of the Berkeley Seismological Laboratory for discussions about earthquake data catalogs.

REFERENCES

Aiken, C., & Peng, Z. (2014). Dynamic triggering of microearthquakes in three geothermal, volcanic regions of California. *Journal of Geophysical Research, Solid Earth, 119,* 6992–7009.

Båth, M. (1965). Lateral inhomogeneities of the upper mantle. *Tectonophysics, 2,* 483–514.

Beall, J. J., Wright, M. C., Pingol, A. S., & Atkinson, P. (2010). Effect of high rate injection on seismicity in The Geysers. *Geothermal Resources Council Transaction, 34,* 1203–1208.

Boyle, K., Jarpe, S., Hutchings, L., Saltiel, S., Peterson, J., & Majer, E. (2011). Preliminary investigation of an aseismic 'doughnut hole' region in the northwest Geysers, California.

Thirty-Sixth Workshop on Geothermal Reservoir Engineering, 2, 595–601.

Brodsky, E. (2006). Long-range triggered earthquakes that continue after the wave train passes. *Geophysical Research Letters, 33*(L15313), 1–5.

Chéry, J. (2008). Geodetic strain across the San Andreas Fault reflects elastic plate thickness variations (rather than fault slip rate). *Earth and Planetary Science Letters, 269,* 352–365.

Dawson, T.E., & Weldon, R.J. (2013). Uniform California Earthquake Rupture Forecast, Version 3 (UCERF3)—The Time-Independent Model, Appendix B, Geological slip rate data and geologic deformation model, USGS Open-File Report 2013-1165.

Edwards, B., & Douglas, J. (2014). Magnitude scaling of induced earthquakes. *Geothermics, 52,* 132–139.

Freed, A. M. (2005). Earthquake triggering by static, dynamic, and postseismic stress transfer. *Annual Review of Earth and Planetary Sciences, 33,* 335–367.

Guilhem, A., Hutchings, L., Dreger, D., & Johnson, L. (2014). Moment tensor inversions of $M \sim 3$ earthquakes in The Geysers geothermal fields, California. *Journal of Geophysical Research, 119,* 2121–2137.

Gutenberg, B., & Richter, C. (1954). *Seismicity of the earth and associated phenomena.* Princeton: Princeton University Press.

Hanks, T. C., & Kanamori, H. (1979). A moment–magnitude scale. *Journal Geophysical Research, 84,* 2348–2350.

Hill, D. P., Reasenberg, P. A., Michael, A., Arabaz, W. J., Beroza, G., Brumbaugh, D., et al. (1993). Seismicity remotely triggered by the magnitude 7.3 Landers, California, earthquake. *Science, 260,* 1617–1623.

Johnson, L. R. (2014a). A source model for induced earthquakes at The Geysers geothermal reservoir. *Pure and Applied Geophysics, 171,* 1625–1640.

Johnson, L. R. (2014b). Source mechanisms of induced earthquakes at The Geysers geothermal reservoir. *Pure and Applied Geophysics, 171,* 1641–1668.

Jolivet, R., Bürgmann, R., & Houlié, N. (2009). Geodetic exploration of the elastic properties across and within the northern San Andreas Fault zone. *Earth and Planetary Science Letters, 288,* 126–131.

Majer, E., Baria, R., Stark, M., Oates, S., Bommer, J., Smith, B., et al. (2007). Induced seismicity associated with enhanced geothermal systems. *Geothermics, 36,* 185–222.

NCEDC (2014), Northern California Earthquake Data Center. UC Berkeley Seismological Laboratory. Dataset. doi:10.7932/NCEDC.

Prejean, S., Hill, D. P., Brodsky, E. E., Hough, S. E., Johnston, M. J. S., Malone, S. D., et al. (2004). Remotely triggered seismicity on the United States west coast following the M_w 7.9 Denali fault earthquake. *Bulletin of the Seismological Society of America, 94,* S348–S359.

Prentice, C., Larsen, M., Kelsey, H., & Zachariasen, J. (2014). Late Holocene slip rate and ages of prehistoric earthquakes along the Maacama Fault near Willits, Mendocino County, northern California. *Bulletin of the Seismological Society of America, 104,* 2966–2984.

Prescott, W. H., Savage, J. C., Svarc, J. L., & Manaker, D. (2001). Deformation across the Pacific-North America plate boundary near San Francisco, California. *Journal of Geophysical Research: Solid Earth, 106,* 6673–6682.

Schmitt, A., Grove, M., Harrison, T., Lovera, O., Hulen, J., & Walters, M. (2003). The Geysers-Cobb Mountain Magma System, California (Part 1): U-Pb zircon ages of volcanic rocks, conditions of zircon crystallization and magma residence times. *Geochimica et Cosmochimica Acta, 67,* 3423–3442.

Shcherbakov, R., & Turcotte, D. (2004). A modified form of Båth's law. *Bulletin of the Seismological Society of America, 94,* 1968–1975.

Shcherbakov, R., Turcotte, D. L., & Rundle, J. B. (2006). Scaling properties of the Parkfield aftershock sequence. *Bulletin of the Seismological Society of America, 96,* 5379–5384.

Stark, M., & Davis, S. (1996). Remotely triggered micro-earthquakes at The Geysers geothermal field, California. *Geophysical Research Letters, 23,* 945–948.

Stimac, J., Goff, F., & Wohletz, K. (2001). Thermal modeling of the Clear Lake magmatic-hydrothermal system, California, USA. *Geothermics, 30,* 349–390.

Utsu, T. (1961). A statistical study on the occurrence of aftershocks. *Geophysical Magazine, 30,* 521–605.

Utsu, T. (2002a). Relationships between magnitude scales. In W. H. K. Lee, H. Kanamori, P. C. Jennings, & C. Kisslinger (Eds.), *International handbook of earthquake and engineering seismology, part A* (pp. 733–746). Amsterdam: Academic Press.

Utsu, T. (2002b). Relationships between magnitude scales. In W. H. K. Lee, H. Kanamori, P. C. Jennings, & C. Kisslinger (Eds.), *International handbook of earthquake and engineering seismology, part A* (pp. 733–746). Amsterdam: Academic Press.

(Received May 25, 2016, revised May 5, 2017, accepted May 8, 2017, Published online May 12, 2017)